Welding

Level One

SIXTH EDITION

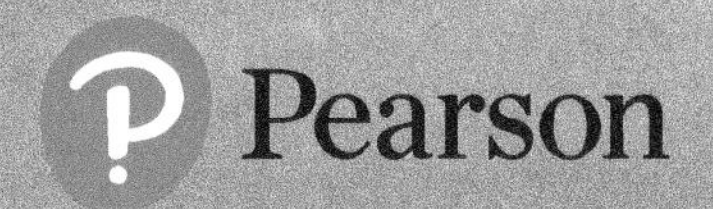

NCCER
President and Chief Executive Officer: Boyd Worsham
Vice President of Innovation and Advancement: Jennifer Wilkerson
Chief Learning Officer: Lisa Strite
Welding Project Manager: Dario VanHorne
Senior Manager of Projects: Chris Wilson
Senior Manager of Production: Erin O'Nora
Testing/Assessment Project Manager: Elizabeth Schlaupitz
Testing and Assessment Project Coordinator: Chelsi Santana
Lead Technical Writer: Don Congdon
Art Manager: Carrie Pelzer
Production Artist: Judd Ivines
Permissions Specialist: Amanda Smith
Desktop Publishing Manager: James McKay, Denise Baco
Production Specialists: Gene Page, Eric Caraballoso, Daphney Milian
Digital Content Coordinators: Rachael Downs, Yesenia Tejas
Managing Editor: Graham Hack
Editors: Karina Kuchta, Hannah Murray, Zi Meng

Pearson
Manager of Project Management: Vanessa Price
Senior Content Producer: Alexandrina B. Wolf
Employability Solutions Coordinator: Monica Perez
Content Producers: Alma Dabral

Composition: NCCER
Content Technologies: Gnostyx
Printer/Binder: Lakeside Book Company
Cover Printer: Lakeside Book Company
Text Fonts: Palatino LT Pro and Helvetica Neue

Cover Image
Cover photo provided by: Used with permission from MSA.

3 2023

Scout Automated Print Code

Paperback
ISBN-10: 0-13-92453-4
ISBN-13: 978-0-13-792453-0

Hardcover
ISBN-10: 0-13-792454-2
ISBN-13: 978-0-13-792454-7

PREFACE

To the Trainee

Despite our advanced welding processes, welding can trace its historic development to ancient times. The earliest examples of welding range from pressure-welded lap joints from the Bronze age to tools made in Egypt during the Iron Age. By 1920, automatic welding was introduced. This eventually led to the creation of plasma arc welding in 1957. Arc welding, which uses electric currents to create heat and bond metals together, is the most common type of welding today. As years go by, welders in the industry constantly find new approaches to refine the craft and increase productivity.

Almost every manufactured product uses welding, either directly or indirectly. From pipelines, aircraft carriers, car racing, and even national defense, the welding trade impacts virtually every industry.

Few career choices offer as many opportunities for employment, entrepreneurship and growth. According to the Department of Labor, the basic skills of welding are similar across industries, so welders can easily shift from one industry to another, repositioning themselves to succeed according to market demand. Welders will continue to be in high demand. In addition to emerging technologies, there is a continual need to rebuild aging infrastructure. The construction of new power generation facilities will yield new job prospects for trained and skilled welders.

Welding Level One introduces the fundamentals of the welding trade and establishes best practices for personal and team safety. The four levels of this curriculum present an apprentice approach and will help you be knowledgeable, safe, and effective on the job. The sixth edition *Welding* curriculum has been revised by industry subject matter experts from across the nation who have incorporated the latest advancements of the trade.

New with *Welding Level One*

The sixth edition *Welding Level One* presents each revised module through a user friendly instructional design which organizes the material efficiently and in a logical sequence. NCCER has drastically overhauled the material and design to make reading and understanding easier for the trainee. The images and diagrams have also been updated to exemplify the most current welding practices, with special emphasis on safety.

Welding Level One aligns with the current standards in American Welding Society's School Excelling through National Skills Education (SENSE) EG2.0 guidelines for Entry Welder. This means that, in addition to conforming to NCCER guidelines for credentialing through its Registry, this program can also be used to meet guidelines provided by AWS for Entry Welder training. For more information on the AWS SENSE program, contact AWS at 1-800-443-9353 or visit **www.aws.org.**

We wish you success as you progress through this training program. If you have any comments on how NCCER might improve upon this textbook, please complete the User Update form using the QR code on this page. NCCER appreciates and welcomes its customers' feedback. You may submit yours by emailing **support@nccer.org**. When doing so, please identify feedback on this title by listing #WeldingL1 in the subject line.

Our website, **www.nccer.org**, has information on the latest product releases and training.

Your feedback is welcome. You may email your comments to **curriculum@nccer.org** or send general comments and inquiries to **info@nccer.org**.

NCCER Standardized Curricula

NCCER is a not-for-profit 501(c)(3) education foundation established in 1996 by the world's largest and most progressive construction companies and national construction associations. It was founded to address the severe workforce shortage facing the industry and to develop a standardized training process and curricula. Today, NCCER is supported by hundreds of leading construction and maintenance companies, manufacturers, and national associations. The NCCER Standardized Curricula was developed by NCCER in partnership with Pearson, the world's largest educational publisher.

Some features of the NCCER Standardized Curricula are as follows:

- An industry-proven record of success
- Curricula developed by the industry, for the industry
- National standardization providing portability of learned job skills and educational credits
- Compliance with the Office of Apprenticeship requirements for related classroom training (*CFR 29:29*)
- Well-illustrated, up-to-date, and practical information

NCCER also maintains the NCCER Registry, which provides transcripts, certificates, and wallet cards to individuals who have successfully completed a level of training within a craft in NCCER's Curricula. *Training programs must be delivered by an NCCER Accredited Training Sponsor in order to receive these credentials.*

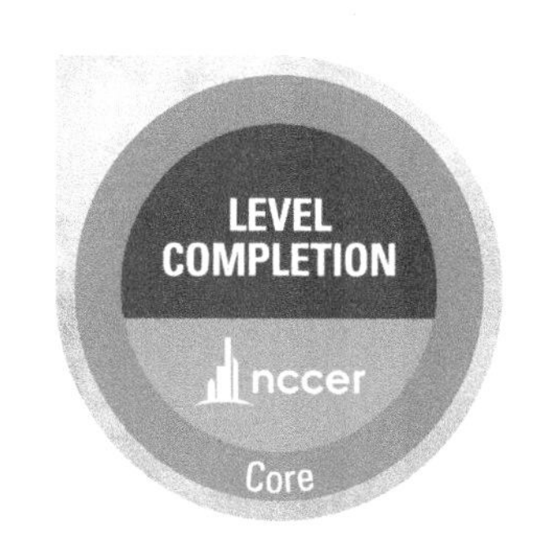

Online Badges

Show off your industry-recognized credentials online with NCCER's digital badges!

NCCER is now providing online credentials. Transform your knowledge, skills, and achievements into badges that you can share across social media platforms, send to your network, and add to your resume. For more information, visit **www.nccer.org**.

Cover Image Provider

MSA Safety Incorporated is a global leader in the development, manufacture and supply of safety products that protect people and facility infrastructures.

DESIGN FEATURES

Content is organized and presented in a functional structure that allows trainees to acces the information where they need it.

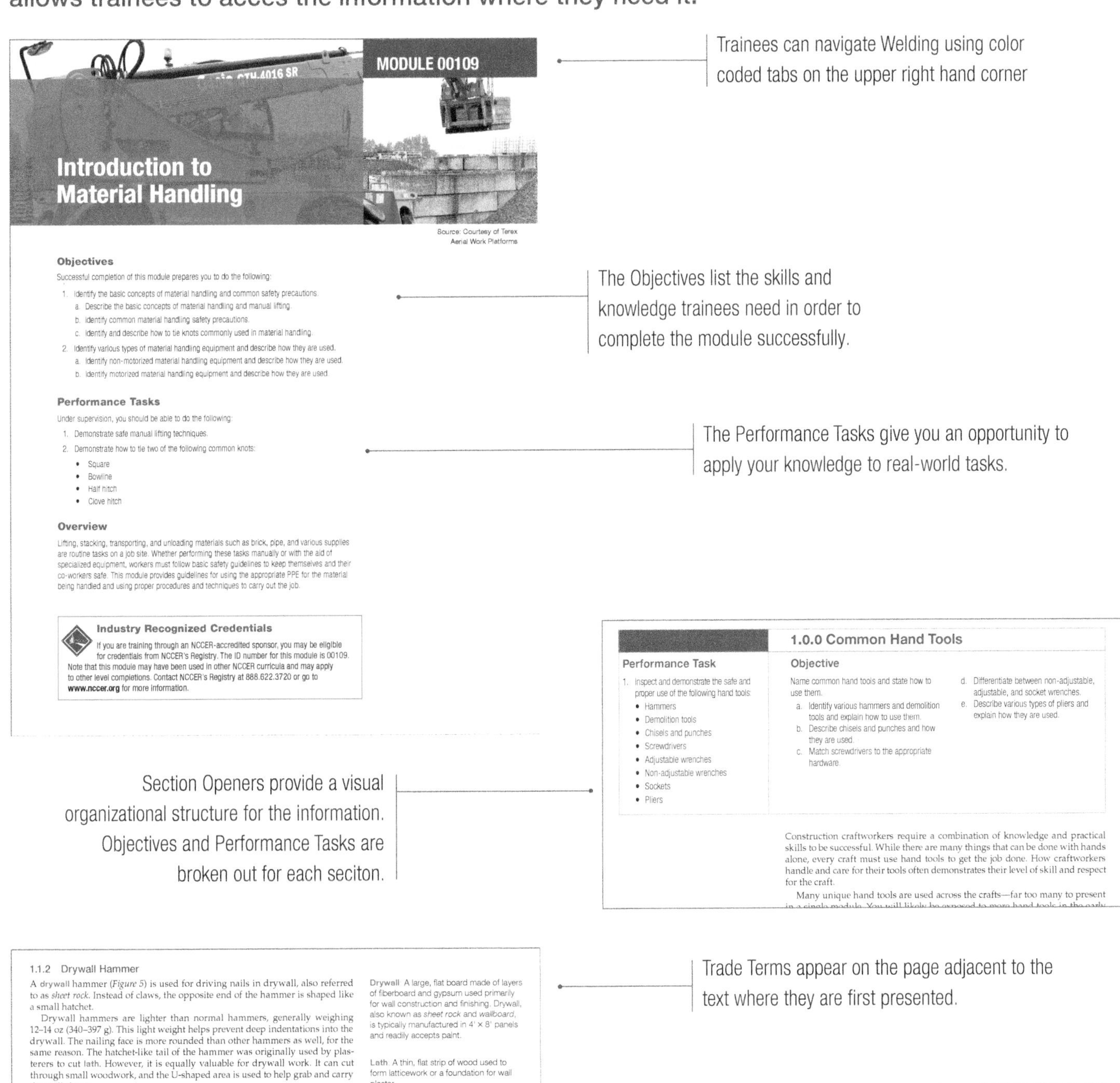

Step-by-step math equations help make the concepts clear and easy to grasp.

1.2.0 Adding and Subtracting Whole Numbers

Sum The result of an addition problem. For example, in the problem 7 + 8 = 15, 15 is the sum.

To add means to combine the values of two or more numbers together into one sum or total. To add whole numbers, perform the following steps:

Step 1 Line up the digits in the top and bottom numbers by place value columns.

$$\begin{array}{r} 723 \\ +\ 84 \\ \hline \end{array}$$

Step 2 Beginning at the right side, add the numbers in the ones column (3 and 4) together first.

$$\begin{array}{r} 723 \\ +\ 84 \\ \hline 7 \end{array}$$

Step 3 Continue adding the digits in each column, moving from right to left, one column at a time. In this example, adding the 2 and 8 in the tens column gives 10. This requires carrying the 1 from the tens column over to the next column on the left. To do so, place the 0 in the tens column and carry the 1 over to the top of the hundreds column as shown. Always add the carried-over number to the rest of the digits in that column.

$$\begin{array}{r} \overset{1}{7}23 \\ +\ 84 \\ \hline 07 \end{array}$$

Step 4 Add the 7 already in the hundreds column to the 1 carried over. The resulting sum is 807.

$$\begin{array}{r} \overset{1}{7}23 \\ +\ 84 \\ \hline 807 \end{array}$$

QR codes link trainees directly to videos that highlight current content.

SUCCESS STORIES: *Credentials Matter: Start Building Your Career Today*

Scan this code using the camera on your phone or mobile device to view this video.

Important information is highlighted, illustrated, and presented to facilitate learning.

Placement of images near the text description and details such as callouts and labels help trainees absorb information.

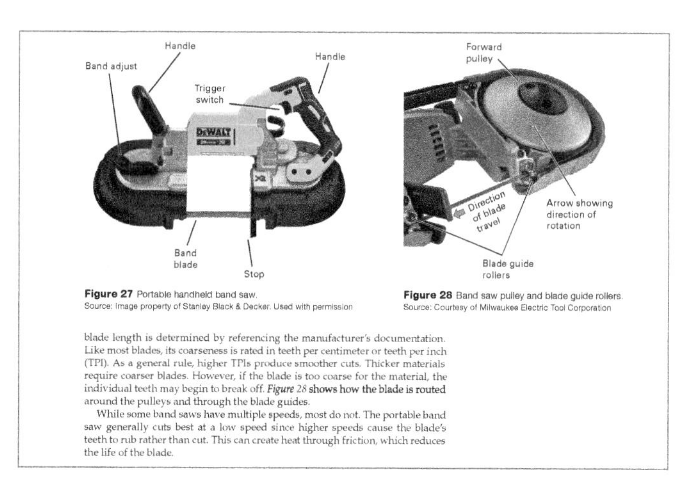

Figure 27 Portable handheld band saw.
Source: Image property of Stanley Black & Decker. Used with permission

Figure 28 Band saw pulley and blade guide rollers.
Source: Courtesy of Milwaukee Electric Tool Corporation

blade length is determined by referencing the manufacturer's documentation. Like most blades, its coarseness is rated in teeth per centimeter or teeth per inch (TPI). As a general rule, higher TPIs produce smoother cuts. Thicker materials require coarser blades. However, if the blade is too coarse for the material, the individual teeth may begin to break off. *Figure 28* shows how the blade is routed around the pulleys and through the blade guides.

While some band saws have multiple speeds, most do not. The portable band saw generally cuts best at a low speed since higher speeds cause the blade's teeth to rub rather than cut. This can create heat through friction, which reduces the life of the blade.

Preparing Drills with Keyless Chucks

Most cordless drills use a keyless chuck. While the steps for preparing a cordless drill are similar, there are some small differences. Follow the steps below when preparing to use drills with keyless chucks:

Step 1 Disconnect the drill from its power source by removing the battery pack before loading a bit.

Step 2 As shown in (*Figure 7A*), open the chuck by turning it counterclockwise until the jaws are wide enough to insert the bit shank.

Step 3 Insert the bit shank into the chuck opening (*Figure 7B*). Keeping the bit centered in the opening, turn the chuck by hand until the jaws grip the bit shank.

Step 4 Tighten the chuck securely with your hand so that the bit does not move (*Figure 7C*). You are now ready to use the cordless drill.

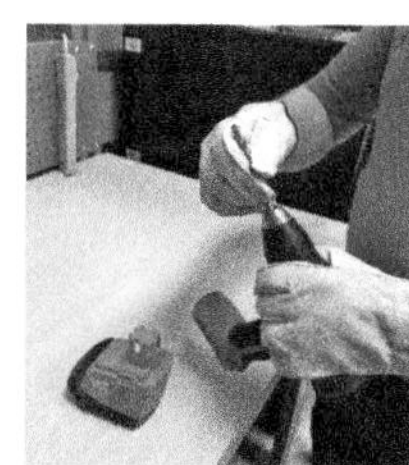

(A) Insert the Bit Shank.

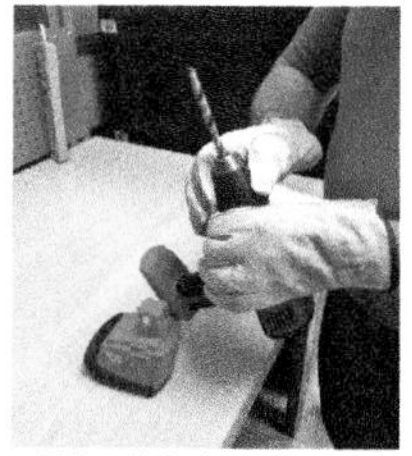

(B) Keep Bit Straight and Partially Tighten the Chuck.

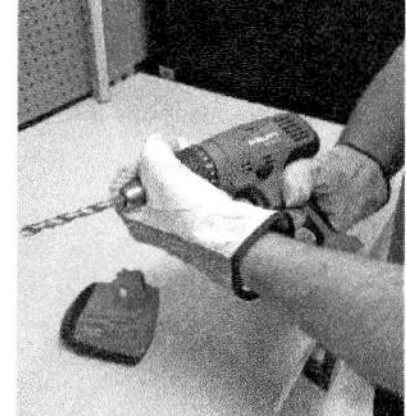

(C) Tighten the Chuck Securely.

Figure 7 Loading the bit on a keyless chuck.
Source: Cianbro Corporation

New boxes highlight safety and other important information for trainees. Warning boxes stress potentially dangerous situations, while Caution boxes alert trainees to dangers that may cause damage to equipment. Note boxes provide additional information on a topic.

WARNING!

A portable band saw always cuts in the direction of the user. For that reason, workers must be especially careful to avoid injury when using this type of saw. Always wear appropriate PPE and stay focused on the work.

CAUTION

Never assume anything. It never hurts to ask questions, but disaster can result if you don't ask. For example, do not assume that an electrical power source is turned off. First ask whether the power is turned off, then check it yourself to be completely safe.

NOTE

This training alone does not provide any level of certification in the use of fall arrest or fall restraint equipment. Trainees should not assume that the knowledge gained in this module is sufficient to certify them to use fall arrest equipment in the field.

Did You Know?

Louis Henry Sullivan, an American architect in the late 19th century, created a new style of architecture that resulted in buildings that were tall but still considered beautiful, a unique concept at the time. Called the "Father of Skyscrapers," he is most known for his design of the Wainwright Building in St. Louis.

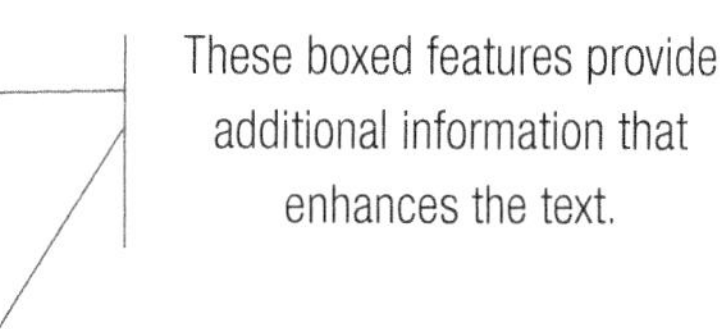

These boxed features provide additional information that enhances the text.

Orthographic Drawings

Orthographic drawings are used for elevation drawings. They show straight-on views of the different sides of an object with dimensions that are proportional to the actual physical dimensions. In orthographic drawings, the designer draws lines that are scaled-down representations of real dimensions. Every 12 inches, for example, may be represented by ¼ inch on the drawing. Similarly, in an example using metric measurements with a ratio of 1:2, every 30 millimeters may be represented by 15 millimeters on the drawing.

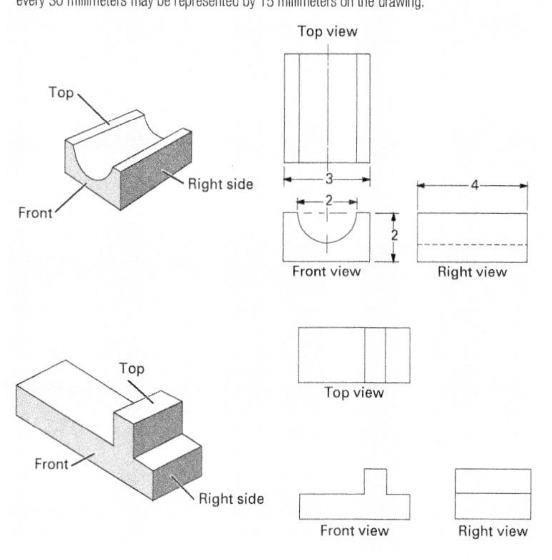

Around the World

GOST

While OSHA serves to protect workers by setting safety standards in the United States, other systems are used internationally. One such set of technical standards used on a regional basis is known as GOST. GOST standards are more far-reaching than OSHA standards, as they cover a much broader range of topics than worker safety alone. The first set of GOST standards were published in 1968 as state standards for the former Soviet Union. After the Soviet Union was dismantled, GOST became a regional standard used by many previous members of the Soviet Union. Although countries may also have some standards of their own, countries such as Belarus, Moldova, Armenia, and Ukraine continue to use GOST standards as well. The standards are no longer administered by Russia, however. Today, the standards are administered by the Euro-Asian Council for Standardization, Metrology and Certification (EASC).

Going Green

Biodiesel

Cranes and other equipment used in rigging operations consume lots of fuel—just like all the other pieces of equipment at a typical job site. Most large trucks and construction equipment run on diesel fuel. These vehicles and machines could go green and use biodiesel instead. Biodiesel is a plant oil based fuel made from soybeans, canola, and other waste vegetable oils. It is even possible to make biodiesel from recycled frying oil from restaurants. Biodiesel is considered a green fuel since it is made using renewable resources and waste products. Biodiesel can be combined with regular diesel at any ratio or be run completely on its own. This means any combination of biodiesel and regular petroleum diesel can be used or switched back and forth as needed.

But what benefits does biodiesel have over traditional fuels?

- It's environmentally friendly. Biodiesel is sustainable and a much more efficient use of our resources than diesel.
- It's non-toxic. Biodiesel reduces health risks such as asthma and water pollution linked with petroleum diesel.
- It produces lower greenhouse gas emissions. Biodiesel is almost carbon-neutral, contributing very little to global warming.
- It can improve engine life. Biodiesel provides excellent lubricity and can significantly reduce wear and tear on your engine.

Think about the environmental impact that would occur if every vehicle and piece of equipment at every job site were converted to biodiesel. The use of biodiesel also continues to increase in Europe, where Germany produces the majority of these fuels. However, even tiny countries such as Malta and Cyprus have some level of production.

Going Green looks at ways to preserve the environment, save energy, and make good choices regarding the health of the planet.

Review questions at the end of each section and module allow trainees to measure their progress.

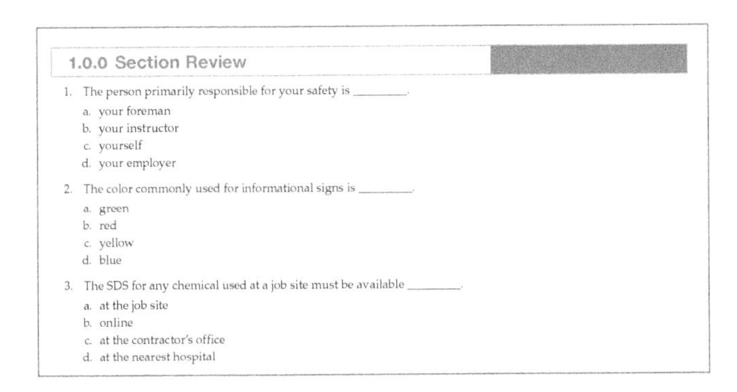
1.0.0 Section Review

1. The person primarily responsible for your safety is ________.
   a. your foreman
   b. your instructor
   c. yourself
   d. your employer
2. The color commonly used for informational signs is ________.
   a. green
   b. red
   c. yellow
   d. blue
3. The SDS for any chemical used at a job site must be available ________.
   a. at the job site
   b. online
   c. at the contractor's office
   d. at the nearest hospital

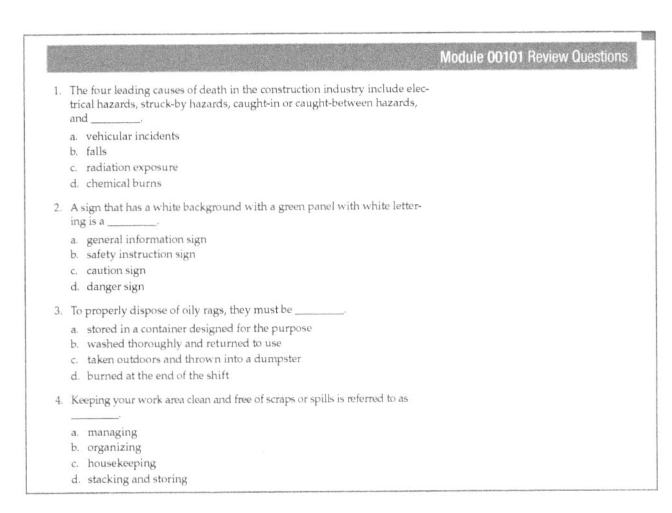
Module 00101 Review Questions

1. The four leading causes of death in the construction industry include electrical hazards, struck-by hazards, caught-in or caught-between hazards, and ________.
   a. vehicular incidents
   b. falls
   c. radiation exposure
   d. chemical burns
2. A sign that has a white background with a green panel with white lettering is a ________.
   a. general information sign
   b. safety instruction sign
   c. caution sign
   d. danger sign
3. To properly dispose of oily rags, they must be ________.
   a. stored in a container designed for the purpose
   b. washed thoroughly and returned to use
   c. taken outdoors and thrown into a dumpster
   d. burned at the end of the shift
4. Keeping your work area clean and free of scraps or spills is referred to as ________.
   a. managing
   b. organizing
   c. housekeeping
   d. stacking and storing

NCCERCONNECT

This interactive online course is a unique web-based supplement that provides a range of visual, auditory, and interactive elements to enhance training. Also included is a full eText.

Visit **www.nccerconnect.com** for more information!

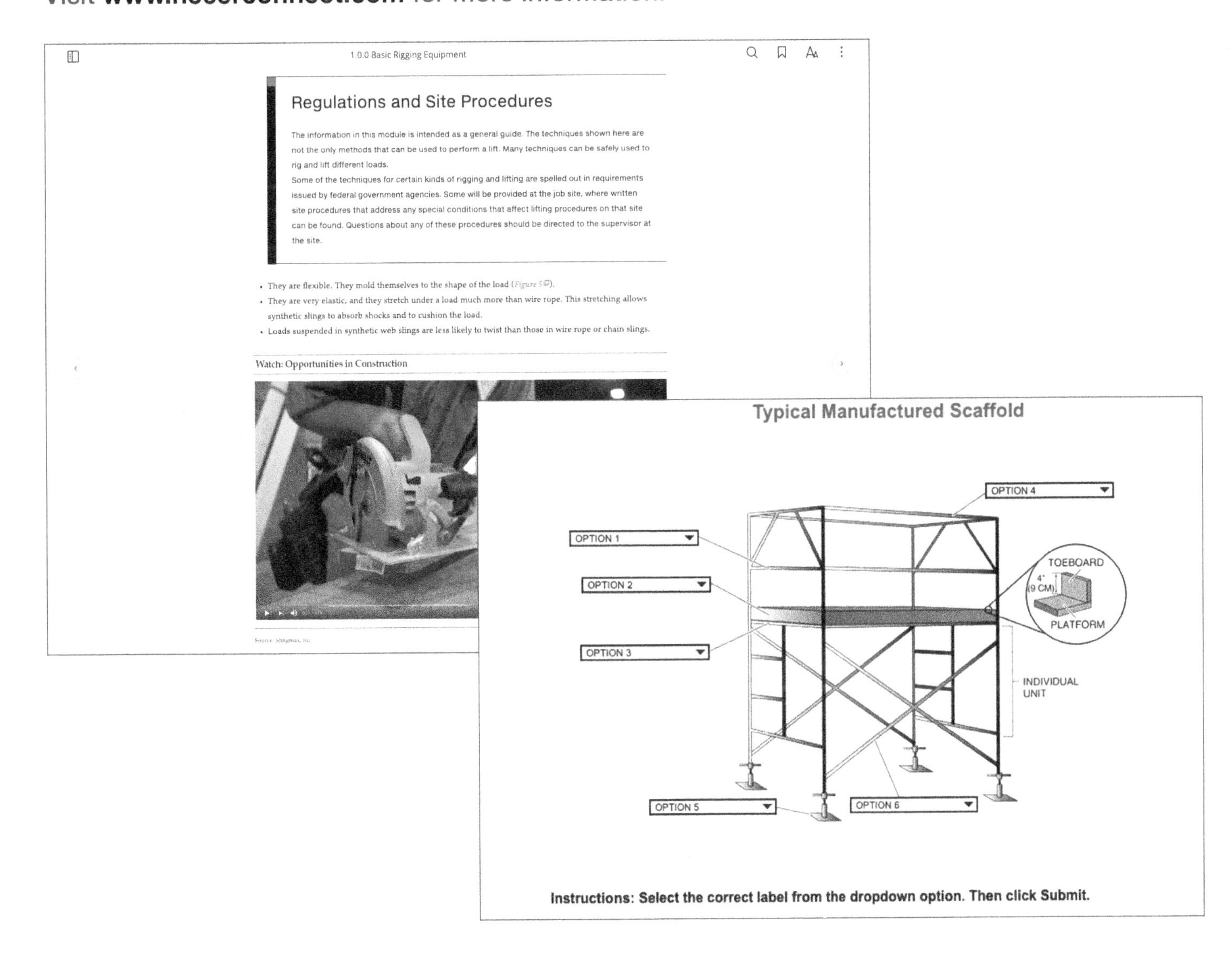

ACKNOWLEDGMENTS

This curriculum was revised as a result of the vision and leadership of the following sponsors:

All Things Metal
Bechtel
College of Southern Nevada
Kentucky Welding Institute
Lee College
Mesa Community College
Miller Electric
NW Florida State College
Raycap
United Group Services
Windham School District
Yates

This curriculum would not exist without the dedication and unselfish energy of those volunteers who served on the Authoring Team. A sincere thanks is extended to the following:

Jim Casey
Ben Pletcher
John Elliott
Matthew Aschoff
Donald "Greg" MacLiver
Brian Dennis
Jimmy Perry
Richard Samanich
Scottie Smith
Don Herron
Ashley Applegate
Curtis Casey

A final note: This book is the result of a collabortive effort involving the production, editorial, and development staff at Pearson Education, Inc., and NCCER. Thanks to all of the dedicated people involved in the many stages of this project.

NCCER PARTNERS

To see a full list of NCCER Partners, please visit:

www.nccer.org/about-us/partners.

You can also scan this code using the camera on your phone or mobile device to view these partnering organizations.

CONTENTS

Module 29101 Welding Safety

Module 29102 Oxyfuel Cutting

Module 29103 Plasma Arc Cutting

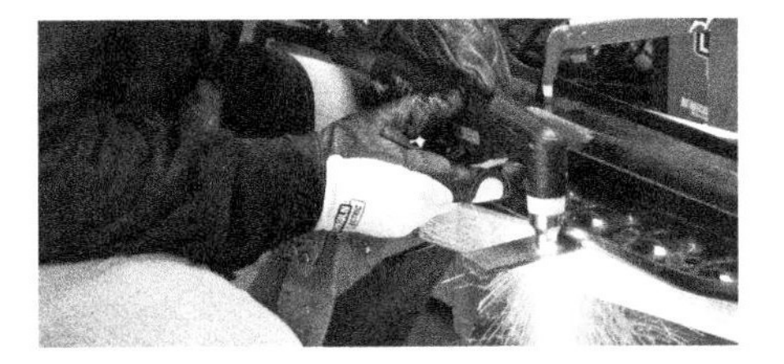

Module 29104 Air-Carbon Arc Cutting

Module 29105 Base Metal Preperation

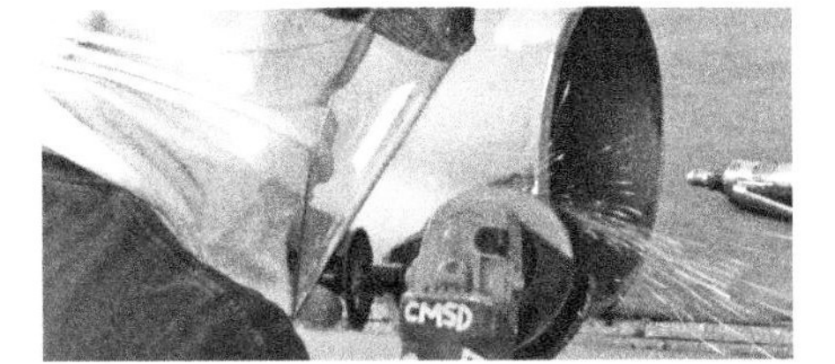

Module 29106 Weld Quality

Module 29108 SMAW Electrodes

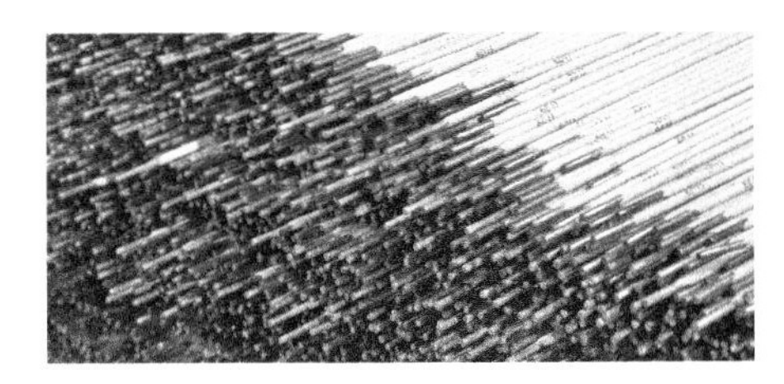

Module 29109 SMAW – Beads and Fillet Welds

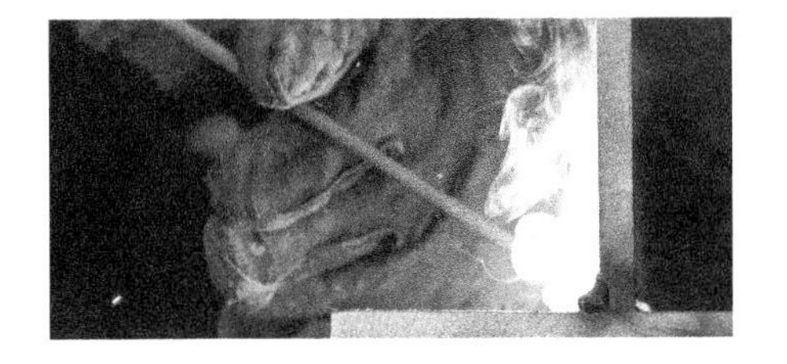

Module 29110 Joint Fit-Up and Alignment

Module 29112 SMAW – Open-Root Groove Welds – (Plate)

Welding Appendixes

Source: © Miller Electric Mfg. LLC

Objectives

Successful completion of this module prepares you to do the following:

1. Describe the welding craft and its apprenticeship program.
   a. Define welding and its common forms.
   b. Summarize the NCCER training and apprenticeship program.
2. List a welder's personal protective equipment (PPE) and describe its role.
   a. Identify and describe body, foot, and hand PPE.
   b. Identify and describe head PPE.
3. Summarize welding environment hazards and associated safety practices.
   a. Outline welding safety and the factors that contribute to accidents.
   b. List general workplace safety practices.
   c. Describe fire safety, hot work permits, and fire watch requirements.
   d. Describe confined spaces and their safety practices.
   e. Summarize welding equipment safety practices.
   f. Explain respiratory hazards, ventilation, and associated safety equipment.
   g. Describe using a Safety Data Sheet (SDS) to handle materials safely.

Performance Tasks

This is a knowledge-based module. There are no Performance Tasks.

Overview

Today's industrial and technological world needs qualified welders. Many work in heavy industry, assembling oil and natural gas pipelines or building ships. Others construct and repair the machinery that keeps power plants, refineries, chemical plants, and manufacturing facilities running smoothly. Still others work in the quieter world of high-tech shops, building precision equipment from exotic metals joined by sophisticated techniques. As you'll learn here, welding offers an excellent career path with solid pay, plenty of challenges, and work to suit almost every taste. You'll also discover, however, that succeeding in welding requires not just skill but a commitment to working safely.

NOTE

This module uses US standard and metric units in up to three different ways. This note explains how to interpret them.

Exact Conversions

Exact metric equivalents of US standard units appear in parentheses after the US standard unit. For example: "Measure 18" (45.7 cm) from the end and make a mark."

Approximate Conversions

In some cases, exact metric conversions would be inappropriate or even absurd. In these situations, an approximate metric value appears in parentheses with the ~ symbol in front of the number. For example: "Grip the tool about 3" (~8 cm) from the end."

Parallel but not Equal Values

Certain scenarios include US standard and metric values that are parallel but not equal. In these situations, a slash (/) surrounded by spaces separates the US standard and metric values. For example: "Place the point on the steel rule's 1" / 1 cm mark."

Digital Resources for Welding

Scan this code using the camera on your phone or mobile device to view the digital resources related to this craft.

Industry Recognized Credentials

If you are training through an NCCER-accredited sponsor, you may be eligible for credentials from NCCER's Registry. The ID number for this module is 29101. Note that this module may have been used in other NCCER curricula and may apply to other level completions. Contact NCCER's Registry at 1.888.622.3720 or go to **www.nccer.org** for more information.

You can also show off your industry-recognized credentials online with NCCER's digital badges. Transform your knowledge, skills, and achievements into badges that you can share across social media platforms, send to your network, and add to your resume. For more information, visit **www.nccer.org**.

1.0.0 The Welding Craft

Performance Tasks

There are no Performance Tasks in this section.

Objective

Describe the welding craft and its apprenticeship program.

a. Define welding and its common forms.

b. Summarize the NCCER training and apprenticeship program.

Welding: Processes that use heat to join materials by softening or melting them, so they blend and form a strong mechanical bond as they cool.

This section introduces the **welding** craft and the most common welding techniques. As you'll discover, welders have many tools at their disposal to join materials. This section also introduces apprenticeship, one of the best ways to become a welding craft professional.

1.1.0 Welding Work

Electric arc: An electric current flow across a gap.

Welding is a family of processes that join materials together through heat. Heat from an **electric arc**, oxyfuel torch, laser, or even friction melts or softens the materials at the joint. The materials flow together and cool to form a strong mechanical bond. Welding typically involves metals, either identical or different, but it works with some plastics and glasses too.

Welding differs from soldering and brazing. Although they too use heat, they don't melt the *base metals*—the metals being joined. Instead, they melt a separate *filler metal* that flows into the joint, cools, and hardens. Soldering and brazing usually use a torch to supply the heat.

Electric arc welding is the most common and familiar welding technology. An electric current heats the joint to a temperature between 6,000°F and 10,000°F (~3,300°C and 5,500°C). This melts the base metals that will form the joint. Often, the welder adds a filler metal to the joint to improve its strength. The finished joint is as strong as, or even stronger than, the joined base metals. Welding commonly unites **carbon steel**, **stainless steel**, and aluminum. It also joins more exotic materials like titanium, cobalt, copper, and specialized alloys.

Carbon steel: An iron alloy containing a small amount of carbon to increase its strength.

Stainless steel: An iron alloy that doesn't rust because it contains chromium.

Welders support many industries, including building and bridge construction, as well as ship and oil rig construction. They repair industrial machines and pipelines. Some build the precision equipment used in scientific research, chemical processing, and nuclear power. Welding requires extensive knowledge and skill, as well as a commitment to perform high-quality work. Good welders also commit themselves to working safely by following codes and workplace standards.

Welding involves much more than simply striking an arc. Welders must know the materials' properties and the correct techniques to join them. They must know how to handle and store base metals and filler metals, as well as how to prepare them for welding. Welders read welding symbols, mechanical drawings, and project specifications to produce quality work that lasts and operates safely.

1.1.1 Common Welding Types

Welders can choose from several methods when they join materials (*Figure 1*). The most popular is shielded metal arc welding (SMAW), also called *stick welding*. In this method, a rod-shaped, consumable **electrode** forms the weld. The electrode's **flux** coating vaporizes, shielding the joint from moisture, oxygen, and contaminants. Welders use SMAW to join iron, steel, and stainless steel.

Gas metal arc welding (GMAW), sometimes called *MIG (metal inert gas) welding*, uses a consumable wire electrode fed through the welding gun to make the weld. It uses a **shielding gas** to protect the joint from the oxygen and moisture in the air. GMAW joins steel and other metals. It's popular in many industrial applications, including robotic automobile manufacturing.

Flux-cored arc welding (FCAW) resembles GMAW but has a few differences. Instead of using a solid wire electrode, it uses a hollow wire electrode containing a flux core. The core melts and flows into the joint, protecting it from oxygen and moisture. It also cleans the joint by dissolving surface contaminants. This makes the joint stronger and more reliable. FCAW welding may or may not use a shielding gas since the flux core provides most of the protection.

Gas tungsten arc welding (GTAW), sometimes called *TIG (tungsten inert gas) welding*, produces the weld with a non-consumable tungsten electrode. It requires a shielding gas. Welders use GTAW to join stainless steel and light metals such as aluminum and magnesium.

Electrode: A metal rod or wire that carries electric current and forms the welding or cutting arc.

Flux: A material that dissolves or inhibits oxides and protects a weld joint from the atmosphere.

Shielding gas: A gas such as argon, helium, or carbon dioxide that protects the welding electrode and weld zone from contamination during certain types of welding.

1.1.2 Cutting

Welders use heat to cut materials as well as join them. Technologies like oxyfuel, air-carbon arc, or plasma arc cut steel and other metals (*Figure 2*). Oxyfuel cutting, the most common method, produces intense heat by burning a fuel and the metal itself in oxygen. Acetylene fuel gas produces the hottest flame. Other common fuels include propane, propylene, and natural gas.

A plasma arc tool cuts metals with a jet of superheated gas produced by an electric arc. Air-carbon arc cutting and gouging melts metals by forming an electric arc between a carbon electrode and the material. Welders use it to prepare a base metal for welding or to expose cracks for repair work.

1.1.3 Inspection

Since welders work on critical structures such as bridges, high-rise buildings, ships, pipelines, refineries, and automobiles, defective welds can have serious consequences. For this reason, welding requires strong quality control and careful inspections. Critical applications require intense inspections that often involve X-ray or ultrasonic testing.

1.2.0 NCCER Standardized Training

Becoming a welder requires extensive training and many hours of practice. This program will prepare you to weld plate and pipe in a variety of positions and to cut carbon steel plate and pipe. You'll learn to use common equipment and techniques.

Welders have wide career choices with many possible paths. After completing your training and becoming a journey-level welder, you can advance in several ways (*Figure 3*). The more training and education you have, the greater your opportunities will be. NCCER standardized training is just the beginning.

1.2.1 NCCER

NCCER is a not-for-profit education foundation established by the nation's leading construction companies. It provides the industry with standardized construction education materials (the NCCER Standardized Curricula) and a system for tracking and recognizing students' training accomplishments (the NCCER Registry).

(A) SMAW

(B) GMAW

(C) GTAW

Figure 1 Welding methods.
Sources: Justin Poland, Robins and Morton (1A); The Lincoln Electric Company, Cleveland, OH, USA (1B–1C)

NCCER also offers accreditation, instructor certification, and craft assessments. It commits itself to developing and maintaining an internationally recognized, standardized, portable, and competency-based training process.

By partnering with industry and academia, NCCER has developed a system for program accreditation like those found in higher education. Its accreditation process ensures that students receive quality training based on uniform standards and criteria. NCCER Accredited Training Sponsors must follow all standards outlined in NCCER's *Accreditation Guidelines*.

Numerous training centers are proud to be NCCER Accredited Training Sponsors / Accredited Assessment Centers. Millions of craft professionals and construction managers have received quality construction education through NCCER's network of Accredited Training Sponsors. Every year the number of NCCER Accredited Training Sponsors increases significantly.

A craft instructor is a journey-level craft professional or a career technical educator trained and certified to teach the NCCER Standardized Curricula. This network of certified instructors ensures that NCCER training programs meet the standards set by the industry. Visit **www.nccer.org** for more information.

NCCER's *Welding* curriculum provides trainees with portable, industry-recognized credentials. These credentials include transcripts, certificates, and wallet cards tracked through NCCER's Registry.

The curriculum provides trainees with industry-driven training and education using a competency-based learning approach. This means that trainees must show the instructor that they possess the knowledge and skills needed to safely perform the hands-on tasks covered in each module.

Instructors use written exams and hands-on performance testing to assess trainees. When the instructor is satisfied that a trainee has demonstrated the required knowledge and skills for a given module, that information is sent to NCCER and kept in the Registry. The NCCER Registry allows employers to confirm workers' training and skills as they move around the country.

1.2.2 Apprenticeship

Apprentice training goes back thousands of years and continues to follow many of the same traditional principles. A person entering a craft learns from those who have mastered it. Apprentices learn by doing. While modern apprenticeships include some classroom time, their ultimate goal is hands-on proficiency in real settings.

The Office of Apprenticeship from the US Department of Labor sets the minimum standards for training programs across the country. These programs include both mandatory classroom instruction and on-the-job learning (OJL). They require at least 144 hours of classroom instruction plus 2,000 hours of OJL per year. In a typical four-year apprenticeship program, trainees spend 576 hours in the classroom and 8,000 hours in OJL. They then receive certificates issued by the registered apprenticeship program.

(A) Oxyfuel Cutting

(B) Manual Plasma Arc Cutting

(C) Air-Carbon Arc Gouging

Figure 2 Cutting methods.
Sources: The Lincoln Electric Company, Cleveland, OH, USA (2A); © Miller Electric Mfg. LLC (2B); Zachry Industrial, Inc (2C)

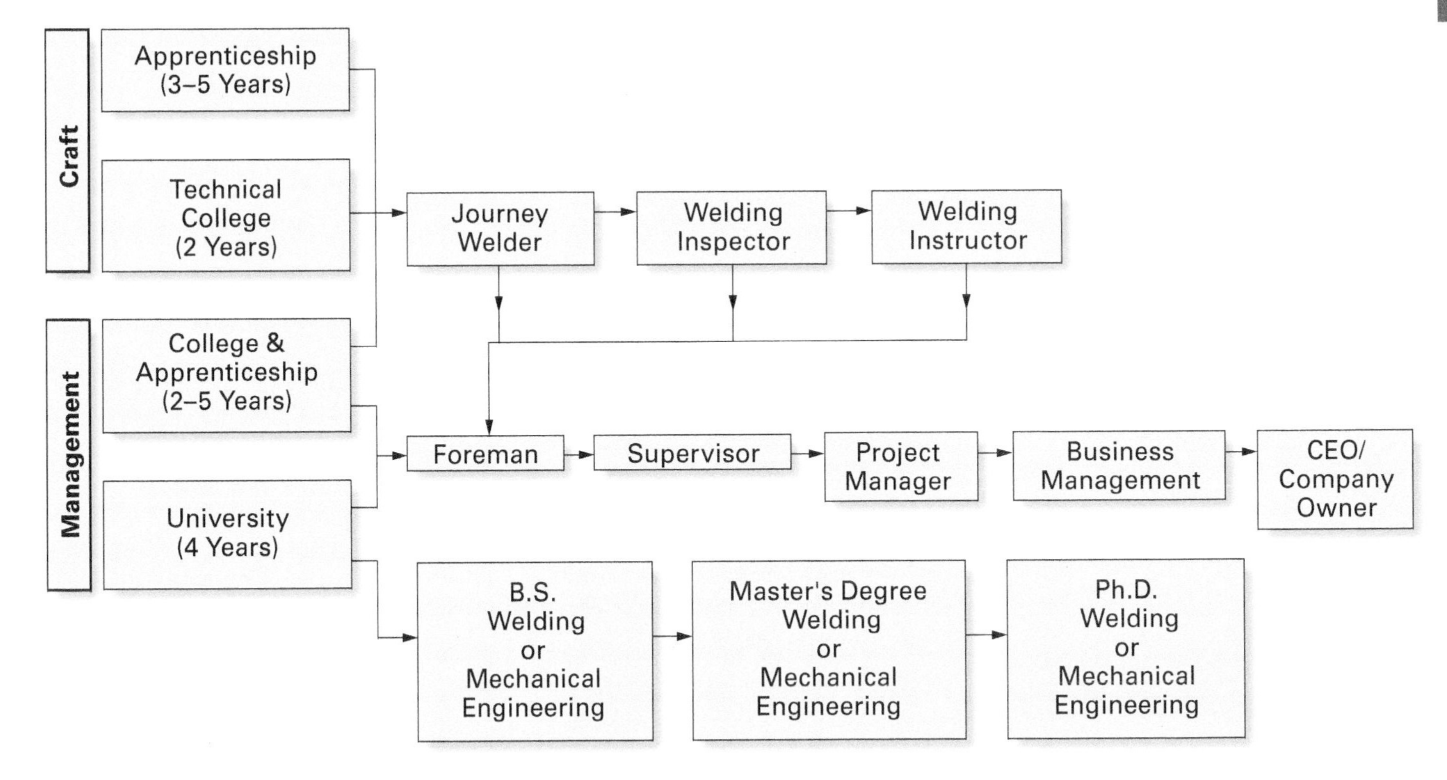

Figure 3 Opportunities in the welding craft.

NCCER uses the Department of Labor apprenticeship standards as the foundation for a comprehensive training program that provides trainees with in-depth classroom instruction and thorough OJL experience.

All apprenticeship standards prescribe specific OJL tasks. The apprentice receives hands-on training in each of these. Each task requires a specific number of hours. In a competency-based program, it may be possible to shorten the time required by testing out of specific tasks through written and performance tests.

In a traditional program, the apprentice may acquire the required OJL in increments of 2,000 hours per year. Layoffs or illness may affect the duration. The apprentice must log all work time and turn the log in to the apprenticeship committee so it can maintain accurate records.

Classroom instruction and OJL don't always happen at the same time. A lot depends on the work environment and the jobs available. Sometimes, apprentices with work experience or other classroom history may count these towards their classroom requirement. Program regulators handle these situations on a case-by-case basis.

Informal, employer-provided OJL is usually less thorough than that from a formal apprenticeship program. Quality depends on the employer and its trainers. A small company may offer training in only one area, while a large company may offer more choice.

Technical courses provided by a high school or community college, such as mechanical drawing, shop, or math, are always useful to an apprentice. Manual dexterity, good physical condition, and quick reflexes are important too. Solving problems quickly and accurately, working well with others, and communicating effectively are essential skills. Above all, apprentices must respect and follow safe workplace practices and standards.

NOTE

Some companies have physical activity requirements that apprentices must meet. These vary from company to company.

1.2.3 Youth Training and Apprenticeship Programs

Youth apprenticeship programs let trainees begin their apprenticeship while still in high school. A student entering the program in the 11th grade may complete as much as one year of the training program before graduation. Local employers may offer students jobs so they can earn money while still in school. After graduation, they can enter the industry at a higher level and with more pay than someone just starting a training program.

Did You Know?

Apprenticeships in the United States

In 2020, the US Department of Labor reported that there were more than 636,000 apprentices receiving registered apprenticeship training in 26,000 programs in the United States.

Did You Know?

An Ancient Apprenticeship Contract

Historical apprenticeship records include a contract between a young Greek named Heracles and a master weaver. In exchange for food, a tunic (a long shirt), and 20 holidays a year, Heracles worked for the weaver for five years. Halfway through this period, he received a salary of 12 drachmas. In his final year, his pay doubled to 24 drachmas.

As he learned the weaving craft, Heracles also had to clean the shop, build fires, pick up supplies, make deliveries, and do whatever other jobs the master weaver required. It's also likely that he had to pay for any damage he caused.

Perhaps you're wondering whether 12 drachmas was a decent salary. There's no way to know for sure. It probably wasn't very impressive, but Heracles wasn't working for a salary. He was learning to become a journeyman weaver. Someday, he could expect to become a master weaver with his own shop, a good salary, and apprentices of his own.

High school graduates who participated in apprenticeship training may enter their training program at its second level. They may also apply for credit at some two- or four-year colleges that offer certificate programs in their field.

1.0.0 Section Review

1. Two welding methods that use a consumable wire electrode are FCAW and _____.
   a. SMAW
   b. GMAW
   c. GTAW
   d. carbon arc

2. Before graduation, a high school student entering an apprenticeship program in the 11th grade may complete _____.
   a. as much as one year of the program
   b. nearly two years of the program
   c. more than two years of the program
   d. three years of the program

2.0.0 Personal Protective Equipment

Performance Tasks

There are no Performance Tasks in this section.

Objective

List a welder's personal protective equipment (PPE) and describe its role.

a. Identify and describe body, foot, and hand PPE.
b. Identify and describe head PPE.

In the workplace, everyone wears clothing appropriate to the work environment. Those working with hot materials require extra protection. A welder's personal protective equipment (PPE) depends on the task and the equipment he or she works with. Each piece of PPE protects a specific body part from specific injuries.

2.1.0 Body, Foot, and Hand Protection

Ultraviolet (UV) radiation: Invisible light rays that can burn skin and damage the eyes.

Welding produces molten metal, sparks, and heat. Chipping produces pieces of hot slag. A welding or cutting arc also emits invisible but dangerous **ultraviolet (UV) radiation**. It produces skin burns much like a bad sunburn. Body, foot, and hand PPE guards against these hazards. Welders choose from these based on the tasks they're doing and applicable safety codes (*Figure 4*).

Figure 4 Welders with proper PPE.
Source: © Miller Electric Mfg. LLC

2.1.1 Body Protection

Proper welding clothing protects from sparks, heat, and UV radiation. Shirts have a tight weave, long sleeves, and pocket flaps. Welders always button their collars to protect their necks. Pants don't have cuffs and hang straight down the leg, touching the shoe tops without creases. Cuffs and creases can catch sparks, which can cause fires.

Polyester or other common synthetic fibers melt, which can cause burns. To avoid this, welders choose natural fibers, like cotton and wool, or special synthetics specifically designed for welding clothing.

Some materials are naturally *fire-resistant*, meaning that they won't catch fire and burn. They may smolder but they won't burst into flames. Others, like cotton, require treating with a *fire retardant* before they're safe. When purchasing protective clothing, confirm from the label that it's either made from fire-resistant materials or treated with a fire retardant. Don't trust the package alone.

WARNING!

Laundering clothing treated with a fire retardant gradually removes the protective chemicals, eventually making the clothing unsafe. Generally, treated clothing must be replaced or re-treated after 50 to 100 washes. Never use starch, fabric softener, or bleach on treated clothing as these can make the clothing flammable or remove the fire retardant. Fire-resistant fabrics, which are not treated, have fewer restrictions. Always consult the clothing label or the manufacturer's instructions for general guidance and before laundering. Another helpful resource is ASTM F2757, *Standard Guide for Home Laundering Care and Maintenance of Flame Resistant or Arc Rated Clothing*.

Tasks like cutting, overhead welding, or welding that produces molten metal spatters and large amounts of heat require additional protective coverings. *Figure 5* shows a welder wearing clothing called *leathers*. As the name suggests, leathers are usually made from leather because it's thick and durable. Unfortunately, it's also hot and heavy.

Some manufacturers offer protective clothing made from welding-appropriate synthetic materials that work well but are lightweight and cooler. *Figure 6* shows leather and fire-resistant protective items that cover different body parts.

Figure 5 Body protection PPE.
Source: The Lincoln Electric Company, Cleveland, OH, USA

Full jackets offer the most protection. A cape is cooler but covers less. Leather pants combined with a jacket offer good body and leg protection. In hot working conditions, leggings, called *chaps*, offer an alternative. They strap on the legs and are open at the back.

An apron alone is cooler and protects the welder's lap and most of the leg area when squatting, sitting, or bending over a table. When welding at or below waist level, full- or half-sleeve arm covers and an apron may be sufficient. The welding position and type usually determines the amount and type of PPE required. Of course, relevant safety codes and company policy ultimately govern PPE choice.

WARNING!

Welding sparks and molten metal spatter can be as hot as 7,000°F (~4,000°C). They sometimes travel a long way and land on people and/or flammable materials. To protect against these, welders and nearby workers must wear suitable PPE. Prevent fires by covering flammable materials with fire-resistant or fireproof blankets before starting to work.

(A) Leather Jacket

(B) Leather Apron

(C) Fire-Resistant Arm Cover

(D) Leather Boot or Shoe Protection (Spats)

(E) Cape (Fastens to an Apron)

(F) Leather Chaps

Figure 6 Leather and fire-resistant PPE.
Sources: The Lincoln Electric Company, Cleveland, OH, USA (6A–6C)

Hybrid Jackets and Specialized Clothing

Today, welders have a lot of choice in their protective clothing. Hybrid welding jackets have standard leather sleeves combined with a body section made from cooler, welding-safe synthetic materials. Since many women work in the welding craft, several manufacturers produce clothing specifically for them.

Source: Image courtesy of Black Stallion (Revco Industries)

2.1.2 Foot Protection

The Occupational Safety and Health Administration (OSHA) requires all workers to wear protective footwear wherever falling, rolling, or sharp objects could injure feet. Workers must also wear suitable protective footwear when working around electrical hazards. Always confirm that safety shoes comply with the standards that govern the working conditions. Ordinary street or athletic shoes are never acceptable in the industrial workplace.

Welders must also protect their feet from sparks and droplets of molten metal. High-top boots or safety shoes with leather-reinforced soles or flexible metal inner soles provide the right protection. Pant legs should cover the tongue and lace area. If the pants don't cover this area, wear leather *spats* to protect this part of the feet (*Figure 6*).

WARNING!

When purchasing safety footwear, confirm that it meets the most recent *ANSI/AWS Z49.1* guidelines. Always follow your workplace's safety guidelines as well.

2.1.3 Hand Protection

Depending on their type, gloves protect the hands from abrasions, cuts, burns, and electrical shock. Select gloves based on the task and its accompanying hazards. Wear gauntlet-type welding gloves when welding or cutting to protect against heat and UV radiation. *Figure 7* shows several glove styles, including special heavy-duty gloves for unusually hot conditions.

WARNING!

Don't weld or cut metal with gloves that have petroleum products on them. **They could catch fire, particularly if you're working around oxygen**. They can also contaminate the welding electrode, which could affect the weld quality.

2.2.0 Head Protection

Your head is particularly vulnerable to injury. Above all, protect your eyesight since it's your most vital sense. Don't forget your skin and hearing either. Proper PPE keeps your head and those crucial senses safe.

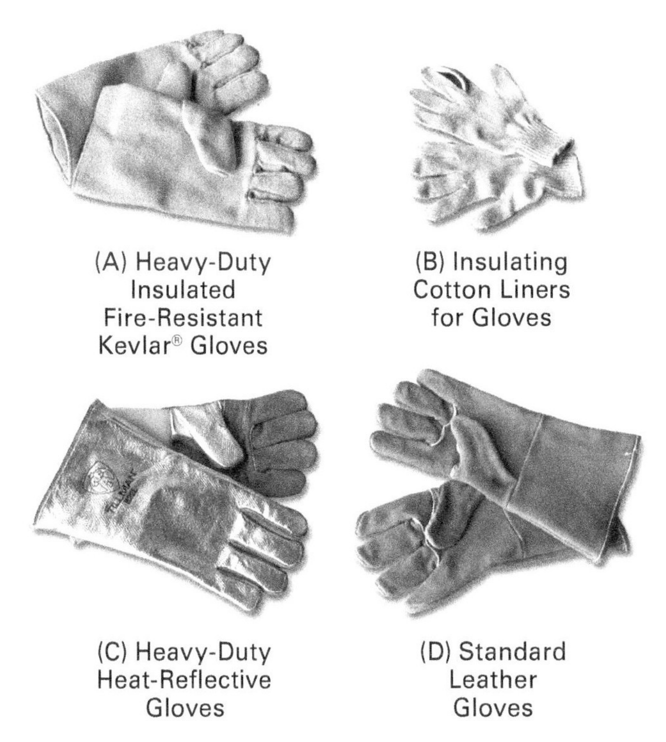

Figure 7 Welding gloves.

Lightweight Welding Gloves

Some operations, such as GMAW, GTAW, brazing, soldering, and oxyfuel welding, don't produce as much heat and spatter as others. For these, soft, lightweight gloves may be a better choice. They give you greater dexterity and make operating the equipment easier.

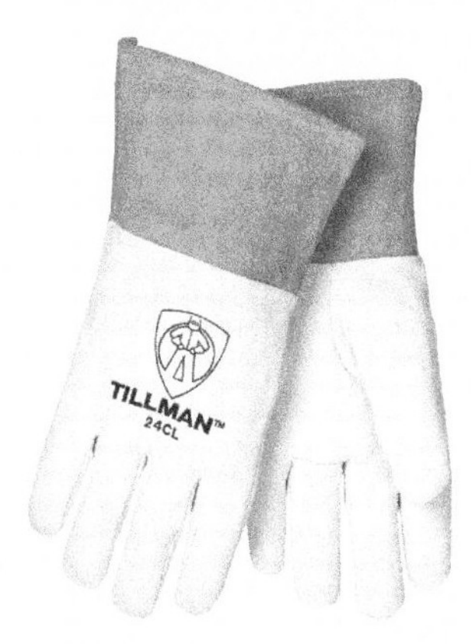

Source: John Tillman Company

2.2.1 Eye and Face Protection

Any industrial environment provides numerous opportunities for eye injuries. Flying parts, metal fragments, abrasive dust, chemical splashes, and sparks can damage the eyes either temporarily or permanently. Since all craftworkers rely on their eyesight, protecting it is a top priority. For this reason, all craftworkers must wear appropriate eye protection on the jobsite. It may be regular safety glasses or a highly specialized piece of PPE.

Welders experience more head hazards than many other workers as they do their jobs. Sparks, molten metal, and metal fragments can fly into the eyes or injure the face during welding and cutting. Even worse, however, is the visible and invisible light that a welding arc, as well as some cutting equipment, produces. The visible light is extremely bright and can strain or injure the eyes over time.

Welding arcs, cutting arcs, and flames also produce invisible light that can harm the eyes and skin. All emit infrared (IR) radiation, which your body senses as heat. Too much can burn your skin. Welding and cutting arcs also emit UV radiation, which is especially dangerous. Viewing an arc without the right eye protection can cause both short- and long-term damage (**flash burn**). UV radiation also burns skin (**arc burn**) and can cause skin cancer over time.

Flash burn: A burn to the eyes produced by brief exposure to intense heat and ultraviolet light. Also called *welder's flash*.

Arc burn: A burn to the skin produced by brief exposure to intense heat and ultraviolet light.

WARNING!

Never look directly or indirectly at an arc without suitable eye protection. Follow industry guidelines when selecting equipment. Always check for unprotected nearby workers before beginning a task.

As *ANSI/AWS Z49.1* specifies, welders must wear protective spectacles with side shields, goggles, or other approved eye protection in addition to a helmet or face shield with a filtered lens. The spectacles or goggles may have either clear or filtered lenses, depending on the operation and welding/cutting type.

For overhead oxyfuel cutting operations, a leather hood may be used in place of a face shield to provide protection from sparks and molten metal. Combine goggles with a leather soft-cap when cutting and burning. A face shield with flip-up lenses can be used at some sites if the lens is dark enough for the application. *Figure 8* shows typical eye and face protection for oxyfuel work.

NOTE

Good-quality welding hoods usually include a manufacturer's guarantee that they block UV radiation.

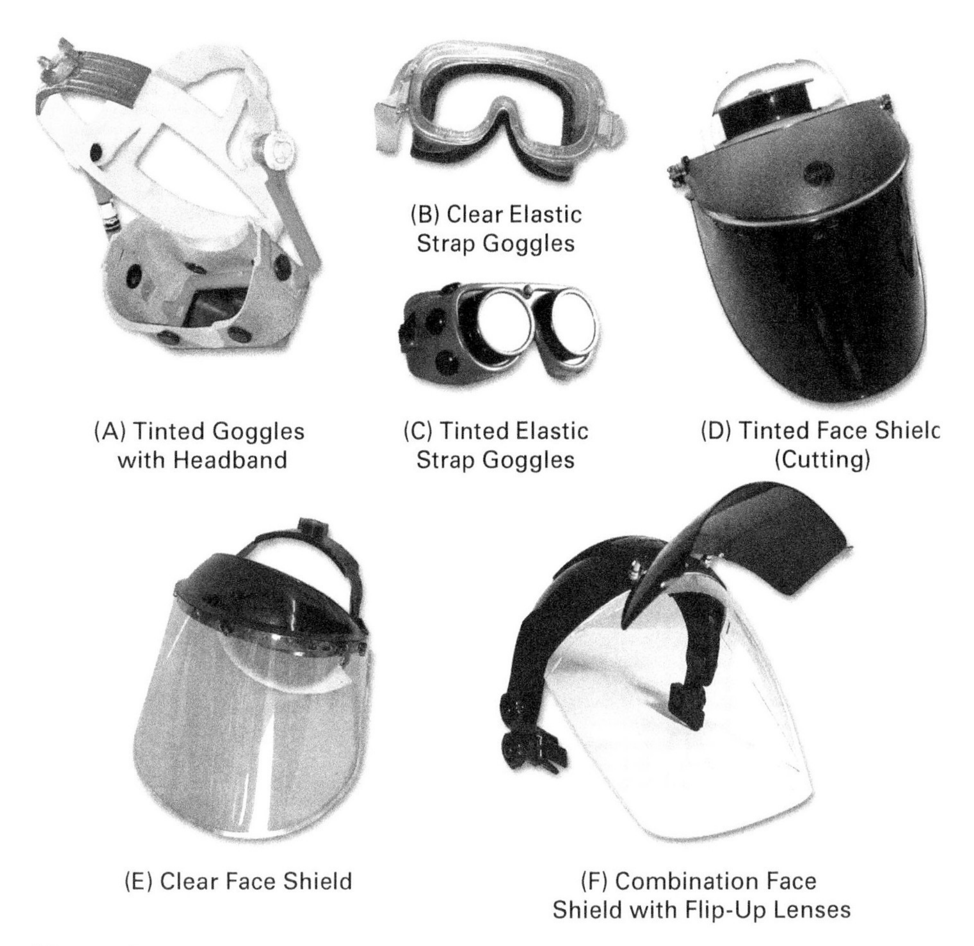

Figure 8 Oxyfuel eye and face PPE.
Source: Sellstrom Manfacturing (8F)

(A) Standard Size Flip-Lens Faceplate

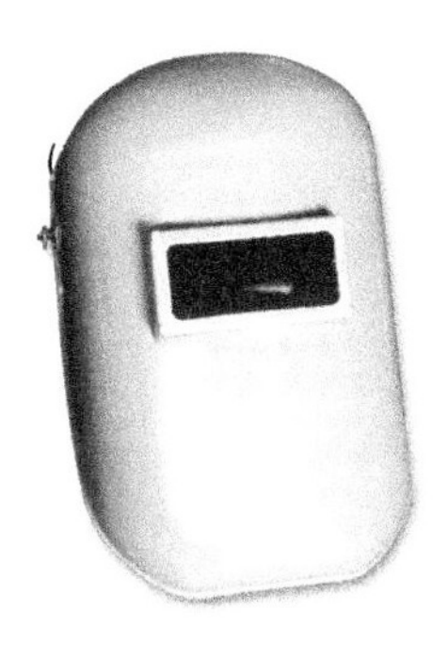

(B) Standard Size Fixed-Lens Faceplate (Standard Helmet)

(C) Large-Lens Faceplate

Figure 9 Arc-welding helmets.

For arc-based operations, welders must wear a helmet with a suitably tinted lens over their safety glasses. Welders can choose from several helmet styles (*Figure 9*). Many have tinted lenses that flip up, leaving a clear safety lens to protect the eyes. This style lets the welder keep the helmet on when chipping and grinding. Other helmets have additional side lenses in a lighter tint so welders can sense activities beside them.

Figure 10 summarizes the correct shade number for common cutting and welding operations and technologies.

NOTE

Manufacturers rate tinted lenses with a standardized *shade number* ranging from 2 to 14. The higher the number, the darker the lens. When selecting a lens, choose the darkest shade possible that still allows you to see adequately. While numbers as low as 7 can work with certain types of arc welding, shade numbers between 9 and 14 are more typical. Cutting produces less light, so lower shade numbers are more appropriate.

WARNING!

When purchasing helmets and lenses, always confirm that they comply with *ANSI/ISEA Z87.1*, *Occupational and Educational Personal Eye and Face Protection Devices*. Using non-compliant eye protection can lead to injury and damaged eyesight.

2.2.2 Hearing Protection

Industrial environments can be very noisy, which damages hearing over time. Hot sparks can burn the ears and enter the ear canals during overhead welding. Such burns are extremely painful, so it is important to wear earplugs when welding (*Figure 11*). Wearing a leather hood also reduces the chances of burns.

Disposable earplugs protect hearing by reducing noise to a safe level. They usually have an outer soft foam layer and an inner fiber core that filters out harmful noise but allows normal conversation to pass through. To use disposable earplugs, roll each plug into a cone and insert the tapered end into the ear canal while pulling up on the upper part of the ear. The earplugs will expand, filling the ear canal and creating a proper fit.

WARNING!

Never use plain cotton or bathroom tissue for ear protection. Always use approved ear protection devices.

Arc Welding Processes

Process	Electrode	Amperage	Lens Shade Numbers (2 3 4 5 6 7 8 9 10 11 12 13 14)
Shielded Metal Arc Welding (SMAW)	< 3/32" (2.4 mm)	< 60A	7
	3/32" – 5/32" (2.4 mm – 4.0 mm)	60A – 160A	8 9 10
	5/32" – 1/4" (4.0 mm – 6.4 mm)	160A – 250A	10 11 12
	> 1/4" (6.4 mm)	250A – 550A	11 12 13 14
Gas Metal Arc Welding (GMAW) and Flux Cored Arc Welding (FCAW)		< 60A	7
		60A – 160A	10 11
		160A – 250A	10 11 12
		250A – 500A	10 11 12 13 14
Gas Tungsten Arc Welding (GTAW)		< 50A	8 9 10
		50A – 150A	8 9 10 11 12
		150A – 500A	10 11 12 13 14
Plasma Arc Welding (PAW)		< 20A	6 7 8
		20A – 100A	8 9 10
		100A – 400A	10 11 12
		400A – 800A	11 12 13 14
Carbon Arc Welding (CAW)			14

Arc Cutting Processes

Process	Electrode	Amperage	Lens Shade Numbers (2 3 4 5 6 7 8 9 10 11 12 13 14)
Plasma Arc Cutting (PAC)		< 20A	4
		20A – 40A	5
		40A – 60A	6
		60A – 80A	8
		80A – 300A	8 9
		300A – 400A	9 10 11 12
		400A – 800A	10 11 12 13 14
Air-Carbon Arc Cutting (A-CAC)	Light	< 500A	10 11 12
	Heavy	500A – 1,000A	11 12 13 14

Gas Processes

Process	Type	Plate Thickness	Lens Shade Numbers (2 3 4 5 6 7 8 9 10 11 12 13 14)
Oxyfuel Welding (OFW)	Light	< 1/8" (3 mm)	4 5
	Medium	1/8" – 1/2" (3 mm – 13 mm)	5 6
	Heavy	> 1/2" (13 mm)	6 7 8
Oxyfuel Cutting (OC)	Light	< 1" (25 mm)	3 4
	Medium	1" – 6" (25 mm – 150 mm)	4 5
	Heavy	> 6" (150 mm)	5 6
Torch Brazing (TB)			3 4
Torch Soldering (TS)			2

Lens shade values are based on OSHA minimum values and ANSI/AWS suggested values. Choose a lens shade that's too dark to see the weld zone. Without going below the required minimum value, reduce the shade value until it's visible. If possible, use a lens that absorbs yellow light when working with fuel gas processes.

Figure 10 Guide to Lens Shade Numbers.

Auto-Darkening Helmets

In the 1980s, a new welding technology came on the market as an alternative to the traditional *passive helmet*—the *auto-darkening welding helmet*. Today, this technology is mature and affordable. It's also extremely convenient and something to consider when choosing a helmet.

Auto-darkening helmets have clear lenses that filter out both infrared and UV radiation. You can see through them easily when performing non-welding tasks. As soon as you start to weld, however, electronic sensors on the helmet trigger the lens, which darkens to the correct shade within less than $^1/_{20,000}$ of a second. When the arc stops, the lens returns to clear.

Helmets run on batteries, although some also have a small solar panel that extends the battery life. Many models have a mode switch that selects between several lens shade ranges. For example, in the *Weld* mode, the lens darkens to between 8 and 13, while in *Cutting* mode, it darkens to between 5 and 8.

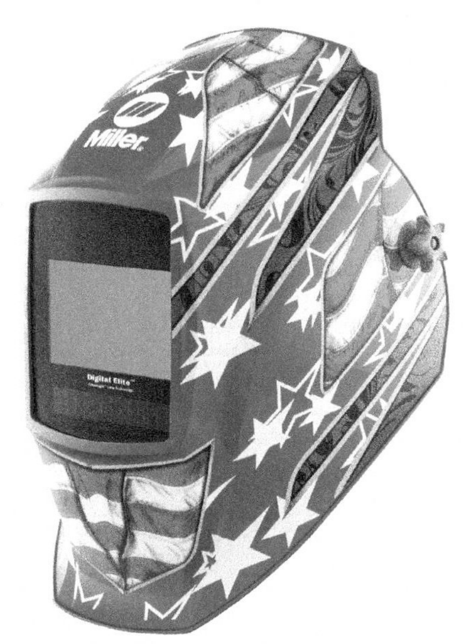

Source: © Miller Electric Mfg. LLC

More expensive models have extra features like multiple sensors. They may also include adjustments to tweak how quickly they darken when they see the arc. This feature is handy since helmets without this feature can be fooled by sunlight and darken unexpectedly.

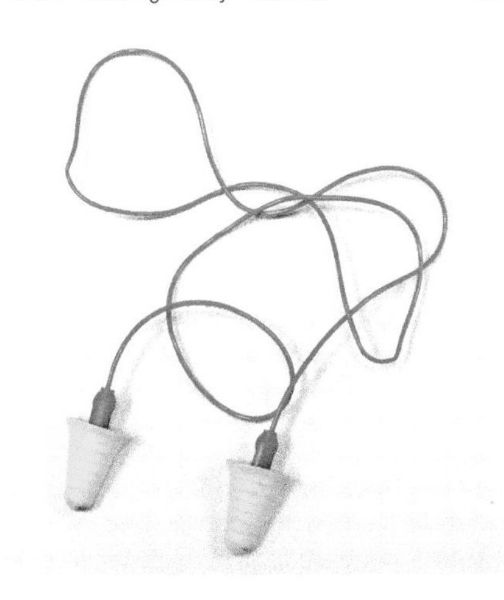

Figure 11 Earplugs.

2.0.0 Section Review

1. Which of the following fabrics is the *best* choice for welding clothing?
   a. Rayon
   b. Polyester
   c. Nylon
   d. Cotton

2. Manufacturers rate a welding helmet lens' light-blocking ability with a(n) _____.
   a. shade number
   b. UV factor
   c. density value
   d. IR/UV index

3.0.0 Welding Safety Practices

Objective

Summarize welding environment hazards and associated safety practices.

a. Outline welding safety and the factors that contribute to accidents.
b. List general workplace safety practices.
c. Describe fire safety, hot work permits, and fire watch requirements.
d. Describe confined spaces and their safety practices.
e. Summarize welding equipment safety practices.
f. Explain respiratory hazards, ventilation, and associated safety equipment.
g. Describe using a Safety Data Sheet (SDS) to handle materials safely.

Performance Tasks

There are no Performance Tasks in this section.

Welding has many associated hazards, including intense heat, flames, and high-intensity electric arcs. Cutting brings additional hazards to the workplace, including explosive and flammable gases, as well as intense flames. These conditions require specialized safety practices to protect both the welder and those working nearby.

3.1.0 Welding Safety Hazards

NOTE

The American Welding Society (AWS) publishes many welding-related standards as well. Trainee welders should become familiar with these too.

Confined space: A potentially hazardous space, not meant to be continuously occupied, that has limited entry and exit routes.

The welding industry has a well-developed safety standard: *ANSI/AWS Z49.1, Safety in Welding, Cutting, and Allied Processes.* All welders should be familiar with this standard and consult it when necessary. It outlines PPE, ventilation, fire prevention, and welding in a **confined space**. It provides precautions for welders, supervisors, and managers, as well as specific welding safety procedures.

Accidents have many causes, but they usually divide into two categories: *personal* and *physical*. As they work, welders must think ahead, spotting factors that could lead to accidents and then correcting them.

3.1.1 Personal Factors That Cause Accidents

Many aspects of your daily life can lead to accidents. Some are harmless by themselves but become significant risks when you're doing a demanding job. For example, a minor illness is normally no more than a nuisance. But an ill person may struggle to concentrate or have trouble handling strenuous work. Ill people performing dangerous tasks tend to have accidents more often than healthy people.

Mental stress reduces concentration too, which also leads to accidents. Someone who is worrying about a personal problem won't focus fully on the task. If you find yourself in this situation, you need to stop and deal with the situation. Distractions can interfere with your ability to concentrate on work tasks, making accidents and injuries far more likely.

Age and inexperience contribute to accidents too. Insurance company studies show that young, inexperienced workers take more risks or don't recognize a hazard until it's too late. Experienced workers, on the other hand, see problems coming and deal with them. There's no fast way to become experienced. Instead, recognize your limitations and constantly remind yourself to stop and think before acting. Learn from more experienced workers by asking questions or asking for help if you're uncertain.

Alcohol and illegal drugs have no place in the workplace. They dull or distort the senses and prevent a worker from doing jobs safely and accurately. Hangovers can be almost as risky since they distract people from paying attention to their job. Never work when you're under the influence and refuse to work with those who are.

Remember that over-the-counter drugs and legal prescriptions can be hazardous too. Some cold and allergy medicines can make you sleepy or sluggish. Many prescriptions include warnings about driving, operating equipment, or performing dangerous tasks. If you're taking one of these, don't perform any hazardous work. Speak to your supervisor. Many companies switch employees taking medications to light duties.

Many people feel sleepy at specific times in the day or after eating. Learn to recognize these factors or conditions and be extra careful. Working overtime too many days in a row can lead to excessive fatigue or a sleep debt. If you feel that the schedule or working conditions are preventing you from working safely, speak to your supervisor.

3.1.2 Physical Factors That Cause Accidents

Jobsite conditions contribute to accidents. Noise, vehicle traffic, crowded spaces, or too many people working together can make it harder to work safely. Many people hurry or cut corners as quitting time approaches. Remind yourself not to develop this pattern in your own work life and be careful around others who do.

Clean and organized workspaces are safer. Put away unneeded tools and clean up as you work. Keep your tools together rather than scattering them. Coil up cords and hoses so they don't trip people or become damaged. Handle solvents and flammable materials carefully. Dispose of rags and waste in approved containers. Since welding involves a lot of heat, think twice when working with flammable materials. Always follow your company's safety and hazardous waste rules.

Using damaged or defective equipment is a bad idea. A frayed or damaged welding cable could give someone a shock or start a fire. Don't use defective equipment. Tag it for repair and get a replacement immediately. Always inspect your equipment at the start of every shift. Follow company maintenance schedules or the manufacturer's guidelines.

3.2.0 Work Area Safety

Working in a safe environment means fewer accidents. Sometimes, you have little control over it, particularly if you're working in the field. But a welding shop is another matter. You can control that environment and make it a safe and productive place to work.

3.2.1 Welding in the Shop

A good welding shop helps prevent fires. Ideally, it has bare concrete floors, as well as bare metal walls and ceilings. These are **electrically grounded** and **bonded**. Never keep solvents or other flammable chemicals in the welding shop. Store rags, waste, scrap wood, and other flammable materials elsewhere.

Electrically grounded: Connected to the earth or a conducting body that behaves like the earth electrically.

Bonded: Permanently joining metallic parts to form a conductive path that can safely carry a specific electric current.

WARNING!

Never weld or cut over wood floors, as this can start fires. When you can't avoid welding or cutting over a wood floor or wood planks, cover everything with fireproof blankets.

Keep the shop clean and organized. Put away tools and keep the floors free from clutter. Clean up grease and oil. Sweep benches and the floor to keep metal fragments under control. Organize stock and scrap materials rather than letting them pile up. Keep hoses and cables coiled up and hung on hooks. Collect electrode stubs in a caddy or steel bucket rather than throwing them on the floor.

Set aside a space for cutting and grinding. Welding generates hot metal spatter, which could melt holes in cutting torch hoses. Keeping cutting equipment away from welding prevents this problem. Separate workspaces also keep craftworkers from interfering with each other, leading to a safer environment.

Every welding shop should have a first aid kit or be close to a designated first aid station. It should also have appropriate fire extinguishers and fire-prevention equipment, like fireproof blankets. Signs should clearly identify safety equipment, emergency procedures, and emergency telephone numbers.

NOTE

Every company has its own safety protocols. Know yours thoroughly. Regularly go over emergency procedures and remind yourself of what to do for each emergency type. Know your company's safety officer or manager and know how to contact that person.

3.2.2 Welding in the Field

Welding in the field or outside the shop creates more opportunities for accidents. You can't control the environment as well, and there are more distractions and other workers nearby. Constantly watch for conditions that could lead to an accident.

If you must weld in an enclosed building, try to eliminate anything that could trap sparks. They can smolder for hours and then burst into flames. Always have a charged fire extinguisher nearby. If you must leave a piece of equipment or hot metal unattended, write the word HOT on it with soapstone before leaving.

NOTE

Besides being a good safety practice, clearly identifying hot metal or equipment is also an OSHA requirement.

Protect other workers from welding glare and UV radiation by putting up portable welding screens (*Figure 12*). These also control drafts, which can

Figure 12 Welding screen.
Source: Sellstrom Manufacturing

Figure 13 Welding blanket.

interfere with the welding arc. Cover material and equipment that could be burned or damaged with welding blankets (*Figure 13*).

3.2.3 General Safety Reminders

The following are useful workplace safety reminders:

- Eliminate trip hazards by coiling cables and keeping clamps and other tools off the floor.
- Don't let yourself get tangled in cables, loose wires, or clothing while you work. You must be able to move freely, especially if your clothing ignites or an accident happens.
- Clean up oil, grease, and other lubricants that could ignite and splatter off surfaces.
- Shut off welding machines and disconnect their power before servicing them or performing maintenance. Hard-wired machines should have a dedicated circuit breaker or disconnect switch. Follow lockout/tagout (LOTO) procedures. Unplug all other machines.
- Don't throw electrode stubs on the floor. Collect them in a caddy or steel bucket.
- Always remove SMAW electrodes from their holders before laying the holder down. Leaving a stub in the electrode holder could cause an arc to a nearby surface.
- Work in a dry area, booth, or other shielded area whenever possible.
- Check for open doors or windows through which sparks could travel and start fires.
- Identify the nearest fire extinguisher before starting work.
- Don't wrap wires, hoses, or cables around your body, especially when working on scaffolding.
- Jobsite conditions can change quickly! Pay attention to your surroundings. Modify your work practices as required to stay safe.

3.3.0 Fire Safety and Hot Work

Welding and cutting are *hot work*, which means that fire is a regular hazard. When a task poses a fire risk, the site manager must issue a *hot work permit* (*Figure 14*). It confirms that the site has been surveyed and is safe for welding or cutting. Hot work permits also provide documentation and accountability. They include the time, location, and type of work.

WARNING!

Cutting and welding tanks and containers carry several extra risks beyond those of other hot work. The tank may contain flammable or toxic vapors, liquids, or residues. Performing hot work could start a fire, cause an explosion, or release dangerous substances into the work environment. Obtaining a hot work permit for a tank or container requires a qualified person to inspect it and certify that it's safe for hot work. If it's not, a qualified person must render it safe before work can begin.

The hot work permit system promotes better fire safety. Permits help managers keep track of hot work, so they know which workers are doing it at a particular time. This information helps coordinate emergency responses and evacuations during fires.

NOTE

OSHA requires both hot work permits and fire watches. Many companies severely discipline employees who bypass these safety protocols.

3.3.1 Fire Watches

A *fire watch* requires a specific worker to constantly monitor the jobsite for fires during hot work. Fire watch personnel perform no other duties, nor do they assist with the work. They must be ready to use a fire extinguisher or activate an alarm as required.

HOT WORK PERMIT

For Cutting, Welding, or Soldering with Portable Gas or ARC Equipment

(References: 1997 Uniform Fire Code Article 49 & National Fire Protection Association Standard NFPA 51B.)

Job Date____________ Start Time____________ Expiration_____________ WO #_____________

Name of Applicant__________________________ Company_____________ Phone_____________

Supervisor_______________________________ Phone_______________

Location / Description of work ____

IS FIRE WATCH REQUIRED?

1. _______ (yes or no) Are combustible materials in building construction closer than 35 feet to the point of operation?
2. _______ (yes or no) Are combustibles more than 35 feet away but would be easily ignited by sparks?
3. _______ (yes or no) Are wall or floor openings within a 35 foot radius exposing combustible material in adjacent areas, including concealed spaces in floors or walls?
4. _______ (yes or no) Are combustible materials adjacent to the other side of metal partitions, walls, ceilings, or roofs which could be ignited by conduction or radiation?
5. _______ (yes or no) Does the work necessitate disabling a fire detection, suppression, or alarm system component?

YES to any of the above indicates that a qualified fire watch is required.

Fire Watcher Name(s) ______________________________________ Phone_____________

NOTIFICATIONS

Notify the following groups at least 72 hours prior to work and 30 minutes after work is completed.
Write in names of persons contacted.

Notify in person OR by phone ONLY if question #5 above is answered "yes":

- Facilities Management Fire Alarm Supervisor

Notify by phone or in person: (If by phone, write down name of person and send them a completed copy of this permit.)

- Facilities Management Fire Protection Group
- Environmental Health & Safety Industrial Hygiene Group

SIGNATURES REQUIRED

University Project Manager________________________________ Date ___________ Phone______________

I understand and will abide by the conditions described in this permit. I will implement the necessary precautions which are outlined on both sides of this permit form. Thirty minutes after each hot work session, I will reinspect work areas and adjacent areas to which spark and heat might have spread to verify that they are fire safe, and contact Facilities Management Alarm Technicians to have any disabled fire protection systems reactivated.

__________________________ __________________________ Date ____________ Phone______________
Permit Applicant Company or Department

1/17/15

Figure 14 Hot work permit.

Welding, cutting, and heating operations all require a fire watch. Fuel-burning equipment, like torches, poses fire risks beyond that of the work itself. These require additional vigilance and care. Hot work permits and fire watches reduce these dangers to manageable levels.

WARNING!

Never perform heating, cutting, or welding tasks until you have obtained a hot work permit and established a fire watch. If you are unsure of the proper procedure, check with your supervisor. Violating the hot work permit and fire watch procedures can result in serious injury or death. OSHA requires maintaining a fire watch for thirty minutes after finishing a welding or cutting job.

3.3.2 Fire/Explosion Prevention

Most welding and cutting processes involve heat. Welding or cutting a vessel or container that once contained flammable or explosive materials is hazardous. Residues can catch fire or explode. Before welding or cutting vessels, check whether they contained explosive, hazardous, or flammable materials. These include petroleum products, citrus products, or chemicals that release toxic fumes when heated.

American Welding Society (AWS) F4.1, Safe Practices for the Preparation of Containers and Piping for Welding and Cutting, and *ANSI/AWS Z49.1* describe safe practices for these situations. Begin by cleaning the container to remove any residue. Steam cleaning, washing with detergent, or flushing with water are possible methods. Sometimes you must combine these to get good results.

NOTE

A *lower explosive limit (LEL)* gas detector can check for flammable or explosive gases in a container or its surroundings. Gas monitoring equipment checks for specific substances, so be sure to use a suitable instrument. Test equipment must be checked and calibrated regularly. Always follow the manufacturer's guidelines.

WARNING!

Clean containers only in well-ventilated areas. Vapors can accumulate in a confined space during cleaning, causing explosions or toxic substance exposure.

After cleaning the container, you must formally confirm that it's safe for welding or cutting. *American Welding Society (AWS) F4.1, Safe Practices for the Preparation of Containers and Piping for Welding and Cutting*, outlines the proper procedure. The following three paragraphs summarize the process.

Immediately before work begins, a qualified person must check the container with an appropriate test instrument and document that it's safe for welding or cutting. Tests should check for relevant hazards (flammability, toxicity,

Oxyfuel Equipment Safety Precautions

- Identify the nearest fire extinguisher before starting work.
- Always light the torch flame with an approved torch lighter to avoid burning your fingers.
- Never point the torch at anyone when lighting or using it.
- Never point the torch at the cylinders, regulators, hoses, or anything else that may be damaged and cause a fire or explosion.
- A torch that is not in the worker's hands must be extinguished. Never place a lighted torch on a bench or workpiece. Never hang it up while lighted either.
- To prevent fires and explosions, install check valves and flashback arrestors in all oxyfuel equipment.
- When cutting with oxyfuel equipment, clear all combustible materials from the area.
- Liquid oxygen can damage skin by causing frostbite. Handle it carefully.
- Never substitute oxygen for compressed air or use it for blowing the workpiece clean. You could cause a serious fire.
- Set aside any fuel or oxygen cylinder with defective valves or safety features. The supplier should repair or replace them. Don't attempt a repair yourself.
- Never use a hammer or wrench to open valves.

etc.). During work, repeated tests must confirm that the container and its surroundings remain safe.

Alternatively, fill the container with an inert material ("inerting") to drive out any hazardous vapors. Water, sand, or an inert gas like argon meets this requirement. Water must fill the container within 3" (~7.5 cm) or less of the work location. The container must also have a vent above the water so air can escape as it expands (*Figure 15*). Sand must completely fill the container.

When using an inert gas, a qualified person must supervise, confirming that the correct amount of gas keeps the container safe throughout the work. Using an inert gas also requires additional safety procedures to avoid accidental suffocation.

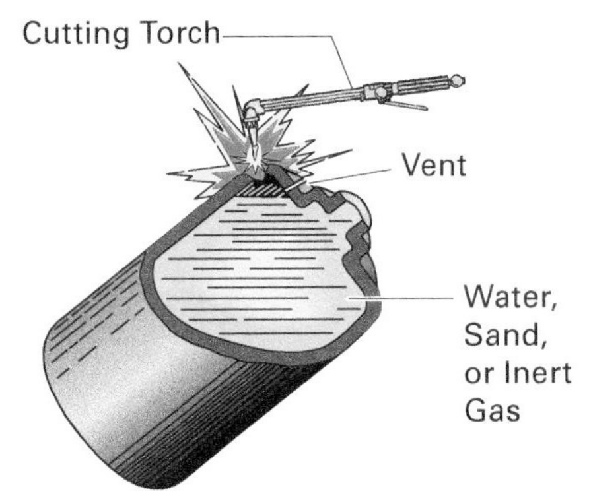

Note: *ANSI Z49.1* and AWS standards should be followed.

Figure 15 "Inerting" a container.

WARNING!

Never weld or cut drums, barrels, tanks, vessels, or other containers until they have been emptied and cleaned thoroughly. Residues, such as detergents, solvents, greases, tars, or corrosive/caustic materials, can produce flammable, toxic, or explosive vapors when heated. Never assume that a container is clean and safe until a qualified person has checked it with a suitable test instrument. Never weld or cut in places with explosive vapors, dust, or combustible products in the air.

3.3.3 Fire Extinguisher Guidelines

A fire extinguisher is your main tool for dealing with small fires. All welding sites must have fire extinguishers nearby. These must be fully charged and have current inspection tags. All workers should know how to operate them correctly. Before starting work, confirm that you know the nearest one's location.

To use a fire extinguisher, remember the **PASS** method:

- **P**ull the pin from the handle.
- **A**im the nozzle at the base of the fire while standing 8' to 10' away.
- **S**queeze the discharge handle.
- **S**weep the nozzle back and forth at the base of the fire.

NOTE

NCCER Module 00101, *Basic Safety*, introduces fire extinguishers. Review the relevant section if you're uncertain about fire extinguisher types (A, B, C, A-B-C, and D) or how to use one.

WARNING!

Flammable metals like magnesium can ignite during welding or cutting. They burn very intensely and are hard to extinguish. Never pour water on them. Never use Class A, Class B, Class C, or A-B-C fire extinguishers. Water, CO_2, and the dry powder in regular extinguishers will make the fire burn faster and more intensely. **Extinguish metal fires with a Class D fire extinguisher only.** If you're working with flammable metals, be sure that you have one in your work area.

3.4.0 Confined Spaces

A confined space is a relatively small or restricted space with limited ways to enter and exit it. They're not meant to be occupied, except for specific work tasks. Storage tanks, boilers, pressure vessels, small compartments, and underground utility vaults are all confined spaces. Some small rooms or even unventilated corners or alcoves in a larger room may also qualify as confined spaces.

OSHA 29 CFR 1910.146 defines a confined space as a space that

- is large enough and so configured that an employee can bodily enter and perform assigned work.
- has a limited or restricted means of entry or exit, for example, tanks, vessels, silos, storage bins, hoppers, vaults, and pits.
- is not designed for continuous employee occupancy.

WARNING!

This module is *not* a confined spaces training program. Completing it does *not* qualify you for working in confined spaces.

Confined spaces can be hazardous since they're more difficult to exit quickly. Most are poorly ventilated and have limited oxygen capacity. They can trap fumes, particles, and gases. Some are small enough that they can trap workers. Working in a confined space requires careful thought and planning to avoid unnecessary risks. Before working in a confined space, workers must get a confined space entry permit (*Figure 16A* and *Figure 16B*).

OSHA 29 CFR 1910.146 further defines a permit-required confined space as a space that

- contains or has the potential to contain a hazardous atmosphere.
- contains a material that has the potential for engulfing an entrant.
- has an internal configuration such that an entrant could be trapped or asphyxiated by inwardly converging walls or by a floor that slopes downward and tapers to a smaller cross-section.
- contains any other recognized serious safety or health hazard.

3.4.1 Working in Confined Spaces

Maintaining a breathable atmosphere in the confined space is a top priority. Normal air contains about 21.5 percent oxygen by volume. The oxygen level in the confined space's air must stay between 19.5 and 23.5 percent by volume. Anything outside this range is hazardous.

Levels below 19.5 percent are *deficient* and cause confusion, poor performance, unconsciousness, and death. Levels above 23.5 percent are *enriched* and dangerous for a different reason. Too much oxygen makes fires start more easily. Once started, they'll burn hotter and more aggressively. If oxygen saturates clothing, it will burn readily and may be difficult to extinguish.

Table 1 summarizes the effects of different oxygen levels in the air.

TABLE 1 Effects of Different Oxygen Levels in Air

Oxygen Level	Effects
> 23.5%	Easy ignition of flammable material such as clothing
19.5% to 21.5%*	Normal levels
17%	Deterioration of night vision, increased breathing rate, accelerated heartbeat
14% to 16%	Very poor muscular coordination, rapid fatigue, intermittent respiration
6% to 10%	Nausea, vomiting, inability to perform, unconsciousness
< 6%	Spasmodic breathing, convulsive movements, and death within minutes

*There is a safety zone of 2%. Above 23.5% is considered enriched.

WARNING!

Never ventilate a confined space with oxygen. It can enrich the atmosphere to the point that fires will start easily and burn vigorously. Oxygen will saturate the worker's clothing, making it a fire hazard.

Before starting work in a confined space, block portable equipment wheels to prevent accidental movement. Work out a plan to remove the welder quickly in an emergency (*Figure 17*). When using safety belts and lifelines, be sure they won't make the welder too large to exit the space. Another worker stationed

Master Card No.______________

1. Work Description

Area____________________ Equipment Location____________________

Work to be done:

2. Gas Test		Results	Recheck	Recheck
Required	☐ Instrument Check			
☐ Yes	☐ Oxygen % 20.8 Min.			
☐ No	☐ Combustible % LFL			
		Date/Time/Sig.	Date/Time/Sig.	Date/Time/Sig.

3. Special Instructions: ☐ Check issuer before beginning work ☐ None

4. Hazardous Materials: ☐ None What did the line / equipment last contain?

5. Special Protection Required: ☐ None ☐ Forced Air Ventilation

☐ Avoid Skin Contact ☐ Gloves ____________ ☐ Suit ____________________

☐ Goggles or Face Shield ☐ Respirator ________________ ☐ Safety Harness

☐ Self-Contained Breathing Equipment ☐ Hoseline Breathing Equipment

☐ Other, Specify ________ ☐ Standby - Name: ______________________________

6. Fire Protection Required: ☐ None ☐ Portable Fire Extinguisher ☐ Fire Hose and Nozzle

☐ Fire Watch ☐ Other, specify:

7. Condition of Area and Equipment

Required Yes	No	THESE KEY POINTS MUST BE CHECKED
		a. Lines disconnected & blinded or where disconnecting is not possible, blinds installed? (Includes drains, vents and instrument leads) and appropriate valves locked out?
		b. Equipment cleaned, washed, purged, ventilated?
		c. Low voltage or GFCI-protected electrical equipment provided?
		d. Explosion-proof electrical equipment provided?
		e. Life lines required to be attached to safety harnesses?

Comments

rev.6/10/15

Figure 16A Confined space entry permit (1 of 2).

8. Approval			Permit Authorization Area Supervisor	Permit Acceptance Maint./Contractor Supervisor/Engineer	Date	Time
	Date	Time				
Issued by						
Endorsed by						
Endorsed by						
Endorsed by						

9. Individual Review

I have been instructed on proper Safety Procedures and proper Confined Space Entry Procedures. I have signed in on the appropriate Master Card and have affixed personal locks on energy isolation devices as appropriate.

Signature of all personnel covered by this permit.

Forward to Production Superintendent 7 days after completion of work.

rev.6/10/15

Figure 16B Confined space entry permit (2 of 2).

outside the confined space must monitor the welder constantly and be ready to initiate a rescue.

During lunch, significant breaks, or overnight, remove electrodes from their holders. Place electrode holders in a safe position where they can't arc. Disconnect welding machines from the power source. Prevent fuel gas or oxygen from accumulating in the space by closing the supply valves. Release the regulators and bleed the lines. Shut off the torch valve. When feasible, remove the torch and its hoses from the confined space.

If other workers will enter the confined space immediately after the welder leaves, mark all hot metal or post a clear warning. Don't leave unnecessary equipment in the space as it will make others' jobs more difficult.

WARNING!

Shielding gases and **purge gases** are not toxic, but they cannot support life either. If they accumulate in a confined space, they can push the oxygen-containing air out. Since they are odorless, the welder won't notice them and may pass out and die. When using *any* gas in a confined space, be extremely cautious and follow all safety protocols.

Purge gases: Inert gases, such as argon, that drive oxygen-containing air away from a weld zone to protect it from defects.

3.5.0 Welding Equipment Safety

Welding equipment carries a variety of hazards and safety risks. Welders must be especially aware of electrical dangers. Some welding machines are large and heavy, creating fall and crush risks. Engine-powered equipment requires special safety precautions. *Figure 18* shows example welding machine safety notices.

Figure 17 Worker in a confined space.

Did You Know?

Oxygen-Deficient Air

Most people don't realize that entering an oxygen-deficient space will lead to death within a very short time. After just a few breaths, most people become confused and uncoordinated. They lose consciousness in less than a minute. Within eight to ten minutes, they're dead. Those with underlying health problems can die even more quickly.

At a refinery in Corpus Christi, Texas, a welder's helper entered a pipe while the welder was at lunch. Argon had been used as a shielding gas, so much of the oxygen-containing air had been displaced. When the welder returned, he found the helper dead.

In the still environment of a confined space, nitrogen and helium rise to the top of the air. Argon and carbon dioxide sink to the bottom. Opening the top of a confined space makes it easier to remove light gases. Opening the bottom makes removing heavy ones easier. In either case, however, powered ventilation equipment is the best way to flush out the gases and restore normal air.

INSTALLATION SAFETY PRECAUTIONS

⚠ WARNING

ELECTRIC SHOCK CAN KILL.

- Do not touch electrically live parts or electrode with skin or wet clothing.
- Insulate yourself from work and ground.
- Always wear dry insulating gloves.

FUMES AND GASES CAN BE DANGEROUS.

- Keep your head out of fumes.
- Use ventilation or exhaust to remove fumes from breathing zone.

WELDING SPARKS CAN CAUSE FIRE OR EXPLOSION.

- Keep flammable material away.
- Do not weld on closed containers.

ARC RAYS CAN BURN EYES AND SKIN

- Wear eye, ear, and body protection.

Figure 18 Safety notices.

3.5.1 Electrical Hazards

Electrical shock from welding and cutting equipment can cause death, painful injury, or severe burns. Electrical shocks also cause secondary injuries when workers fall or slam into equipment. The shock's dangerousness, as well as its short- and long-term consequences, depends on the body parts that the electrical current passes through (*Figure 19*). The amount of current—measured in amperes (A) or milliamperes (mA)—also determines its consequences.

The Construction Employers Association has issued the following guidelines regarding the seriousness and consequences of electrical shock:

Greater than 3 mA — Indirect accident

Greater than 10 mA — Muscle contraction

Greater than 30 mA — Lung paralysis, usually temporary

Greater than 50 mA — Possible **ventricular fibrillation**

100 mA to 4A — Certain ventricular fibrillation

Greater than 4A — Heart paralysis, severe burns

15A, 20A, and 30A — Common household electrical circuits

Ventricular fibrillation: A lethal heart rhythm in which the heart muscle writhes rather than pumps.

Generally, higher voltages cause higher current flows, leading to more serious shocks. But even relatively low voltages can deliver significant current if the body's electrical resistance is low. Wet or cut skin has a much lower resistance than dry, intact skin. Similarly, standing in water or on a damp surface lowers electrical resistance.

Even relatively large currents won't cause death if they flow through a non-vital body part. These shocks, however, are more insidious since they damage internal tissue along the current's path. Workers may not realize that they're badly injured until several days later when the tissue dies and puts them in the hospital.

All craftworkers who use electrical equipment must receive general training in electrical safety, as well as in the specific equipment that they use. They should also know how to respond if a coworker is shocked or injured by equipment. Most companies have a safety officer and employee first responders in each part of the facility. They may also have first aid stations that can handle common injuries. Many keep automatic external defibrillators (AEDs) in easily accessible locations.

NOTE

Most companies have strict emergency and accident-handling policies. In general, you should not render first aid unless you are properly trained or a qualified first responder isn't available. Many people take first aid and CPR classes so they can act as volunteer first responders within their companies. You may wish to do this, so you'll be able to help coworkers in emergencies.

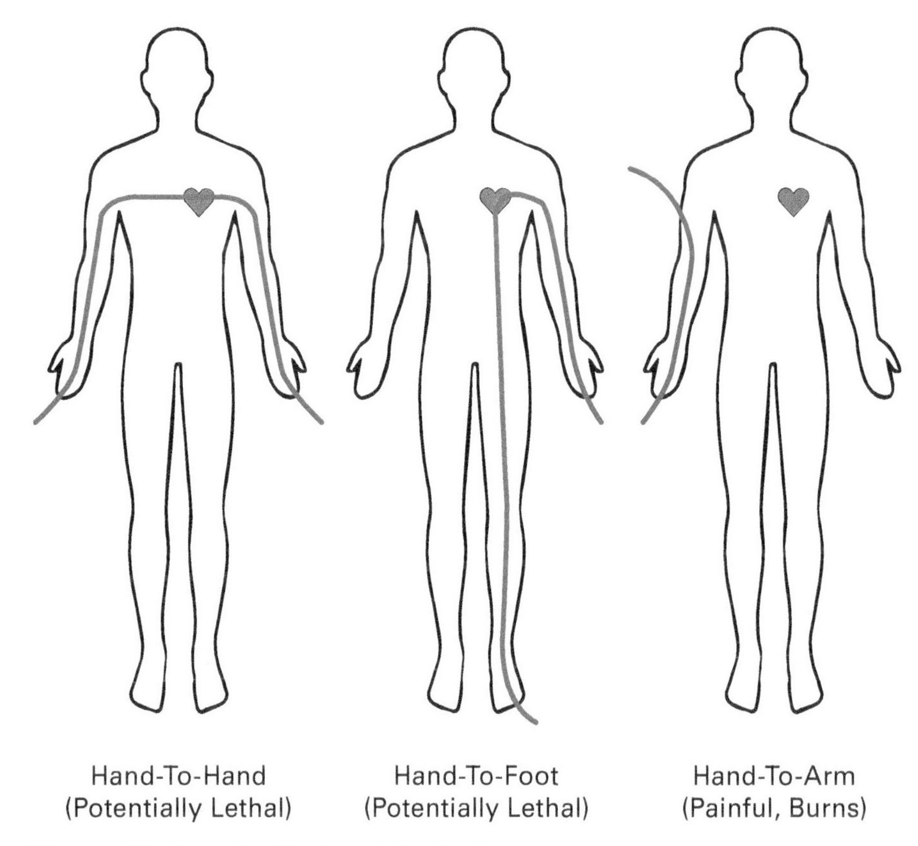

Figure 19 Current paths through the body.

Welding electrical hazards come in two forms. The first comes from the electrical supply that powers the welding equipment. It usually has a voltage between 240V and 480V, which is more than enough to injure and kill. Even when the welding machine is switched off, this voltage will be present inside the machine. Only by unplugging the machine or disconnecting its power at the source will the machine be electrically safe.

WARNING!

Always follow your company's lockout/tagout (LOTO) procedures when removing a welding machine's power. Qualified electricians must install and service permanent welding machines. If a machine develops an electrical problem, refer it to an electrician rather than trying to fix it yourself. Never work on a portable welding machine either.

Welding machines must be properly grounded. Machines that plug into an outlet will be grounded if the building's electrical system is correctly grounded. Permanently connected machines must be wired to a suitable ground by a qualified electrician. Ungrounded or improperly grounded equipment can be a shock hazard.

The welding process itself creates the second kind of electrical hazard. Although the welding machine's output voltage is lower than its supply voltage, it's still potentially dangerous. Leaning against the work while welding can cause a shock. Since the welding machine can deliver a large current, shocks passing through nearby metal objects can heat them dangerously within seconds. For this reason, welders never wear metal jewelry or rings when working.

The following guidelines will help you avoid electrical injuries when welding:

- The electrode voltage is highest when you are not welding. During welding, the voltage drops.
- The electrode holder's jaws and screws, as well as the electrode itself, are electrically energized. Make sure not to touch them or let them touch your clothing, especially if it's damp or wet.

- Never unintentionally touch the electrode to a conductive surface. It will energize the surface and might arc as well.
- Never operate welding equipment on a wet or damp floor.
- Wear dry gloves in good condition when welding. Keep an extra pair on hand.
- Keep dry insulation between your body and the metal you're welding. If the ground or floor is wet, stand on an insulating mat.
- Keep all equipment, including cables and electrode holders, in good condition.
- Ground all metal workbenches.

If you must weld in conditions in which you might touch the work, wear gloves designed to protect against electrical shock. Plug the welder into a ground fault circuit interrupter (GFCI). Using one of the following welding machines reduces the risk too:

- Semi-automatic DC constant-voltage welder
- DC manual (stick) welder
- AC welder with reduced voltage control

3.5.2 Engine-Powered Equipment

When a site has no electrical power, welders must use gasoline- or diesel-powered equipment. These range from small, portable units to large, trailer-mounted machines (*Figure 20*).

Besides the usual electrical hazards, these machines have extra risks associated with their engines:

- Engine-powered machines produce deadly carbon monoxide gas just like car engines. Never operate one indoors or in an enclosed space.
- Carbon monoxide can also enter a confined space if an engine-powered machine is operating upwind of it. Similarly, an engine operating near a ventilation intake can accidentally deliver carbon monoxide to a distant location. Check for these situations very carefully before starting the engine.
- Gasoline and diesel fuel are flammable. Keep them in approved containers away from hot work and running equipment.

4-Wheel Trailer with Multi-Process Welder

Figure 20 Gasoline-powered welding machine.
Source: The Lincoln Electric Company, Cleveland, OH, USA

- When refueling, turn off all welding and cutting flames. Stop the engine and allow it to cool to prevent accidental fires caused by spilling fuel on hot parts.

3.6.0 Fumes and Gases

Welding and cutting produces hazardous gases, fumes, and dust (the **fume plume**). Welders must check for adequate ventilation before starting work to prevent themselves and others from breathing these products. They may choose from one of three ventilation methods:

Fume plume: Smoke, gases, and particles produced from the consumables, metals, and metal coatings during welding.

Natural ventilation — Normal airflow through the workplace

Mechanical ventilation — Fixed or portable fans that increase airflow

Source extraction — A device mounted near the welding arc that removes products as they're produced

As a rule, natural ventilation is adequate if the space meets the following requirements:

- At least 10,000 ft^3 per welder
- A ceiling height no less than 16'
- Cross-ventilation not blocked by partitions, equipment, or other barriers
- Not a designated confined space

If natural ventilation isn't sufficient, welders use one of the other two methods. Many welding shops rely on all three to keep the air as clean as possible. Sometimes, specialists need to determine the correct method for a job.

Regardless of the ventilation methods, never breathe the fume plume. Watch the smoke column and avoid getting too close to it. Some welders use a small fan to redirect it away from their face. Be careful, however, not to let the fan blow on the joint as you weld. It could interfere with the flux gases or shielding gas, causing weld defects that weaken the weld.

WARNING!

The fume plume is dangerous. Overexposure can cause nausea, headaches, dizziness, metal fume fever, and potentially fatal toxic effects. Fumes also irritate eyes, skin, and the respiratory system. In confined spaces, fumes and gases can kill.

Some welding materials, electrodes, and base metals contain toxic substances that require special ventilation. These include the following:

- Barium
- Cadmium
- Chromium
- Cobalt
- Copper
- Manganese
- Nickel
- Paint and other surface coatings
- Silica
- Zinc

Stainless steel contains chromium. When heated, it changes into Cr(VI) (*hexavalent chromium*). It can cause cancer and other health problems. Stainless steel may also contain manganese, another hazardous metal. Some welding electrodes contain manganese too. High exposure to manganese can cause neurological problems, including one that resembles Parkinson's disease.

Galvanized steel: Carbon steel dipped in zinc to inhibit corrosion.

The zinc that coats **galvanized steel** produces flu-like symptoms, often called *metal fume fever.* While these symptoms can be unpleasant, they're usually temporary and go away over time. High exposure may result in more severe symptoms that cause hospitalization.

Some fasteners contain cadmium, which is more toxic. Cadmium exposure initially may produce symptoms like metal fume fever. Over time, however, it accumulates in the body and causes serious organ damage and neurological problems. It can also produce long-term health effects that don't go away, such as kidney damage.

Some materials produce fumes that can cause cancer or serious respiratory illness. Long-term exposure can even result in death. For these reasons, always know what you're welding or cutting. Some products require special ventilation procedures or PPE. Check with the manufacturer or your company's welding protocols before working with these. Always follow the correct procedure and wear the right PPE.

WARNING!

Welding or cutting cadmium-plated fasteners requires appropriate respiratory PPE since cadmium is quite toxic. Unfortunately, cadmium-plated fasteners look like several other kinds of plated fasteners. Never assume that a material is safe unless you can prove it with proper documentation or tests. When in doubt, assume the material is toxic and use appropriate respiratory PPE.

3.6.1 Source Extraction Equipment

Source extraction, also called *local exhaust,* uses a machine to pull fumes away from the weld as they're produced. Units come in many styles (*Figure 21*). The farther the extraction hose is from the arc, the greater is the volume of air required to remove fumes. Source extraction equipment falls into two classes: low vacuum / high volume and high vacuum / low volume.

Low vacuum / high volume equipment uses weak suction to pull a lot of air through a large hose (*Figure 21* [A]). The hose needs to be within 6" to 15" (~15 cm to 38 cm) of the arc to work properly.

High vacuum / low volume systems use strong suction to draw in lower amounts of air through a small hose (*Figure 21* [B]). Because the hose is small, it fits in difficult-to-reach or confined workspaces. The nozzle must be close to the arc—within 2" to 4" (~5 cm to 10 cm). Some welding guns have a built-in extraction nozzle.

When using an extraction system, position the nozzle or hood as close to the weld as practical. Be careful not to disrupt the flux gases or shielding gas. Confirm that the smoke column flows into the extraction system and away from you. If the extraction system alone isn't sufficient to keep fumes under control, the welder will have to use respiratory protection too (*Figure 22*).

3.6.2 Respirators

Welders wear respiratory PPE to avoid toxic or irritating fumes and gases. Respiratory PPE, or *respirators*, divide into three categories:

- Air-purifying respirators
- Supplied-air respirators (SARs)
- Self-contained breathing apparatus (SCBA)

Before using a respirator, you must receive training on its use and maintenance. Employers provide training, as well as other services associated with respiratory safety.

Air-Purifying Respirators

These respirators filter the surrounding air to remove undesirable gases, fumes, and particles. The air must therefore have enough oxygen for safe breathing. The respirator has a filter cartridge attached to a mask. The cartridge contains

(A) Low Vacuum
High Volume

(B) High Vacuum
Low Volume

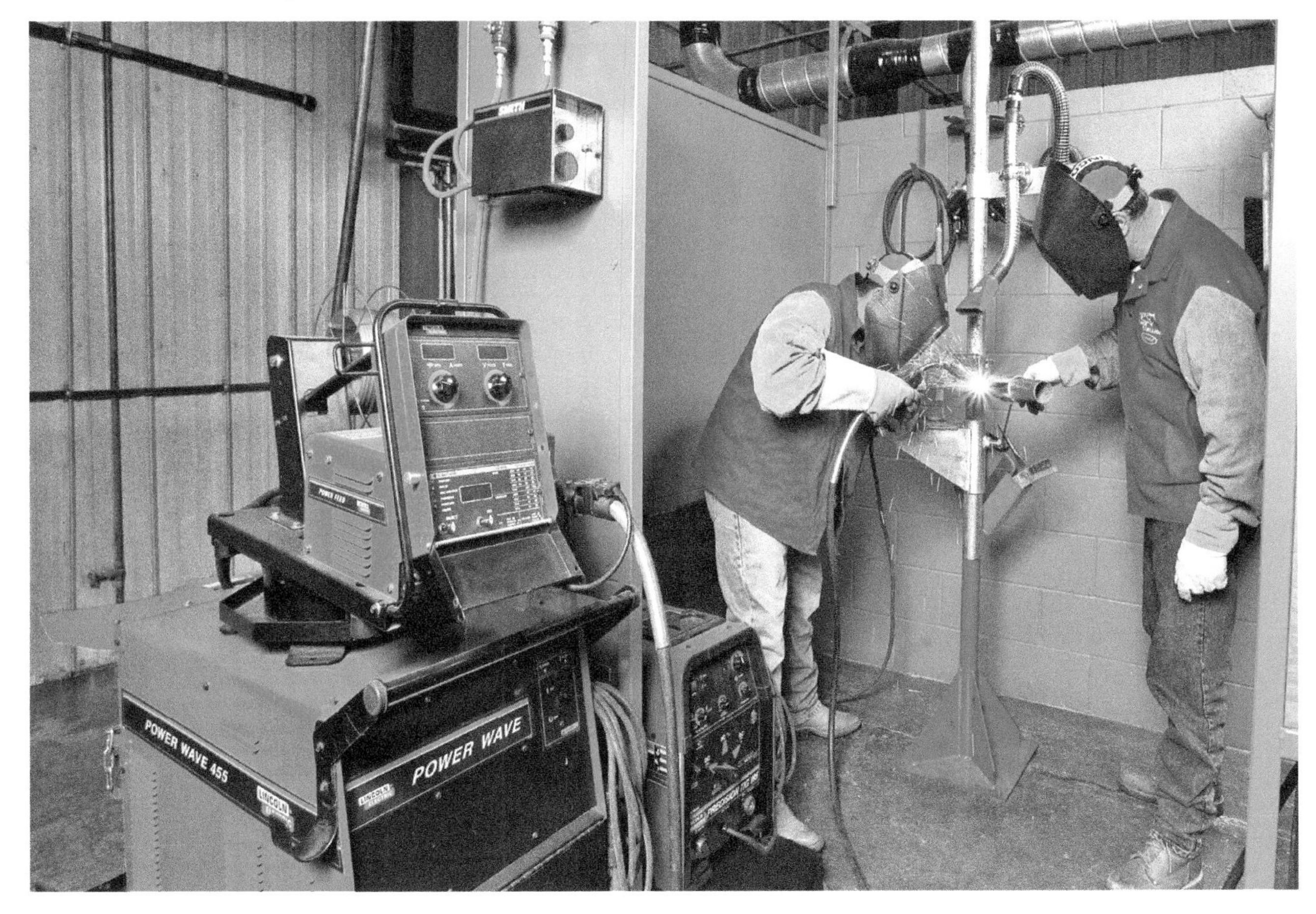

(C) Engineered System

Figure 21 Source extraction equipment.
Sources: © Miller Electric Mfg. LLC (21A); The Lincoln Electric Company, Cleveland, OH, USA (21B–21C)

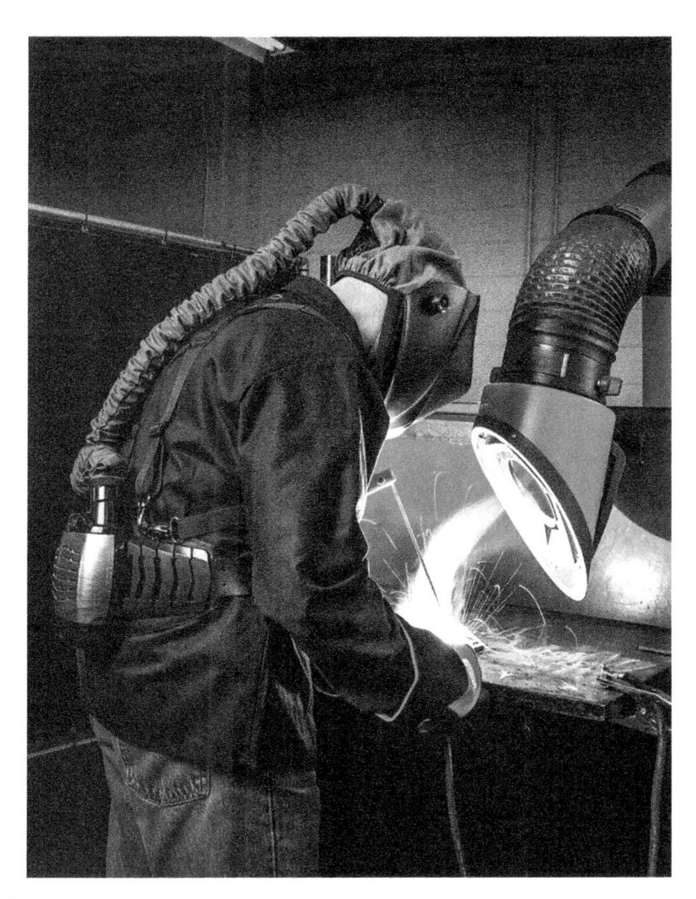

Figure 22 Personal filtration and an extraction system.
Source: The Lincoln Electric Company, Cleveland, OH, USA

chemicals that absorb toxic substances, as well as a soft material that traps particles. Filter cartridges remove only specific substances. Always confirm that your respirator is correct for your work environment. Cartridges should have a National Institute for Occupational Safety and Health (NIOSH) approval label attached. It identifies the filter's rating. Never use an unmarked cartridge.

WARNING!

Filter cartridges eventually stop working. Always replace the cartridge within the time interval specified by the manufacturer. If you begin to taste or smell anything unusual or find it difficult to breath, replace the cartridge immediately!

Air-purifying respirator manufacturers classify their products into four groups:

- No-maintenance
- Low-maintenance
- Reusable
- Powered air-purifying respirators (PAPRs)

No-maintenance and low-maintenance respirators are suitable for light-duty work only. Most cover the mouth and nose rather than the whole face (*Figure 23* [A]). No-maintenance units don't have replaceable filter cartridges, while low-maintenance ones do.

Reusable respirators have replaceable parts and work well for day-to-day use. Some cover the entire face (*Figure 23* [B]). Workers must keep their respirators maintained by checking and replacing parts as they wear out or degrade.

Powered air-purifying respirators come in half-mask, full-mask, and hood versions. They use a battery-powered blower to draw air through the filter. The blower may mount on the mask (*Figure 24*) or the belt (*Figure 22*). The blower makes breathing easier and more comfortable. Most units include alarms that go off when the battery is low or airflow drops below an acceptable level.

(A) Half-Mask (B) Full-Mask

Figure 23 Air-purifying respirators.
Sources: MSA The Safety Company (23A); iStock@NanoStockk (23B)

Figure 24 Full-mask powered air-purifying respirator (PAPR).
Source: MSA The Safety Company

Supplied-Air Respirators

Sometimes, filter-type respirators aren't enough. Perhaps the air doesn't contain enough oxygen, or contains toxic gases, chemicals, or particles. It may even be classified as **immediately dangerous to life or health (IDLH)**. An IDLH environment will kill you or produce irreversible health effects. In these cases, workers can protect themselves with a supplied-air respirator.

Immediately dangerous to life or health (IDLH): As defined by OSHA, an atmosphere that poses an immediate threat to life, would cause irreversible adverse health effects, or would impair an individual's ability to escape from a dangerous atmosphere.

These units deliver air through a hose to a full mask or hood (*Figure 25*). A compressor or bank of compressed air cylinders supplies the air. Respirators used in IDLH environments must include an emergency air tank that can supply air if the main supply fails.

Continuous-flow models deliver a steady airflow to the mask or hood. Usually, the worker can adjust it with a valve. Pressure-demand models deliver air only when the worker inhales. During exhalation, a valve opens and vents the used air outside the mask. Some models support multiple breathing modes.

Self-Contained Breathing Apparatus

Dragging a supplied-air respirator's hose around can be a nuisance in some environments. The self-contained breathing apparatus solves this problem with an air tank carried on the back (*Figure 26*). A regulator delivers air from the tank to a full mask or hood for between 30 and 60 minutes. Workers must watch their air supply and swap tanks as they run out.

3.6.3 Respiratory Programs

In the United States, OSHA regulates respirator protocols and procedures. It requires all companies to have a respiratory program that includes the following features:

- Standard procedures for respirator selection and use
- Respirator training
- Regular respirator cleaning and disinfecting
- Sanitary respirator storage
- Regular respirator inspection
- Annual fit testing
- Employee pulmonary function testing

Always follow OSHA regulations, as well as your company's procedures, when selecting and using a respirator. Choosing the wrong respirator can be extremely dangerous (*Figure 27*). Selection depends on the contaminants present, as well as their concentration. Most companies have a trained safety specialist create their respirator selection protocols. Respirators must fit properly, and employees must follow the manufacturer's guidelines when using them.

Figure 25 Supplied-air respirator (SAR).
Source: MSA The Safety Company

Figure 26 Self-contained breathing apparatus (SCBA).
Source: MSA The Safety Company

Before using a respirator, you must determine the following:

1. The type of contaminant(s) for which the respirator is being selected
2. The concentration level of that contaminant
3. Whether the respirator can be properly fitted on the wearer's face

You must read and understand all respirator instructions, warnings, and use limitations contained on each package before use.

Figure 27 Use the right respirator for the job.

Always wear your respirator whenever you're working in an environment that makes it necessary. If it's an air-purifying model, confirm that it includes the correct filter type. Verify that it's in good condition. If you're working with a new or unusual material and aren't sure what respirator to select, consult your company's safety officer. Never guess. Don't start the work until you are sure.

The following limitations and guidelines apply to all air-purifying respirators:

- They do not supply oxygen. Never use one in an oxygen-deficient atmosphere!
- They reduce rather than fully eliminate contaminants. When they're used within their design guidelines, inhaled contaminants will be within acceptable levels.
- Do not use them with harmful substances that have little or no taste or smell. These make it difficult to tell when the respirator has stopped working.
- Replace the respirator or its filter if you taste or smell anything from the environment.
- Replace the respirator or its filter if you have trouble breathing through it.

WARNING!

If you feel ill, dizzy, nauseated, or have trouble breathing, leave the area immediately. Go to fresh air and seek assistance.

3.6.4 Checking a Respirator

A respirator that doesn't fit properly is useless. Every time you use one, perform both positive and negative fit checks first. Don't start working until you've confirmed a good seal.

WARNING!

Facial hair can prevent a respirator mask from making a tight seal. *OSHA 29 CFR 1910.134(g)(1)(i)(A)* permits facial hair provided it doesn't interfere with a respirator. Well-trimmed moustaches, sideburns, and small beards are usually acceptable. If you have facial hair and cannot form a good seal with your respirator mask, you must trim or remove it.

To perform positive and negative fit checks, do the following:

Step 1 Adjust the mask for the best fit. Adjust the head and neck straps for a firm fit that feels comfortable.

WARNING!

Do not overtighten the head and neck straps. Tighten them only enough for a good fit. Overtightening can cause mask distortion and dangerous leaks.

Step 2 Block the exhalation valve with your hand.

Step 3 Breathe out into the mask to increase the inside pressure.

Step 4 If the mask puffs out slightly without leaking for a few seconds, you have a good seal.

Step 5 Block the inhalation valve with your hand.

Step 6 Attempt to inhale.

Step 7 If the mask caves in slightly without leaking for a few seconds, you have a good seal.

3.7.0 Safety Data Sheets

Your workplace is full of chemicals, materials, consumables, and manufactured products. Some are harmless, but many are toxic, flammable, environmentally hazardous, or can cause health problems over time. How can you stay healthy and safe as you do your job? What should you do in an emergency?

Your first stop is the product's or material's label on which manufacturers must provide basic information and safety guidelines. For complete information, however, there is a second source, the product's Safety Data Sheet (SDS).

Respirator Inspection and Maintenance

To work properly, respirators must be clean and in good condition. If they're not regularly washed and disinfected, bacteria will grow on their surfaces. It could make you sick or damage the respirator's materials. All companies should have a respirator cleaning and disinfection protocol in place to avoid this problem.

Follow these guidelines to inspect and maintain your respirator.

- Inspect your respirator before and after each use.
- Do not wear a respirator if the mask is distorted, cracked, or worn. It won't seal properly.
- Do not wear a respirator if any part is missing. Replace worn straps or missing parts immediately.
- Do not expose respirators to chemicals or excessive heat, cold, or sunlight.
- Clean and wash your respirator after every session. Remove the cartridge and filter. Wash the mask with mild soap and a soft brush. Rinse and let it air dry overnight.
- Sanitize your respirator once per week. Remove the cartridge and filter. Soak the mask in a sanitizing solution for at least two minutes. Thoroughly rinse it with warm water. Let it air dry overnight.
- When not using them, store respirators in a resealable plastic bag. Do not store it face down as that can distort the mask.

NOTE

Prior to 2015, SDSs went by the name Material Safety Data Sheets (MSDSs). They contained essentially the same information but in a different format. Companies were required to switch over to the new format in 2015 and replace all MSDSs with SDSs. Some employees may still refer to an SDS as an MSDS.

All chemical and material manufacturers must prepare and provide SDSs for their products. These contain essential information about the product, including what it contains and the hazards it represents. SDSs also include instructions for accidents, exposure, storage, and cleanup. Manufacturers ship SDSs with their products and also distribute them online.

In the United States, under the "Right to Know" laws, employers must make SDSs readily available to all employees. Every material or substance that the employee could encounter must have a matching SDS. Most companies keep them in binders or specially marked boxes near workspaces (*Figure 28*). They may post them online as well.

WARNING!

By law, companies cannot conceal information about substances that workers encounter in the workplace. SDSs in the employee's language must be available at the worksite. All employees have a right to consult SDSs at any time. If you can't find an SDS binder or storage box near your workspace, or if SDSs are missing, speak to your supervisor at once. If you have questions about interpreting an SDS, speak to your company's safety officer.

3.7.1 Using an SDS

The Hazard Communication (HAZCOM) standard specifies what information an SDS must contain and how it should be presented. Every SDS includes the following information:

Section 1 — Product identification, manufacturer information, recommended uses and restrictions
Section 2 — Hazard identification
Section 3 — Composition/information on ingredients
Section 4 — First aid measures
Section 5 — Firefighting information
Section 6 — Accidental release measures
Section 7 — Handling and storage
Section 8 — Exposure controls / personal protection
Section 9 — Physical and chemical properties
Section 10 — Stability and reactivity
Section 11 — Toxicological properties
Section 12 — Ecological properties
Section 13 — Disposal considerations
Section 14 — Transport information
Section 15 — Regulatory information
Section 16 — Other information

Figure 28 SDS storage box.
Source: Photo courtesy of Justrite Mfg. Co. LLC

Figure 29A and *Figure 29B* show a sample SDS for a solvent cement used with PVC products. When reading an SDS, look for specific hazards, personal protection, handling procedures, and first aid information. Most SDSs have a 24-hour emergency-response number that you can call for questions or extra guidance.

If you were using this product for the first time, you would consult *Section 2* for the hazard list. According to the SDS, this product is flammable as well as an eye and skin irritant that can cause respiratory irritation, dizziness, and drowsiness.

Next, you'd read *Section 7* and *Section 8* to learn how to handle and store the product safely. It produces hazardous vapors, so you need to ensure good ventilation or wear a suitable respirator. You must also wear eye and skin PPE.

To learn how to deal with accidental exposure and injury, read *Section 4*. *Section 5* provides firefighting guidance. It's smart to read these now rather than for the first time during an emergency.

Figure 30A, *Figure 30B*, *Figure 30C*, *Figure 30D*, *Figure 30E*, and *Figure 30F* show a typical welding product SDS for SMAW electrodes. As you can see, it contains the same type of information arranged in the same way.

Whenever you begin a new job or work in a new location, look for the SDS binder or storage box. Check the materials you'll be using. If any are unfamiliar, read their SDSs, so you'll be familiar with the materials' hazards and proper handling. Always ask your supervisor if you're uncertain about something in your workplace.

GHS SAFETY DATA SHEET

WELD-ON® 705™ Low VOC Cements for PVC Plastic Pipe

Date Revised: **DEC 2011**
Supersedes: **FEB 2010**

SECTION I - PRODUCT AND COMPANY IDENTIFICATION

PRODUCT NAME: **WELD-ON® 705™ Low VOC Cements for PVC Plastic Pipe**

PRODUCT USE: Low VOC Solvent Cement for PVC Plastic Pipe

SUPPLIER:

MANUFACTURER: IPS Corporation
17109 South Main Street, Carson, CA 90248-3127
P.O. Box 379, Gardena, CA 90247-0379
Tel. 1-310-898-3300

EMERGENCY: Transportation: CHEMTEL Tel. 800.255-3924, 813-248-0585 (International) **Medical:** Tel. 800.451.8346, 760.602.8703 3E Company (International)

SECTION 2 - HAZARDS IDENTIFICATION

GHS CLASSIFICATION:

Health		Environmental		Physical	
Acute Toxicity:	Category 4	Acute Toxicity:	None Known	Flammable Liquid	Category 2
Skin Irritation:	Category 3	Chronic Toxicity:	None Known		
Skin Sensitization:	NO				
Eye:	Category 2B				

GHS LABEL: OR 

Signal Word: **Danger**

WHMIS CLASSIFICATION: CLASS B, DIVISION 2

Hazard Statements	Precautionary Statements
H225: Highly flammable liquid and vapor	P210: Keep away from heat/sparks/open flames/hot surfaces – No smoking
H319: Causes serious eye irritation	P261: Avoid breathing dust/fume/gas/mist/vapors/spray
H332: Harmful if inhaled	P280: Wear protective gloves/protective clothing/eye protection/face protection
H335: May cause respiratory irritation	P304+P340: IF INHALED: Remove victim to fresh air and keep at rest in a position comfortable for breathing
H336: May cause drowsiness or dizziness	P403+P233: Store in a well ventilated place. Keep container tightly closed
EUH019: May form explosive peroxides	P501: Dispose of contents/container in accordance with local regulation

SECTION 3 - COMPOSITION/INFORMATION ON INGREDIENTS

	CAS#	EINECS #	REACH Pre-registration Number	CONCENTRATION % by Weight
Tetrahydrofuran (THF)	109-99-9	203-726-8	05-2116297729-22-0000	25 - 50
Methyl Ethyl Ketone (MEK)	78-93-3	201-159-0	05-2116297728-24-0000	5 - 36
Cyclohexanone	108-94-1	203-631-1	05-2116297718-25-0000	15 - 30

All of the constituents of this adhesive product are listed on the TSCA inventory of chemical substances maintained by the US EPA, or are exempt from that listing.
* Indicates this chemical is subject to the reporting requirements of Section 313 of the Emergency Planning and Community Right-to-Know Act of 1986 (40CFR372).
# indicates that this chemical is found on Proposition 65's List of chemicals known to the State of California to cause cancer or reproductive toxicity.

SECTION 4 - FIRST AID MEASURES

Contact with eyes: Flush eyes immediately with plenty of water for 15 minutes and seek medical advice immediately.
Skin contact: Remove contaminated clothing and shoes. Wash skin thoroughly with soap and water. If irritation develops, seek medical advice.
Inhalation: Remove to fresh air. If breathing is stopped, give artificial respiration. If breathing is difficult, give oxygen. Seek medical advice.
Ingestion: Rinse mouth with water. Give 1 or 2 glasses of water or milk to dilute. Do not induce vomiting. Seek medical advice immediately.

SECTION 5 - FIREFIGHTING MEASURES

Suitable Extinguishing Media: Dry chemical powder, carbon dioxide gas, foam, Halon, water fog.
Unsuitable Extinguishing Media: Water spray or stream.
Exposure Hazards: Inhalation and dermal contact
Combustion Products: Oxides of carbon, hydrogen chloride and smoke

	HMIS	NFPA	
			0-Minimal
Health	2	2	1-Slight
Flammability	3	3	2-Moderate
Reactivity	0	0	3-Serious
PPE	B		4-Severe

Protection for Firefighters: Self-contained breathing apparatus or full-face positive pressure airline masks.

SECTION 6 - ACCIDENTAL RELEASE MEASURES

Personal precautions: Keep away from heat, sparks and open flame.
Provide sufficient ventilation, use explosion-proof exhaust ventilation equipment or wear suitable respiratory protective equipment.
Prevent contact with skin or eyes (see section 8).
Environmental Precautions: Prevent product or liquids contaminated with product from entering sewers, drains, soil or open water course.
Methods for Cleaning up: Clean up with sand or other inert absorbent material. Transfer to a closable steel vessel.
Materials not to be used for clean up: Aluminum or plastic containers

SECTION 7 - HANDLING AND STORAGE

Handling: Avoid breathing of vapor, avoid contact with eyes, skin and clothing.
Keep away from ignition sources, use only electrically grounded handling equipment and ensure adequate ventilation/fume exhaust hoods.
Do not eat, drink or smoke while handling.
Storage: Store in ventilated room or shade below 44°C (110°F) and away from direct sunlight.
Keep away from ignition sources and incompatible materials: caustics, ammonia, inorganic acids, chlorinated compounds, strong oxidizers and isocyanates.
Follow all precautionary information on container label, product bulletins and solvent cementing literature.

SECTION 8 - PRECAUTIONS TO CONTROL EXPOSURE / PERSONAL PROTECTION

EXPOSURE LIMITS:

Component	ACGIH TLV	ACGIH STEL	OSHA PEL	OSHA STEL:
Tetrahydrofuran (THF)	50 ppm	100 ppm	200 ppm	
Methyl Ethyl Ketone (MEK)	200 ppm	300 ppm	200 ppm	
Cyclohexanone	20 ppm	50 ppm	50 ppm	

Engineering Controls: Use local exhaust as needed.
Monitoring: Maintain breathing zone airborne concentrations below exposure limits.
Personal Protective Equipment (PPE):
Eye Protection: Avoid contact with eyes, wear splash-proof chemical goggles, face shield, safety glasses (spectacles) with brow guards and side shields, etc. as may be appropriate for the exposure.
Skin Protection: Prevent contact with the skin as much as possible. Butyl rubber gloves should be used for frequent immersion.
Use of solvent-resistant gloves or solvent-resistant barrier cream should provide adequate protection when normal adhesive application practices and procedures are used for making structural bonds.
Respiratory Protection: Prevent inhalation of the solvents. Use in a well-ventilated room. Open doors and/or windows to ensure airflow and air changes. Use local exhaust ventilation to remove airborne contaminants from employee breathing zone and to keep contaminants below levels listed above.
With normal use, the Exposure Limit Value will not usually be reached. When limits approached, use respiratory protection equipment.

Figure 29A Solvent Cement SDS (1 of 2).
Source: Weld-On Adhesives, Inc., a division of IPS Corporation

SDS NO: 415884-B-EN-NA
REVISED: October 29, 2018
C5085
Page 2 of 6

B

SAFETY DATA SHEET

INGREDIENT	CAS NO.	EINECS[Γ]	GROUP AND %WEIGHT				GHS Classification(s)	GHS HAZARD STATEMENTS (See Section 16 for Complete Phrases)
			A	B	C	D		
ALUMINUM OXIDE	1344-28-1	215-691-6	<5	<1	<1	---	NONE	
CALCIUM CARBONATE	1317-65-3	215-279-6	---	2-10	2-10	---	NONE	
CELLULOSE	9004-34-6	232-674-9	<5	---	---	<5	NONE	
CHROMIUM (metal)	7440-47-3	231-157-5	---	---	<9	---	NONE	
FLUORSPAR	7789-75-5	232-188-7	---	1-12	4-15	---	NONE	
IRON	7439-89-6	231-096-4	70-90	60-80	60-90	70-90	NONE	
MAGNESIUM CARBONATE	546-93-0	208-915-9	<2	<5	<1	<1	NONE	
MANGANESE	7439-96-5	231-105-1	1-5	1-5	1-5	1-5	- Acute Tox. 4 (Inhalation)(1) - Acute Tox. 4 (Oral) (1) - STOT RE 1(2)	H332 H302 H372
MICA	12001-26-2	None	<5	---	---	---	NONE	
MOLYBDENUM	7439-98-7	231-107-2	---	---	<2	<1	- STOT RE 2(2) - Eye Irrit. 2(3) - STOT SE 3(4)	H373 H319 H335
NICKEL	7440-02-0	231-111-4	---	---	<5	<2	Powder/Element: - Carc. 2(5) - Skin Sens. 1(6) - STOT RE 1(2) - Aquatic Chronic 3	H351 H317 H372 H412
POTASSIUM SILICATE	1312-76-1	215-199-1	<2	<2	<2	<2	NONE	
SILICA	14808-60-7	238-878-4	<7	<8	<7	<7	- STOT RE 2(2) - Carc. 2(5) - Acute Tox. 4 (Inhalation)(1)	H373 H351 H332
(Amorphous Silica Fume)	69012-64-2	273-761-1	---	---	---	---	NONE	
SILICON	7440-21-3	231-130-8	<2	<2	<5	<2	NONE	
SODIUM SILICATE	1344-09-8	215-687-4	<2	<2	<2	<2	NONE	
STRONTIUM CARBONATE	1633-05-2	216-643-7	---	<2	<2	---	NONE	
TITANIUM DIOXIDE	13463-67-7	236-675-5	<14	<10	<5	<5	- Carc. 2(5)	H351
VANADIUM	7440-62-2	231-171-1	---	---	<1	---	- Acute Tox. 4 (Inhalation)(6) - STOT RE 2(7) - Eye Irrit. 1(8) - Aquatic Chronic 2	H332 H373 H318 H411
HEXAVALENT CHROMIUM [CHROMIUM (VI) TRIOXIDE] (Fume constituent)	1333-82-0	215-607-8	Varies	Varies	Varies	Varies	- Ox. Sol. 1(7) - Carc. 1A(5) - Muta. 1B(8) - Repr. Tox 2(9) - Acute Tox. 2 (Inhalation)(1) - Acute Tox. 3 (Skin & Oral)(1) - STOT RE 1(2) - Skin Corr. 1A(10) - Skin Sens. 1(6) - Resp. Sens. 1(11) - Aquatic Acute 1 - Aquatic Chronic 1	H271 H350 H340 H361f H330 H311, H301 H372 H314 H317 H334, H317 H400 H410

--- Dashes indicate the ingredient is not present within the group of products **Γ** – European Inventory of Existing Chemical Commercial Substance Number **(1)** Acute toxicity (Cat. 1, 2, 3 and 4) **(2)** Specific target organ toxicity (STOT) – repeated exposure (Cat. 1 and 2) **(3)** Serious eye damage/eye irritation (Cat. 1 and 2) **(4)** Specific target organ toxicity (STOT) – single exposure ((Cat. 1, 2) and Cat. 3 for narcotic effects and respiratory tract irritation, only) **(5)** Carcinogenicity (Cat. 1A, 1B and 2) **(6)** Skin sensitization (Cat. 1, Sub-cat. 1A and 1B) **(7)** Oxidizing solid (Cat. 1, 2 and 3) **(8)** Germ cell mutagenicity (Cat. 1A, 1B and 2) **(9)** Reproductive toxicity (Cat. 1A, 1B and 2) **(10)** Skin corrosion/irritation (Cat. 1, 1A, 1B, 1C and 2) **(11)** Respiratory sensitization (Cat. 1, Sub-cat. 1A and 1B)

Figure 30B Welding Electrode SDS (2 of 6).
Source: Courtesy of Hobart Brothers LLC

HOBART FILLER METALS

SDS NO: 415884-B-EN-NA
REVISED: October 29, 2018
C5085
Page 3 of 6

B

SAFETY DATA SHEET

SECTION 4 – FIRST AID MEASURES

INGESTION: Not an expected route of exposure. Do not eat, drink, or smoke while welding; wash hands thoroughly before performing these activities. If symptoms develop, seek medical attention at once.
INHALATION during welding: If breathing is difficult, provide fresh air and contact physician. If breathing has stopped, perform artificial respiration and obtain medical assistance at once.
SKIN CONTACT during welding: Remove contaminated clothing and wash the skin thoroughly with soap and water. If symptoms develop, seek medical attention at once.
EYE CONTACT during welding: Dust or fume from this product should be flushed from the eyes with copious amounts of clean, tepid water until victim is transported to an emergency medical facility. Do not allow victim to rub or keep eyes tightly closed. Obtain medical assistance at once.
Arc rays can injure eyes. If exposed to arc rays, move victim to dark room, remove contact lenses as necessary for treatment, cover eyes with a padded dressing and rest. Obtain medical assistance if symptoms persist.

Section 11 of this SDS covers the acute effects of overexposure to the various ingredients within the welding consumable. Section 8 of this SDS lists the exposure limits and covers methods for protecting yourself and your co-workers.

SECTION 5 – FIRE-FIGHTING MEASURES

Fire Hazards: Welding consumables applicable to this sheet as shipped are nonreactive, nonflammable, non-explosive and essentially nonhazardous until welded.

Welding arcs and sparks can ignite combustibles and flammable products. If there are flammable materials, including fuel or hydraulic lines, in the work area and the worker cannot move the work or the flammable material, a fire-resistant shield such as a piece of sheet metal or fire resistant blanket should be placed over the flammable material. If welding work is conducted within 35 feet or so of flammable materials, station a responsible person in the work zone to act as fire watcher to observe where sparks are flying and to grab an extinguisher or sound the alarm if needed.
Unused welding consumables may remain hot for a period of time after completion of a welding process. See American National Standard Institute (ANSI) Z49.1 for further general safety information on the use and handling of welding consumables and associated procedures.
Suitable Extinguishing Media: This product is essentially nonflammable until welded; therefore, use a suitable extinguishing agent for a surrounding fire.
Unsuitable Extinguishing Media: None known.

SECTION 6 - ACCIDENTAL RELEASE MEASURES

In the case of a release of solid welding consumable products, solid objects can be picked up and placed into a disposal container. If airborne dust and/or fume is present, use adequate engineering controls and, if needed, personal protection to prevent overexposure. Refer to recommendations in Section 8. Wear proper personal protective equipment while handling. Do not discard as general trash.

SECTION 7 - HANDLING AND STORAGE

HANDLING: No specific requirements in the form supplied. Handle with care to avoid cuts. Wear gloves when handling welding consumables. Avoid exposure to dust. Do not ingest. Some individuals can develop an allergic reaction to certain materials. Retain all warning and product labels.
STORAGE: Keep separate from acids and strong bases to prevent possible chemical reactions.

SECTION 8 - EXPOSURE CONTROLS AND PERSONAL PROTECTION

Read and understand the instructions and the labels on the packaging. Welding fumes do not have a specific OSHA PEL (Permissible Exposure Limit) or ACGIH TLV (Threshold Limit Value). The OSHA PEL for Particulates – Not Otherwise Regulated (PNOR) is 5 mg/m^3 – Respirable Fraction, 15 mg/m^3 – Total Dust. The ACGIH TLV for Particles – Not Otherwise Specified (PNOS) is 3 mg/m^3 – Respirable Particles, 10 mg/m^3 – Inhalable Particles. The individual complex compounds within the fume may have a lower OSHA PEL or ACGIH TLV than the OSHA PNOR and ACGIH PNOS. An Industrial Hygienist, the OSHA PELs for Air Contaminants (29 CFR 1910.1000), and the ACGIH TLVs should be consulted to determine the specific fume constituents present and their respective exposure limits. All exposure limits are in milligrams per cubic meter (mg/m^3).

INGREDIENT	CAS	EINECS	OSHA PEL	ACGIH TLV
ALUMINUM###	7429-90-5	231-072-3	5 R*, 15 (Dust)	1 R* {A4} 5 (Welding fumes, as Al)
CALCIUM CARBONATE	1317-65-3	215-279-6	5 R*, 5 (as CaO)	3 R*, 2 (as CaO)
CELLULOSE	9004-34-6	232-674-9	5 R*	10 Dust
CHROMIUM#	7440-47-3	231-157-5	1 (Metal) 0.5 (Cr II & Cr III Cpnds) 0.005 (Cr VI Cpnds, Calif. OSHA PEL)	0.5 (Metal) 0.003 (Cr III Cpnds) {A4; DSEN; RSEN} 0.0002 (Cr VI Sol Cpnds) {A1; Skin; DSEN; RSEN} 0.0005 (Cr VI STEL)
FLUORSPAR	7789-75-5	232-188-7	2.5 (as F)	2.5 (as F) {A4}
IRON+	7439-89-6	231-096-4	5 R*	5 R* (Fe_2O_3) {A4}
IRON OXIDE	1309-37-1	215-168-2	10 (Oxide Fume)	5 R* (Fe_2O_3) {A4}
MAGNESIUM CARBONATE+	546-93-0	208-915-9	5 R*	3 R*
MANGANESE#	7439-96-5	231-105-1	5 CL ** (Fume) 1, 3 STEL*** ■	0.1 I* {A4} ♦ 0.02 R* ♦♦
MICA	12001-26-2	None	3 R*■	3 R*
MOLYBDENUM	7439-98-7	231-107-2	5 R*	3 R*; 10 I* (Ele and Insol) 0.5 R* (Sol Cpnds) {A3}
NICKEL#	7440-02-0	231-111-4	1 (Metal) 1 (Sol Cpnds) 1 (Insol Cpnds)	1.5 I* (Ele) {A5} 0.1 I* (Sol Cpnds) {A4} 0.2 I* (Insol Cpnds) {A1}
POTASSIUM SILICATE	1312-76-1	215-199-1	Not established	Not established
SILICA++	14808-60-7	238-878-4	0.05 R*	0.025 R* {A2}
(Amorphous Silica Fume)	69012-64-2	273-761-1	0.8	2 R*
SILICON+	7440-21-3	231-130-8	5 R*	3 R*
SODIUM SILICATE	1344-09-8	215-687-4	Not established	Not established
STRONTIUM CARBONATE+	1633-05-2	216-643-7	5 R*	3 R*
TITANIUM DIOXIDE	13463-67-7	236-675-5	15 (Dust)	10 {A4}
VANADIUM	7440-62-2	231-171-1	Not Established (Ele) 1 TWA, 3 STEL***■ (Ele) 0.1 CL** (Fume as V_2O_5) 0.5 R* CL** (Dust as V_2O_5)	Not Established (Ele) 0.05 R* (Dust as V_2O_5) {A4} 0.05 I* (Fume as V_2O_5) {A3}

R* - Respirable Fraction I* - Inhalable Fraction ** - Ceiling Limit *** - Short Term Exposure Limit + - As a nuisance particulate covered under "Particulates Not Otherwise Regulated" by OSHA or "Particulates Not Otherwise Classified" by ACGIH ++ - Crystalline silica is bound within the product as it exists in the package. However, research indicates silica is present in welding fume in the amorphous (noncrystalline) form #- Reportable material under Section 313 of SARA ## - Reportable material under Section 313 of SARA only in fibrous form ■ - NIOSH REL TWA and STEL ■■ - AIHA Ceiling Limit of 1 mg/m^3 ♦ - Limit of 0.1 mg/m^3 is for Inhalable Mn in 2015 by ACGIH ♦♦ - Limit of 0.02 mg/m^3 is for Respirable Mn in 2015 by ACGIH Ele – Element Sol – Soluble Insol – Insoluble Inorg – Inorganic Cpnds – Compounds NOS – Not Otherwise Specified

Figure 30C Welding Electrode SDS (3 of 6).
Source: Courtesy of Hobart Brothers LLC

SAFETY DATA SHEET

{A1} - Confirmed Human Carcinogen per ACGIH {A2} - Suspected Human Carcinogen per ACGIH {A3} - Confirmed Animal Carcinogen with Unknown Relevance to Humans per ACGIH {A4} - Not Classifiable as a Human Carcinogen per ACGIH {A5} - Not Suspected as a Human Carcinogen per ACGIH (noncrystalline form) DSEN – Dermal Sensitization RSEN – Respiratory Sensitization EINECS – European Inventory of Existing Commercial Chemical Substances OSHA – U.S. Occupational Safety and Health Administration ACGIH – American Conference of Governmental Industrial Hygienists

VENTILATION: Use enough ventilation or local exhaust at the arc or both to keep the fumes and gases below the PEL/TLV in the worker's breathing zone and the general area. Train the welder to keep his head out of the fumes.
RESPIRATORY PROTECTION: Use NIOSH-approved or equivalent fume respirator or air supplied respirator when welding in confined space or where local exhaust or ventilation does not keep exposure below the regulatory limits.
EYE PROTECTION: Wear helmet or use face shield with filter lens for open arc welding processes. As a rule of thumb begin with Shade Number 14. Adjust if needed by selecting the next lighter and/or darker shade number. Provide protective screens and flash goggles, if necessary, to shield others from the weld arc flash.
PROTECTIVE CLOTHING: Wear hand, head and body protection which help to prevent injury from radiation, sparks and electrical shock. See ANSI Z49.1. At a minimum this includes welder's gloves and a protective face shield, and may include arm protectors, aprons, hats, shoulder protection as well as dark non-synthetic clothing. Train the welder not to touch live electrical parts and to insulate himself from work and ground.
PROCEDURE FOR CLEANUP OF SPILLS OR LEAKS: Not applicable
SPECIAL PRECAUTIONS (IMPORTANT): When welding with electrodes that require special ventilation (such as stainless or hardfacing, or other products which require special ventilation, or on lead- or cadmium-plated steel and other metals or coatings like galvanized steel, which produce hazardous fumes) maintain exposure below the PEL/TLV. Use industrial hygiene monitoring to ensure that your use of this material does not create exposures which exceed PEL/TLV. Always use exhaust ventilation. Refer to the following sources for important additional information: American National Standard Institute (ANSI) Z49.1; Safety in Welding and Cutting published by the American Welding Society, 8669 NW 36 Street, # 130, Miami, Florida 33166-6672, Phone: 800-443-9353 or 305-443-9353; and OSHA Publication 2206 (29 CFR 1910), U.S. Government Printing Office, Washington, DC 20402.

SECTION 9 – PHYSICAL AND CHEMICAL PROPERTIES

Welding consumables applicable to this sheet as shipped are nonreactive, nonflammable, non-explosive and essentially nonhazardous until welded.
PHYSICAL STATE: Solid
APPEARANCE: Cored Wire/Coated Rod
COLOR: Gray
ODOR: Not Applicable
ODOR THRESHOLD: Not Applicable
pH: Not Applicable
MELTING POINT/FREEZING POINT: Not Available
INITIAL BOILING POINT AND BOILING RANGE: Not Available
FLASH POINT: Not Available
EVAPORATION RATE: Not Applicable
FLAMMABILITY (SOLID, GAS): Not Available
UPPER/LOWER FLAMMABILITY OR EXPLOSIVE LIMITS: Not Available
VAPOR PRESSURE: Not Applicable
VAPOR DENSITY: Not Applicable
RELATIVE DENSITY: Not Available
SOLUBILITY(IES): Not Available
PARTITION COEFFICIENT: N-OCTANOL/WATER: Not Applicable
AUTO-IGNITION TEMPERATURE: Not Available
DECOMPOSITION TEMPERATURE: Not Available
VISCOSITY: Not Applicable

SECTION 10 – STABILITY AND REACTIVITY

GENERAL: Welding consumables applicable to this sheet are solid and nonvolatile as shipped. This product is only intended for use per the welding parameters it was designed for. When this product is used for welding, hazardous fumes may be created. Other factors to consider include the base metal, base metal preparation and base metal coatings. All of these factors can contribute to the fume and gases generated during welding. The amount of fume varies with the welding parameters.
STABILITY: This product is stable under normal conditions.
REACTIVITY: Contact with acids or strong bases may cause generation of gas.

SECTION 11 – TOXICOLOGICAL INFORMATION

SHORT-TERM (ACUTE) OVEREXPOSURE EFFECTS: Welding Fumes - May result in discomfort such as dizziness, nausea or dryness or irritation of nose, throat or eyes. **Aluminum Oxide** - Irritation of the respiratory system. **Calcium Oxide** - Dust or fumes may cause irritation of the respiratory system, skin and eyes. **Chromium** - Inhalation of fume with chromium (VI) compounds can cause irritation of the respiratory tract, lung damage and asthma-like symptoms. Swallowing chromium (VI) salts can cause severe injury or death. Dust on skin can form ulcers. Eyes may be burned by chromium (VI) compounds. Allergic reactions may occur in some people. **Fluorides** - Fluoride compounds evolved may cause skin and eye burns, pulmonary edema and bronchitis. **Iron, Iron Oxide** - None are known. Treat as nuisance dust or fume. **Magnesium, Magnesium Oxide** - Overexposure to the oxide may cause metal fume fever characterized by metallic taste, tightness of chest and fever. Symptoms may last 24 to 48 hours following overexposure. **Manganese** - Metal fume fever characterized by chills, fever, upset stomach, vomiting, irritation of the throat and aching of body. Recovery is generally complete within 48 hours of the overexposure. **Mica** - Dust may cause irritation of the respiratory system, skin and eyes. **Molybdenum** - Irritation of the eyes, nose and throat. **Nickel, Nickel Compounds** - Metallic taste, nausea, tightness in chest, metal fume fever, allergic reaction. **Potassium Silicate** - Dust or fumes may cause irritation of the respiratory system, skin and eyes. **Silica (Amorphous)** - Dust and fumes may cause irritation of the respiratory system, skin and eyes. **Sodium Silicate** - Dust or fumes may cause irritation of the respiratory system, skin and eyes. **Strontium Compounds** - Strontium salts are generally non-toxic and are normally present in the human body. In large oral doses, they may cause gastrointestinal disorders, vomiting and diarrhea. **Titanium Dioxide** - Irritation of respiratory system. **Vanadium** - Overexposure to the oxide causes green tongue, cough, metallic taste, throat irritation and eczema.

LONG-TERM (CHRONIC) OVEREXPOSURE EFFECTS: Welding Fumes - Excess levels may cause bronchial asthma, lung fibrosis, pneumoconiosis or "siderosis." Studies have concluded that there is sufficient evidence for ocular melanoma in welders. **Aluminum Oxide** - Pulmonary fibrosis and emphysema. **Calcium Oxide** - Prolonged overexposure may cause ulceration of the skin and perforation of the nasal septum, dermatitis and pneumonia. **Chromium** - Ulceration and perforation of nasal septum. Respiratory irritation may occur with symptoms resembling asthma. Studies have shown that chromate production workers exposed to hexavalent chromium compounds have an excess of lung cancers. Chromium (VI) compounds are more readily absorbed through the skin than chromium (III) compounds. Good practice requires the reduction of employee exposure to chromium (III) and (VI) compounds. **Fluorides** - Serious bone erosion (Osteoporosis) and mottling of teeth. **Iron, Iron Oxide Fumes** - Can cause siderosis (deposits of iron in lungs) which some researchers believe may affect pulmonary function. Lungs will clear in time when exposure to iron and its compounds ceases. Iron and magnetite (Fe_3O_4) are not regarded as fibrogenic materials. **Magnesium, Magnesium Oxide** - No adverse long term health effects have been reported in the literature. **Manganese** - Long-term overexposure to manganese compounds may affect the central nervous system. Symptoms may be similar to Parkinson's disease and can include slowness, changes in handwriting, gait impairment, muscle spasms and cramps and less commonly, tremor and behavioral changes. Employees who are overexposed to manganese compounds should be seen by a physician for early detection of neurologic problems. Overexposure to manganese and manganese compounds above safe exposure limits can cause irreversible damage to the central nervous system, including the brain, symptoms of which may include slurred speech, lethargy, tremor, muscular weakness, psychological disturbances and spastic gait. **Mica** - Prolonged overexposure may cause scarring of the lungs and pneumoconiosis characterized by cough, shortness of breath, weakness and weight loss. **Molybdenum** - Prolonged overexposure may result in loss of appetite, weight loss, loss of muscle coordination, difficulty in breathing and anemia. **Nickel, Nickel Compounds** - Lung fibrosis or pneumoconiosis. Studies of nickel refinery workers indicated a higher incidence of lung and nasal cancers. **Potassium Silicate** - Prolonged overexposure may cause ulceration of the skin and perforation of the nasal septum, dermatitis and pneumonia. **Silica (Amorphous)** - Research indicates that silica is present in welding fume in the amorphous form. Long term overexposure may cause

Figure 30D Welding Electrode SDS (4 of 6).
Source: Courtesy of Hobart Brothers LLC

SAFETY DATA SHEET

pneumoconiosis. Noncrystalline forms of silica (amorphous silica) are considered to have little fibrotic potential. **Sodium Silicate** - Prolonged overexposure may cause ulceration of the skin and perforation of the nasal septum, dermatitis and pneumonia. **Strontium Compounds** - Strontium at high doses is known to concentrate in bone. Major signs of chronic toxicity, which involve the skeleton, have been labeled as "strontium rickets". **Titanium Dioxide** - Pulmonary irritation and slight fibrosis. **Vanadium** - Prolonged overexposure to vanadium pentoxide can cause nasal catarrh or nose bleeds and chronic respiratory problems.

MEDICAL CONDITIONS AGGRAVATED BY EXPOSURE: Persons with pre-existing impaired lung functions (asthma-like conditions). Persons with a pacemaker should not go near welding and cutting operations until they have consulted their doctor and obtained information from the manufacturer of the device. Respirators are to be worn only after being medically cleared by your company-designated physician.

EMERGENCY AND FIRST AID PROCEDURES: Call for medical aid. Employ first aid techniques recommended by the American Red Cross. If irritation or flash burns develop after exposure, consult a physician.

CARCINOGENICITY: Chromium VI compounds, nickel compounds and silica (crystalline quartz) are classified as IARC Group 1 and NTP Group K carcinogens. Titanium dioxide, nickel metal/alloys, vanadium (V_2O_5) and welding fumes are classified as IARC Group 2B carcinogens.

CALIFORNIA PROPOSITION 65:
⚠**WARNING:** These products can expose you to chemicals, including titanium dioxide and/or chromium and/or nickel, which are known to the State of California to cause cancer, and to carbon monoxide, which is known to the State of California to cause birth defects or other reproductive harm. For more information, go to www.P65Warnings.ca.gov.

INGREDIENT	CAS	IARC^{E}	NTP^{Z}	OSHA^{H}	65$^{\Theta}$
ALUMINUM OXIDE	1344-28-1	---	---	---	---
CALCIUM CARBONATE	1317-65-3	---	---	---	---
CELLULOSE	9004-34-6	---	---	---	---
CHROMIUM	7440-47-3	3^{Σ}, $1^{\Sigma\Sigma}$	$K^{\Sigma\Sigma}$	$X^{\Sigma\Sigma}$	$X^{\Sigma\Sigma}$
FLUORSPAR	7789-75-5	---	---	---	---
IRON	7439-89-6	---	---	---	---
IRON OXIDE	1309-37-1	3	---	---	---
MAGNESIUM CARBONATE	546-93-0	---	---	---	---
MANGANESE	7439-96-5	---	---	---	---
MICA	12001-26-2	---	---	---	---
MOLYBDENUM	7439-98-7	---	---	---	---
NICKEL	7440-02-0	$2B^{\beta}$, $1^{\beta\beta}$	S^{β}, $K^{\beta\beta}$	---	X^{β}, $X^{\beta\beta}$
POTASSIUM SILICATE	1312-76-1	---	---	---	---
SILICA	14808-60-7	1^{Ψ}	K	---	X
(Amorphous Silica fume)	69012-64-2	3	—	---	---
SILICON	7440-21-3	---	---	---	---
SODIUM SILICATE	1344-09-8	---	---	---	---
STRONTIUM CARBONATE	1633-05-2	---	---	---	---
TITANIUM DIOXIDE	13463-67-7	2B	---	---	X
Ultraviolet Radiation	---	1	---	---	---
VANADIUM	7440-62-2	$2B^{\Omega}$	---	---	X^{Ω}
Welding Fumes	--	1	--	---	--

E – International Agency for Research on Cancer (1 – Carcinogenic to Humans, 2A – Probably Carcinogenic to Humans, 2B – Possibly Carcinogenic to Humans, 3 – Not Classifiable as to its Carcinogenicity to Humans, 4 --- Probably Not Carcinogenic to Humans) Z – US National Toxicology Program (K – Known Carcinogen, S – Suspected Carcinogen) H – OSHA Designated Carcinogen List Θ – California Proposition 65 (X – On Proposition 65 list) Σ – Chromium Metal and Chromium III Compounds ΣΣ – Chromium VI β – Nickel metal and alloys ββ -- Nickel compounds Ψ – Silica Crystalline α-Quartz Ω – Vanadium pentoxide --- Dashes indicate the ingredient is not listed with the IARC, NTP, OSHA or Proposition 65

SECTION 12 – ECOLOGICAL INFORMATION

Welding processes can release fumes directly to the environment. Welding wire can degrade if left outside and unprotected. Residues from welding consumables and processes could degrade and accumulate in the soil and groundwater.

SECTION 13 – DISPOSAL CONSIDERATIONS

Use recycling procedures if available. Discard any product, residue, packaging, disposable container or liner in an environmentally acceptable manner, in full compliance with federal, state and local regulations.

SECTION 14 – TRANSPORT INFORMATION

No international regulations or restrictions are applicable. No special precautions are necessary.

SECTION 15 – REGULATORY INFORMATION

Read and understand the manufacturer's instructions, your employer's safety practices and the health and safety instructions on the label and the safety data sheet. Observe all local and federal rules and regulations. Take all necessary precautions to protect yourself and others.
United States EPA Toxic Substance Control Act: All constituents of these products are on the TSCA inventory list or are excluded from listing.
CERCLA/SARA TITLE III: Reportable Quantities (RQs) and/or Threshold Planning Quantities (TPQs):

Ingredient name	**RQ(lb)**	**TPQ (lb)**
Products on this SDS are a solid solution in the form of a solid article.	--	--

Spills or releases resulting in the loss of any ingredient at or above its RQ require immediate notification to the National Response Center and to your Local Emergency Planning Committee.
Section 311 Hazard Class
As shipped: Immediate In use: Immediate delayed
EPCRA/SARA TITLE III 313 TOXIC CHEMICALS: The following metallic components are listed as SARA 313 "Toxic Chemicals" and potentially subject to annual SARA 312 reporting: Aluminum Oxide (fibrous forms),Chromium, Manganese, Nickel and Vanadium. See Section 3 for weight percentage.
CANADIAN WHMIS CLASSIFICATION: Class D; Division 2, Subdivision A
CANADIAN CONTROLLED PRODUCTS REGULATION: This product has been classified in accordance with the hazard criteria of the CPR and the SDS contains all of the information required by the CPR.
CANADIAN ENVIRONMENTAL PROTECTION ACT (CEPA): All constituents of these products are on the Domestic Substance List (DSL).

Figure 30E Welding Electrode SDS (5 of 6).
Source: Courtesy of Hobart Brothers LLC

SDS NO: 415884-B-EN-NA
REVISED: October 29, 2018
C5085
Page 6 of 6

B

SAFETY DATA SHEET

SECTION 16 – OTHER INFORMATION

The following Hazard Statements, provided in the OSHA Hazard Communication Standard (29 CFR Part 1910.1200) correspond to the columns labeled 'GHS Hazard Statements' within Section 3 of this safety data sheet. Take appropriate precautions and protective measures to eliminate or limit the associated hazard.

H271: May cause fire or explosion; strong oxidizer
H301: Toxic if swallowed
H302: Harmful if swallowed
H311: Toxic in contact with skin
H314: Causes severe skin burns and eye damage
H317: May cause an allergic skin reaction
H318: Causes serious eye damage
H319: Causes serious eye irritation
H330: Fatal if inhaled
H332: Harmful if inhaled
H334: May cause allergy or asthma symptoms or breathing difficulties if inhaled
H335: May cause respiratory irritation
H340: May cause genetic defects
H350: May cause cancer
H351: Suspected of causing cancer
H361f: Suspected of damaging fertility or the unborn child
H372: Causes damage to organs through prolonged or repeated exposure
H373: May cause damage to organs through prolonged or repeated exposure
H400: Very toxic to aquatic life.
H410: Very toxic to aquatic life with long lasting effects
H411: Toxic to aquatic life with long lasting effects
H412: Harmful to aquatic life with long lasting effects.

For additional information please refer to the following sources:

USA: **American National Standard Institute (ANSI) Z49.1** "Safety in Welding and Cutting", **ANSI/American Welding Society (AWS) F1.5** "Methods for Sampling and Analyzing Gases from Welding and Allied Processes", **ANSI/AWS F1.1** "Method for Sampling Airborne Particles Generated by Welding and Allied Processes", **AWSF3.2M/F3.2** "Ventilation Guide for Weld Fume", American Welding Society, 8669 NW 36 Street, # 130, Miami, Florida 33166-6672, Phone: 800-443-9353 or 305-443-9353. Safety and Health Fact Sheets available from AWS at www.aws.org.
OSHA Publication 2206 (29 C.F.R. 1910), U.S. Government Printing Office, Superintendent of Documents, P.O. Box 371954, Pittsburgh, PA 15250-7954.
Threshold Limit Values and Biological Exposure Indices, American Conference of Governmental Industrial Hygienists (ACGIH), 6500 Glenway Ave., Cincinnati, Ohio 45211, USA.
NFPA 51B "Standard for Fire Prevention During Welding, Cutting and Other Hot Work" published by the National Fire Protection Association, 1 Batterymarch Park, Quincy, MA 02169.

Canada: **CSA Standard CAN/CSA-W117.2-01** "Safety in Welding, Cutting and Allied Processes".

Hobart Brothers LLC strongly recommends the users of this product study this SDS, the product label information and become aware of all hazards associated with welding. Hobart Brothers LLC believes this data to be accurate and to reflect qualified expert opinion regarding current research. However, Hobart Brothers LLC cannot make any expressed or implied warranty as to this information.

Figure 30F Welding Electrode SDS (6 of 6).
Source: Courtesy of Hobart Brothers LLC

3.0.0 Section Review

1. The ANSI standard for welding and cutting safety is ____.
   a. *Z49.1*
   b. *CFR 1929*
   c. *CFR 1910*
   d. *Z45.5*

2. The *primary* purpose for a welding screen is to ____.
   a. keep non-welders from distracting welders at work
   b. protect others from the UV and glare of the welding arc
   c. prevent insects from affecting the welding arc
   d. keep nearby material from catching fire

3. When a worker is assigned the responsibility for providing a fire watch, the worker _____.
   a. cannot do other tasks
   b. can retrieve additional consumables for the welder from the shop
   c. can also weld or cut to remain productive
   d. may test or inspect welds

4. Argon is sometimes used as a _____.
   a. breathing gas
   b. fuel gas
   c. solvent
   d. purge gas

5. Which of the following is a *correct* statement about electrical hazards associated with welding?
   a. The jaws of the electrode holder do not carry electrical current.
   b. GFCIs should not be used when welding.
   c. Electrical shocks can cause severe burns.
   d. A welding machine's output voltage is low and harmless.

6. Which of these metals is *least* likely to produce toxic fumes when heated?
   a. Manganese
   b. Chromium
   c. Carbon steel
   d. Galvanized steel

7. In which section of an SDS are hazards identified?
   a. Section 1
   b. Section 2
   c. Section 5
   d. Section 12

Module 29101 Review Questions

1. The *most* popular welding type is _____.
   a. GTAW
   b. GMAW
   c. FCAW
   d. SMAW

2. A competency-based curriculum is one in which _____.
   a. the focus is on classroom training
   b. students must demonstrate knowledge and skills
   c. instructors must have four-year degrees
   d. all the training is done on the job

3. Youth apprenticeship programs allow students to begin their apprenticeship or craft training _____.
   a. in middle school
   b. in high school
   c. instead of traditional subjects
   d. instead of English classes

4. While welding, protect your body by wearing _____.
   a. pants with cuffs
   b. polyester clothing
   c. cotton clothing
   d. short-sleeve shirts

5. Which of the following is a *correct* statement about eye protection?
   a. It is not necessary to wear standard eye protection if you are wearing a welding helmet.
   b. The same shade of lens can be used for almost all welding and cutting tasks.
   c. Shade numbers for arc welding can range from 7 to 14.
   d. Safety glasses with side shields should not be used because they block peripheral vision.

6. What is a legitimate safety reason for asking your supervisor to assign you light, non-hazardous work?
   a. You're intoxicated and don't feel that you can handle hazardous work.
   b. You're taking a doctor-prescribed medication that makes you feel a bit slow.
   c. You took a recreational drug several hours ago and think it's still in your system.
   d. You're feeling very ill and can't concentrate properly.

7. An OSHA-required safety practice is _____.
   a. welding near an open window
   b. welding over wooden floors
   c. using cardboard to control welding/grinding sparks
   d. writing HOT on hot metal before leaving it unattended

8. A hot work permit _____.
   a. authorizes performing work that could potentially start a fire
   b. promotes using PPE that reduces burn injuries
   c. records unsafe conditions at a jobsite
   d. helps the manager keep records of hazardous spaces

9. *Never* perform cutting without a _____.
   a. bucket of sand
   b. bucket of water
   c. fire watch
   d. fire hose

10. A confined space is one that _____.
    a. has a flammable atmosphere
    b. is smaller than 10' × 10' × 10'
    c. is designed for continuous employee occupancy
    d. has limited means of entry and exit

11. An atmosphere is oxygen deficient when the oxygen level is below _____.
    a. 19.5 percent
    b. 20 percent
    c. 21.5 percent
    d. 23.5 percent

12. The potential for electrical shock _____.
    a. decreases when the skin is damp
    b. remains the same when the skin is damp
    c. increases when the skin is damp
    d. decreases when the skin is cut

13. Which metal, used as a coating in galvanized steel, can cause metal fume fever?
    a. Zinc
    b. Chromium
    c. Barium
    d. Manganese

14. To make sure that a respirator provides proper protection, what must you do each time you wear it?
    a. Perform only a positive fit check.
    b. Perform both positive and negative fit checks.
    c. Perform only a negative fit check.
    d. Tighten the head and neck straps as much as possible.

15. An SDS is a form used to _____.
    a. file a worker's compensation claim with an insurance company
    b. list the contents, hazards, and precautions that pertain to a chemical or material
    c. record unsafe conditions that exist at a jobsite
    d. recommend changes to an employer safety program

Answers to odd-numbered Module Review Questions are found in Appendix A.

Cornerstone of Craftsmanship

Scottie Smith

Director of Welding Technology

Northwest Florida State College

How did you choose a career in the industry?

I wanted to be like my father, Frank Smith. He was a welder and my hero.

Who inspired you to enter the industry?

My father Frank Smith inspired me to enter the welding industry.

What types of training have you been through?

I enlisted in the Navy in 1991 as a Hull Maintenance Technician (HT). HTs are the welders on board Navy ships. I went through the Navy's weld training so that I was qualified to do repairs on nuclear-powered submarines. I was stationed at Point Loma Subbase in San Diego, CA. I received my associate's degree in Welding Technology from Southern Union State Community College in Opelika, Alabama.

How important is education and training in construction?

Education and training are very important because there are always new techniques and technology being developed. Without my associate's degree in Welding Technology, I would not have my current job as Director of Welding Technology.

What kinds of work have you done in your career?

I have done maintenance work during shutdowns at nuclear power plants, natural gas-fired power plants, and paper mills all across our country. I have also helped construct new power plants as well. I have been a welding instructor at three different colleges in three different states—Southern Union State Community College in Alabama, West Georgia Technical College in Georgia, Northwest Florida State College in Florida.

Tell us about your current job.

I am the Director of Welding Technology at Northwest Florida State College. I teach the daytime welding courses at the main campus in Niceville, Florida. I also oversee our night Welding program and the Welding program at our DeFuniak Springs campus.

What has been your favorite project?

I started the Welding program at Northwest Florida State College from scratch. I really enjoyed seeing the program take shape and helping the program grow to its current size. I enjoy the people I work with at our college.

What do you enjoy most about your job?

The most fulfilling part of my job is seeing students be successful. Knowing I am a part of the process that helps our students gain knowledge and skills that allow them to have a career in welding is a blessing to me.

What factors have contributed most to your success?

I am where I am because of the support of my family and the folks at our college. My wife, Kim Smith, has supported me every step of my career and I would not be where I am without her. In addition, my colleagues at Northwest Florida State College have helped me grow and flourish. They are true professionals. I have reached this height in my career because I am standing on the shoulders of giants.

Would you suggest construction as a career to others? Why?

Absolutely, because in construction you see the fruits of your labor. You get to build things that keep our country running. It is very satisfying work and the pay is very good.

What advice would you give to those new to the field?

I would tell someone starting a new field to never give up. Starting something new is never easy and to be successful, you have to stay with it.

Interesting career-related fact or accomplishment?

I wrote the book *How I Teach Welding: Professional Advice on Implementing a Welding Program.* It explains the teaching techniques I use to teach welding and how I set up the Welding program at Northwest Florida State College.

I got to be the Welding instructor for the program I graduated from at Southern Union State Community College. It was exciting to go back to the shop where my father taught me and continue his work.

I am a Certified Welding Inspector (CWI) and a Certified Welding Educator (CWE) by the American Welding Society. I have been a CWI for 18 years.

How do you define craftsmanship?

Craftsmanship is doing a high-quality job every time, even when no one will see the completed job.

Answers to Section Review Questions

Answer	Section Reference	Objective
Section 1.0.0		
1. b	1.1.1	1a
2. a	1.2.3	1b
Section 2.0.0		
1. d	2.1.1	2a
2. a	2.2.1	2b
Section 3.0.0		
1. a	3.1.0	3a
2. b	3.2.2	3b
3. a	3.3.1	3c
4. d	3.4.1	3d
5. c	3.5.1	3e
6. c	3.6.0	3f
7. b	3.7.1	3g

User Update

Did you find an error? Submit a correction by visiting **https://www.nccer.org/olf** or by scanning the QR code using your mobile device.

Source: The Lincoln Electric Company, Cleveland, OH, USA

Oxyfuel Cutting

Objectives

Successful completion of this module prepares you to do the following:

1. Describe oxyfuel cutting and its safety procedures.
   a. Summarize oxyfuel cutting.
   b. List oxyfuel cutting safety procedures.
2. Summarize oxyfuel cutting equipment.
   a. Identify and describe oxyfuel cutting gases and their storage cylinders.
   b. Describe oxyfuel cutting regulators and hoses.
   c. Describe oxyfuel cutting torches and tips.
   d. Describe specialized oxyfuel cutting equipment.
   e. Describe oxyfuel accessories.
3. Outline safely setting up, lighting, and shutting down oxyfuel equipment.
   a. Explain how to set up oxyfuel equipment.
   b. Explain how to leak test oxyfuel equipment.
   c. Explain how to light, adjust, and shut down oxyfuel equipment.
4. Outline common oxyfuel cutting procedures.
   a. Identify good and bad cuts, as well as what creates them.
   b. Explain how to cut, bevel, wash, and gouge.

Performance Tasks

Under supervision, you should be able to do the following:

1. Set up oxyfuel cutting equipment.
2. Light and adjust an oxyfuel torch.
3. Shut down oxyfuel cutting equipment.
4. Disassemble oxyfuel cutting equipment.
5. Change gas cylinders.
6. Cut instructor-specified shapes from various thicknesses of steel.
7. Wash metal with oxyfuel equipment.
8. Gouge metal with oxyfuel equipment.
9. Cut straight lines and bevels either with a track cutter or manually with/without a guide.

NOTE

This module uses US standard and metric units in up to three different ways. This note explains how to interpret them.

Exact Conversions

Exact metric equivalents of US standard units appear in parentheses after the US standard unit. For example: "Measure 18" (45.7 cm) from the end and make a mark."

Approximate Conversions

In some cases, exact metric conversions would be inappropriate or even absurd. In these situations, an approximate metric value appears in parentheses with the ~ symbol in front of the number. For example: "Grip the tool about 3" (~8 cm) from the end."

Parallel but not Equal Values

Certain scenarios include US standard and metric values that are parallel but not equal. In these situations, a slash (/) surrounded by spaces separates the US standard and metric values. For example: "Place the point on the steel rule's 1" / 1 cm mark."

Digital Resources for Welding

Scan this code using the camera on your phone or mobile device to view the digital resources related to this craft.

Overview

Welders cut metals using several different technologies. Oxyfuel cutting, an extremely versatile method, produces a very hot flame by burning a fuel gas with pure oxygen. Since the method involves flammable gases and open flames, however, it's quite hazardous. This module introduces the method, its equipment, procedures, and safety practices.

Industry Recognized Credentials

If you are training through an NCCER-accredited sponsor, you may be eligible for credentials from NCCER's Registry. The ID number for this module is 29102. Note that this module may have been used in other NCCER curricula and may apply to other level completions. Contact NCCER's Registry at 1.888.622.3720 or go to **www.nccer.org** for more information.

You can also show off your industry-recognized credentials online with NCCER's digital badges. Transform your knowledge, skills, and achievements into badges that you can share across social media platforms, send to your network, and add to your resume. For more information, visit **www.nccer.org**.

1.0.0 Oxyfuel Cutting Basics

Performance Tasks

There are no Performance Tasks in this section.

Objective

Describe oxyfuel cutting and its safety procedures.

a. Summarize oxyfuel cutting.

b. List oxyfuel cutting safety procedures.

Oxyfuel cutting is an open-flame process by which welders cut certain metals. It's versatile and requires no electrical power. Some oxyfuel outfits are portable too. The process relies on a fuel gas and oxygen, both stored in cylinders and delivered to the cutting torch by hoses. Using this technology safely requires great care, however, as it has many potential hazards.

1.1.0 The Oxyfuel Cutting Process

Oxyfuel cutting, also called *flame cutting* or *burning*, uses the flame produced by a torch to cut **ferrous metals**. The process, however, is more complex than it seems. Mixing fuel gas with oxygen instead of air produces a much hotter flame. This *preheat flame* rapidly heats the metal, usually carbon steel, to its *kindling temperature*, the temperature at which it can catch fire and burn.

When the metal reaches this temperature, it turns a cherry-red color. The craftworker then turns on an oxygen jet that flows from the torch's center. The flame can't consume the extra oxygen, so the oxygen causes the metal to burn (**oxidize** rapidly) instead. The burning metal provides much of the cutting heat. Oxidized metal, along with molten metal, forms a waste product called **dross**. The oxygen jet blasts the dross from the **kerf** (cut). *Figure 1* shows oxyfuel equipment in use.

Oxyfuel cutting works only on metals that oxidize rapidly, limiting it mostly to carbon steels (which also rust easily). For nonferrous metals, stainless steel, and difficult steel alloys, craftworkers must use other cutting technologies. These include plasma arc cutting (PAC) and air-carbon arc cutting (A-CAC). NCCER Module 29103, *Plasma Arc Cutting*, and NCCER Module 29104, *Air-Carbon Arc Cutting and Gouging*, introduce these.

NOTE

It surprises many people to learn that metals can burn. Some burn more easily than others. Metals burn when they combine rapidly with oxygen. Burning metals can release significant heat, making some metal fires hard to extinguish. Metals that rust combine with oxygen too, but the process is much slower than burning.

Ferrous metals: Metals containing iron.

Oxidize: To chemically combine with oxygen either quickly or slowly.

Dross: Oxidized and molten metal expelled by a thermal cutting process. Sometimes called *slag*.

Kerf: The slot produced by a cutting process.

Figure 1 Oxyfuel cutting.
Source: © Miller Electric Mfg. LLC

1.2.0 Safety Practices

Oxyfuel torches generate high temperatures by burning a flammable gas. They also use oxygen, which makes many things burn easily. Both gases are stored under pressure in cylinders. The equipment, as well as the cutting process itself, is hazardous and requires following careful safety procedures. This section summarizes good practices but isn't a complete oxyfuel safety training. Always complete all training that your workplace requires. Follow all workplace safety guidelines and wear appropriate PPE.

NOTE

Be sure that you've completed NCCER Module 29101, *Welding Safety*, before continuing.

1.2.1 Protective Clothing and Equipment

Welding and cutting tasks are dangerous. Unless you wear all required PPE, you're at risk of serious injury (*Figure 2*). The following guidelines summarize PPE for oxyfuel cutting:

- Always wear safety glasses, plus a full face shield, helmet, or hood. The glasses, face shield, or helmet/hood lens must have the proper shade value for oxyfuel cutting work (*Figure 3*). Never cut without using the proper lens.
- Wear protective leather and/or flame-retardant clothing along with welding gloves that will protect you from flying sparks and molten metal.
- Wear high-top safety shoes or boots. Make sure the pant leg covers the tongue and lace area. If necessary, wear leather spats or chaps for protection. Boots without laces or holes are better for welding and cutting.
- Wear a 100 percent cotton cap with no mesh material in its construction. The bill should point to the rear or the side with the most exposed ear. If you must wear a hard hat, use one with a face shield.

WARNING!

Never wear a cap with a button on top. The conductive metal under the fabric is a safety hazard.

- Wear earplugs to protect your ear canals from sparks. Since the torch makes a continuous noise, always wear hearing protection.

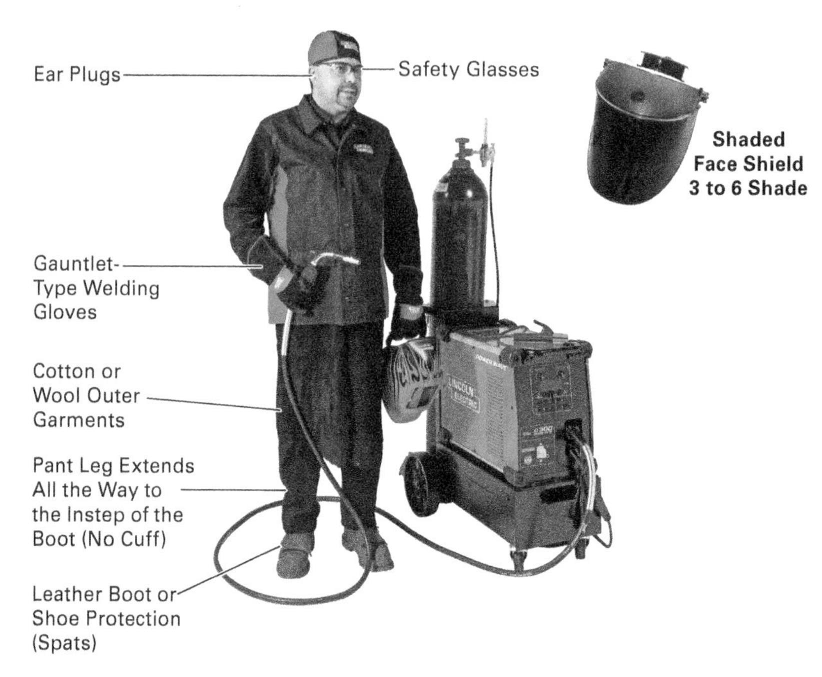

Figure 2 Welding PPE suitable for oxyfuel cutting.
Source: The Lincoln Electric Company, Cleveland, OH, USA

Gas Processes

Process	Type	Plate Thickness	Lens Shade Numbers (2 3 4 5 6 7 8 9 10 11 12 13 14)
Oxyfuel Welding (OFW)	Light	< 1/8" (3 mm)	4 5
	Medium	1/8" – 1/2" (3 mm – 13 mm)	5 6
	Heavy	> 1/2" (13 mm)	6 7 8
Oxyfuel Cutting (OC)	Light	< 1" (25 mm)	3 4
	Medium	1" – 6" (25 mm – 150 mm)	4 5
	Heavy	> 6" (150 mm)	5 6
Torch Brazing (TB)			3 4
Torch Soldering (TS)			2

Lens shade values are based on OSHA minimum values and ANSI/AWS suggested values. Choose a lens shade that's too dark to see the weld zone. Without going below the required minimum value, reduce the shade value until it's visible. If possible, use a lens that absorbs yellow light when working with fuel gas processes.

Figure 3 Oxyfuel Cutting – Guide to Lens Shade Numbers.

1.2.2 Fire/Explosion Prevention

Oxyfuel work causes most welding environment fires. Operators must be properly trained in oxyfuel equipment before using it. When possible, do cutting operations in a dedicated welding shop space. This area should have a concrete floor, protective drapes, and appropriate fire extinguishers. When you must cut outside the shop, remove all flammables from the work area. If this isn't possible, set up protective screens and cover flammables with welding blankets (*Figure 4*). Most companies require posting at least one fire watch with a fire extinguisher during oxyfuel operations.

Oxyfuel cutting produces extremely high temperatures to cut metal. Welding or cutting a vessel or container that once contained combustible, flammable, or explosive materials is hazardous. Residues can catch fire or explode. Before welding or cutting vessels, check whether they contained explosive, hazardous, or flammable materials. These include petroleum products, citrus products, or chemicals that release toxic fumes when heated.

Figure 4 Welding blanket.

American Welding Society (AWS) F4.1, Safe Practices for the Preparation of Containers and Piping for Welding and Cutting, and *ANSI/AWS Z49.1* describe safe practices for these situations. Begin by cleaning the container to remove any residue. Steam cleaning, washing with detergent, or flushing with water are possible methods. Sometimes you must combine these to get good results.

WARNING!

Clean containers only in well-ventilated areas. Vapors can accumulate in a confined space during cleaning, causing explosions or toxic substance exposure.

After cleaning the container, you must formally confirm that it's safe for welding or cutting. *American Welding Society (AWS) F4.1, Safe Practices for the Preparation of Containers and Piping for Welding and Cutting*, outlines the proper procedure. The following three paragraphs summarize the process.

Immediately before work begins, a qualified person must check the container with an appropriate test instrument and document that it's safe for welding or cutting. Tests should check for relevant hazards (flammability, toxicity, etc.). During work, repeated tests must confirm that the container and its surroundings remain safe.

Alternatively, fill the container with an inert material ("inerting") to drive out any hazardous vapors. Water, sand, or an inert gas like argon meets this requirement. Water must fill the container to within 3" (~7.5 cm) or less of the work location. The container must also have a vent above the water so air can escape as it expands (*Figure 5*). Sand must completely fill the container.

When using an inert gas, a qualified person must supervise, confirming that the correct amount of gas keeps the container safe throughout the work. Using an inert gas also requires additional safety procedures to avoid accidental suffocation.

NOTE

A *lower explosive limit (LEL)* gas detector can check for flammable or explosive gases in a container or its surroundings. Gas monitoring equipment checks for specific substances, so be sure to use a suitable instrument. Test equipment must be checked and calibrated regularly. Always follow the manufacturer's guidelines.

WARNING!

Never weld or cut drums, barrels, tanks, vessels, or other containers until they have been emptied and cleaned thoroughly. Residues, such as detergents, solvents, greases, tars, or corrosive/caustic materials, can produce flammable, toxic, or explosive vapors when heated. Never assume that a container is clean and safe until a qualified person has checked it with a suitable test instrument. Never weld or cut in places with explosive vapors, dust, or combustible products in the air.

The following general fire- and explosion-prevention guidelines apply to oxyfuel cutting work:

- Never carry matches or gas-filled lighters in your pockets. Sparks can ignite the matches or cause the lighter to explode.
- Always comply with all site and employer requirements for hot work permits and fire watches.
- Never use oxygen in place of compressed air to blow off the workpiece or your clothing. Never release large amounts of oxygen. Oxygen makes fires start readily and burn intensely. Oxygen trapped in fabric makes it ignite easily if a spark falls on it. Keep oxygen away from oil, grease, and other petroleum products.
- Remove all flammable materials from the work area or cover them with a fire-resistant blanket.
- Before welding, heating, or cutting, confirm that an appropriate fire extinguisher is available. It must have a valid inspection tag and be in good condition. All workers in the area must know how to use it.

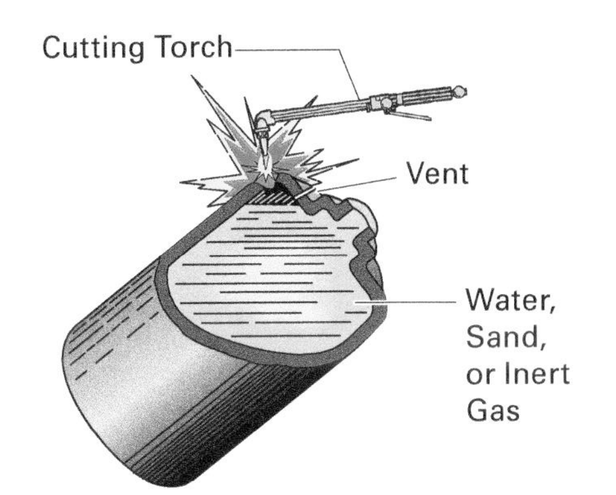

Note: *ANSI Z49.1* and AWS standards should be followed.

Figure 5 "Inerting" a container.

- Never release a large amount of fuel gas into the work environment. Some gases (propane and propylene) accumulate on the floor, while others (acetylene and natural gas) accumulate near the ceiling. Either can ignite far from the release point. Acetylene will explode at lower concentrations than other fuel gases.
- Prevent fires by maintaining a neat and clean work area. Confirm that metal scrap and slag are cold before disposing of them.

1.2.3 Work Area Ventilation

Vapors and fumes fill the air around their sources. Welding and cutting fumes can be hazardous. Welders often work above the area from which fumes come. Good ventilation and source extraction equipment help remove the vapors and protect the welder. The following lists good ventilation practices to use when welding and cutting:

- Always weld or cut in a well-ventilated area. Operations involving hazardous materials, such as metal coatings and toxic metals, produce dangerous fumes. You must wear a suitable respirator when working with them. Follow your workplace's safety protocols and confirm that your respirator is in good condition.

WARNING!

Cadmium, mercury, lead, zinc, chromium, and beryllium produce toxic fumes when heated. When cutting these materials, wear appropriate respiratory PPE. Supplied-air respirators (SARs) work best for long-term work.

- Never weld or cut in a confined space without a confined space permit. Prepare for working in the confined space by following all safety protocols.
- Set up a confined space ventilation system before beginning work.
- Never ventilate with oxygen or breathe pure oxygen while working. Both actions are very dangerous.

WARNING!

When working around other craftworkers, always be aware of what they're doing. Take precautions to keep your work from endangering them.

1.2.4 Cylinder Handling and Storage

Craftworkers must understand the hazards associated with compressed gases and their storage cylinders. Cylinders, pressure regulators, fittings, and hoses must match the gas. Never use equipment in poor condition or that doesn't work correctly.

Report to the gas supplier any leaky cylinders or those with bad valves or fittings. Tag a faulty cylinder or write the problem on it with **soapstone**. If it leaks, move it outdoors to a safe place. Post a warning sign and slowly release the pressure.

Soapstone: A soft, white stone used to mark metal.

Fuel gases vary in their hazardousness. Acetylene, the most useful, is extremely hazardous. It explodes if compressed in cylinders at too great a pressure. For this reason, it's stored in special cylinders that require careful handling. Other gases, such as oxygen, have their own unique needs and precautions. The next section introduces these in detail.

WARNING!

Every gas has its own hazards, storage requirements, and safety procedures. If you aren't familiar with a gas, don't use it until you've received proper training. Misusing a gas can easily cause an accident, a fire, or a fatality.

The following are general safety precautions for working with compressed gas cylinders:

- Always keep cylinders upright and securely chained to a support. This keeps them from being knocked over or falling. Secure cylinders attached to a gas distribution manifold too.
- Store cylinders in a dedicated room. Keep those holding combustible gases separate from other gases. Keep them away from welding, cutting, heat sources, and flames.
- Never store cylinders in halls, stairwells, or near exits. In an emergency, they could prevent people from escaping.
- Keep cylinder storerooms locked. Post signs reading "Danger—No Smoking, Matches, or Open Flames" on the door and inside.
- Store oxygen cylinders in a separate space free from flammables. Never store oxygen near combustible gas cylinders. Storage areas must be at least 20' (6.1 m) apart. Alternatively, a firewall 5' (1.6 m) high with a 30-minute burn rating is acceptable.
- Inert or nonflammable gas cylinders require fewer precautions. You may store them with fuel gas or oxygen cylinders.
- When stored together, keep empty cylinders separate from full ones.
- Store all cylinders upright with their safety caps in place. Don't move a cylinder without a safety cap.
- Never touch a cylinder with a welding electrode, an electrode holder, or an electrically energized part.
- When opening a cylinder valve, stand with the regulator and outlet facing *away* from you.
- Use warm, *not* boiling, water to free a cylinder frozen to the ground.

Safety caps keep the valve from snapping off if the cylinder falls over. A broken valve could turn the cylinder into a small rocket, possibly causing an injury or death. Keep safety caps in place except when using the cylinder. Never lift the cylinder by its safety cap or the valve. Never drop cylinders or handle them roughly.

WARNING!

Use a strap wrench to remove a stuck safety cap. Call the gas supplier if you can't remove it.

Oxygen Consumption

Even though oxyfuel torches have an oxygen supply, they consume oxygen from the surrounding air too. In a heavily occupied or confined workspace, they could reduce breathable oxygen below safe levels.

NOTE

The separation or wall keeps a small fire from heating an oxygen cylinder, causing its safety valve to open. Oxygen released near a small fire will quickly turn it into an inferno.

1.0.0 Section Review

1. The oxyfuel cutting process creates a very hot flame and oxidizes the metal with a jet of _____.
   a. nitrogen
   b. oxygen
   c. carbon dioxide
   d. water vapor

2. Argon is suitable for _____.
   a. mixing with a fuel gas to make it burn hotter
   b. oxidizing dross to create "recycled" slag
   c. keeping acetylene from exploding
   d. "inerting" a container

2.0.0 Oxyfuel Cutting Equipment

Performance Tasks

There are no Performance Tasks in this section.

Objective

Summarize oxyfuel cutting equipment.

a. Identify and describe oxyfuel cutting gases and their storage cylinders.
b. Describe oxyfuel cutting regulators and hoses.
c. Describe oxyfuel cutting torches and tips.
d. Describe specialized oxyfuel cutting equipment.
e. Describe oxyfuel accessories.

Oxyfuel cutting equipment includes gas cylinders, regulators, hoses, and a cutting torch (*Figure 6*). Numerous accessories and specialized components customize the outfit for many tasks. Understanding each part's role and hazards will help you use oxyfuel equipment safely and competently.

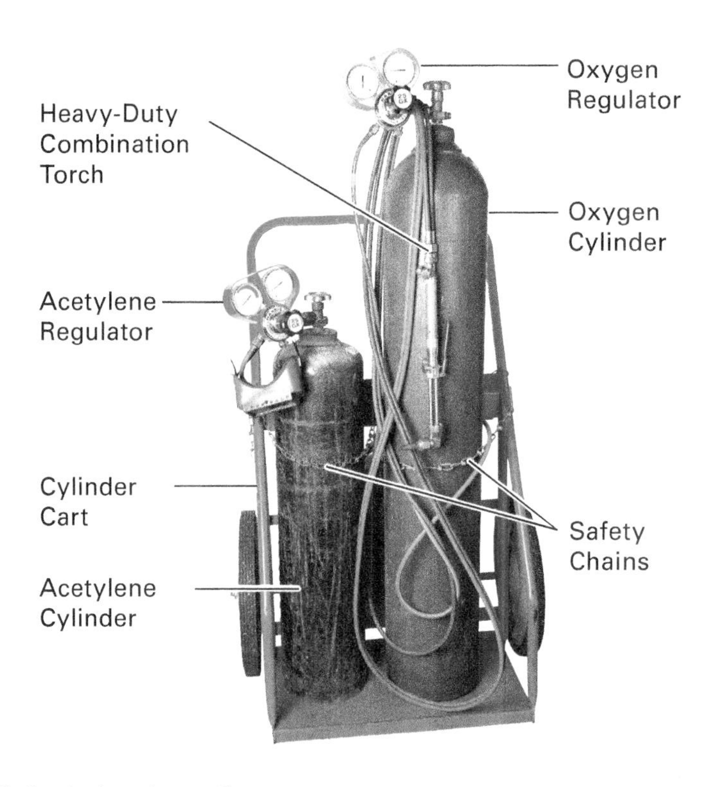

Figure 6 Oxyfuel cutting outfit.
Source: The Lincoln Electric Company, Cleveland, OH, USA

2.1.0 Cutting Gases

Many oxyfuel cutting outfits use oxygen and acetylene. This mix produces the hottest flame. Other fuel gases work for certain applications, however, and are safer. Each gas has its own unique qualities and safety guidelines. The following sections introduce each gas.

2.1.1 Gas Pressure Units

Oxyfuel equipment requires fuel gas and oxygen at the correct pressure. Pressure gauges on each cylinder report the cylinder's internal pressure and the pressure delivered to the torch. Gauges made in the United States display pressure in *pounds per square inch (psi)*. Gauges from metric-speaking countries report pressure in *kilopascals (kPa)* or *bars (bar)*. Sea-level air pressure is about 14.7 psi, 101 kPa, and 1.01 bar.

A gauge can have either an *absolute pressure* scale or a *gauge pressure* scale. They differ only by what's considered "zero pressure." An absolute pressure gauge displays 0 only when it's in a vacuum (a space with no gas at all). An absolute pressure gauge scaled in psi and sitting on a table at sea level will display 14.7 psi. Sea-level pressure is 14.7 psi greater than a vacuum's 0 psi pressure.

A gauge with a gauge pressure scale behaves differently. If it's sitting on a table at sea level, it displays 0. That's because it measures pressure relative to the surrounding air pressure. If it's connected to a tank and displays 100 psi, the pressure in the tank is 100 psi above the outside air pressure. If the tank is at sea level, its internal absolute pressure is 114.7 psi (100 psi + 14.7 psi).

To prevent confusion, pressure gauges with absolute scales place an "A" after the pressure unit. Pressure gauges with gauge pressure scales place a "G" after the unit. Typically, psi becomes *psia* or *psig*. The metric units become kPaA, kPaG, bar(a), and bar(g). *Figure 7* compares absolute and gauge pressures.

NOTE

Most industries and crafts use gauge pressures rather than absolute. However, it's best to always read specifications and gauge faces carefully to be sure!

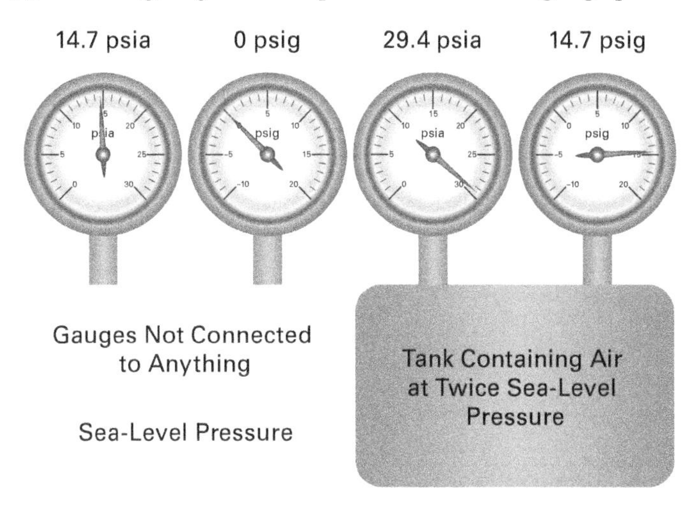

Figure 7 Absolute vs. gauge pressures.

2.1.2 Oxygen

Contrary to what many believe, oxygen (O_2) isn't flammable. Instead, flammable substances require it to burn. Oxygen does, however, make combustion happen more easily and aggressively. It's colorless, odorless, and tasteless. Mixing fuel gas with oxygen instead of air produces a hotter flame.

Oxygen comes in hollow steel cylinders at a pressure of 2,000 psig or more (~14,000 kPaG). International standards specify cylinder sizes. In the United States, the US Department of Transportation (DOT) regulates cylinders. *Figure 8* summarizes oxygen cylinder markings and capacities based on US DOT specifications.

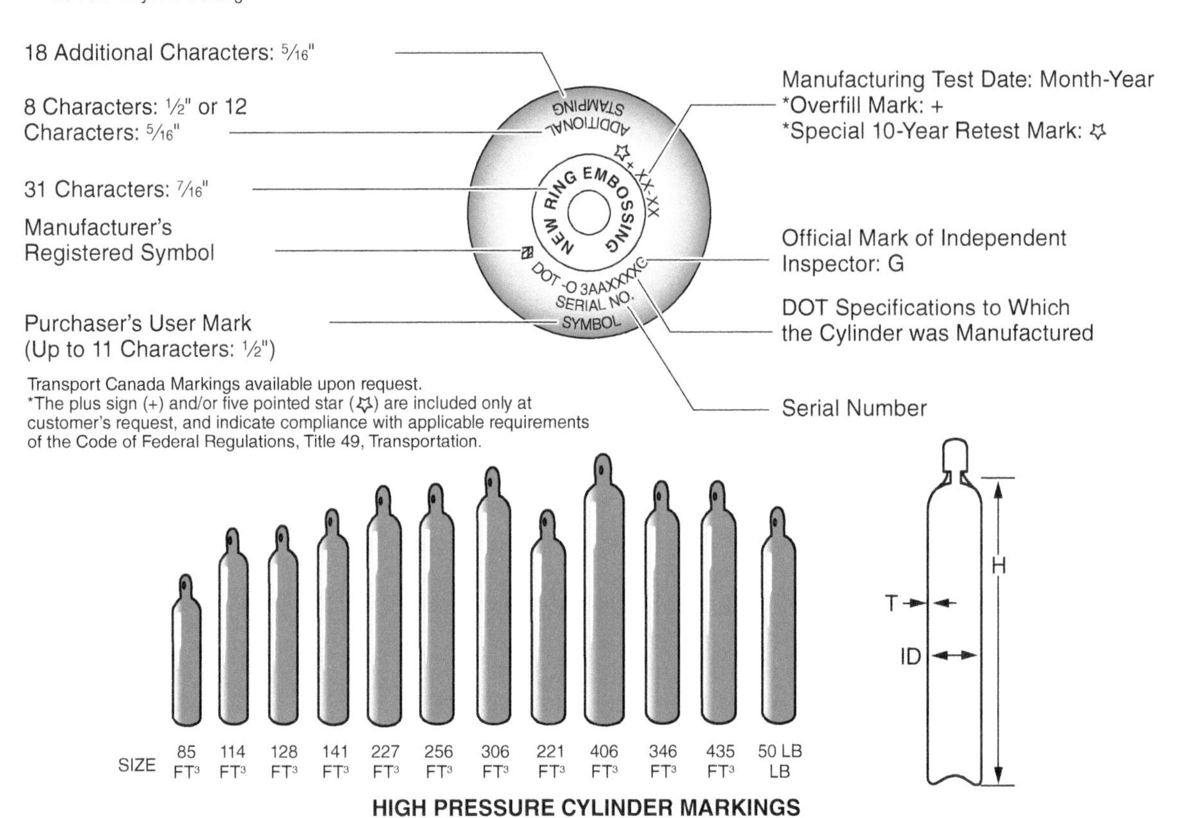

DOT SPECIFICATIONS	O_2 CAPACITY (FT³)		WATER CAPACITY (IN³)		NOMINAL DIMENSIONS (IN)			NOMINAL WEIGHT (LB)	PRESSURE (PSI)	
	AT RATED SERVICE PRESSURE	AT 10% OVERCHARGE	MINIMUM	MAXIMUM	AVG. INSIDE DIAMETER "ID"	HEIGHT "H"	MINIMUM WALL "T"		SERVICE	TEST
STANDARD HIGH PRESSURE CYLINDERS[1]										
3AA2015	85	93	960	1040	6.625	32.50	0.144	48	2015	3360
3AA2015	114	125	1320	1355	6.625	43.00	0.144	61	2015	3360
3AA2265	128	140	1320	1355	6.625	43.00	0.162	62	2265	3775
3AA2015	141	155	1630	1690	7.000	46.00	0.150	70	2015	3360
3AA2015	227	250	2640	2710	8.625	51.00	0.184	116	2015	3360
3AA2265	256	281	2640	2710	8.625	51.00	0.208	117	2265	3775
3AA2400	306	336	2995	3060	8.813	55.00	0.226	140	2400	4000
3AA2400	405	444	3960	4040	10.060	56.00	0.258	181	2400	4000
ULTRALIGHT® HIGH PRESSURE CYLINDERS[1]										
E-9370-3280	365	NA	2640	2710	8.625	51.00	0.211	122	3280	4920
E-9370-3330	442	NA	3181	3220	8.813	57.50	0.219	147	3330	4995
ULTRA HIGH PRESSURE CYLINDERS[2]										
3AA3600	347[3]	374	2640	2690	8.500	51.00	0.336	170	3600	6000
3AA6000	434[3]	458	2285	2360	8.147	51.00	0.568	267	6000	10000
E-10869-4500	435[3]	NA	2750	2890	8.813	51.00	0.260	148	4500	6750
E-10869-4500	485[3]	NA	3058	3210	8.813	56.00	0.260	158	4500	6750

1. Regulators normally permit filling these cylinders with 10% overcharge, provided certain other requirements are met.
2. Under no circumstances are these cylinders to be filled to a pressure exceeding the marked service pressure at 70°F.
3. Nitrogen capacity at 70°F.

All cylinders normally furnished with 3/4" NGT internal threads, unless otherwise specified.
Nominal weights include neck ring but exclude valve and cap, add 2 lb (.91 kg) for cap and 1½ lb (.8 kg) for valve.
Cap adds approximately 5" (127 mm) to height.
Cylinder capacities are approximately 5" (127 mm) to height.
Cylinder capacities are approximately at 70°F (21°C).

Figure 8 US DOT specifications for oxygen cylinders.

The smallest standard cylinder holds about 85 ft^3 (2.4 m^3), while the largest holds about 485 ft^3 (13.7 m^3). The most common size for cutting equipment holds 227 ft^3 (6.4 m^3). It's about 4' (1.2 m) tall and 9" (~22 cm) in diameter.

An oxygen cylinder has a bronze cylinder valve on top (*Figure 9*). The valve controls the oxygen flow from the cylinder. A safety plug (rupture disk) on the valve side blows out and releases the oxygen if the tank's pressure rises dangerously high. Escaping oxygen is hazardous, but an explosion is worse.

Oxygen cylinders have Compressed Gas Association (CGA) Type 540 valves for pressures up to 3,000 psig (~20,700 kPaG). CGA Type 577 valves handle up to 4,000 psig (~27,600 kPaG), while CGA Type 701 handles up to 5,500 psig (~37,900 kPaG). Always keep the valve's safety cap in place when you're not using the cylinder (*Figure 10*).

WARNING!

Always secure the cylinder before removing its safety cap. If the cylinder falls over without its cap, the valve could break off. A high-pressure cylinder with a broken valve is extremely dangerous.

NOTE

In the United States, owners must have their cylinders tested and re-certified every 10 years. Other countries may have different rules.

Other Standards

In Canada, Transport Canada (TC) regulates gas cylinders. The Department for Transport (DfT) manages European cylinder standards.

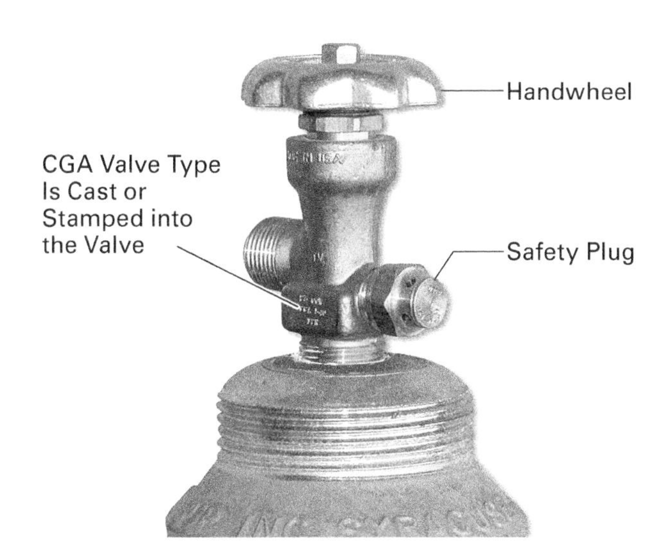

Figure 9 Oxygen cylinder valve.

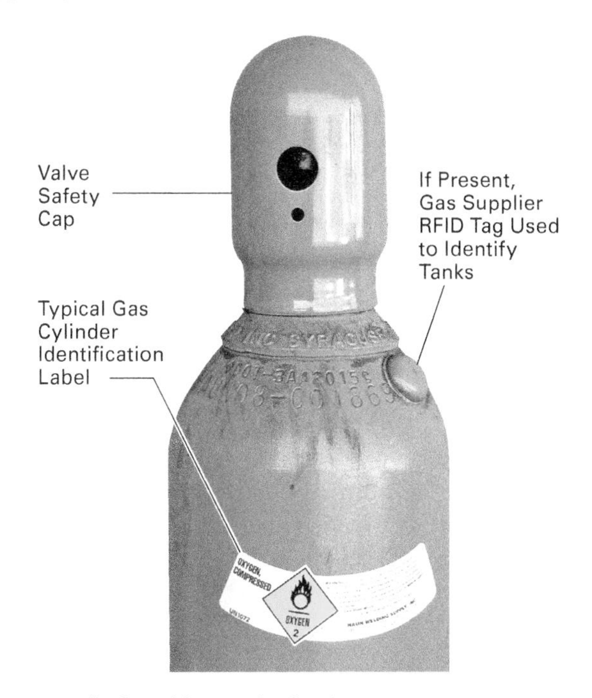

Figure 10 Oxygen cylinder with standard safety cap.

Cylinder Management

Some suppliers track their cylinders with a radio frequency identification (RFID) tag. When scanned, the tag returns a unique number. Using this number, a database supplies information about the cylinder, including its purchaser and maintenance history. Some RFID tags look like buttons. Others fit into the tank walls or attach by a collar around the valve.

Source: Xerafy

Clamshell Safety Cap

Some high-pressure cylinders come with a clamshell safety cap. It splits in half, giving access to the cylinder valve. When closed, it fits around the regulator fittings. This arrangement protects the valve, so you may be able to move the cylinder with the regulator attached. The safety cap threads into place around the valve. A latch pin or padlock secures the clamshell halves. Never move the cylinder with the regulator attached unless the cylinder is secured in an approved carrier. Also confirm that state and local regulations permit it. When in doubt, remove the regulator.

Clamshell Open to Allow Cylinder Valve Operation

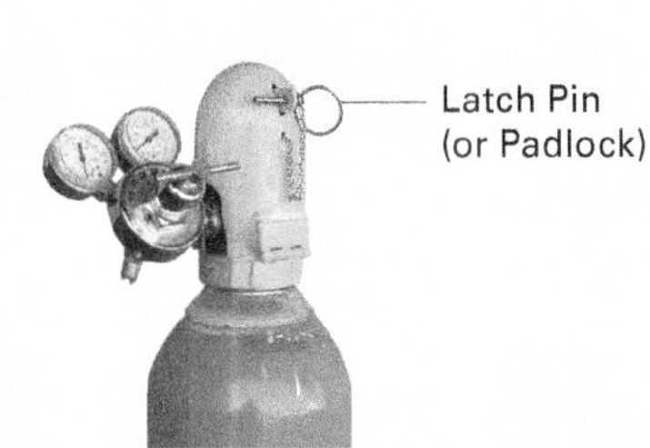

Clamshell Closed for Movement or Padlocked to Prevent Operation of Cylinder Valve

Clamshell Closed for Transport

2.1.3 Acetylene

Acetylene (C_2H_2) is lighter than air and has a garlic odor. When mixed with oxygen, it burns hotter than 5,500°F (~3,040°C). Acetylene flames can cut, weld, harden, and relieve stress. But this useful gas is extremely hazardous and unstable. It easily explodes and is difficult to store safely.

Acetylene explodes when compressed above 15 psig (103 kPaG). Unlike many other gases, regular high-pressure storage isn't an option. Instead, gas suppliers use a special cylinder to overcome this problem. It contains a porous material soaked with liquid acetone. The compressed acetylene gas dissolves in the acetone, which keeps it from exploding. Opening the tank valve causes the acetylene to leave the acetone.

Because the cylinder contains a liquid, it must stay upright. If it tips over, turn it upright. Wait at least one hour before using it. Liquid acetone must not enter the valves or regulator, as it can clog them. It will also make the torch operate erratically.

Never let the acetylene gas pressure rise higher than 15 psig (103 kPaG) as it leaves the cylinder. The flow rate per hour must stay below $^1/_{10}$ the cylinder's capacity. A torch with a large nozzle used with a small tank can easily exceed this limit. High flow rates may draw liquid acetone into the gas stream.

Acetylene cylinders have safety plugs at the top and bottom (*Figure 11*). During a fire, they will melt and release the cylinder's pressure, preventing an explosion. Plugs on acetylene tanks melt at 212°F (100°C).

NOTE

Unlike pressure relief valves, safety plugs are "one-shot" devices. After they melt and open, you must replace them. True pressure relief valves close once the pressure falls below their opening value.

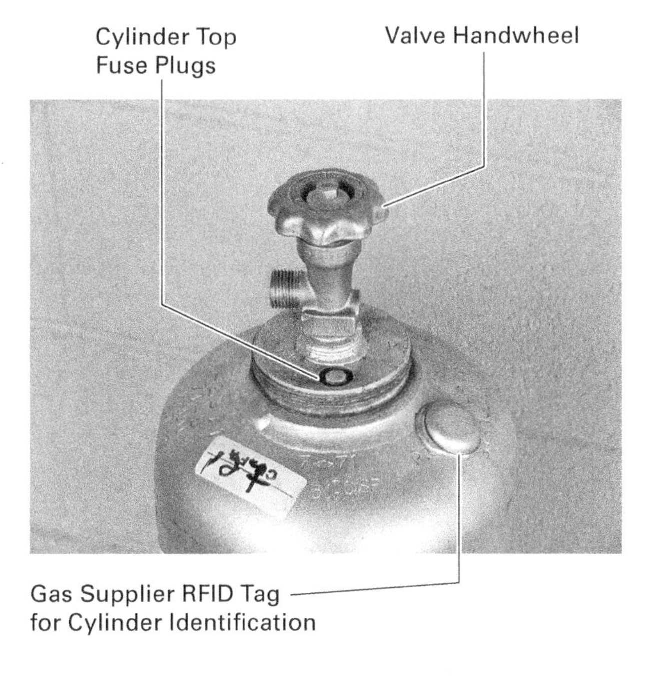

Figure 11 Acetylene cylinder valve and safety plugs.

Ring Guard Safety Cap

Acetylene and propane cylinders may come with a ring-shaped safety cap. It protects the cylinder valve and can fit around an installed pressure regulator. With the ring in place and the valve closed, it may be permissible to move the cylinder. Never move the cylinder with the regulator attached unless the cylinder is secured in an approved carrier. Also confirm that state and local regulations permit it. When in doubt, remove the regulator.

As with oxygen cylinders, international standards specify acetylene cylinder sizes. In the United States, the US DOT regulates them. *Figure 12* summarizes acetylene cylinder markings and capacities based on US DOT specifications.

NOTE

In the United States, owners must have their cylinders tested and re-certified every 10 years. Other countries may have different rules.

The smallest standard cylinder holds about 10 ft^3 (0.3 m^3), while the largest holds about 420 ft^3 (11.9 m^3). An extra-large 850 ft^3 (24.1 m^3) version is sometimes available.

An acetylene cylinder has a brass cylinder valve on top. For larger cylinders, the CGA Type 510 and Type 300 are standard. Smaller cylinders have the CGA Type 520 or Type 200. Always keep the valve's safety cap in place when you're not using the cylinder (*Figure 13*).

WARNING!

Always secure the cylinder before removing its safety cap. If the cylinder falls over without its cap, the valve could break off. Acetylene is explosive, so a broken valve is extremely dangerous.

NOTE

Leaking fuel gas is a serious hazard. Since some have little or no natural odor, manufacturers usually add an *odorant* to them. This gives the gas an unpleasant rotten-egg smell that's easily noticed.

2.1.4 Other Fuel Gases

While acetylene offers the highest flame temperature, it's a dangerous and relatively costly gas. When a lower temperature is acceptable, welders may choose safer and cheaper alternatives. *Table 1* lists these and compares them to acetylene.

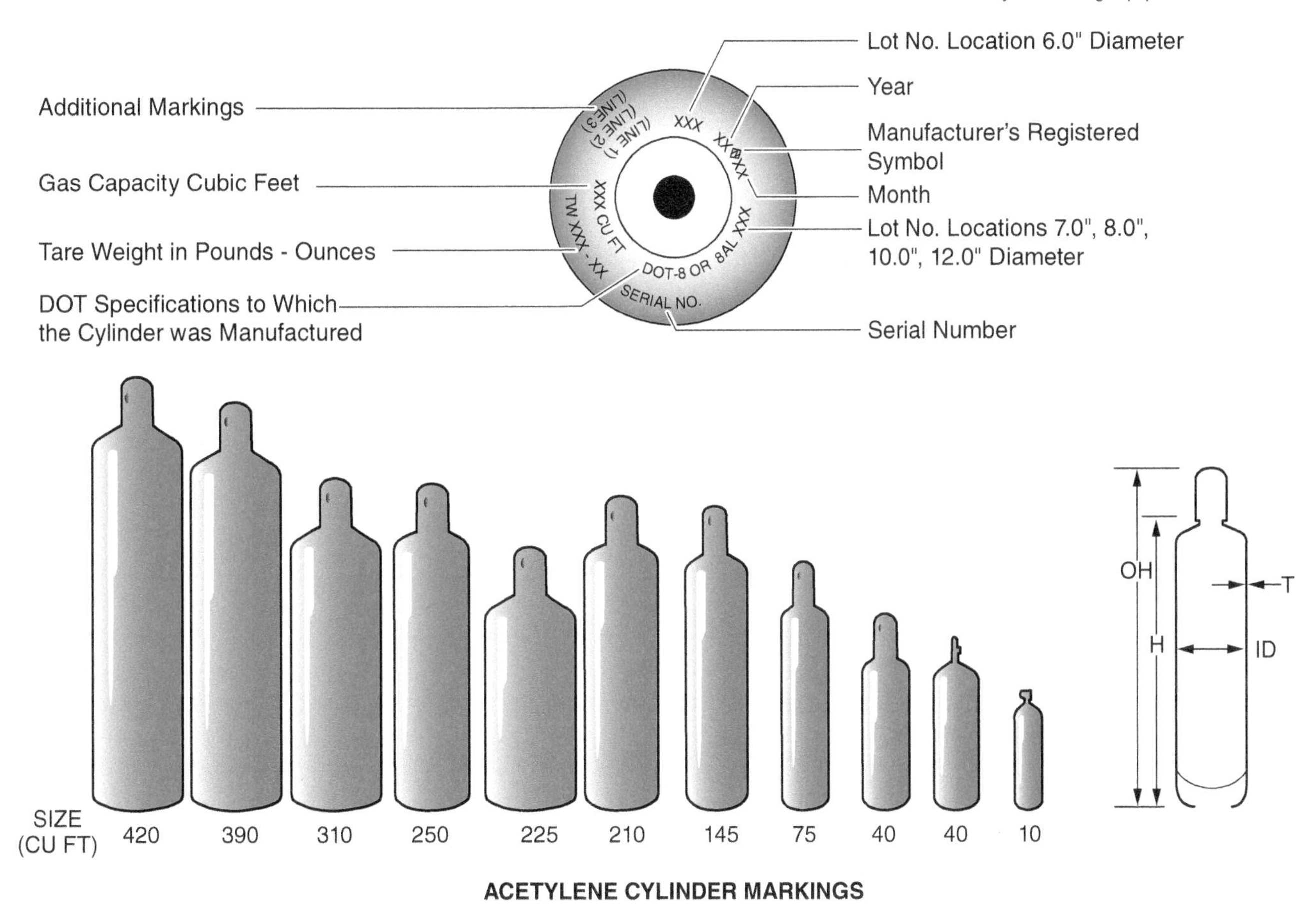

DOT SPECIFICATIONS	CAPACITY: ACETYLENE (FT³)	CAPACITY: MIN. WATER (IN³)	CAPACITY: MIN. WATER (LB)	NOMINAL DIMENSIONS (IN): AVG. INSIDE DIAMETER "ID"	NOMINAL DIMENSIONS (IN): HEIGHT W/OUT VALVE OR CAP "H"	NOMINAL DIMENSIONS (IN): HEIGHT W/VALVE AND CAP "OH"	NOMINAL DIMENSIONS (IN): MINIMUM WALL "T"	ACETONE (LB - OZ)	APPROXIMATE TARE WEIGHT WITH VALVE WITHOUT CAP (LB)
8 AL[1]	10	125	4.5	3.83	13.1375	14.75	0.0650	1-6	8
8[1]	40	466	16.8	6.00	19.8000	23.31	0.0870	5-7	25
8[2]	40	466	16.8	6.00	19.8000	28.30	0.0870	5-7	28
8[3]	75	855	30.8	7.00	25.5000	31.25	0.0890	9-8	45
8	100	1055	38.0	7.00	30.7500	36.50	0.0890	12-2	55
8	145	1527	55.0	8.00	34.2500	40.00	0.1020	18-10	76
8	210	2194	79.0	10.00	32.2500	38.00	0.0940	25-13	105
8AL	225	2630	94.7	12.00	27.5000	32.75	0.1280	29-6	110
8	250	2606	93.8	10.00	38.0000	43.75	0.0940	30-12	115
8AL	310	3240	116.7	12.00	32.7500	38.50	0.1120	39-5	140
8AL	390	4151	150.0	12.00	41.0000	46.75	0.1120	49-14	170
8AL	420	4375	157.5	12.00	43.2500	49.00	0.1120	51-14	187
8	60	666	24.0	7.00	25.79 OH		0.0890	7-11	40
8	130	1480	53.3	8.00	36.00 OH		0.1020	17-2	75
8AL	390	4215	151.8	12.00	46.00 OH		0.1120	49-14	180

1. Tapped for ⅜" valve but are not equipped with valve protection caps.
2. Includes valve protection cap.
3. Can be tared to hold 60 ft³ (1.7 m³) of acetylene gas.

Standard tapping (except cylinders tapped for ⅜") ¾"-14 NGT.

Weight includes saturation gas, filler, paint, solvent, valve, fuse plugs.
Does not include cap of 2 lb (.91 kg.)
Cylinder capacities are based upon commercially pure acetylene gas at 250 psi (17.5 kg/cm²), and 70°F (15°C).

Figure 12 US DOT specifications for acetylene cylinders.

Figure 13 Acetylene cylinder with standard safety cap.

TABLE 1 Fuel Gas Comparison

Gas	Flame Temperature (with Oxygen)
Acetylene	More than 5,500°F (~3,040°C)
Propylene	5,130°F (~2,830°C)
Natural gas	4,600°F (~2,540°C)
Propane	4,580°F (~2,530°C)

Several of these gases come as liquids stored under pressure in cylinders (*Figure 14*). A layer of gas sits on top of the liquid. Opening the tank valve releases the gas and causes the liquid to start vaporizing.

Changing from a liquid to a gas absorbs heat from the surroundings, so cylinders get cold as they release gas. Frost may form on the outside, or the regulator may ice up. Some high-flow regulators have built-in electric heaters to prevent this problem.

WARNING!

Never deal with icing by applying heat directly to a cylinder or regulator. This can create a pressure buildup and possibly an explosion.

The gas pressure inside the cylinder depends on the surrounding temperature. In cold weather, the pressure will be lower. For this reason, cylinder pressure doesn't indicate how much fuel the cylinder contains. Weight is the best way to evaluate its condition.

Liquefied gas cylinders come in sizes holding 30 lb to 225 lb (14 kg to 102 kg) of fuel. They come with CGA Type 510, 350, or 695 valves, depending on the storage pressure.

WARNING!

Always secure the cylinder before removing its safety cap. If the cylinder falls over without its cap, the valve could break off. Fuel gases are flammable and sometimes explosive, so a broken valve is extremely dangerous. Liquid fuel can spurt from a broken valve and ignite.

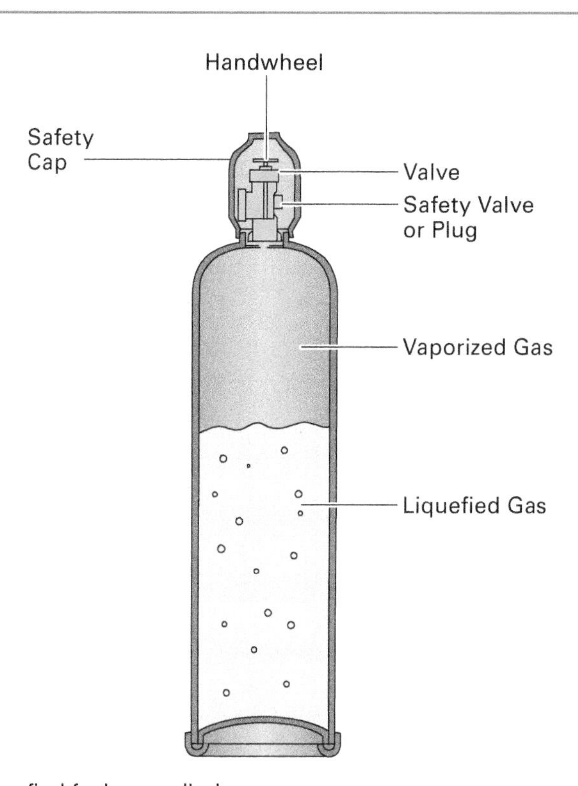

Figure 14 Liquefied fuel gas cylinder.

The following three paragraphs summarize the common acetylene alternatives used in oxyfuel cutting.

Propylene (Propene)

Liquid propylene mixtures are stable, shock-resistant, and relatively safe. They're sold under trade names such as High Purity Gas (HPG™), Apachi™, and Prestolene™. They burn hotter than natural gas and propane. Welders use propylene mixtures for flame cutting, scarfing, heating, stress relieving, brazing, and soldering.

Natural Gas

Natural gas is stable, shock-resistant, and relatively safe. It's common as a cooking and heating fuel. Local gas companies deliver it by pipeline to a facility. For this reason, welders use it in the shop at fixed workstations. It burns hotter than propane but cooler than propylene. It can **backfire** and produce **flashback** in oxyfuel equipment.

Backfire: A loud snap or pop that happens upon extinguishing a torch.

Flashback: A hissing or whistling sound caused by the flame burning back into the torch, hose, or regulator.

Propane (LP)

Propane (liquefied petroleum) is stable, shock-resistant, and relatively safe. It's common as a cooking and heating fuel. Plumbers use small propane torches for soldering. Among the fuel gases, it burns at the lowest temperature. It can backfire and produce flashback in oxyfuel equipment. Welders use it for cutting.

Handling and Storing Liquefied Gas Cylinders

Handle liquefied gas cylinders carefully because they contain flammable gas at a high pressure. Never drop the cylinder or hit it with a heavy object. Keep them upright, so the liquid doesn't enter the valve. Liquefied fuel gas cylinders have a safety valve built into the cylinder valve. If the internal pressure becomes unsafe, the valve opens and releases gas. Always keep the valve's safety cap in place when you're not using the cylinder.

Used Cylinders

Buying used cylinders can save money. But a cylinder in poor condition is no bargain. Before buying used cylinders, ask a few questions:

- Is the cylinder in good condition?
- Is the valve and its threads in good condition?
- Does the cylinder include the valve safety cap?
- When was the cylinder last inspected and certified? Who did it?
- Is the correct certification paperwork included?

2.2.0 Regulators and Hoses

Oxyfuel cutting torches require fuel gas and oxygen at specific pressures. Cylinder pressures are too high, so a suitable *pressure regulator* first reduces the pressure. Safety devices keep gas flowing in the right direction and prevent flame from traveling backwards. Hoses deliver the fuel gas and oxygen to the torch.

2.2.1 Pressure Regulators

Every oxyfuel outfit has two regulators, one for the oxygen and one for the fuel gas (*Figure 15*). Two gauges per regulator show the cylinder (inlet) pressure and torch delivery (outlet) pressure. A knob or screw adjusts the outlet pressure. Turning it clockwise increases pressure, while turning it counterclockwise reduces it.

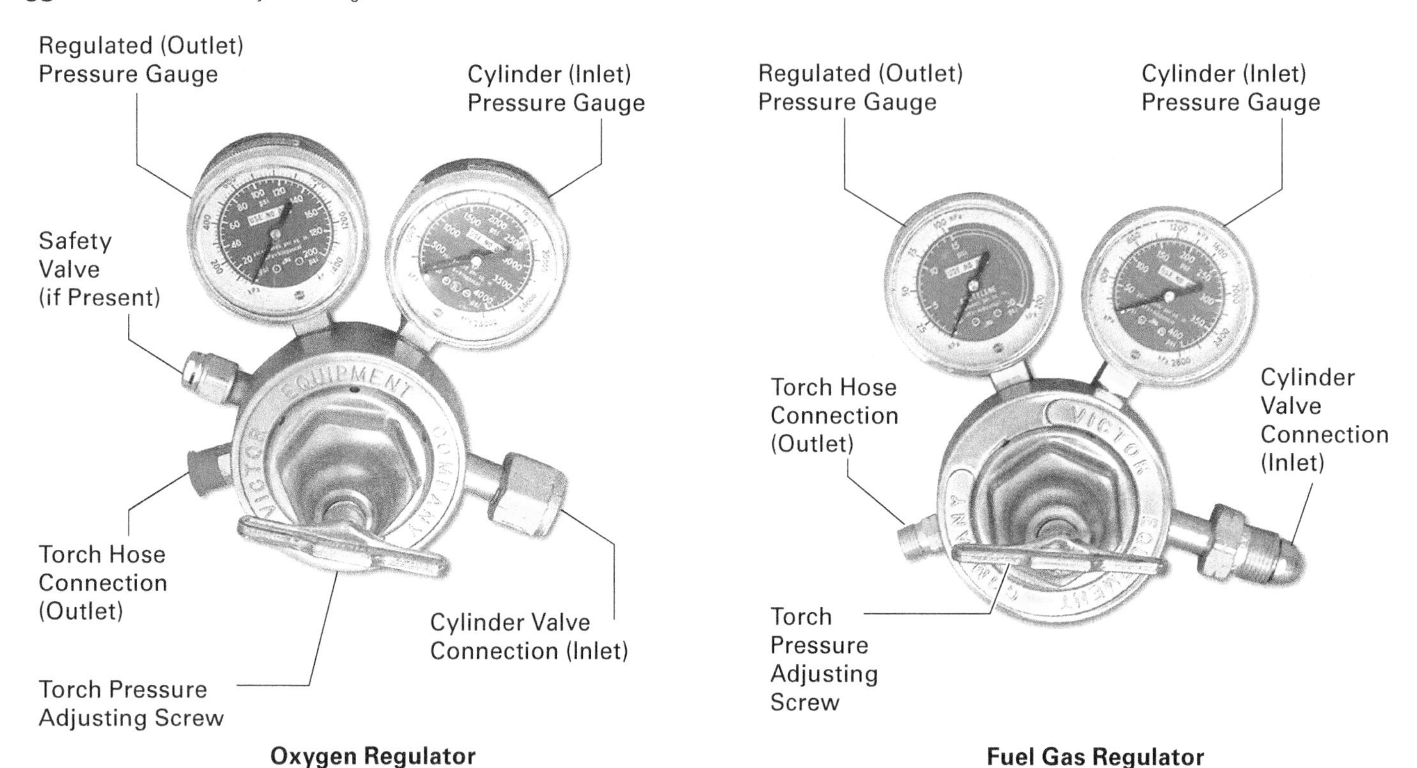

Figure 15 Oxygen and fuel gas regulators.

Turning the adjustment too far counterclockwise unscrews it from the internal mechanism. In this *fully released* condition it feels loose, and no gas will flow. But don't use the regulator adjustment to stop the gas flow. Instead, shut off the gas with the cylinder valve. You should fully release the regulator adjustment, however, before re-opening the cylinder valve.

Oxygen regulators have green markings and right-hand (RH) threads on all connections. The regulator's inlet gauge generally reads up to 3,000 psig (~20,700 kPaG). The outlet gauge may read up to 100 psig (~700 kPaG) or higher.

Fuel gas regulators have red markings and left-hand (LH) threads on all the connections. Some fitting nuts have notches cut on their corners as a reminder. The regulator's inlet gauge generally reads up to 400 psig (~2,760 kPaG). The outlet gauge may read up to 40 psig (~280 kPaG). Acetylene gauges are always red-lined at 15 psig (103 kPaG). This reminds the craftworker that acetylene explodes above this pressure.

Pressure regulators divide into two types, *single-stage* and *two-stage*. Single-stage regulators reduce the pressure in one step. They're simpler and therefore less expensive. But the craftworker must keep adjusting the outlet pressure as the cylinder pressure falls. This is a minor nuisance.

Two-stage regulators contain two pressure regulators, one fixed and one adjustable (*Figure 16*). The first stage reduces the cylinder pressure to a lower, intermediate value. The second stage lets the craftworker control the outlet pressure. Two-stage regulators maintain a constant outlet pressure as the cylinder pressure falls.

A two-stage regulator is more complex, so it's more expensive. It usually can't deliver as high a flow rate as a single-stage unit. Expensive, high-flow versions are available for continuous cutting.

2.2.2 Regulator Safety

The following general safety guidelines apply to all gas regulators:

- Regulators are sensitive. Never shake or jar them.
- Fully release (unscrew counterclockwise) the regulator adjustment before opening the cylinder valve. Fully release the regulator when finished.

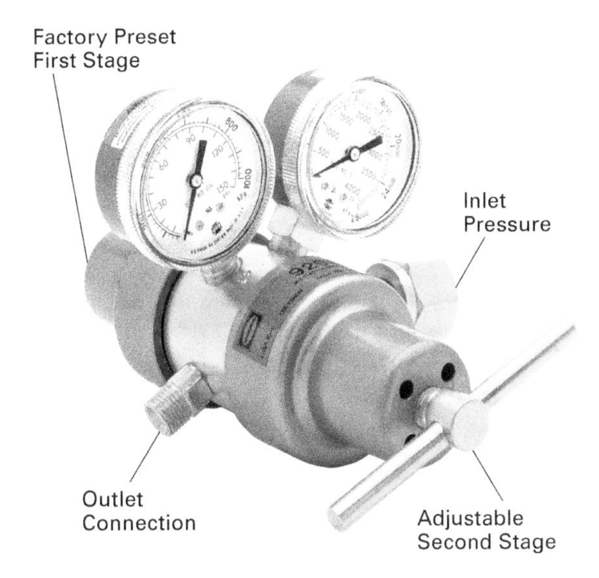

Figure 16 Two-stage regulator.
Source: The Lincoln Electric Company, Cleveland, OH, USA

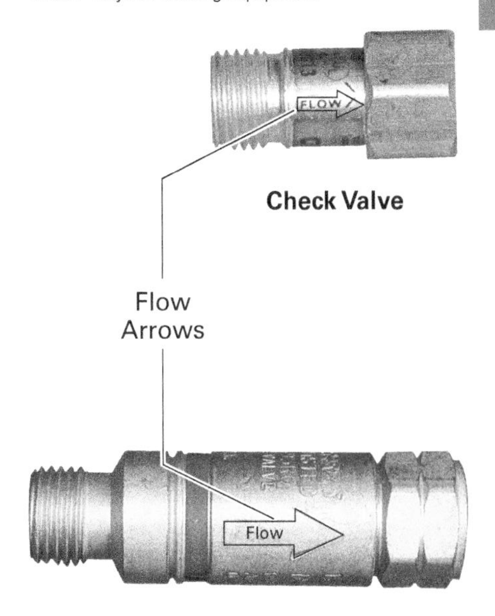

Figure 17 Add-on check valve and flashback arrestor.

- Open cylinder valves slowly. Stand opposite the regulator and outlet. Make sure no one stands directly in line with the regulator or outlet.
- Never lubricate a regulator or gauge with oil or any petroleum-based lubricant. It can explode during operation.
- Never swap fuel gas and oxygen regulators. Use them only with their specified gas.
- Never use a defective regulator. Replace it, or have a qualified technician repair it.
- Install and remove regulators with approved wrenches. Never use large wrenches, pipe wrenches, pliers, vise grips, or slip-joint pliers.

2.2.3 Check Valves and Flashback Arrestors

Check valves keep gas flowing in one direction only. They're an important safety device because they keep gases from flowing backwards at startup. Without them, fuel gas and oxygen may mix in the hoses. A check valve contains a ball held in place with a spring. When the gas flows in the correct direction, it pushes the ball back against the spring, opening the valve. Gas flowing in the wrong direction closes the valve.

Flashback arresters prevent the flame from moving backwards into the hoses or regulators. They contain a filter that blocks the flame. Some also contain a check valve. *Figure 17* shows both devices.

Add-on check valves and flashback arrestors attach to the regulator outlets or the torch connections. Every oxyfuel outfit should have a flashback arrester and check valve attached to each torch connection. A second pair attached to each regulator outlet is also a good idea. Check valves and flashback arrestors have arrows indicating the gas flow direction. When installing them, align the arrow with the flow direction.

2.2.4 Fittings and Wrenches

Oxyfuel fittings are brass or bronze. Some include flexible O-ring seals. Overtightening damages the O-rings, so follow the manufacturer's guidelines. Some fittings are "hand-tighten only." Others require a wrench. Never use regular wrenches on fittings. Instead, use a *torch wrench* (gang wrench). Some have multiple cutouts for different fittings and standard CGA components (*Figure 18*). Never extend the handle with a cheater bar or pipe. Too much force can damage seals and fittings.

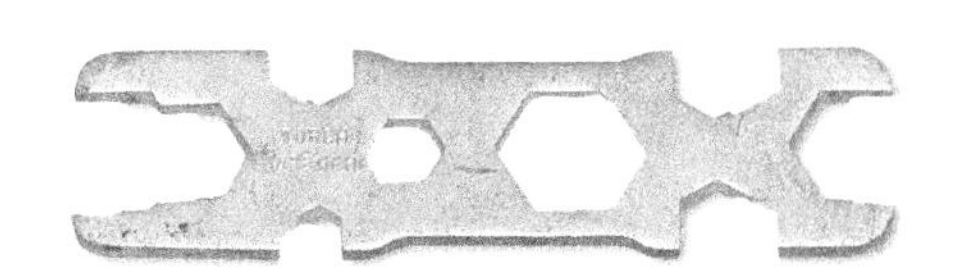

Figure 18 Universal torch wrench.

WARNING!

Never lubricate O-rings with oil or petroleum-based lubricants. These can explode if they're used around oxygen. Always follow the equipment manufacturer's recommendations for installing and maintaining O-rings and other seals.

2.2.5 Gas Distribution Manifolds

Some industrial installations have many fixed stations for cutting and fabrication. Equipping each with individual gas cylinders is inefficient. Instead, companies deliver oxyfuel gases through a pipeline to the workstations. Gas cylinders at a central location connect to a *manifold* with short *pigtail hoses*. The manifold delivers the gas to the supply pipeline. Workers connect their torch hoses to the pipeline (*Figure 19*).

Pressure regulators at both the supply and workstation ends ensure the correct pressure. Workers can adjust the pressure to their torches individually (*Figure 20*). Gauges report the pressure and alert them to leaks. Long hoses may need to be of a larger diameter to avoid unnecessary pressure drops between the connection and workstation.

OSHA specifies the following rules for manifold systems:

- Fuel gas and oxygen manifolds must clearly identify the gas they carry.
- Fuel gas and oxygen manifolds must not be in a confined space. They must be in safe, well-ventilated, accessible locations.
- Hose connections must prevent the wrong hoses from being connected. Workers may not defeat this safety feature with adapters.
- Hose connections must be free from grease and oil.
- Unused connections must have caps.
- The manifold, as well as its valves, must have nothing attached that could damage it or interfere with the valves in an emergency.

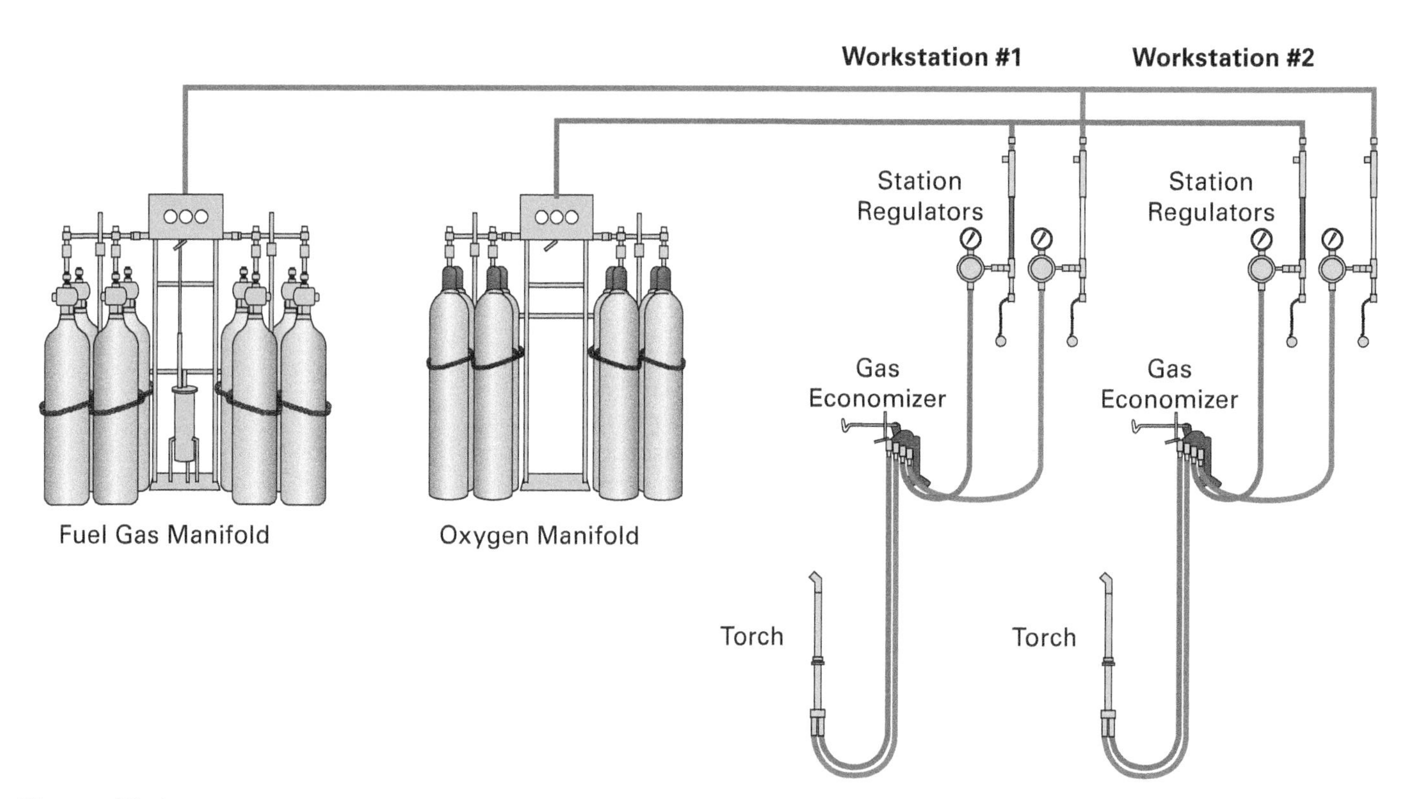

Figure 19 Gas distribution system.

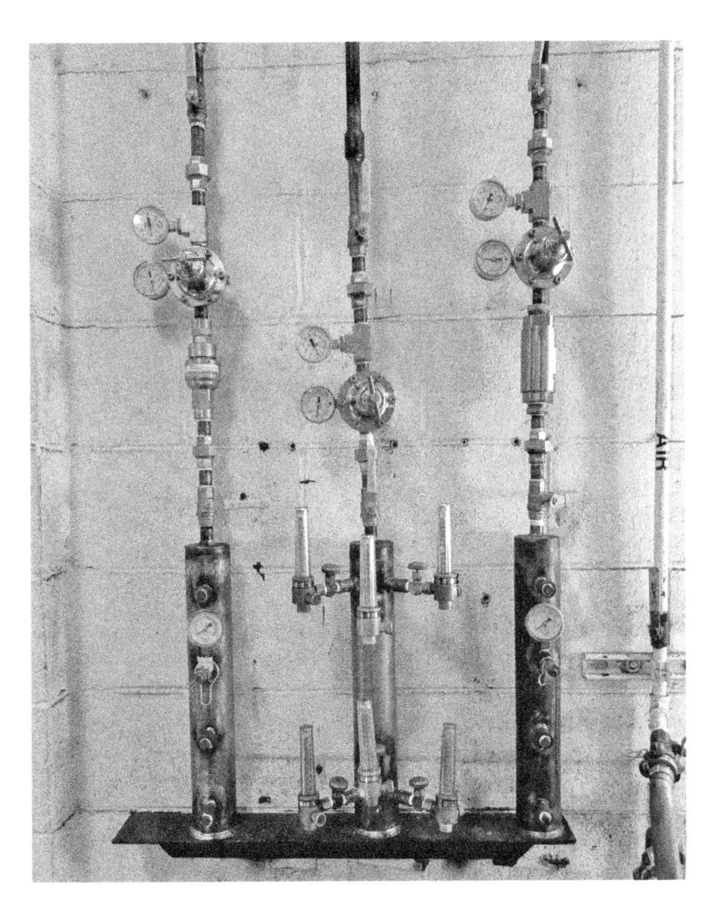

Figure 20 Operator hose connection points.

Keeping Cylinders Separate

Fixed installation gas delivery systems have multiple cylinders in one place. For safety, *ANSI/AWS Z49.1* requires fuel gas and oxygen cylinders to be at least 20' (6.1 m) apart. Alternatively, a firewall 5' (1.6 m) or higher with a 30-minute burn rating, may separate them.

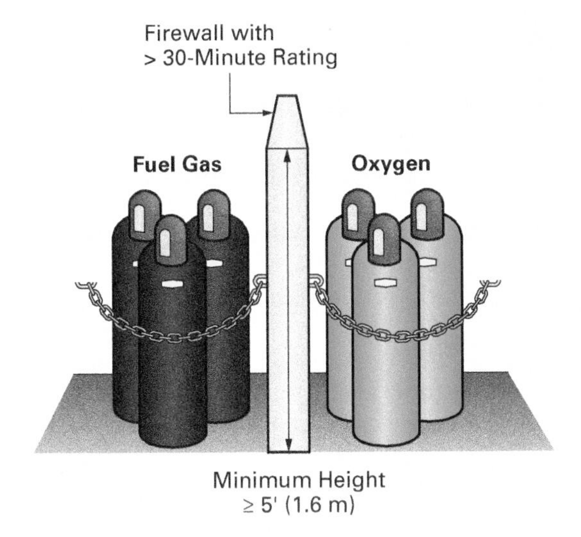

2.2.6 Hoses

Hoses link the torch to the regulator outlet. Oxygen hoses are green and have RH threads. Fuel gas hoses are red and have LH threads. Their fittings also have notches to remind you that their threads are left-hand.

Welding and cutting operations can easily damage hoses. Inspect them regularly for damage. Properly caring for hoses extends their lifespan. The following guidelines apply to all oxyfuel hoses:

- Protect hoses from molten metal and sparks. Some hoses have a fire-retardant outer layer, but molten metal can still damage it.
- Don't let hoses hang under the metal being cut. Hot metal will fall on them. Keep hoses away from the cutting zone.
- Regularly inspect hoses for damage (cuts, burns, abrasions, cracks, and damaged fittings). Replace damaged hoses.
- Never use pipe compounds or lubricants on hose connections. Oil or grease in the compound can catch fire or explode in an oxygen atmosphere.
- Always confirm that the hose is correct for the gas you plan to use. Propane and propylene require hoses designed for them. If you're uncertain about a hose, check with the manufacturer.

2.3.0 Torches and Tips

NOTE

Some torches come with built-in check valves and/or flashback arrestors. Always verify the torch's features before using it. If necessary, connect add-on check valves and/or flashback arrestors to make the oxyfuel outfit safe.

Cutting torches mix the fuel gas with oxygen to produce the preheat flame. They also deliver the cutting oxygen stream that oxidizes the metal. Controls on the torch adjust both the flame and the cutting oxygen stream. Torches come in several styles for different tasks. They accept different tips for specific work.

2.3.1 One-Piece Torches

As the name suggests, one-piece torches have everything in a single package. These torches, sometimes called *demolition torches*, perform heavy-duty cutting (*Figure 21*). They're often very long so the operator isn't too close to the heat. Some can cut steel 12" (~30 cm) or thicker. They have valves on the handle to control the preheat flame. A separate spring-loaded lever controls on the oxygen cutting jet. The hoses connect behind the valves.

Torches mix the fuel gas with oxygen in two possible ways. *Tip-mixing torches* combine the gases at the torch's tip. They have three tubes running from the handle to the tip. One carries fuel gas. The other two carry oxygen. One supplies the preheat flame, while the other delivers the cutting oxygen stream. *Figure 21* is a three-tube torch.

Supply tube mixing torches combine the gases as they travel to the tip. They have two tubes. One carries fuel gas and oxygen for the preheat flame. The other carries the cutting oxygen stream.

One-piece torches further divide into two categories. *Positive-pressure torches* require regulated, pressurized oxygen and fuel gas. *Injector torches* require regulated, pressurized oxygen. But they can handle fuel gas at very low pressures. A vacuum created by the flowing oxygen sucks in the low-pressure fuel gas. Injector torches work well with pipeline-supplied natural gas fuel.

2.3.2 Combination Torches

These torches are more versatile than one-piece torches because they split into two parts. The *handle* connects to the gas supplies and has valves controlling each. The *torch attachment* screws onto the handle. Cutting torch attachments handle medium- or light-duty tasks. Manufacturers also offer brazing and heating torch attachments for non-cutting tasks.

Cutting torch attachments have two additional valves. A *knob* adjusts the preheat flame's oxygen. A *lever* controls the cutting oxygen stream. *Figure 22* shows a combination torch.

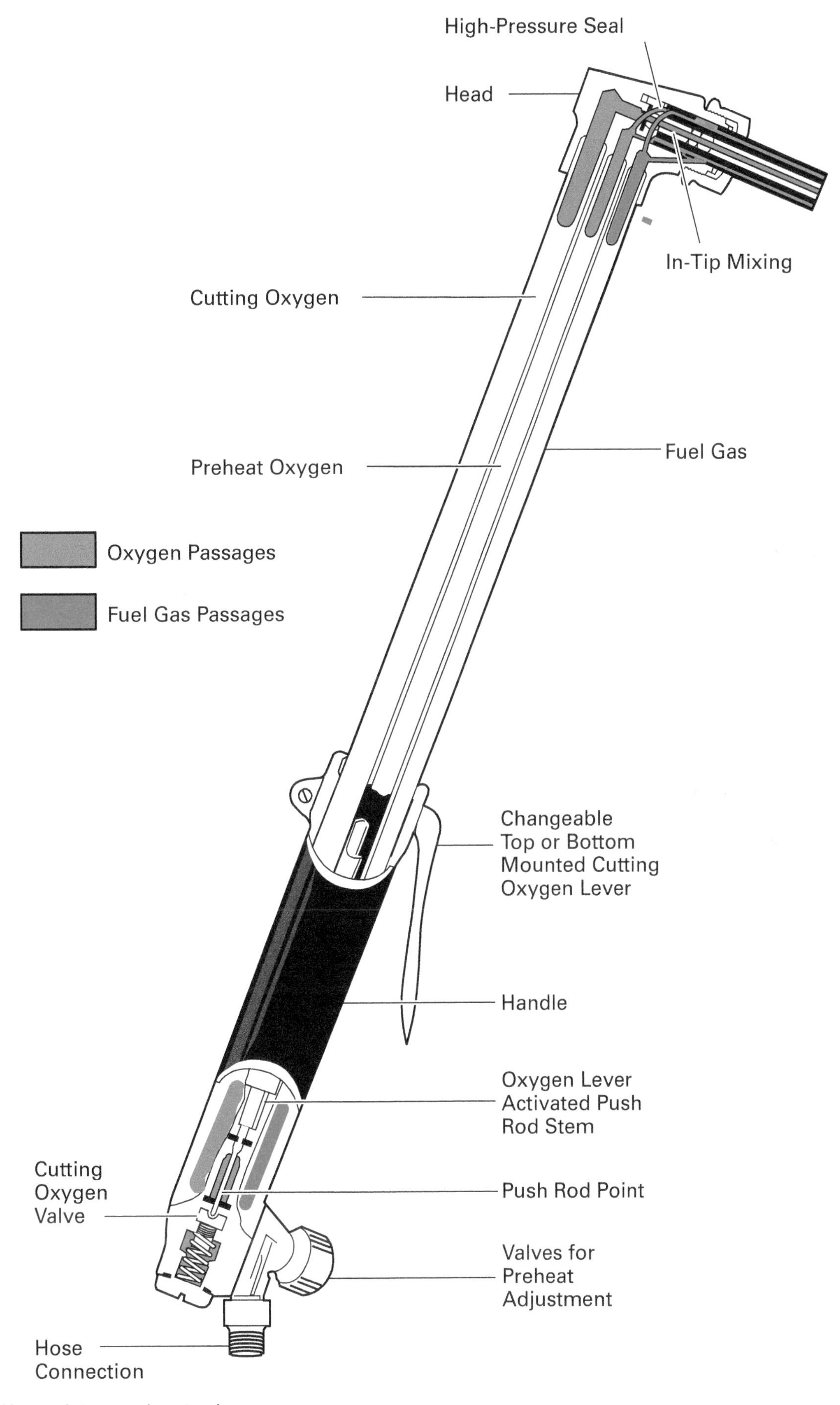

Figure 21 Heavy-duty one-piece torch.

2.3.3 Standard Cutting Tips

Cutting torch tips (nozzles) fit onto the cutting torch end. Some screw into place, while others have a securing nut. Tips come in many styles for different tasks. Broadly, they divide into one-piece and two-piece designs (*Figure 23*). One-piece tips are solid copper. Two-piece tips have a sleeve and an internal section. Tips for acetylene cutting are almost always one-piece. Tips for other gases come in both styles.

The tip end has a single, central orifice (hole) for the cutting oxygen stream. Several smaller orifices surround it. These generate the preheat flame. Acetylene tips have four, six, or eight preheat orifices. Tips for other gases have six or more, because those gases burn at lower temperatures. *Figure 24* shows different tips styles and orifice arrangements.

The tip that you select depends on the fuel gas and the base metal thickness. Tip manufacturers provide tables to guide you. *Table 2* shows an example for acetylene cutting.

CAUTION

Normally, tips and torches must come from the same manufacturer. While other tips might fit, they probably won't seal correctly. Improper seals can cause a flashback or a fire. Other tips may also require different flow rates and pressures than the torch delivers. Unless a tip manufacturer specifically indicates that its tips are compatible with the torch, don't use them. Similarly, never use another manufacturer's tip table for guidance. Every tip brand has its own unique operating behaviors.

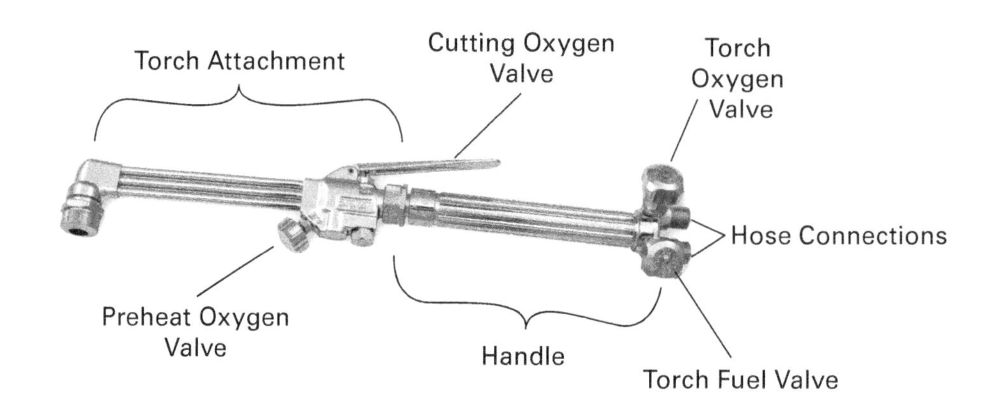

Figure 22 Combination cutting torch.
Source: © Miller Electric Mfg. LLC

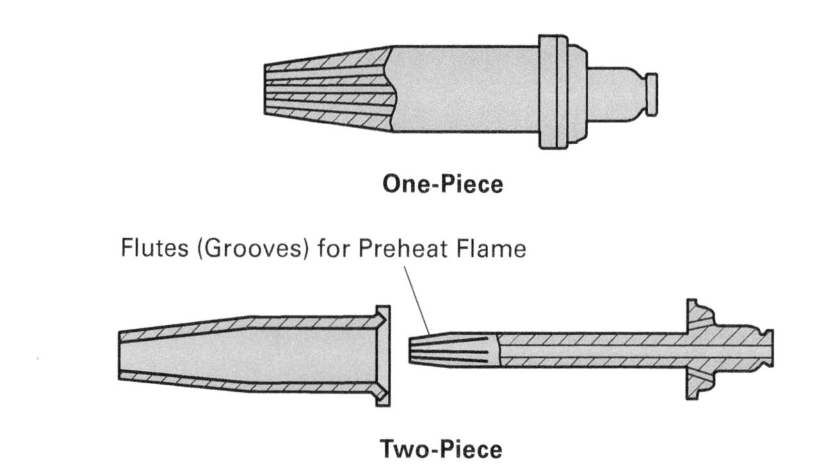

Figure 23 One- and two-piece cutting tips.

TABLE 2 Sample Acetylene Cutting Tip Chart

Cutting Tip Series 1-101, 3-101, and 5-101												
Metal Thickness		**Tip Size**	**Cutting Oxygen Pressure***		**Preheat Oxygen***		**Acetylene Pressure***		**Speed**		**Kerf Width**	
(in)	**(mm)**		**(psig)**	**(kPaG)**	**(psig)**	**(kPaG)**	**(psig)**	**(kPaG)**	**(in/min)**	**(cm/min)**	**(in)**	**(mm)**
1/8	3.18	000	20–25	138–172	3–5	21–34	3–5	21–34	20-30	51-76	0.04	01.02
1/4	6.35	00	20–25	138–172	3–5	21–34	3–5	21–34	20-28	51-71	0.05	01.27
3/8	9.52	0	25–30	172–207	3–5	21–34	3–5	21–34	18-26	46-66	0.06	01.52
1/2	12.70	0	30–35	207–241	3–6	21–41	3–5	21–34	16-22	41-56	0.06	01.52
3/4	19.05	1	30–35	207–241	4–7	28–48	3–5	21–34	15-20	38-51	0.07	1.78
1	25.40	2	35–40	241–276	4–8	28–55	3–6	21–41	13-18	33-46	0.09	02.29
2	50.80	3	40–45	276–310	5–10	34–69	4–8	28–55	10-12	25-30	0.11	02.79
3	76.20	4	40–50	276–345	5–10	34–69	5–11	34–76	8-10	20-25	0.12	03.05
4	101.60	5	45–55	310–379	6–12	41–83	6–13	41–90	6-9	15-23	0.15	03.81
6	152.40	6	45–55	310–379	6–15	41–103	8–14	55–97	4-7	10-18	0.15	03.81
10	254.00	7	45–55	310–379	6–20	41–138	10–15	69–103	3-5	8-13	0.34	08.64
12	304.80	8	45–55	310–379	7–25	48–172	10–15	69–103	3-4	8-10	0.41	10.41

*The lower side of the pressure listings is for hand cutting and the higher side is for machine cutting.

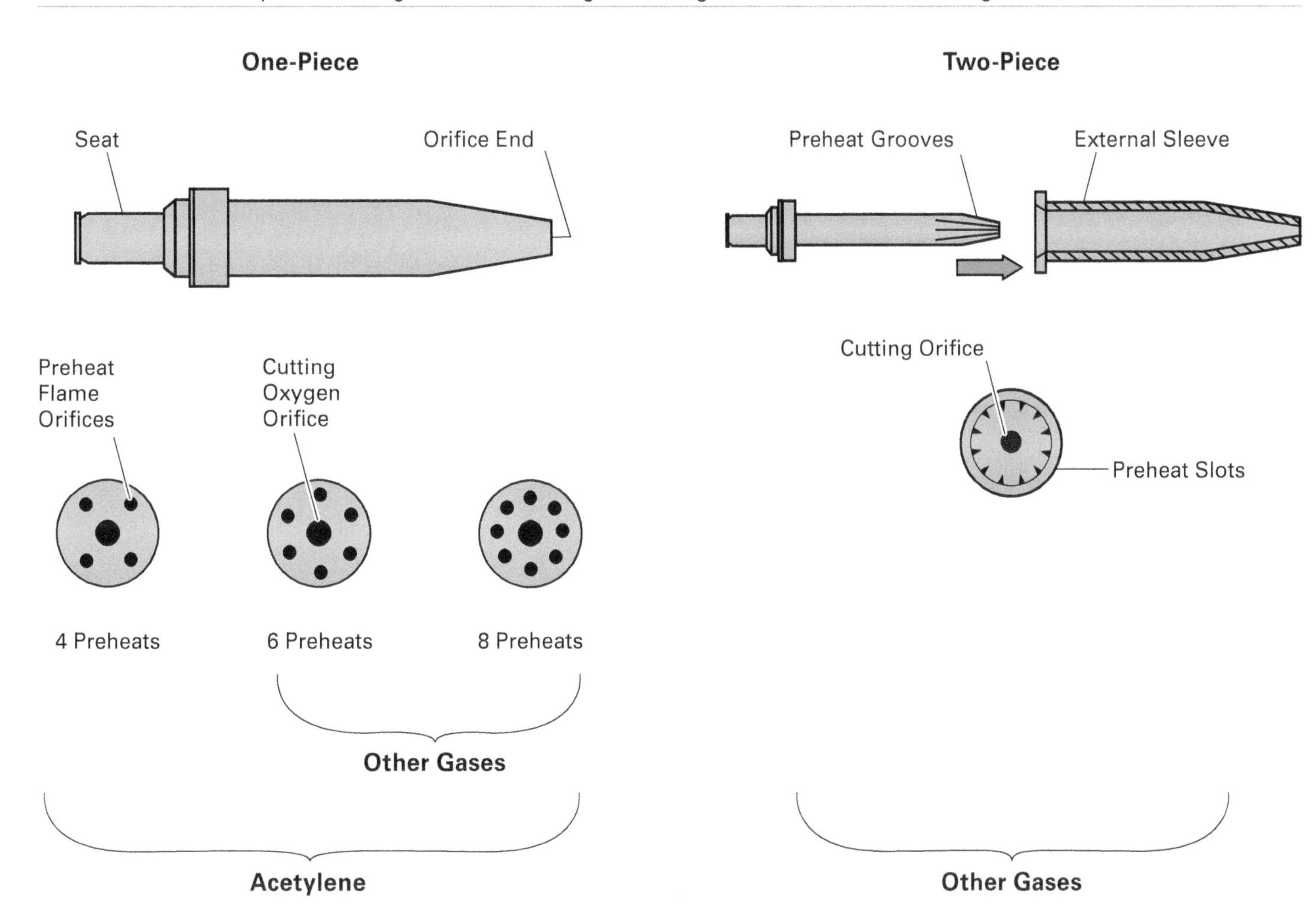

Figure 24 Tip styles and orifices.

Washing: Cutting off bolts, rivets, and other projections from a metal surface.

Gouging: Cutting a groove into a surface.

2.3.4 Specialized Cutting Tips

Some cutting jobs require special tips. These cut sheet metal, rivets, risers, or flues. **Washing** and **gouging** require special tips too. *Figure 25* shows several specialized cutting tips. Notice that they have unusual shapes, as well as different orifice arrangements.

Sheet Metal Cutting Tips

These tips have only one preheat orifice. This keeps the heat low to avoid distorting the thin metal. These tips work with motorized cutting machines, as well as hand cutting.

Rivet and Riser Cutting Tips

These tips are flat on one side, so they can cut parallel to a surface. They cut off rivet heads, bolt heads, and nuts. Riser cutting tips have more preheat orifices to cut risers, flanges, and angle legs faster.

Rivet Blowing and Metal Washing Tips

These tips are heavy duty to withstand high heat. They perform coarse cutting and remove clips, angles, and brackets.

Gouging Tips

These tips cut grooves in metal to prepare it for welding.

Flue Cutting Tips

These tips cut flues inside boilers. They also work for general cutting operations in tight spaces.

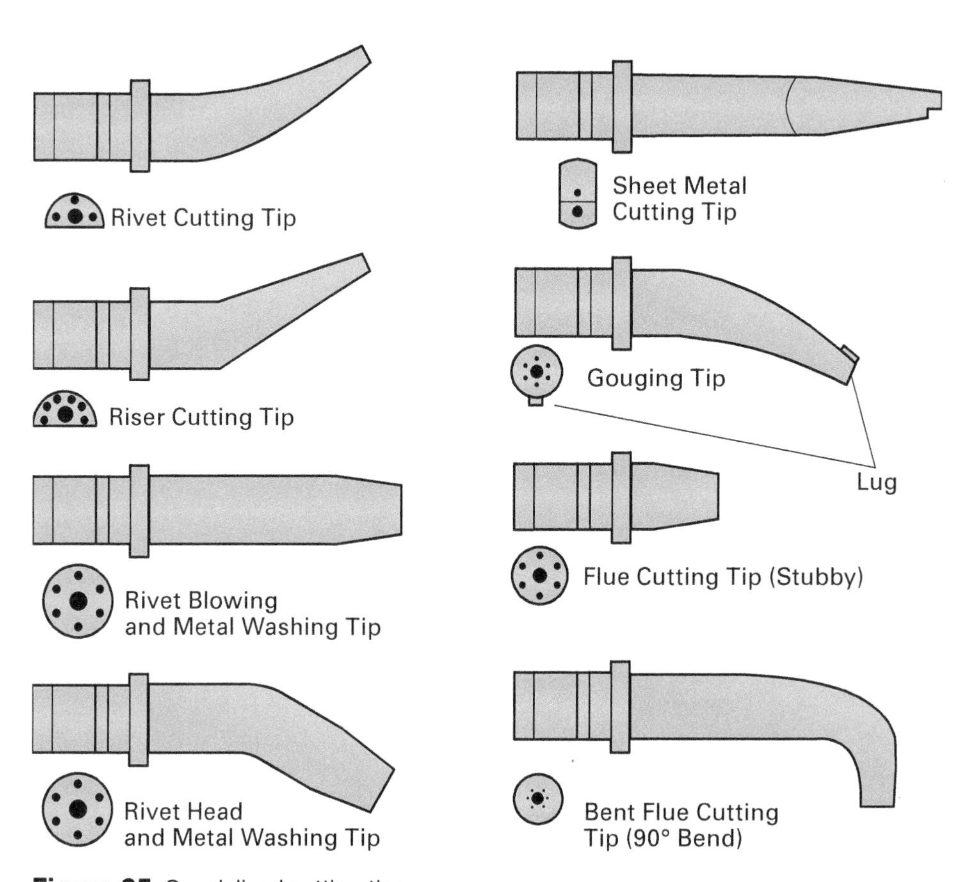

Figure 25 Specialized cutting tips.

2.3.5 Cutting Tips and Acetylene Safety

As you've learned, acetylene is unstable and dangerous. It explodes easily, requiring special cylinders containing acetone to store it. The acetylene dissolves in the acetone, leaving it when you open the cylinder valve. If you release the gas too quickly, however, liquid acetone will come along with it. To keep the acetone in the cylinder where it belongs, you must control the acetylene's flow rate.

To be safe, never use a gas flow that removes more than $^1/_{10}$ of the acetylene in the cylinder per hour. If you remove gas at a higher rate, you'll draw acetone along with it. This problem happens most often with small cylinders or high-flow tips.

Tip manufacturers specify the maximum fuel gas flow rate at the recommended pressure for each acetylene tip. Before cutting, confirm that the tip's maximum flow rate per hour doesn't exceed $^1/_{10}$ of the acetylene cylinder's capacity.

When a job requires a higher flow rate than one cylinder can safely supply, welders connect multiple tanks to a manifold. The combined acetylene flows keep the individual cylinder flow rates below the safe maximum value.

WARNING!

If the preheat flame turns purple, acetone is mixing with the fuel and oxygen. Extinguish the torch immediately.

2.4.0 Specialized Oxyfuel Cutting Equipment

While many tasks require nothing more than an oxyfuel hand torch, some require specialized cutting tools. These include simple guides to make hand cutting more accurate. But they also include specialized cutting torches and sophisticated cutting machines.

2.4.1 Mechanical Guides

Freehand cutting a long line, a circle, or an irregular path can be quite challenging. Welders often use special tools or guides to produce better results. A wheeled guide tool makes a big difference for many cuts. It holds the torch the proper distance from the work and rolls along the surface. Combining this tool with a clamped straightedge makes cutting long lines easier. It also works well with templates for cutting complicated shapes.

A circle guide makes cutting accurate circles simple (*Figure 26*). It works like a beam compass. One end rests in a drilled or pierced hole. The other holds the torch the correct distance from the surface. The worker rotates the tool around the center, cutting a perfect circle.

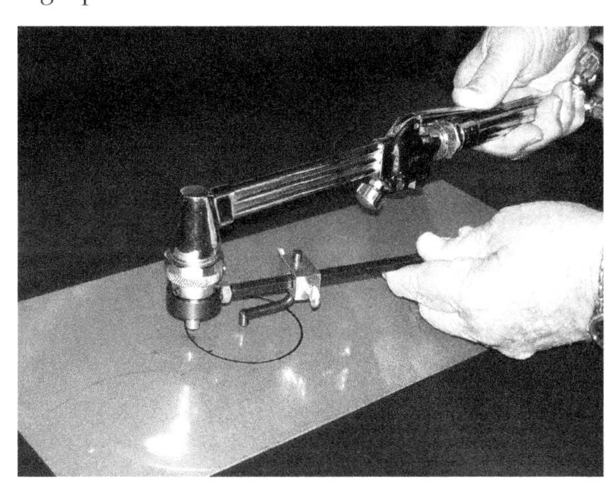

Figure 26 Circle cutting guide.
Source: © Miller Electric Mfg. LLC

Straight Line Cutting Guide

To cut a straight line, clamp a piece of angle iron to the workpiece. It acts as a straightedge to guide the torch. Wrap a hose clamp around the torch tip to control the height. Rest the clamp on the angle iron, start the cut, and move forward at the proper speed.

2.4.2 Computer-Controlled Cutting Machines

Industrial fabricators save a lot of time by using a motor-driven cutting machine (*Figure 27*). These provide precise, repeatable results. An industrial computer operates a motorized carriage that carries the torch over the workpiece. Some use a programmed path to guide the torch. Others have a sensor that follows a line on the workpiece, a separate drawing, or a template.

Sophisticated motorized cutters can handle multiple torches. These reduce the cutting time or make multiple copies at once. Some machines can enlarge or reduce the pattern by a specific percentage. Most motorized cutters work with both oxyfuel torches and plasma arc cutters.

2.4.3 Track Cutting Machines

Track cutting machines semi-automatically produce straight, curved, or beveled cuts. Some can even cut circles. The machine carries the oxyfuel torch in a motorized carriage that runs on a track. A variable-speed electric motor drives the carriage at whatever speed the operator selects.

The machine shown in *Figure 28* has the following features:

- Cuts straight-line cuts of any length
- Cuts circles up to 96" (~244 cm) in diameter
- Cuts bevels or chamfers
- Cuts in either direction
- Has high- and low-speed modes
- Offers cutting speeds between 1" (~2.5 cm) and 110" (~280 cm) per minute

As *Figure 29* shows, the machine's controls are quite simple. The Power control turns the machine on or off. The Speed Range selector chooses between high- and low-speed modes. The Direction switch selects the travel direction. The Speed knob adjusts the cutting speed smoothly from a crawl to the range's top speed.

Figure 27 Computer-controlled cutting machine.
Source: Kolke Aronson Inc. – Worldwide manufacturer of cutting, welding and positioning equipment.

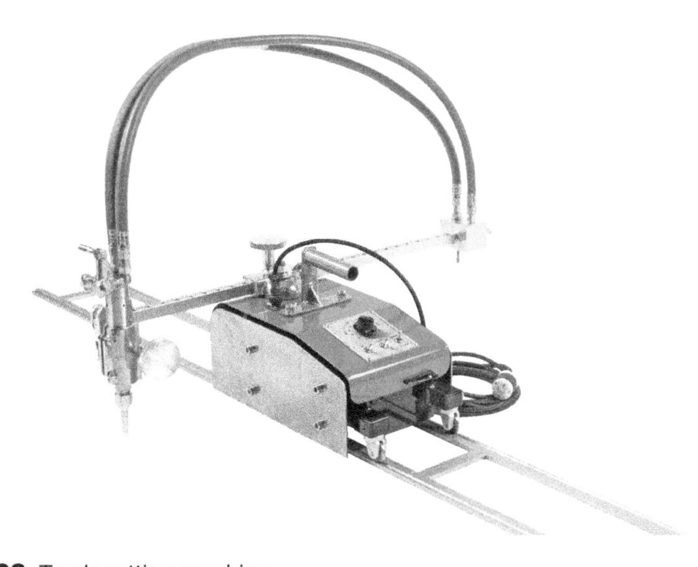

Figure 28 Track cutting machine.
Source: The Lincoln Electric Company, Cleveland, OH, USA

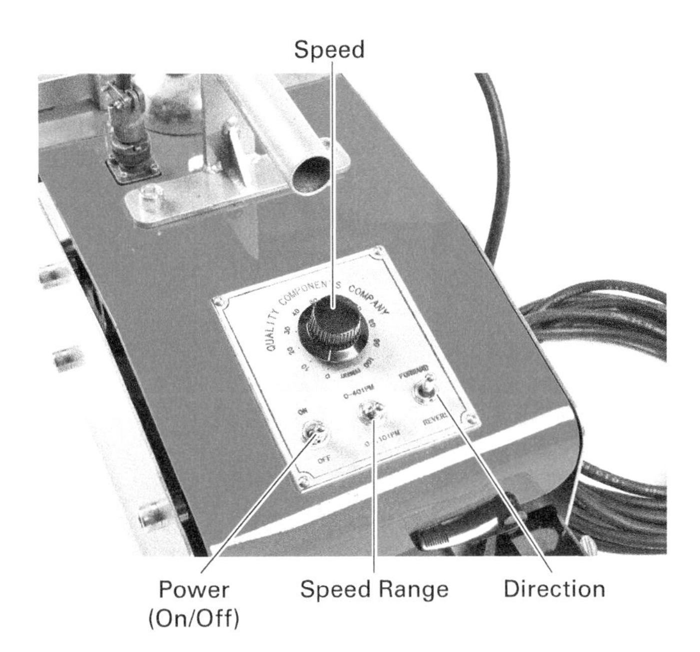

Figure 29 Track cutting machine controls.
Source: The Lincoln Electric Company, Cleveland, OH, USA

2.4.4 Pipe-Cutting Machines

Pipe-cutting machines cut and bevel pipes in the shop or field. Most work with both oxyfuel torches and plasma arc cutters. Some have a motor, while others are cranked. The band cutter version attaches with a stainless steel band that wraps around the pipe (*Figure 30*). A chain-and-sprocket drive pulls the torch around the pipe, producing the cut.

The ring gear version clamps onto the pipe (*Figure 31*). Gears connected to the crank or motor rotate the torch carriage and the ring gear at different speeds. This double motion carries the torch completely around the pipe, producing the cut.

Figure 30 Motorized band-type pipe-cutting machine.
Source: Courtesy of H&M Pipe Beveling Machine Company, Inc.

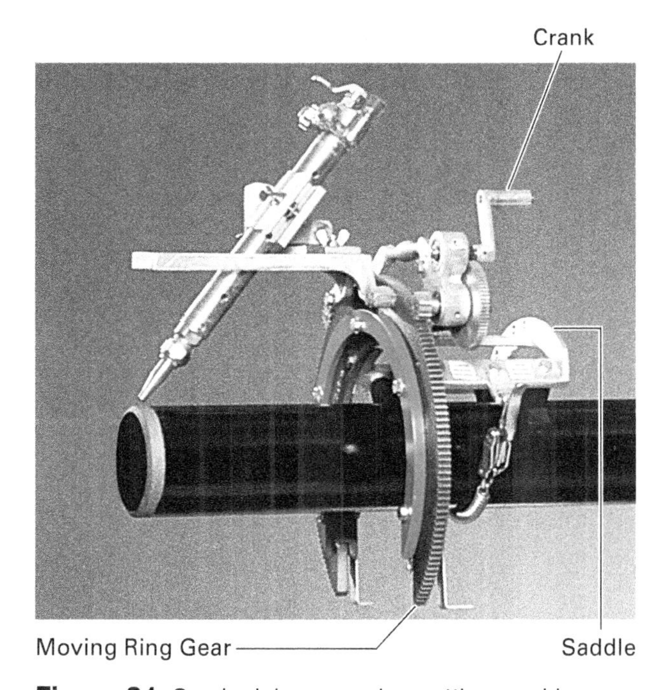

Figure 31 Cranked ring gear pipe-cutting machine.
Source: Courtesy of H&M Pipe Beveling Machine Company, Inc.

2.4.5 Thermic Lances

These unusual cutting tools can burn their way through almost any material. They're perfect for large industrial jobs, demolition, and underwater work. The lance produces extremely high temperatures by burning metal rods inside a steel tube. Oxygen flowing through the tube raises the temperature to between 7,000°F and 8,000°F (~3,900°C and 4,500°C).

The lance, which burns itself up, comes in several sizes. Small ones are about 20" (~50 cm) long and 0.25" (~6 mm) in diameter. Large ones are up to 10' (~3 m) long and 1" (~25 mm) in diameter. The outer tube contains wires or rods made from iron, steel, aluminum, or magnesium.

The lance attaches to a handle that supplies a steady flow of oxygen (*Figure 32*). An acetylene torch preheats the lance to its ignition temperature. Then, the operator turns on the oxygen, which ignites the lance. Turning off the oxygen extinguishes it.

2.5.0 Oxyfuel Accessories

While oxyfuel cutting equipment is relatively simple, it requires a lot of supporting accessories. These keep it working properly or make tasks easier and more efficient.

2.5.1 Tip Cleaners and Tip Drills

Eventually, torch tips become covered with molten metal or carbon. Then they stop producing clean, even cuts. Dross will build up around the kerf. Tip cleaners and drills restore the tip to good working condition (*Figure 33*).

A tip cleaner is a tiny round file that fits into the orifices. Kits usually include several sizes. Some kits include a flat file for smoothing the tip face. They're softer than regular files since tips are copper, a soft metal. For this reason, never use a regular file on a tip. To use a tip cleaner, insert it into the orifice and move it back and forth a few times.

Tip drills are tiny drill bits that clear plugged orifices. They too come in several sizes, along with a handle that holds the bit. Select the correct size, clamp it in the handle, and carefully drill out the blockage. Tip drills are brittle, so handle them carefully.

CAUTION

Use tip cleaners and drills carefully. They can break off in the orifices. Overly aggressive cleaning or using the wrong size drill will enlarge the orifices. This will cause the torch flame to burn incorrectly. Discard and replace a damaged tip.

2.5.2 Friction Lighters

Lighting an oxyfuel torch with a match or pocket lighter is extremely dangerous. Always use a friction lighter (striker) to ignite the torch. These work by rubbing a flint on a steel surface, producing a shower of sparks (*Figure 34*).

WARNING!

Never light a torch with a match or pocket lighter. You could be burned, or the lighter could explode.

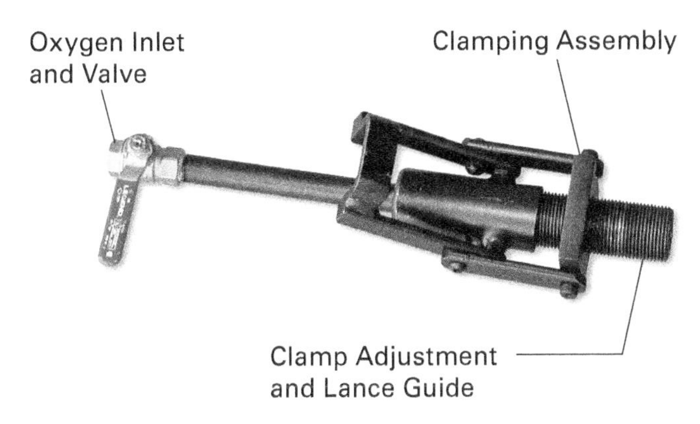

Figure 32 Thermic lance handle.

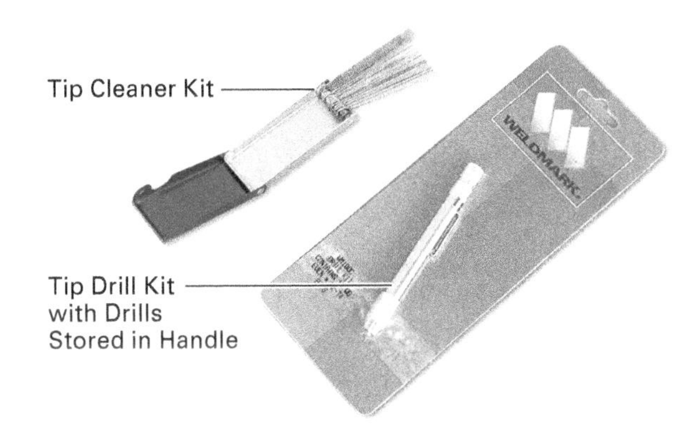

Figure 33 Tip cleaner and drill kits.

Figure 34 Friction lighters.

Figure 35 Cylinder cart.
Source: Vestil Manufacturing

2.5.3 Cylinder Cart

Cylinder (bottle) carts are hand trucks specially equipped to carry gas cylinders safely. They cradle the cylinders and secure them with chains or straps. *Figure 35* shows a cylinder cart suitable for oxyfuel work. It includes a firewall between the oxygen and acetylene cylinders. Some carts have attached tool trays or boxes.

WARNING!

Never carry fuel gas and oxygen cylinders together on a bottle cart without a firewall.

2.5.4 Heat-Resistant Markers

Regular pens and pencils aren't suitable for marking hot surfaces. Their marks are also difficult to see through a tinted lens. Pencils made from soapstone—a soft, greasy mineral—work extremely well. They leave a white, heat-resistant mark that shows up clearly under an arc or cutting flame.

Some craftworkers prefer pencils made from silver or red graphite (*Figure 36*). Both leave heat-resistant marks. Silver graphite shows up well on dark surfaces, while red works on bright metals. Several manufacturers make dyes and paints suitable for hot surfaces.

Using a Cup-Type Striker

To light a torch with a cup-type striker, hold the cup slightly below the tip and to one side. It should be parallel to the gas stream. This prevents the ignited gas from deflecting back from the cup. It also reduces soot buildup on the striker. Replace the striker's flint when it sparks poorly.

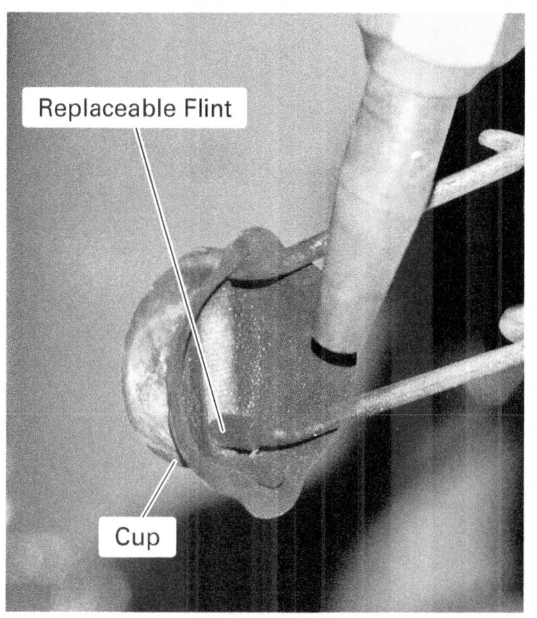

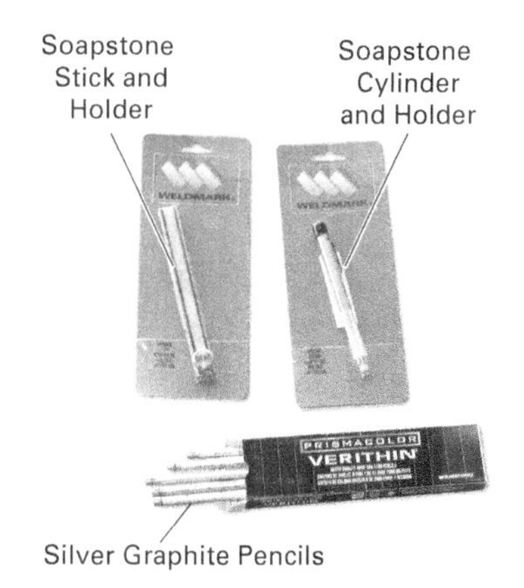

Figure 36 Soapstone and graphite markers.

Sharpening Soapstone Sticks

Sharpen a soapstone marker by shaving one side with a file. Leave the other side flat so it can run along a straightedge when drawing lines.

2.0.0 Section Review

1. Which fuel gas produces the *hottest* oxyfuel flame?
   a. Natural gas
   b. Acetylene
   c. Propane
   d. Propylene

2. To prevent gas from mixing in a torch's hoses or the cutting flame from going back inside, equip the torch with _____.
   a. check valves and flashback arrestors
   b. orifice plates and torch tips
   c. HEPA filters and anti-flame valves
   d. distribution manifolds and friction strikers

3. Which oxyfuel torch type requires pressurized, regulated fuel gas?
   a. A motor-controlled torch
   b. A vacuum-injector torch
   c. A neutral-flame torch
   d. A positive-pressure torch

4. Which oxyfuel tool produces very high temperatures by burning metal rods in oxygen?
   a. A circle cutter
   b. A plasma arc cutter
   c. A thermic lance
   d. A ring gear cutting machine

5. The *most* common material for marking metal with heat-resistant lines is _____.
   a. soapstone
   b. chalk
   c. lead
   d. flint

3.0.0 Using Oxyfuel Equipment

Objective

Outline safely setting up, lighting, and shutting down oxyfuel equipment.

a. Explain how to set up oxyfuel equipment.
b. Explain how to leak test oxyfuel equipment.
c. Explain how to light, adjust, and shut down oxyfuel equipment.

Performance Tasks

1. Set up oxyfuel cutting equipment.
2. Light and adjust an oxyfuel torch.
3. Shut down oxyfuel cutting equipment.
4. Disassemble oxyfuel cutting equipment.
5. Change gas cylinders.

Oxyfuel cutting equipment is relatively simple, but it's also hazardous. If set up or handled improperly, it can start fires, explode, or injure. Before using an oxyfuel outfit, craftworkers must learn how to set it up, test it, and light it. They must also learn how to shut it down properly when they're finished.

WARNING!

Always wear suitable PPE for setting up and using oxyfuel equipment.

3.1.0 Setting Up

NOTE

Some oxyfuel connections contain sealing O-rings. When assembling equipment, always check the O-rings first. Replace them if they're damaged or distorted. Overtightening can damage O-rings, so always follow the manufacturer's guidelines for tightening connections. Never use a wrench if they specify hand tightening only.

Properly prepared oxyfuel equipment works safely. The following sections guide you through the setup process, emphasizing safety at each step.

WARNING!

Do not use plumbing pipe tape or pipe dope on any gas fittings. It won't seal properly and will cause leaks.

3.1.1 Transporting and Securing Cylinders

Transport cylinders to the workstation or worksite with a hand truck or bottle cart. Be sure they have safety caps before moving them. Keep them upright and securely chained in place. An acetylene cylinder not kept upright can't be used for at least an hour.

WARNING!

Handle cylinders with care since they're under high pressure. Never drop them, knock them over, or roll them. Keep them below 140°F (60°C). Moving them without the valve safety cap is extremely dangerous. Never lift them with a sling or electromagnet. Use a *cylinder cage* instead.

At the destination, chain them to a fixed upright. If you moved them with a bottle cart equipped with a firewall, you may leave them secured to it. Remove the safety caps. Inspect the valve outlets for damage. Keep the safety caps near the cylinders.

Hoisting Cylinders

Never lift a cylinder by its safety cap or the valve. Always use a lifting cage. Secure the cylinder in the cage before lifting. Cages designed to lift oxygen and fuel gas cylinders together have a firewall dividing the space. Never lift fuel gas and oxygen cylinders together in a cage not equipped with one.

3.1.2 Cracking the Valves

Cracking the valves blows out dirt or debris in the outlet. Confirm that the cylinder is secure. Stand with the outlet facing away from you. Open the valve briefly to clear the outlet (*Figure 37*). Don't leave it open too long since both oxygen and fuel gases are dangerous.

WARNING!

To avoid being hit by dirt or debris, always stand with the outlet facing *away* from you. Don't point the outlet at anyone else either!

3.1.3 Attaching the Regulators

Inspect each regulator carefully. Never use one with damaged or missing parts. Replace questionable regulators or have a qualified service technician repair them.

Close (fully release) each regulator by turning its adjustment knob or screw counterclockwise until it feels loose. Check the fittings for dirt, oil, and grease (*Figure 38*). Clean off dirt with a clean, oil-free cloth. Follow your company's safety procedures for dealing with oil- or grease-contaminated fittings.

WARNING!

Oil or grease on fittings is a serious safety hazard. It can ignite or explode around oxygen. Contaminated fittings require special cleaning and an inspection before they can return to service. Do not do this yourself unless you're properly qualified. Never clean fittings with an oily cloth!

CAUTION

Never overtighten gas fittings. They're usually made from soft brass or bronze. Overtightening can strip threads or damage seals. Follow the manufacturer's guidelines for tightening. Use only appropriate tools.

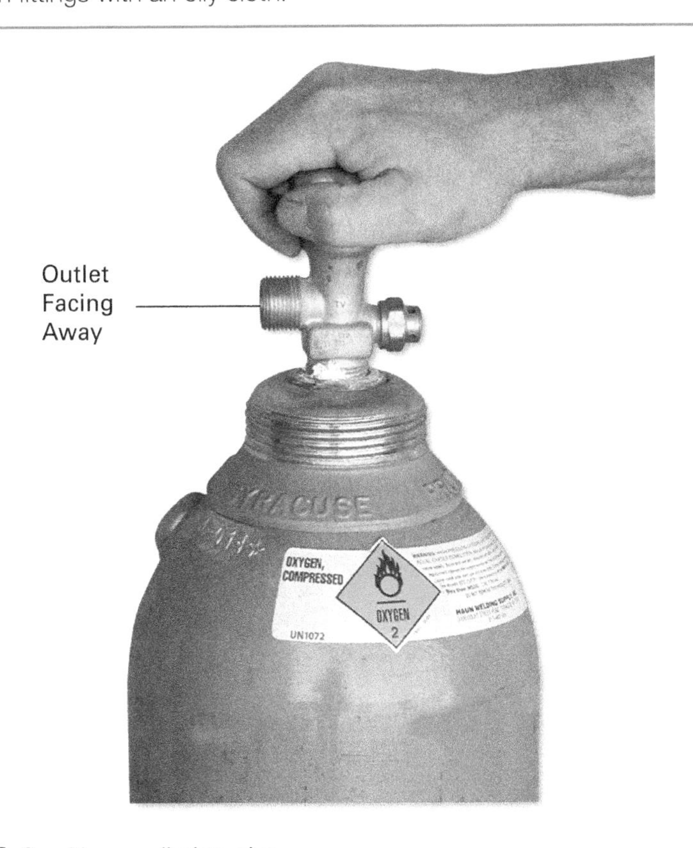

Figure 37 Cracking a cylinder valve.

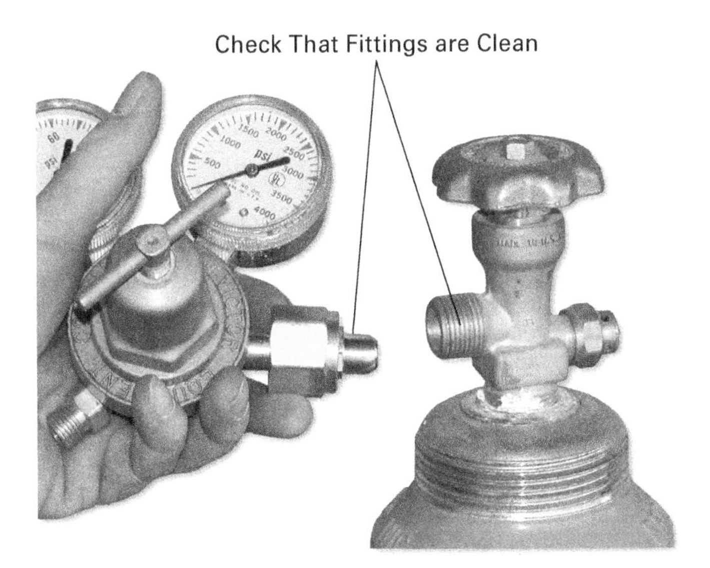

Figure 38 Checking connection fittings.

Connect the oxygen regulator to the oxygen cylinder. Tighten it with a torch wrench (*Figure 39*). Connect the fuel gas regulator to the fuel gas cylinder. Tighten it with a torch wrench. Remember, fuel gas fittings usually have LH threads.

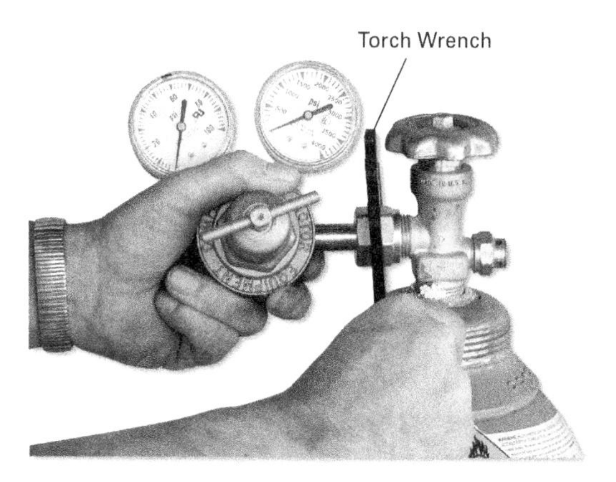

Figure 39 Tightening the regulator.

Clear each regulator's outlet. Crack the cylinder valve slightly. Turn the regulator's knob or adjustment screw clockwise until the gas blows debris from the outlet. Shut the cylinder valve and close the regulator (*Figure 40*).

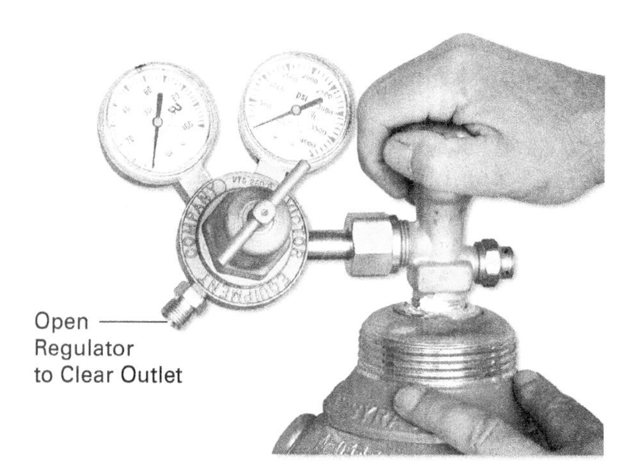

Figure 40 Clearing the regulator.

3.1.4 Installing Flashback Arrestors and/or Check Valves

Flashback arrestors and check valves are essential safety devices. You must have at least one of each per hose. Connect them to the regulator outlet and/or the torch inlet. If you use only one pair, connect it to the torch. Some torches already contain these devices. Know what your equipment includes.

If necessary, connect a flashback arrestor and a check valve to the regulator outlet (*Figure 41*). If possible, use a flashback arrestor with a built-in check valve. Install it with the arrow pointing *away* from the regulator. Tighten it with a torch wrench.

WARNING!

Never operate an oxyfuel outfit without at least one flashback arrestor and one check valve per hose.

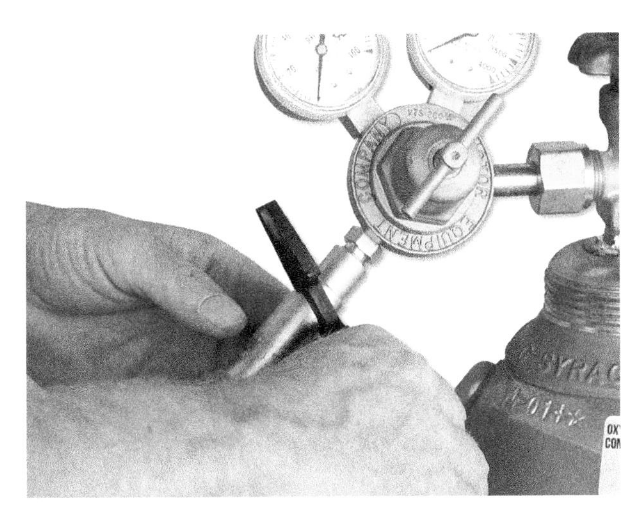

Figure 41 Attaching a flashback arrestor.

3.1.5 Connecting the Hoses

Hoses may contain dirt or debris. New hoses usually have talc and bits of rubber in them. These materials will clog torch valves or tip orifices. Properly preparing hoses prevents this problem.

Inspect both hoses for damage like burns, cuts, or abrasions. Check the fittings for damaged threads, oil, or grease. Replace anything that's questionable. Connect each hose to the correct regulator outlet or flashback arrestor (*Figure 42*). Tighten it with a torch wrench. Blow out the hoses by briefly cracking the cylinder valves and opening the regulators.

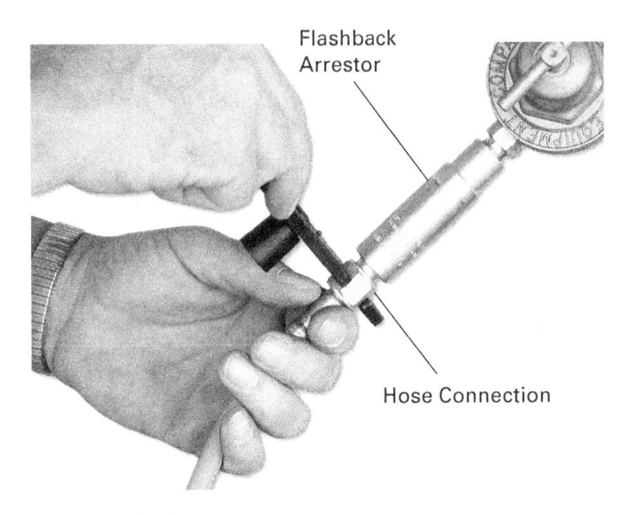

Figure 42 Connecting the hose.

3.1.6 Preparing the Torch

If the torch doesn't include built-in flashback arrestors and check valves, add them to its inlets. Connect them with their arrows pointing *towards* the inlet. Tighten with a torch wrench. Connect each hose to the correct inlet (*Figure 43*). Tighten with a torch wrench.

For a combination torch, connect the cutting torch attachment to the handle. Check its O-ring(s) first. Follow the manufacturer's recommendations for tightening the cutting torch attachment. Most are hand-tighten only.

Select a torch tip based on the cutting job, material thickness, and fuel gas. Consult the manufacturer's tip selection chart for guidance. Check the tip for damage. Confirm that the orifices are clear and that the tip is free from dross. If required, use a tip cleaner or drill. Inspect the sealing surface. Discard and replace damaged tips.

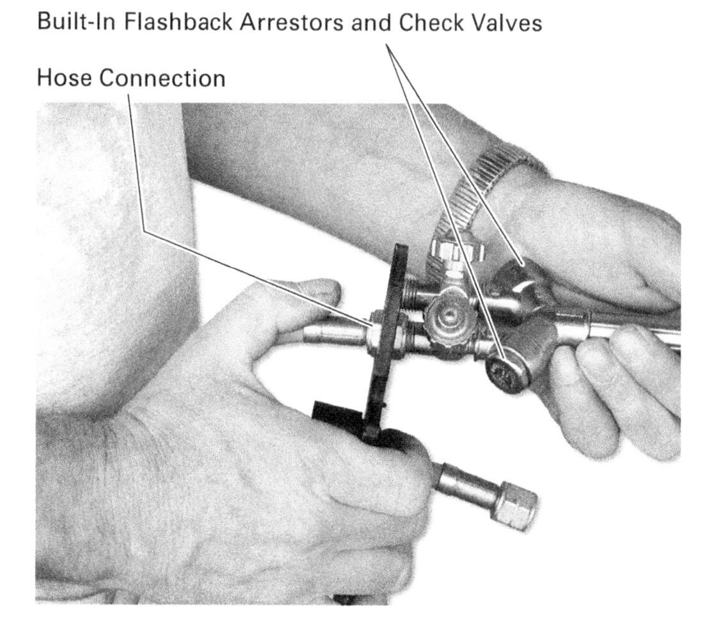

Figure 43 Connecting the hose to the torch.

WARNING!

If you're going to use acetylene, consult the manufacturer's selection chart to determine the tip's maximum hourly flow rate. Confirm that it's less than $1/10$ of the cylinder's capacity. Higher values will draw liquid acetone into the torch.

CAUTION

Overtightening connections can damage the threads and seals. Tighten by following the manufacturer's recommendations. If a torch valve leaks, lightly tighten the packing nut. If a cutting torch attachment or torch tip leaks, disassemble and check the sealing surfaces and/or O-rings. Replace damaged or distorted O-rings. With new equipment, briefly overtighten the connection to seat the seals. Then loosen and re-tighten the correct amount. Replace leaking components that you cannot fix.

Follow the torch manufacturer's instructions to install the tip. Always tighten using the specified method (*Figure 44*).

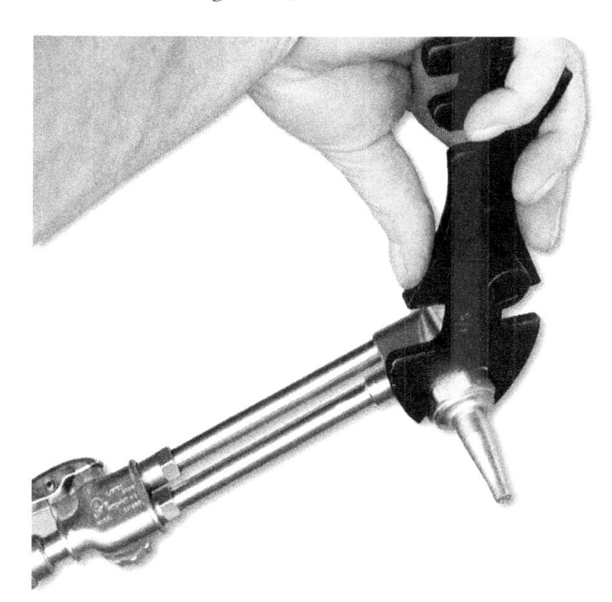

Figure 44 Installing the tip.

3.1.7 Readying the Torch

Close both torch valves. Turn both regulator adjustment knobs or screws counterclockwise until they're loose (*Figure 45*). Closing the regulators protects them from damage when you open the cylinder valves.

Figure 45 Torch valves and regulator adjustment knobs.
Source: Zachry Industrial, Inc.

Readying the Gas Supply

Stand opposite the oxygen regulator and *not* in line with the adjustment knob or screw. Also face away from the pressure gauges. Make sure that no one else is standing in line either. Crack open the oxygen cylinder valve until the inlet pressure gauge rises and stops (*Figure 46*). Now open the cylinder valve all the way until the valve seats at the top. This position prevents high-pressure leaks at the valve stem.

WARNING!

Opening the cylinder valve too quickly can damage the regulator or pressure gauge. The regulator adjustment screw could blow out or the gauge could explode. Standing with these components facing away from you prevents injury if this happens.

Figure 46 Cylinder valve and gauges.
Source: Zachry Industrial, Inc.

Stand opposite the fuel gas regulator and not in line with the adjustment knob or screw. Also face away from the pressure gauges. Make sure that no one else is standing in line either. Crack open the fuel gas cylinder valve until the inlet pressure gauge rises and stops. Now open the cylinder valve only $1\frac{1}{2}$ turns. This valve position makes it easy to close the valve quickly in an emergency—especially important when using acetylene.

WARNING!

To avoid an explosion, purge in a well-ventilated area with no nearby heat sources or open flames.

Setting the Working Pressure and Purging the Torch (Oxygen)

Fully open the torch's oxygen valve. Depress and hold the cutting oxygen lever. Slowly turn the oxygen regulator's adjustment knob or screw clockwise. Watch the outlet (working) pressure gauge. Adjust the regulator until the gauge shows the correct working pressure with the oxygen flowing. Let the oxygen flow for 5–10 seconds. This purges the air from the hoses and torch. Release the cutting lever and close the torch's oxygen valve.

NOTE

With single-stage regulators, the outlet pressure gauge will rise when you release the cutting lever. This is normal. Adjust the oxygen regulator to deliver the correct outlet pressure when the gas is flowing.

Setting the Working Pressure and Purging the Torch (Fuel Gas)

Open the torch's fuel gas valve about $\frac{1}{8}$ of a turn. Slowly turn the fuel gas regulator's adjustment knob or screw clockwise. Watch the outlet (working) pressure gauge. Adjust the regulator until the gauge shows the correct working pressure with the fuel gas flowing. Let the fuel gas flow for 5–10 seconds. This purges the air from the hoses and torch. Close the torch's fuel gas valve.

If you're using acetylene, watch the outlet pressure gauge while you close the torch's fuel gas valve. If the pressure rises above 15 psig (103 kPaG), *immediately* re-open the torch's fuel gas valve. Acetylene can explode inside the hose if its pressure rises above this point! With the gas flowing, adjust the pressure regulator to reduce the outlet pressure. Then, re-close the torch fuel gas valve, watching the outlet pressure gauge as you do so.

WARNING!

With single-stage regulators, the outlet pressure gauge will rise when you close the torch's fuel gas valve. This is normal. If you're using acetylene, however, the outlet pressure with the torch valve closed must *never* rise above 15 psig (103 kPaG). Adjust the regulator to guarantee this essential safety condition.

Old Cylinder Valves

Some cylinders have an obsolete valve style that doesn't have a handwheel. These require a special T-shaped wrench. Always leave the wrench in place so you can quickly turn off the gas in an emergency.

3.2.0 Leak Testing

After setup, immediately test oxyfuel equipment for leaks. Check the equipment periodically, and check the torch before each use. Take leaks seriously as they can cause a fire or an explosion.

WARNING!

To avoid an explosion, do leak testing in a well-ventilated area with no nearby heat sources or open flames.

3.2.1 General Procedure

To identify leaks, apply a commercial leak-test solution to each possible leak point. A water/detergent mix is an acceptable, but poorer, substitute. Bubbles will form if gas is leaking (*Figure 47*).

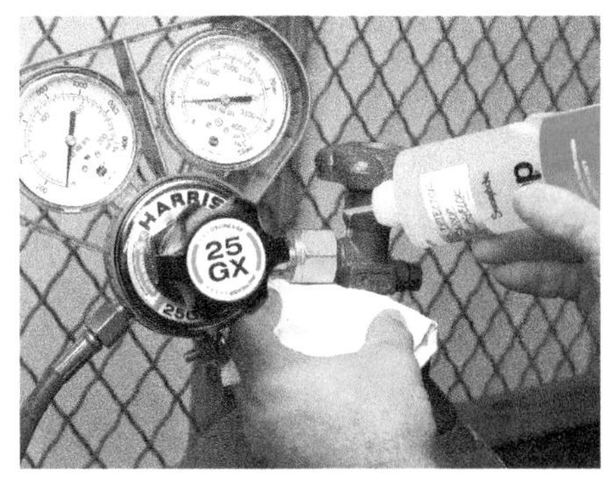

Applying the Solution

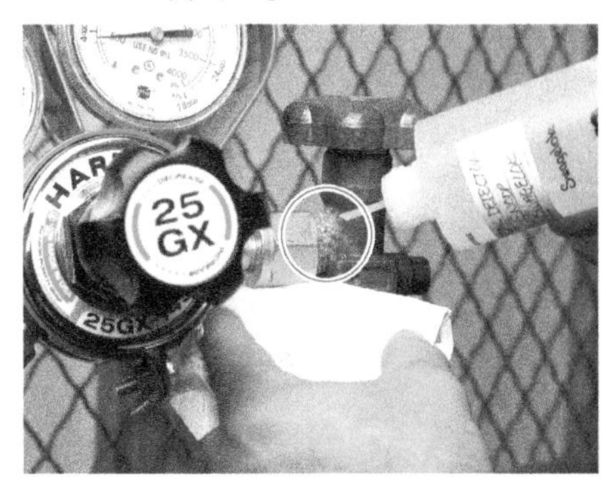

Identifying a Leak

Figure 47 Leak-test solution.
Source: The Lincoln Electric Company, Cleveland, OH, USA

WARNING!

Never use an open flame or sparks from a friction lighter to check for leaks.

WARNING!

Never use detergents containing oils as a leak-test solution substitute. Oils can explode or cause a fire around oxygen.

Check the following points for leaks:

- Oxygen cylinder valve
- Fuel gas cylinder valve
- Oxygen regulator adjustment, as well as the regulator inlet and outlet connections
- Fuel gas regulator adjustment, as well as the regulator inlet and outlet connections
- Flashback arrester / check valve connections
- Hose connections at each end
- Torch valves and cutting lever oxygen valve
- Cutting torch attachment connection (combination torches only)
- Cutting tip seal

Tighten a leaking cylinder valve's packing gland nut. If this doesn't stop the leak, close the valve. Mark the cylinder and move it to a safe, well-ventilated place. Notify the gas supplier. For other leaks, tighten the connections slightly. If this doesn't stop the leak, turn off the gas, open all connections, and inspect the fitting for damage. Replace questionable O-rings.

WARNING!

Never use plumbing pipe tape or pipe dope to fix leaking gas fittings.

3.2.2 Initial/Periodic Leak Testing

Perform leak tests after the initial setup and periodically as required. Set the equipment to the correct working pressures with the torch valves closed. Apply a leak-test solution to the cylinder valves, regulator adjustments, regulator relief ports, and regulator gauge connections (*Figure 48*). Check for leaks at the hose connections, regulator connections, and flashback arrestor / check valve connections.

Portable Oxyfuel Equipment

Most oxyfuel equipment is heavy and bulky. Moving it requires a bottle cart or hand truck. For small tasks or those in the field, a portable outfit is very handy. As you can see, it includes everything needed for small cutting jobs.

Source: © Miller Electric Mfg. LLC

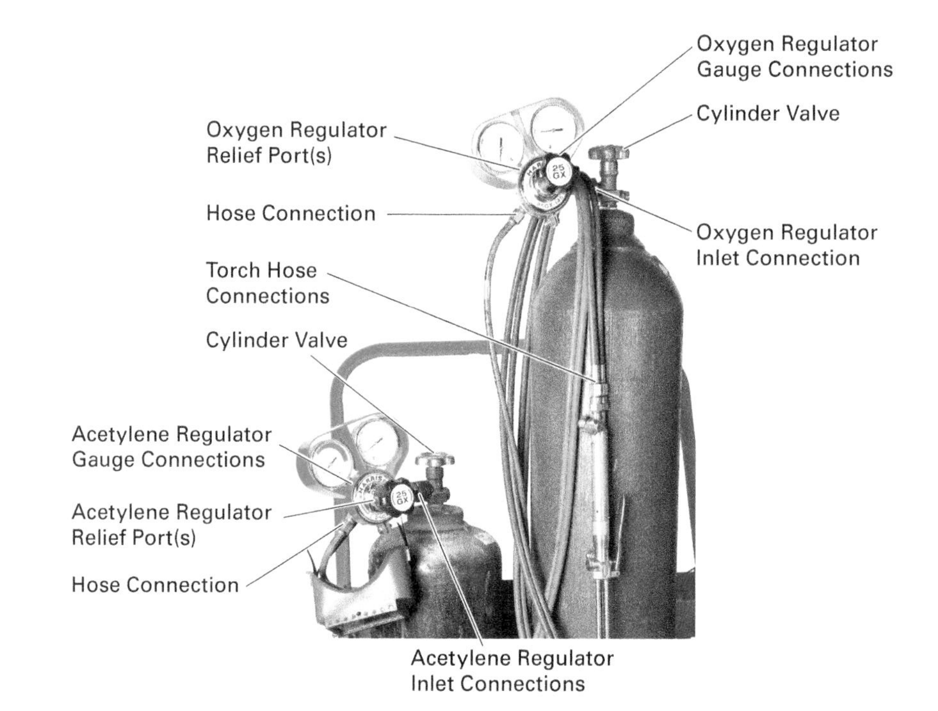

Figure 48 Initial/periodic leak-test points.
Source: The Lincoln Electric Company, Cleveland, OH, USA

3.2.3 Leak-Down Testing

Before lighting the torch, perform a leak-down test. It checks the regulators, flashback arrestors / check valves, hoses, and torch. Set the equipment to the correct working pressures with the torch valves closed. Then close both cylinder valves. Turn the regulator adjustment knobs or screws $^1/_2$ turn counterclockwise. Note the inlet and outlet pressures on each regulator.

Wait several minutes. Read all pressure gauges. If any show a drop, it indicates a leak. Track it down with leak-test solution. Once the oxyfuel outfit passes the leak-down test, set the equipment to the correct working pressures.

Daily Leak Testing

At the start of each day, always leak test regulators, hoses, and the torch. Leak test the torch after changing the torch attachment (combination torches only) or the tip. Leaks can cause fires and explosions. Workers don't always notice fires immediately because they're looking at the workpiece through a tinted lens.

3.2.4 Torch Leak Testing

Torches have multiple possible leak points (*Figure 49*). Check these with leak-test solution by following the manufacturer's recommendations. Correct any problems that you discover. Replace components that continue to leak. Never work with a leaky torch.

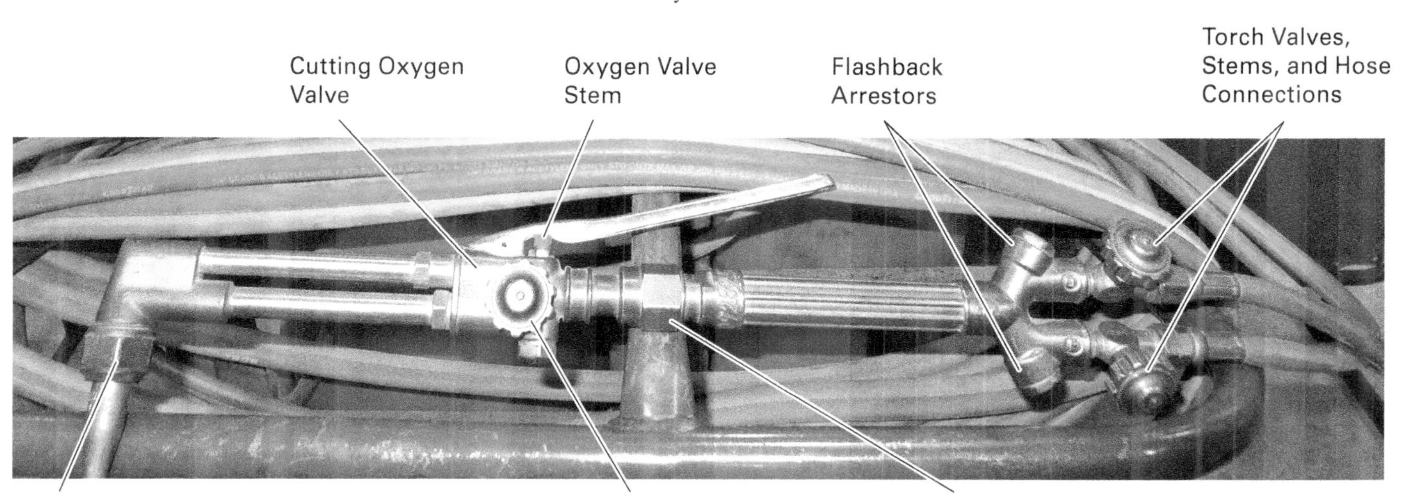

Figure 49 Torch leak points.
Source: Zachry Industrial, Inc.

3.3.0 Lighting, Adjusting, and Shutting Down

Operating an oxyfuel torch requires careful attention to detail. From the moment you light it until the moment you extinguish it, stay alert. You'll have to watch the flame and adjust the torch controls to make it do what you require.

3.3.1 Understanding the Flame

Successfully using an oxyfuel torch depends on understanding the flame. Its appearance and behavior tell you a lot about the gas mixture and the torch. Interpreting the flame helps you adjust the torch valves and deal with problems.

Oxyfuel flames divide into three types: **neutral flame**, **carburizing flame**, and **oxidizing flame**. *Figure 50* shows examples for both acetylene and propane (LP).

Neutral flame: A flame burning with the best mix of fuel gas and oxygen.

Carburizing flame: A flame burning with an excess amount of fuel. Also called a *reducing flame.*

Oxidizing flame: A flame burning with an excess amount of oxygen.

- *Neutral flame* — This flame has the perfect mix of fuel and oxygen. The inner cones are light blue. The outer is a darker blue. Workers use the neutral flame for most tasks.
- *Carburizing flame* — This flame has too much fuel gas. It has a white feather created by the excess fuel. The carburizing flame is brighter and longer than the neutral flame. Length depends on the amount of excess fuel. Carburizing flames deposit carbon from the unconsumed fuel gas in the metal, making it hard and brittle. Workers use this flame for special heating applications.
- *Oxidizing flame* — This flame has too much oxygen. The inner cones are shorter, bluer, and more pointed than those in neutral flames. The outer envelope is very short and often fans out at the ends. Oxidizing flames are the hottest. The extra oxygen in the flame can combine with metals, forming a hard, brittle, low-strength oxide. The preheat flames of a properly adjusted cutting torch will be slightly oxidizing after you shut off the cutting oxygen. Some specialized fuel gases also require a slightly oxidizing flame.

WARNING!

If you're using acetylene and the flame burns purple, acetone from the cylinder is entering the torch. You may be using too high a flow rate. Shut down the torch immediately and investigate.

3.3.2 Backfires and Flashbacks

Sometimes the torch will go out with a loud pop or snap—a backfire. Touching the tip to the work surface or a bit of dross on the tip can cause backfires. When they occur, relight the torch immediately if the torch doesn't relight itself.

Investigate unexplained backfires. Possible causes include the following:

- Improper operating pressures
- A loose tip
- Dirt in the tip seat or a bad seat

Flashbacks are much more serious. The flame retracts into the torch with a hissing or whistling sound. If this happens, immediately shut off the torch oxygen valve. This extinguishes the flame inside the torch. Flashbacks can cause fires, explosions, or melt the torch end. Possible causes include the following:

- Equipment malfunction
- Overheated tip
- Dross or spatter sticking to the torch tip
- Too large a tip for the gas flow rate

Acetylene Burning in Atmosphere
Open fuel gas valve until smoke clears from flame.

Carburizing Flame (Excess acetylene with oxygen)
Preheat flames require more oxygen.

Neutral Flame (Acetylene with oxygen)
Temperature 5,589°F (3,087°C).
Proper preheat adjustment when cutting.

Neutral Flame with Cutting Jet Open
Cutting jet must be straight and clean.
If it flares, the pressure is too high for the tip size.

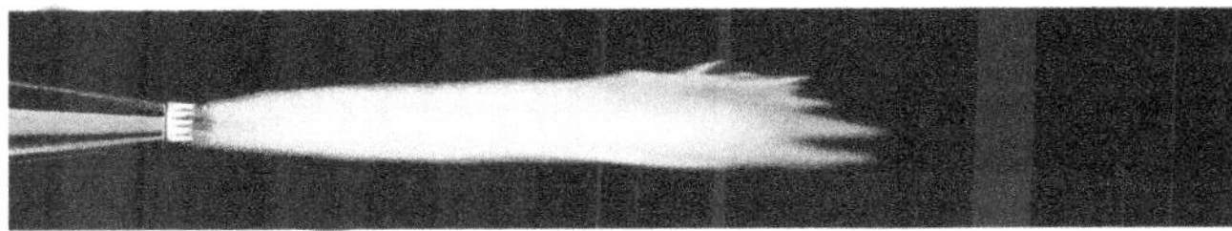

Oxidizing Flame (Acetylene with excess oxygen)
Not recommended for average cutting. However, if the preheat flame is adjusted for neutral with the cutting oxygen on, then this flame is normal after the cutting oxygen is off.

OXYACETYLENE FLAME

LP Gas Burning in Atmosphere
Open fuel gas valve until flame begins to leave tip end.

Reducing Flame (Excess LP gas with oxygen)
Not hot enough for cutting.

Neutral Flame (LP gas with oxygen)
For preheating prior to cutting.

Oxidizing Flame with Cutting Jet Open
Cutting jet stream must be straight and clean.

Oxidizing Flame without Cutting Jet Open
(LP gas with excess oxygen)
The highest temperature flame for fast starts and high cutting speeds.

OXYPROPANE FLAME

Figure 50 Acetylene and propane flames.
Source: © Miller Electric Mfg. LLC

After stopping a flashback, let the torch cool down. Blow oxygen alone through the torch for several seconds to remove soot produced by the flashback. Relight the torch. If it makes a hissing or whistling sound, or if the flame is abnormal, shut off the torch immediately and have it serviced by a qualified technician.

3.3.3 Igniting the Torch and Adjusting the Flame

Once you've assembled the oxyfuel equipment and tested it for leaks, you're ready to ignite the torch. Confirm that you know the correct pressure and flow rate for the tip you've installed. Consult the manufacturer's tip selection chart for guidance.

WARNING!

Always wear appropriate PPE when using oxyfuel torches.

Confirm that the oxygen and fuel gas regulators are set to the correct working pressures. Don't depress the oxygen cutting lever. Make sure a one-piece torch's oxygen valve is closed. Close a combination torch's preheat oxygen valve and open its torch oxygen valve.

Open the torch fuel gas valve about ¼ turn. Hold a friction lighter to the tip's side toward the front. Ignite the torch.

WARNING!

Never light a torch with a match or pocket lighter. You could be burned, or the lighter could explode. Holding the friction lighter to one side prevents the ignited gas from deflecting back at you.

Once the torch ignites, adjust the flame with the torch fuel gas valve. Increase the fuel gas flow until the flame stops smoking or pulls slightly away from the tip. Reduce the flow until the flame returns to the tip. Open a one-piece torch's oxygen valve very slowly. Open a combination torch's preheat oxygen valve very slowly. Increase the oxygen until the flame becomes neutral.

Maximum Fuel Gas Flow

Set the maximum fuel flow with the following procedure. Increase the fuel gas flow until the flame pulls away from the tip. Decrease it until the flame just returns to the tip.

Press the cutting oxygen lever all the way down and watch the flame. It should have a long, thin, high-pressure oxygen cutting jet up to 8" (~20 cm) long, extending from the cutting orifice in the tip center. If it doesn't, do the following:

- Use the manufacturer's tip selection chart to verify the working pressures.
- Shut down the torch and clean the tip. Replace the tip if this doesn't fix the problem.

Watch the preheat flame with cutting oxygen on. If it changes to a carburizing flame, increase the preheat oxygen until the flame is neutral. After this adjustment, the preheat flame will be slightly oxidizing when you shut off the cutting oxygen. This is acceptable.

3.3.4 Shutting Off the Torch

Shut off the torch by releasing the cutting oxygen lever. For a one-piece torch, shut off the torch oxygen valve. For a combination torch, shut off the preheat oxygen valve. Close the torch fuel gas valve to extinguish the flame.

WARNING!

Always turn off the oxygen flow *first* to prevent flashback.

3.3.5 Shutting Down the Equipment

When you're finished with oxyfuel equipment, shut it down in the correct order (*Figure 51*). The following steps outline the process:

Step 1 Close the fuel gas and oxygen cylinder valves. Leave the regulator adjustments alone.

Step 2 Open the torch fuel gas valve. This releases pressure in the hose and regulator. Close it when the pressure has bled off. Repeat the process with the torch oxygen (or preheat) valve.

NOTE

If you've relieved the pressure properly, all regulator gauges should read 0. Don't continue until they do.

Step 3 Turn the fuel gas and oxygen regulator adjustment knobs or screws counterclockwise until they're loose.

Step 4 Coil and secure the hoses and torch.

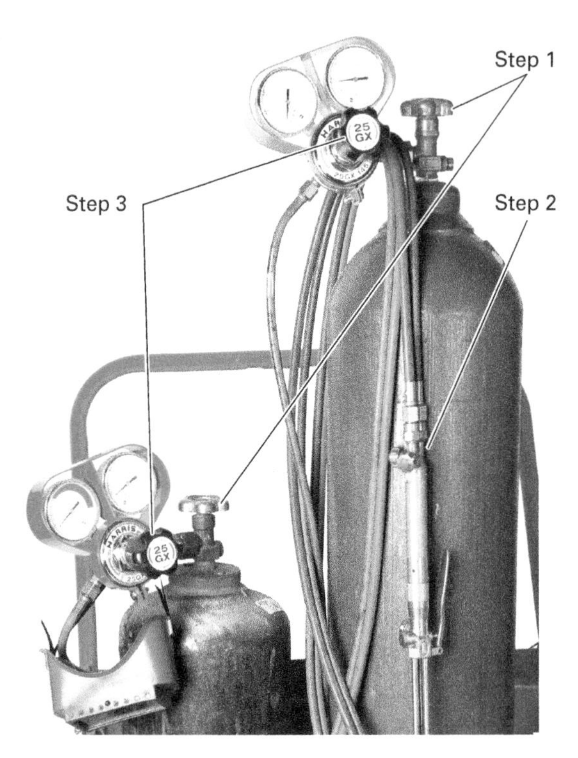

Figure 51 Shutting down oxyfuel cutting equipment.
Source: The Lincoln Electric Company, Cleveland, OH, USA

3.3.6 Disassembling the Equipment

Disassemble oxyfuel equipment that you won't use for a while. Begin by checking that it's properly shut down. All valves should be closed. All pressure gauges should read 0. Disconnect the hoses from the torch and the regulators. Remove the regulators. Replace the safety caps over the cylinder valves. Secure the cylinders to a bottle cart or hand truck and move them to an appropriate storage area.

WARNING!

Always keep cylinders upright during transport and in storage. Never move a cylinder without first securing it to the cart. Always use a bottle cart with a suitable firewall to move oxygen and fuel gas cylinders together. Treat empty cylinders by the same rules as full ones.

3.3.7 Changing Cylinders

Treat gas cylinders as "empty" before they're truly empty. Once the oxygen cylinder's working pressure drops below 25 psig (~170 kPaG), the torch won't operate reliably. Gases can flow backwards if the pressure drops too low. To prevent these problems, replace the cylinder before it reaches this point.

The following steps outline changing a cylinder:

Step 1 Shut down the oxyfuel equipment properly. All valves should be closed. All pressure gauges should read 0.

Step 2 Remove the hose and then the regulator from the empty cylinder. Replace the safety cap over the cylinder valve.

Step 3 Mark "MT" ("empty") and the date in soapstone near the cylinder top (*Figure 52*). If your company has a different identification procedure, follow it.

Step 4 Properly transport the cylinder from the workstation to the correct storage area. Place the cylinder in the storage area's empty cylinder section.

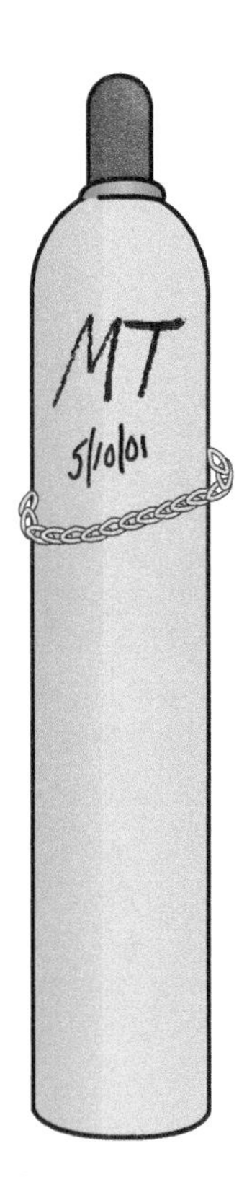

Figure 52 Empty cylinder marking.

WARNING!

Always keep cylinders upright during transport and in storage. Never move a cylinder without first securing it to the cart. Always use a bottle cart with a suitable firewall to move oxygen and fuel gas cylinders together. Treat empty cylinders by the same rules as full ones.

Marking and Tagging Cylinders

Mark cylinders with soapstone or another temporary marking tool. Never use permanent markers. Tag defective cylinders with a warning sign.

3.0.0 Section Review

1. *Always* lift cylinders with a _____.
   a. strong electromagnet
   b. cylinder cage
   c. sling through the safety cap slots
   d. cable wrapped around the valve

2. After setup, as well as periodically, oxyfuel equipment requires _____.
   a. gas purging
   b. carburizing
   c. backfiring
   d. leak testing

3. During shutdown, *after* bleeding the pressure from oxyfuel equipment, turn the regulator adjustment knobs or screws _____.
   a. clockwise until they're tight
   b. halfway between open and closed
   c. until the pressure gauges move
   d. counterclockwise until they're loose

4.0.0 Oxyfuel Cutting Procedures

Objective

Outline common oxyfuel cutting procedures.

a. Identify good and bad cuts, as well as what creates them.

b. Explain how to cut, bevel, wash, and gouge.

Performance Tasks

6. Cut instructor-specified shapes from various thicknesses of steel.
7. Wash metal with oxyfuel equipment.
8. Gouge metal with oxyfuel equipment.
9. Cut straight lines and bevels either with a track cutter or manually with/without a guide.

Good oxyfuel cutting requires practice and attention to detail. Understanding what produces good and bad cuts is essential. Using the right technique for each operation is vital too.

4.1.0 Producing Good Cuts

Clean cuts don't happen by chance. Experienced craftworkers understand that good preparation backs up their skills, helping them produce the best possible cut. Planning is the first step towards precise cutting. Identifying the reason for a poor cut turns mistakes into learning experiences.

4.1.1 Preparing to Cut

Prepare the metal by removing rust, scale, and foreign matter from the surface. Position the work at a comfortable height. Mark the cut lines with soapstone or a scriber. Select the correct cutting tip for the fuel gas and metal thickness. Consult the manufacturer's tip selection chart for guidance.

4.1.2 Good and Bad Cuts

Good craftworkers recognize a good cut. Not only can they spot a bad cut, but they can explain *why* it's bad. *Figure 53* shows examples of good and bad cuts. The following list summarizes each:

Drag lines: Parallel, nearly vertical lines on oxyfuel-cut edges produced by the oxygen stream.

- *Good cut* — This cut has a square top edge. It's sharp, straight, and not jagged. The bottom edge may have some dross adhering to it. It shouldn't be excessive, however, and can be easily removed with a chipping hammer. The **drag lines** are regular, nearly vertical, and not too pronounced.
- *Erratic cut* — Holding or moving the torch unsteadily produced a jagged kerf with irregular drag lines.
- *Cutting speed too slow* — The kerf has bad gouging at the bottom and almost vertical drag lines. The top is "rough."
- *Cutting speed too fast* — The kerf is irregular, with slag at the bottom. The drag lines are curved.
- *Preheat too low* — The bottom is badly gouged because of the slow cutting speed required.
- *Preheat too high* — The top edge is melted, giving an irregular line with excessive dross. Low cutting pressure produces a similar effect.
- *Poor restarts* — Carelessly restarting the cut produced bad gouges at each start point.
- *Cutting pressure too high* — The kerf is ugly and uneven with erratic drag lines because the pressure was too high for the tip size.

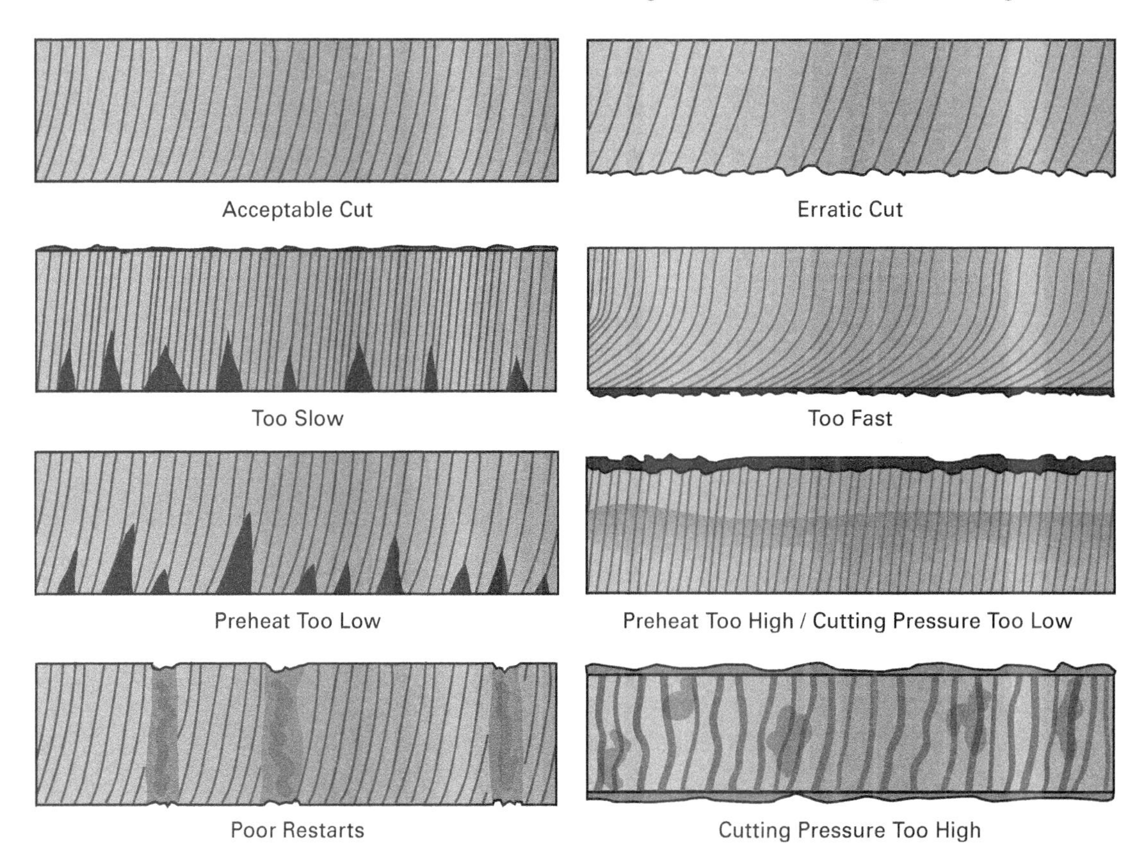

Figure 53 Cut examples.

Some inspectors evaluate cuts with a *surface roughness gauge*. It's a plastic plate with four numbered surfaces (*Figure 54*). The inspector compares the cut with the tool and assigns it a number based on the surface that most closely matches the cut's appearance. If the cut receives a #1, it automatically fails. Some job specifications include a roughness value. *AWS C4.1, Criteria for Describing Oxygen-Cut Surfaces, and Oxygen Cutting Surface Roughness Gauge*, explains how to use the tool.

NOTE

The Performance Accreditation Tasks at the end of this module require excellent cuts. They must have square kerf faces with minimal notching, not exceeding $^{1}/_{16}$" (1.6 mm) deep. Practice each task until you're confident you can do it well. Your instructor will confirm when you're ready to move on to other tasks.

4.2.0 Cutting Steel

Oxyfuel equipment cuts carbon steel extremely well. While it can cut thick or thin metal easily, the technique for each differs. Cut quality depends on the metal's thickness, the selected tip, and the craftworker's skill. Besides cutting through plates, oxyfuel equipment cuts bevels and grooves. It also removes bolts and rivets.

WARNING!

Always wear appropriate PPE when cutting with oxyfuel equipment.

4.2.1 Cutting Thin Steel

Thin steel is $^{3}/_{16}$" (~5 mm) thick or less. Heat warps and distorts thin materials. Controlling the heat is therefore crucial. Cut quickly without producing a jagged kerf. The following steps outline the process:

Step 1 Prepare the surface. Light the torch and adjust the flame.

Step 2 Hold the torch so the tip points in the travel direction at a 15- to 20-degree angle. Center one preheat orifice, as well as the cutting orifice, on the cut line (*Figure 55*).

Step 3 Preheat the metal to a dull red. Don't overheat the steel or it will warp.

Step 4 Rest the tip lightly on the surface. Press the cutting oxygen lever to start the cut. Move along the line as quickly as you can without losing the cut.

NOTE

Don't hold the tip perpendicular (vertical) to the surface, or you'll overheat the steel.

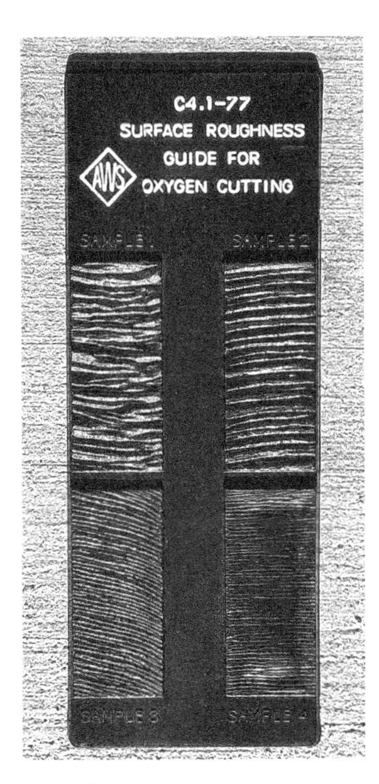

Figure 54 Surface roughness gauge.
Source: Rich Samanich, M.Ed., SCWI

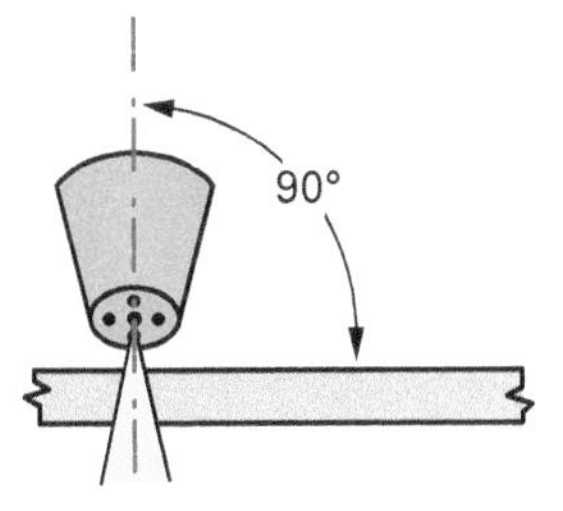

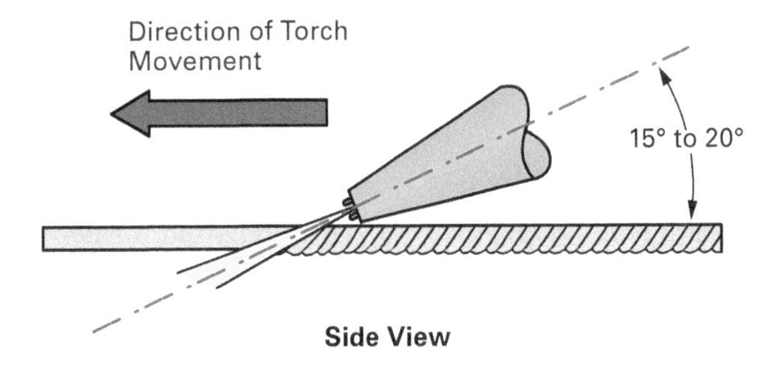

Figure 55 Cutting thin steel.

4.2.2 Cutting Thick Steel

Thick steel is anything greater than $^{3}/_{16}$" (~5 mm) thick. It warps and distorts far less than thin metal, so it's easier to cut. The following steps outline the process:

Step 1 Prepare the surface. Light the torch and adjust the flame.

Step 2 Follow the sequence shown in *Figure 56*. You can cut from left to right or from right to left. Select the direction that gives the best visibility. When you start cutting, hold the preheat orifices $^{1}/_{16}$" to $^{1}/_{8}$" (~2 mm to 3 mm) above the workpiece.

NOTE

For steel up to $^{3}/_{8}$" (~10 mm) thick, you can omit the first and third steps shown in *Figure 56*.

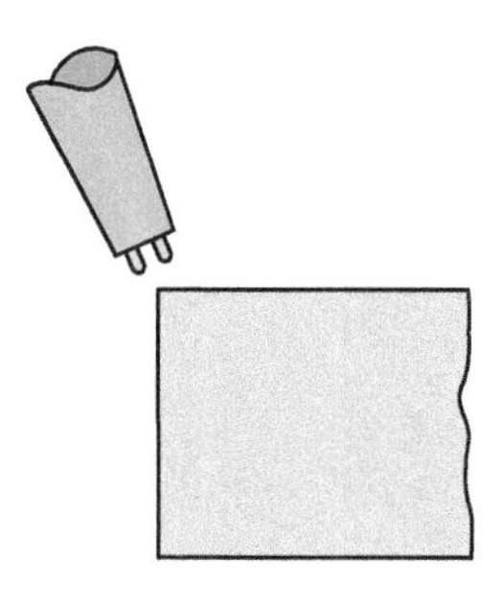

1. Start to preheat; point tip at angle on edge of plate.

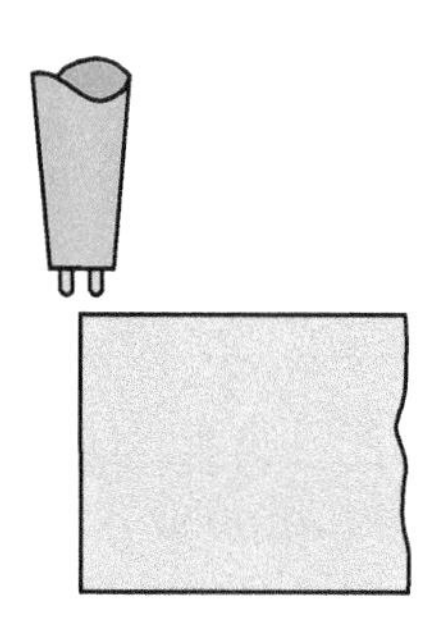

2. Rotate tip to upright position.

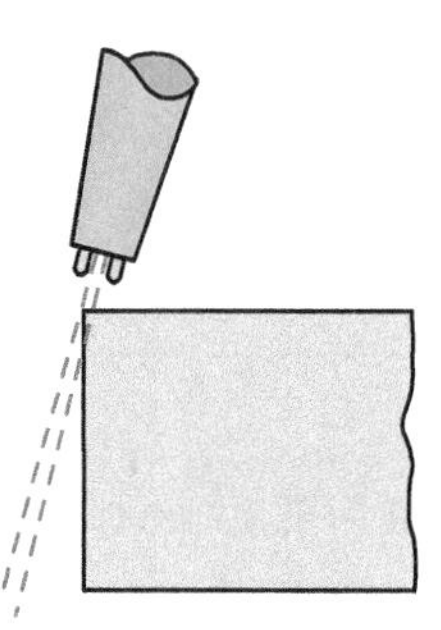

3. Press cutting oxygen valve slowly; as cut starts, rotate tip backward slightly.

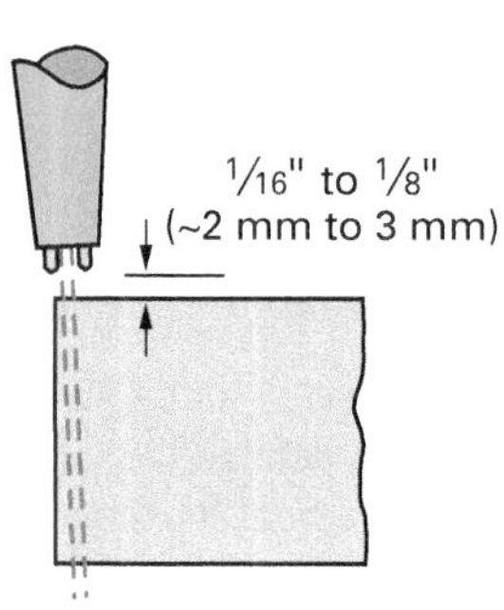

4. Rotate to upright position without moving tip forward.

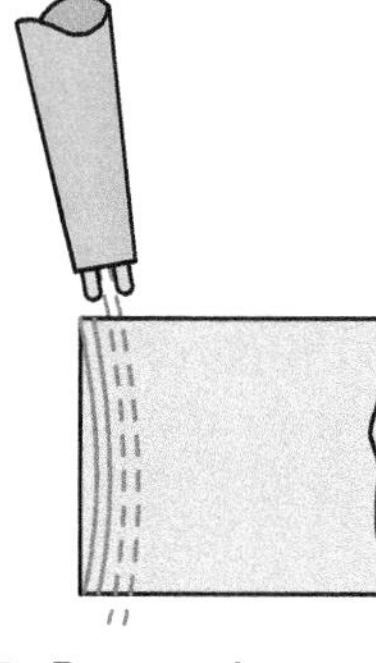

5. Rotate tip more to point slightly in direction of cut.

6. Advance as fast as good cutting action will permit.

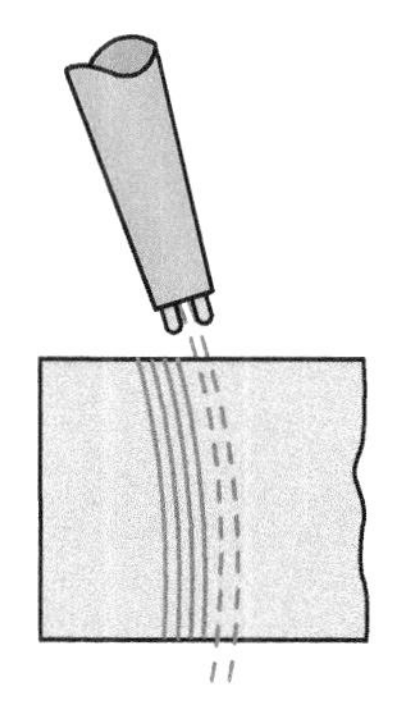

7. Do not jerk; maintain slight leading angle toward direction of cut.

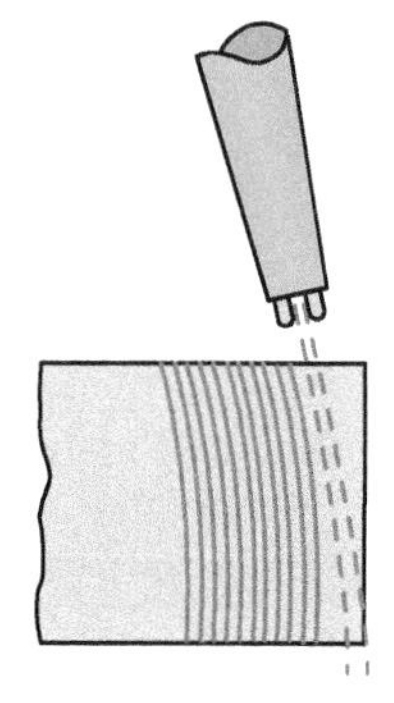

8. Slow down; let cutting stream sever corner edge at bottom.

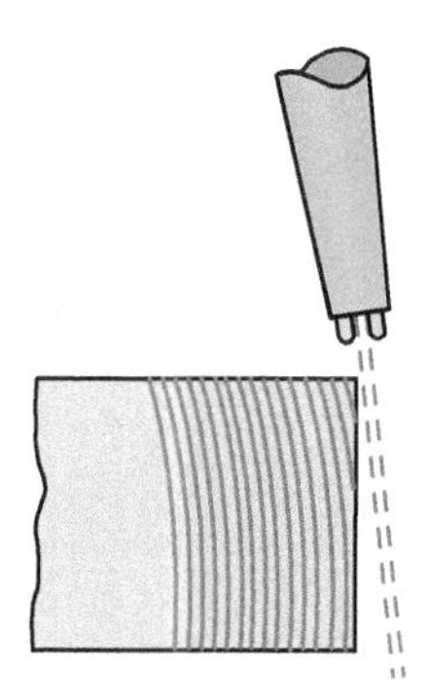

9. Continue steady forward motion until tip has cleared end.

Figure 56 Cutting thick steel.

4.2.3 Piercing a Plate

Piercing: To penetrate a metal plate with a cutting torch.

Piercing—creating a small hole in a plate—is the start of a larger hole or slot. Beginning in the middle requires more preheating, so you must choose a larger tip. For plates thicker than 3" (~8 cm), preheating the bottom side first is wise. The following steps outline the process (*Figure 57*):

Step 1 Prepare the surface. Light the torch and adjust the flame.

Step 2 Hold the tip ¼" to 5⁄16" (~6 mm to 8 mm) above the surface until it's a bright cherry red.

Step 3 Raise the tip to about ½" (~13 mm) above the surface and tilt the torch slightly. This angle prevents molten metal from splashing the tip when you start the oxygen stream.

Step 4 Press the cutting oxygen lever.

Step 5 Maintain the angled position until you break through.

Step 6 Rotate the torch to the vertical (perpendicular) position.

Step 7 Lower the torch to ¼" to 5⁄16" (~6 mm to 8 mm). Cut outwards, following the hole or slot's outline.

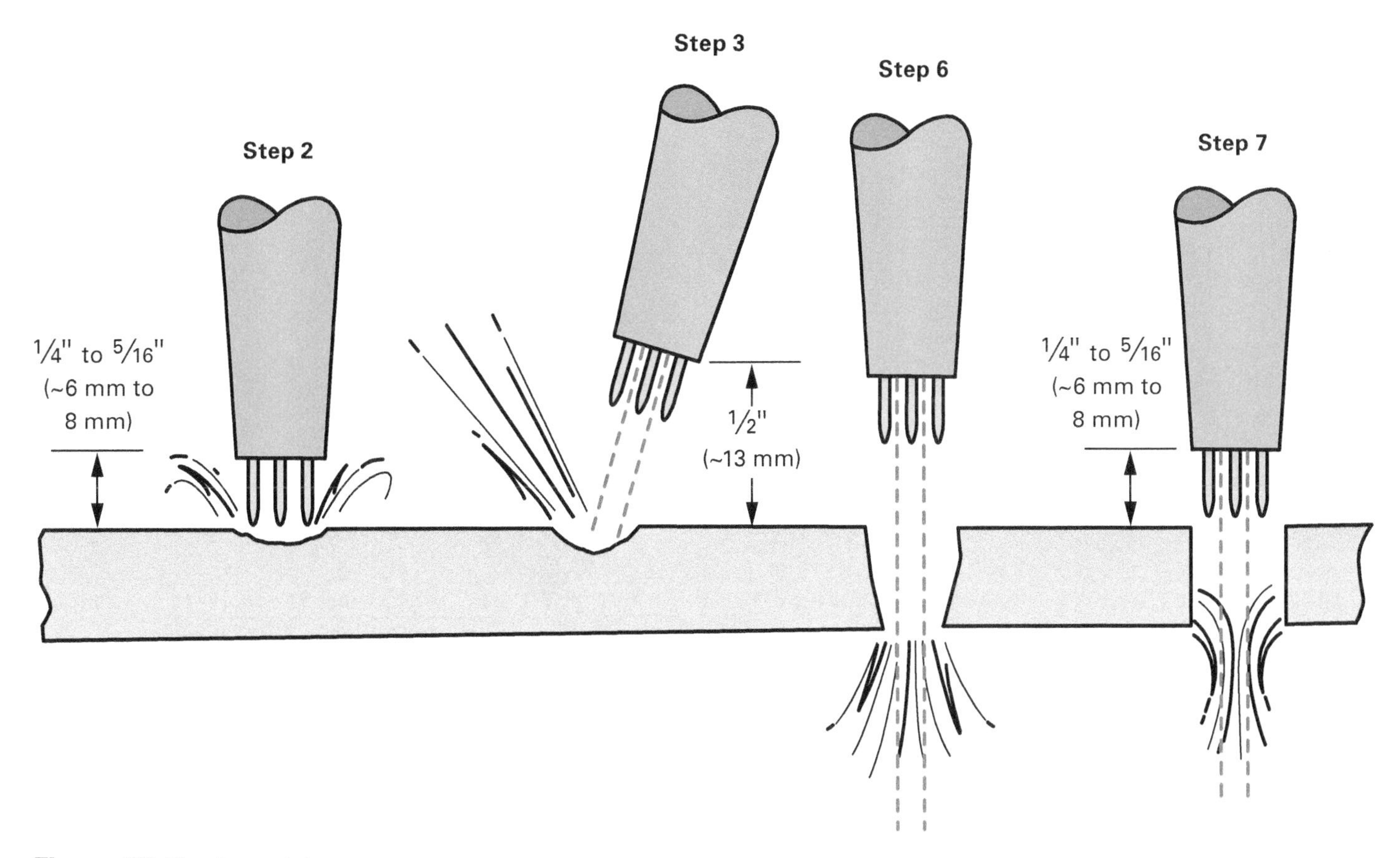

Figure 57 Piercing a plate.

4.2.4 Cutting Bevels

Beveling a steel plate's edge prepares it for welding. The following steps outline the process (*Figure 58*):

Step 1 Prepare the surface. Light the torch and adjust the flame.

Step 2 Hold the torch so the tip faces the metal at the desired bevel angle.

Step 3 Preheat the edge to a bright cherry red.

Step 4 Press the cutting oxygen lever.

Step 5 Move the tip steadily along the bevel line. If you're not using a guide, keep the tip angle as consistent as possible.

NOTE

A piece of angle iron will hold the torch at a 45-degree angle. Clamped in place, it makes an excellent torch guide.

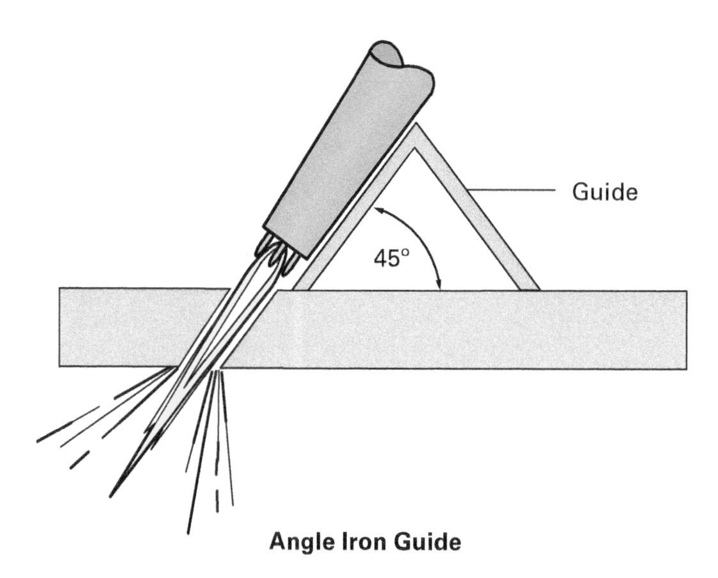

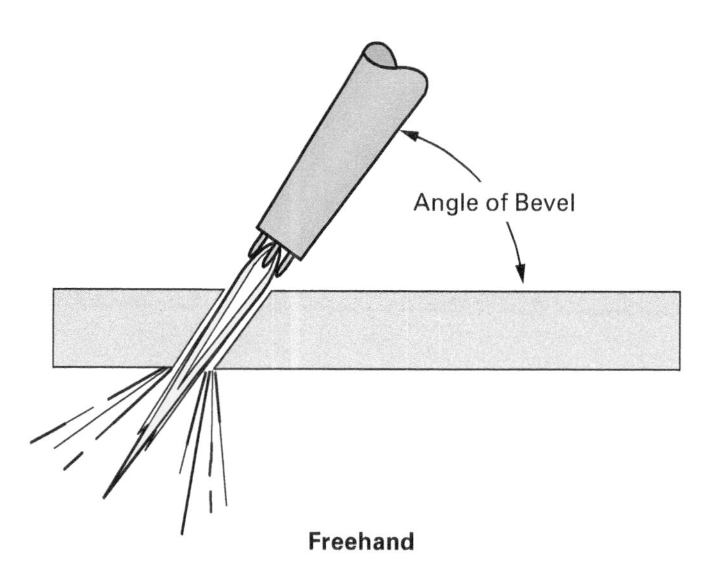

Figure 58 Cutting a bevel.

4.2.5 Washing

Washing refers to cutting bolts, rivets, welded parts, and projections from a surface. It requires a special tip with a large cutting orifice. This configuration produces a low-velocity cutting stream that doesn't damage the surrounding metal. The following steps outline the process (*Figure 59*):

Step 1 Prepare the surface. Light the torch and adjust the flame.

Step 2 Preheat the projection to a bright cherry red.

Step 3 Rotate the tip so it's at a 55-degree angle to the surface.

Step 4 Press the cutting oxygen lever.

Step 5 Beginning at the top, start cutting away the projection. Continue moving back and forth across the material. Rotate the tip parallel to the projection. Move toward the surface as you cut away the projection. Don't cut the surface itself!

CAUTION

If the surrounding metal becomes too hot, you could damage the surface. Cut away the projection as quickly as possible. If necessary, stop and let the surface cool.

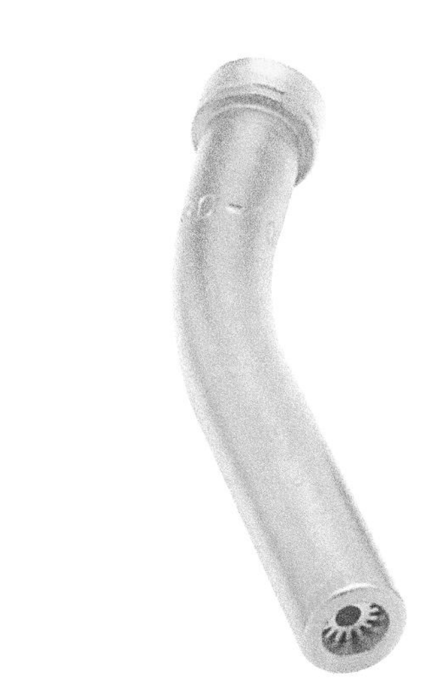

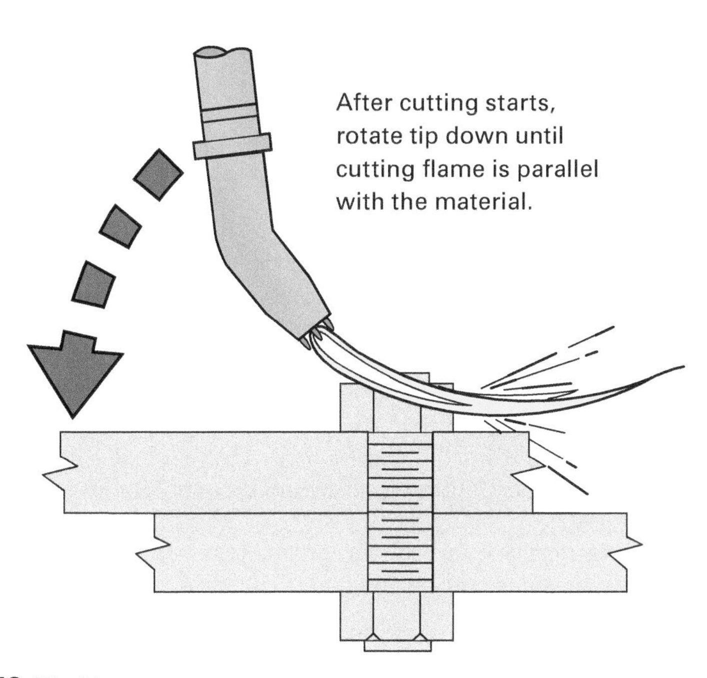

Figure 59 Washing.
Source: The Lincoln Electric Company, Cleveland, OH, USA (59A)

4.2.6 Gouging

Gouging cuts a groove in a surface (*Figure 60*). Welders do this to remove cracks and defects before welding. A gouging tip is curved and directs a low-velocity oxygen stream upwards. This helps the craftworker control the groove's width and depth. Gouging tips can also wash a surface, although they can't handle large projections.

Producing a smooth groove with a consistent depth takes practice. Travel speed and angle are crucial. The following steps outline the process:

Step 1 Prepare the surface. Light the torch and adjust the flame.

Step 2 Hold the torch so the preheat holes point directly at the surface. Preheat the surface to a bright cherry red.

Step 3 When the surface is ready, slowly roll the torch away. Tilt it to an angle that will produce the correct groove depth. While rolling the torch away, depress the cutting oxygen lever gradually.

Step 4 Move the tip along the groove line. Rock it back and forth to cut the correct width and depth.

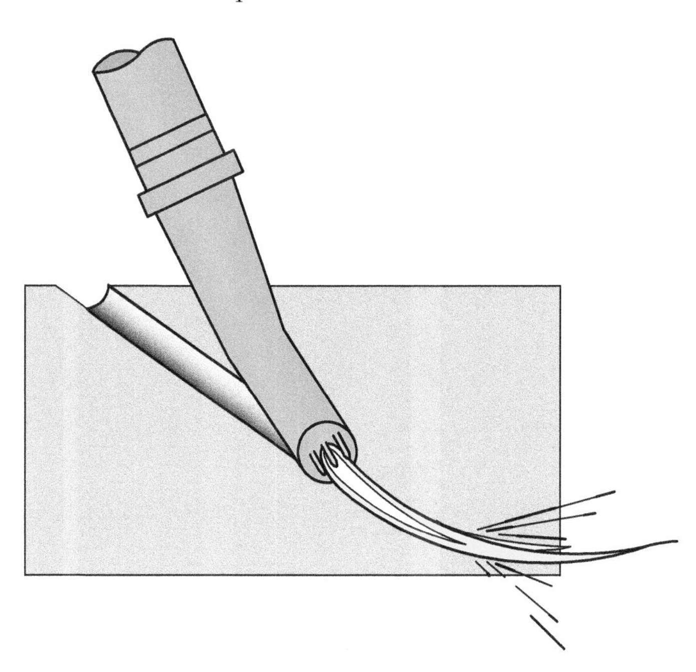

Figure 60 Gouging.

4.2.7 Using a Track Cutting Machine

An oxyfuel track cutting machine produces precision cuts semi-automatically (*Figure 61*). These machines are especially useful for production tasks with repeated cuts. Many models are portable, so you can use them in the field.

An electric motor propels the *tractor unit* along the track, producing the cut. The *torch holder* positions the torch vertically. The *rack assembly* moves the torch holder towards or away from the tractor unit. The *bevel adjustment* holds the torch at the required angle. Clamping screws lock everything in position for the cut.

All track cutting machines work similarly. Consult the manual for specific details. The following steps outline the general process:

Step 1 Place the track on the workpiece in the correct position. If necessary, add extra sections. Support the track properly. If necessary, clamp it at both ends so the tractor will move smoothly.

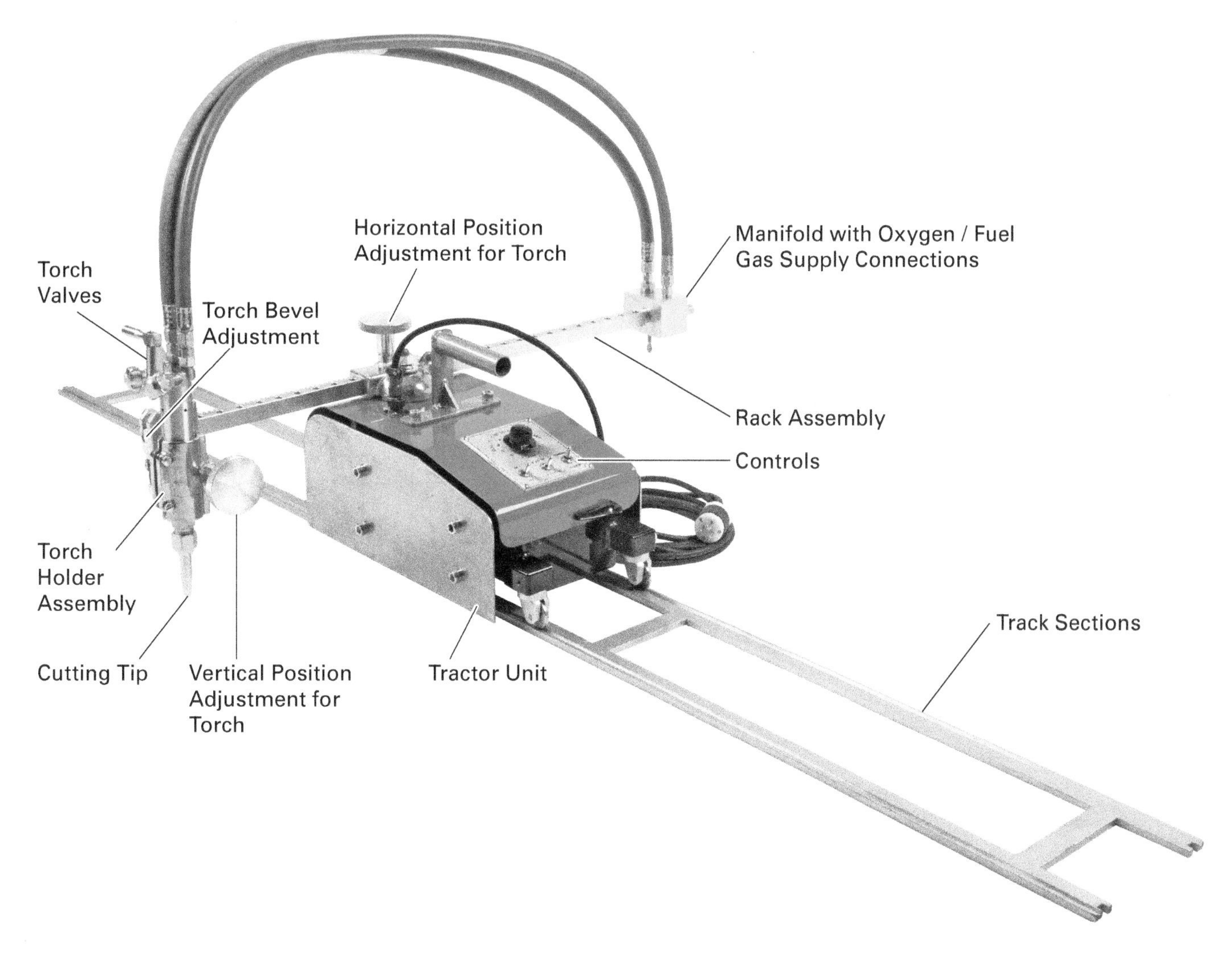

Figure 61 Portable track cutting machine.
Source: The Lincoln Electric Company, Cleveland, OH, USA

WARNING!

Many cutting machines can't detect the track end and will fall off. Never leave an operating track cutting machine unattended.

Step 2 Place the tractor on the track. Confirm that the power line and hoses are long enough for the cut.

Step 3 Position the tractor at the starting position. Confirm that the power switch is off. Select the cut direction, speed range, and travel speed. Plug in the tractor's power cord.

Step 4 Install the torch. Make all required adjustments (height, position, and angle). Tighten the clamping screws.

Step 5 Light the torch and adjust the flame.

Step 6 Preheat the start of the cut.

Step 7 Turn on the cutting oxygen. Switch on the tractor and start it moving.

Step 8 When the machine completes the cut, switch off the tractor and extinguish the torch.

4.0.0 Section Review

1. An oxyfuel cut has gouging at the bottom, a pronounced break in the drag lines, and an irregular kerf. What *most* likely caused this poor-quality cut?
   a. Travel speed was too fast.
   b. Preheat was insufficient.
   c. Cutting oxygen pressure was too low.
   d. Cutting oxygen pressure was too high.

2. A washing tip has a large cutting orifice that produces a _____.
   a. high-velocity oxygen stream
   b. grooved cut in the surface
   c. cleansing layer of dross
   d. low-velocity oxygen stream

Module 29102 Review Questions

1. Oxyfuel cutting works *best* with _____.
   a. copper
   b. ferrous metals
   c. nonferrous metals
   d. stainless steel

2. In oxyfuel cutting, what produces much of the cutting heat?
   a. The burning fuel gas
   b. The exploding fuel gas
   c. The expanding oxygen
   d. The burning metal

3. When cutting with oxyfuel equipment, use eye protection with a shade value of _____.
   a. 1 to 2
   b. 3 to 6
   c. 7 to 8
   d. 9 to 10

4. If an acetylene cylinder tips on its side, what should you do?
   a. Contact the gas supplier for instructions.
   b. Release a small amount of gas to purge the valve.
   c. Stand it upright and wait at least one hour.
   d. Inject acetone into the tank with a recharge valve.

5. Above what pressure will acetylene explode?
   a. 5 psig (~34 kPaG)
   b. 10 psig (~69 kPaG)
   c. 15 psig (103 kPaG)
   d. 25 psig (~172 kPaG)

6. Most pressure regulators include two pressure gauges. One shows the cylinder pressure, while the other shows the _____.
   a. torch's gas flow rate
   b. regulator's outlet pressure
   c. amount of gas left in the cylinder
   d. temperature inside the regulator

7. An oxyfuel gas connection has notches on the corners and left-hand threads. What type of connection is it?
   a. An oxygen fitting
   b. An inert gas fitting
   c. A fuel gas fitting
   d. A backlash fitting

8. To prevent the torch flame from going inside the torch and hose, install a(n) _____.
   a. flashback arrestor
   b. check valve
   c. anti-siphon valve
   d. ball valve

9. Which torch can work with low-pressure fuel gas because it draws it in with the vacuum produced by the flowing oxygen?
   a. A siphon torch
   b. An inert torch
   c. An injector torch
   d. A suspension torch

10. Which cutting machine semi-automatically cuts lines, bevels, and curves?
    a. A track cutting machine
    b. A ring gear pipe cutter
    c. A motorized band cutter
    d. A thermic lance

11. A thermic lance generates intense cutting heat _____.
    a. by burning hydrogen and oxygen
    b. with a plasma arc flame
    c. by burning metal rods
    d. with molten sodium and oxygen

12. What must a bottle cart have to carry an oxygen and a fuel gas cylinder together?
    a. A fire extinguisher
    b. A firewall
    c. A valve safety cap
    d. A tool tray

13. Cracking the cylinder valves during oxyfuel equipment setup _____.
    a. tests the regulators for leaks
    b. equalizes the cylinder pressures
    c. clears dirt from the outlets
    d. purges air from the cylinders

14. Check oxyfuel equipment for leaks by _____.
    a. smelling for gas near the outlets
    b. applying a solution that bubbles
    c. listening for hissing near connections
    d. sparking a friction lighter near connections

15. An oxyfuel cutting flame that has too much fuel gas is a(n) _____.
    a. enriched flame
    b. neutral flame
    c. oxidizing flame
    d. carburizing flame

16. Before removing the hoses from an oxyfuel outfit, what should you do?
    a. Verify that all pressure gauges read 0.
    b. Remove the torch tip.
    c. Press the torch oxygen valve lever.
    d. Open both torch valves.

17. A good oxyfuel cut has drag lines that are _____.
    a. at a 30-degree angle
    b. at a perfect 45-degree angle
    c. almost horizontal
    d. nearly vertical

18. Moving an oxyfuel cutting torch unsteadily produces _____.
    a. a badly contaminated kerf
    b. a jagged kerf
    c. excessive dross
    d. nearly horizontal drag lines

19. When cutting thin steel, the crucial factor is preventing _____.
    a. oxidation
    b. sparking
    c. distortion
    d. scarring

20. If you must cut off several large bolts projecting from a metal plate, you should select a _____.
    a. beveling tip
    b. gouging tip
    c. piercing tip
    d. washing tip

Answers to odd-numbered Module Review Questions are found in *Appendix A*.

Cornerstone of Craftsmanship

Richard Samanich
Lead Welding Instructor
College of Southern Nevada

How did you choose a career in the industry?

When I was a young jet engine mechanic in the US Air Force, I purchased a used motorcycle. What I didn't know was that the frame was cracked; it was on the verge of breaking. While telling my fellow airmen about my bad luck, one of our supervisors overhead me. He said it was no big deal. We could take it to the auto-hobby shop on base and weld it. We did, and I was amazed by it, and asked him where he learned to weld. He said at the community college. So, I enrolled as soon as I could. Welding became my main interest, so I applied to cross train to be a Metals Processing Specialist in the air force. The application was approved and got me started in the welding and heat-treating profession—all because of a used motorcycle.

Who inspired you to enter the industry?

One of my supervisors when I was a jet engine mechanic in the air force. He was responsible for my interest in welding and my desire to take classes at a community college.

What types of training have you been through?

My first welding training began at a community college welding program, in 1979. After that I was trained in the US Air Force in oxyfuel welding, SMAW, GTAW, basic heat treating and electroplating of metals, welding high- and low-pressure lines for aircraft. Over the years, I took several courses from the American Welding Society, as well as Lincoln Electric. I became an AWS Certified Welding Inspector in 1999 and currently hold a Senior Certified Welding Inspector certification. I have held certifications in several welding processes and on numerous types of metals. I have previously held these certifications from the International Code Counsel: Structural Steel and Welding Special Inspector, Structural Steel & Bolting Special Inspector, Spray Applied Fireproofing Special Inspector, and Reinforced Concrete Special Inspector. Also, I've held the Concrete Field-Testing Technician certification from the American Concrete Institute. I was a company Level II NDT technician in dye penetrant and magnetic particle testing for aircraft and construction.

I have two associate degrees in applied science (aircraft powerplant technology and metals technology), also a bachelor's degree in professional aeronautics from Embry-Riddle Aeronautical University, and a master's degree in career and technical education from Concordia University.

How important is education and training in construction?

Education and training are extremely important, especially when it comes to safety. Working in construction is very dangerous and requires attention on everyone's part. Being well educated in your field is also very important. When issues arise on a job, your education can help you come up with a solution that saves the day and possibly saves the project money.

What kinds of work have you done in your career?

Most of my career has been working in aviation, mostly military aircraft, but also working on civilian aircraft. I also worked for Carroll Shelby at Shelby American when we started making the CSX-4000 Cobras in 1997. We had to reverse engineer most of the parts and welding fixtures/jigs.

I also worked on test and evaluation of aircraft and weapon systems, both classified and unclassified. Some of the aircrafts I worked on are in the National Museum of the US Air Force, the Threat Training Facility at Nellis Air Force Base, the Evergreen Aviation & Space Museum in Oregon, and a few other museums. Makes me feel old now that I see those in museums.

Tell us about your current job.

I am currently the Lead Welding Instructor at the College of Southern Nevada in Henderson, Nevada. Teaching basic and advanced GTAW and blueprint reading for welders. I keep my hand in the welding industry by helping to weld and provide consulting for local machine shops, transmission shops and aviation companies.

What has been your favorite project?

This is a difficult question, as I have had the honor of working on many rewarding projects. Some were very difficult and frustrating at the time, but being a part of a team that built something successful was satisfying. In construction, my favorite project was the Cleveland Clinic—Lou Ruvo Center for Brain Health. This was a very interesting project due to many reasons. It was designed by Pritzker Prize-winning architect Frank Gehry. The structural engineer for the curved stainless steel Keep Memory Alive Event Center was in Germany, the fabrication shop was in China, and the building was erected in Las Vegas, Nevada. The different time zones, use of metric measurements, and ISO Standards all presented their own issues in the construction of this building.

What do you enjoy most about your job?

In my current position, I really enjoy seeing students that have never struck an arc develop into qualified entry-level welders. Hearing from previous students on how we have helped them become successful welders is very rewarding. In my previous jobs, I enjoy seeing the completed project doing what it was designed for.

What factors have contributed most to your success?

Always wanting to learn more and being a team worker. Being able to communicate and helping to resolve technical issues that arise.

Would you suggest construction as a career to others? Why?

Yes, I would—it is a very rewarding career. You can make a good income for sure, but being able to look at the finished product and know you were a member of the team that made it, is a great reward.

What advice would you give to those new to the field?

Put a sincere effort in learning more about welding then just being able to lay down beads. Also, when working in construction, it is a good idea to calculate your living expenses for a month. Then save enough money in a separate account to cover a minimum of six months of your expenses.

If you are interested in becoming an inspector, make sure you are open to learning and getting other inspection certifications. These will make you more marketable and more valuable to your employer.

Interesting career-related fact or accomplishment?

First of two welders to receive the American Welding Society's Distinguished Welder Award.

How do you define craftsmanship?

Taking pride in doing high-quality work. Not settling for "it's good enough" or "it looks good from my house." Going beyond just getting the job done, collecting the check, and moving on to the next job. After you complete a weld, ask yourself if you would pay for that weld. If not, then the weld is not good enough.

Answers to Section Review Questions

Answer	Section Reference	Objective
Section 1.0.0		
1. b	1.1.0	1a
2. d	1.2.2	1b
Section 2.0.0		
1. b	2.1.4; *Table 2*	2a
2. a	2.2.3	2b
3. d	2.3.1	2c
4. c	2.4.5	2d
5. a	2.5.4	2e
Section 3.0.0		
1. b	3.1.1	3a
2. d	3.2.0	3b
3. d	3.3.5	3c
Section 4.0.0		
1. a	4.1.2	4a
2. d	4.2.5	4b

User Update

Did you find an error? Submit a correction by visiting **https://www.nccer.org/olf** or by scanning the QR code using your mobile device.

Plasma Arc Cutting

Source: The Lincoln Electric Company, Cleveland, OH, USA

Objectives

Successful completion of this module prepares you to do the following:

1. Describe plasma arc cutting and its safety procedures.
   a. Summarize plasma arc cutting.
   b. List plasma arc cutting safety procedures.
2. Summarize plasma arc cutting equipment.
   a. Describe plasma arc cutting power units.
   b. Describe plasma arc cutting gases and gas control devices.
   c. Describe plasma arc cutting torches and accessories.
3. Outline setting up, safely operating, and maintaining plasma arc cutting equipment.
   a. Explain how to set up a plasma arc cutting project.
   b. Explain how to operate plasma arc cutting equipment safely.
   c. Explain how to maintain plasma arc cutting equipment.

Performance Tasks

Under supervision, you should be able to do the following:

1. Set up plasma arc cutting equipment.
2. Select amperage, gas pressure, and flow rate appropriate to the metal type and thickness.
3. Square-cut metal with plasma arc cutting equipment.
4. Bevel-cut metal with plasma arc cutting equipment.
5. Pierce and cut slots in metal with plasma arc cutting equipment.
6. Dismantle and store plasma arc cutting equipment.

Overview

Welders cut metals using several different technologies. Almost nothing works as well as plasma arc cutting. This method rapidly and precisely cuts metals with a jet of superheated, electrically charged gas. This module introduces the method, its equipment, procedures, and safety practices.

NOTE

This module uses US standard and metric units in up to three different ways. This note explains how to interpret them.

Exact Conversions

Exact metric equivalents of US standard units appear in parentheses after the US standard unit. For example: "Measure 18" (45.7 cm) from the end and make a mark."

Approximate Conversions

In some cases, exact metric conversions would be inappropriate or even absurd. In these situations, an approximate metric value appears in parentheses with the ~ symbol in front of the number. For example: "Grip the tool about 3" (~8 cm) from the end."

Parallel but not Equal Values

Certain scenarios include US standard and metric values that are parallel but not equal. In these situations, a slash (/) surrounded by spaces separates the US standard and metric values. For example: "Place the point on the steel rule's 1" / 1 cm mark."

Digital Resources for Welding

Scan this code using the camera on your phone or mobile device to view the digital resources related to this craft.

Industry Recognized Credentials

If you are training through an NCCER-accredited sponsor, you may be eligible for credentials from NCCER's Registry. The ID number for this module is 29103. Note that this module may have been used in other NCCER curricula and may apply to other level completions. Contact NCCER's Registry at 1.888.622.3720 or go to **www.nccer.org** for more information.

You can also show off your industry-recognized credentials online with NCCER's digital badges. Transform your knowledge, skills, and achievements into badges that you can share across social media platforms, send to your network, and add to your resume. For more information, visit **www.nccer.org**.

1.0.0 Plasma Arc Cutting

Performance Tasks

There are no Performance Tasks in this section.

Objective

Describe plasma arc cutting and its safety procedures.

a. Summarize plasma arc cutting.

b. List plasma arc cutting safety procedures.

Plasma: A superheated gas that's electrically charged (ionized).

Dross: Oxidized and molten metal expelled by a thermal cutting process. Sometimes called *slag*.

Kerf: The slot produced by a cutting process.

Plasma arc cutting (PAC) does its work with a superheated stream of gas. This gas doesn't behave like an everyday gas. Instead, it's been changed into a **plasma**, which is extremely hot and electrically conductive (ionized). The plasma rapidly melts almost any metal, blowing it away from the worksite at the same time.

Plasma arc cutting is both versatile and adaptable to many cutting applications (*Figure 1*). Craftworkers can cut metals with a hand torch, or CNC machines can do it under computer control. This method works for most metals, both ferrous and nonferrous. It produces little **dross**, rapidly cutting a narrow **kerf**. Since it mostly heats the metal near the cut, it doesn't distort the surrounding metal.

1.1.0 The PAC Process

A plasma arc cutting setup has an electrical power supply, a cutting gas delivery system, and a torch (*Figure 2*). Some systems use a second gas for shielding or cooling. PAC torches change the cutting gas into a plasma by passing it through an electric arc inside the torch. The superheated, conductive gas exits through a small orifice (opening) at high speed. An electrical current flowing through the plasma stream keeps it at the required temperature.

Plasma can reach temperatures of 30,000°F (~17,000°C) or higher. This is far hotter than other cutting methods like oxyfuel cutting. The plasma jet melts the metal rapidly and blows it away from the cut, hole, or groove.

WARNING!

Molten material can travel as far as 40' (~12 m) from the workpiece. Take precautions to protect nearby workers and to prevent fires or other damage.

Figure 1 Plasma arc cutting (PAC).
Source: © Miller Electric Mfg. LLC

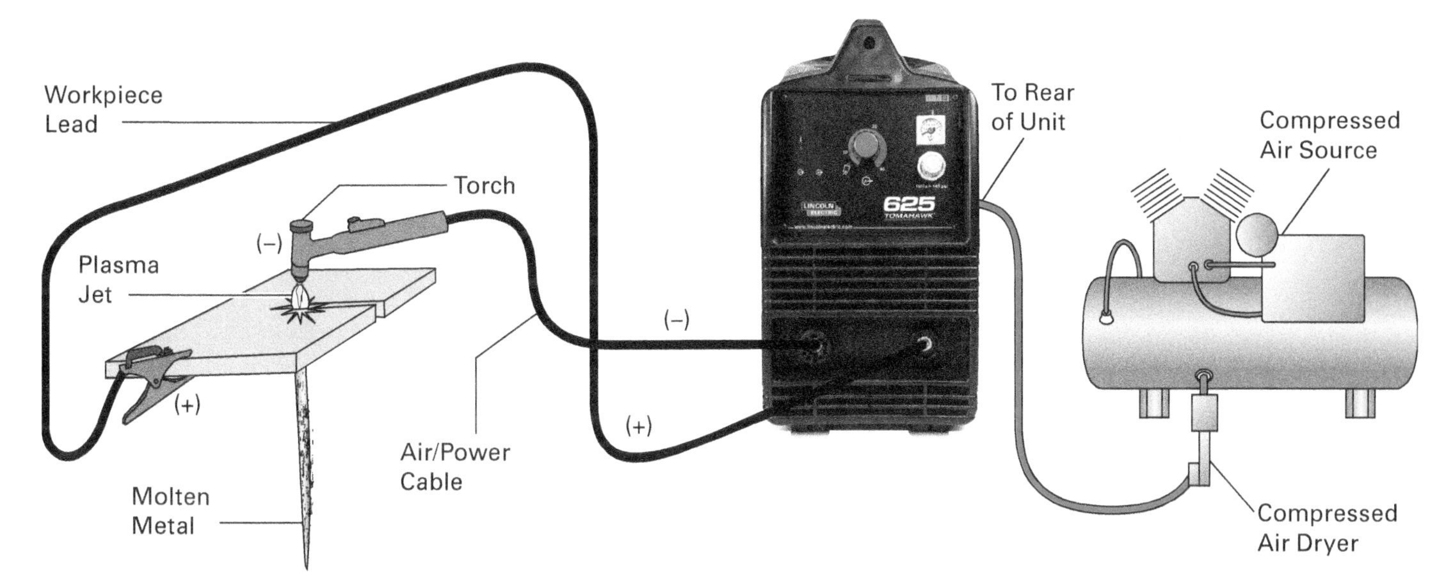

Figure 2 Plasma arc cutting setup.
Source: The Lincoln Electric Company, Cleveland, OH, USA

Plasma arc cutters divide into two types: *transferred arc* and *non-transferred arc*. Transferred arc is more common. It only works with electrically conductive materials like metals. Non-transferred arc cutters can cut non-conductive materials like ceramics and concrete. Facilities that cut only metals usually don't use them.

A Different Kind of Cut

Unlike oxyfuel cutting methods, PAC doesn't depend on oxidation. For this reason, it can cut metals that oxyfuel methods can't. Welders treat each cutting method that they learn as yet another tool to help them solve problems.

1.1.1 Transferred Arc Cutters

A transferred arc cutter creates an electrical circuit between the torch and the conductive workpiece. The plasma flows from the torch to the workpiece, carrying an electric current that maintains the plasma at its high temperature.

Creating the plasma inside the torch so it can transfer the arc to the workpiece requires an *arc initiation* step. Cutters use one of several methods. The most common starts a small electric arc, called the *pilot arc*, between the torch electrode and the nozzle. The pilot arc heats the cutting gas, turning it into a plasma.

When the craftworker brings the torch near the workpiece, the plasma flowing from the torch touches the workpiece, forming an electrical circuit. The current flowing through the plasma heats it to the cutting temperature. The pilot arc then shuts off because it's no longer needed.

To create the required electrical circuit, the craftworker connects one power supply lead to the workpiece by its attached clamp. The torch electrode connects to the other power supply lead. *Figure 3* shows a transferred arc cutter diagram.

1.1.2 Non-Transferred Arc Cutters

Non-transferred arc cutters work with non-conductive materials, so they generate the plasma-forming arc within the torch. The nozzle and the electrode form the electrical circuit, bringing the cutting gas up to the right temperature. The hot plasma exits the nozzle and cuts the workpiece. *Figure 4* shows a non-transferred arc cutter diagram.

NOTE

During startup, the nozzle and the electrode erode slightly. Eventually, they'll wear out and require replacing. For this reason, they're often called *consumables*.

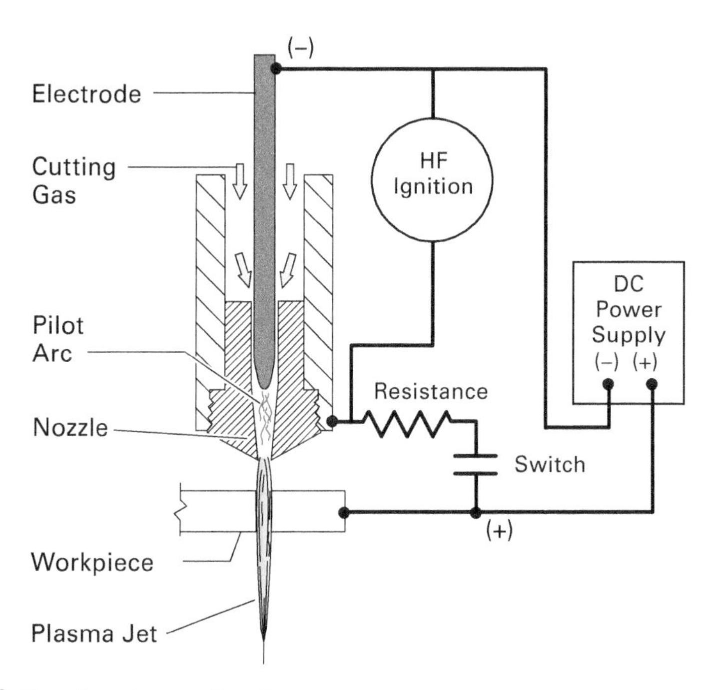

Figure 3 Transferred arc cutter diagram.

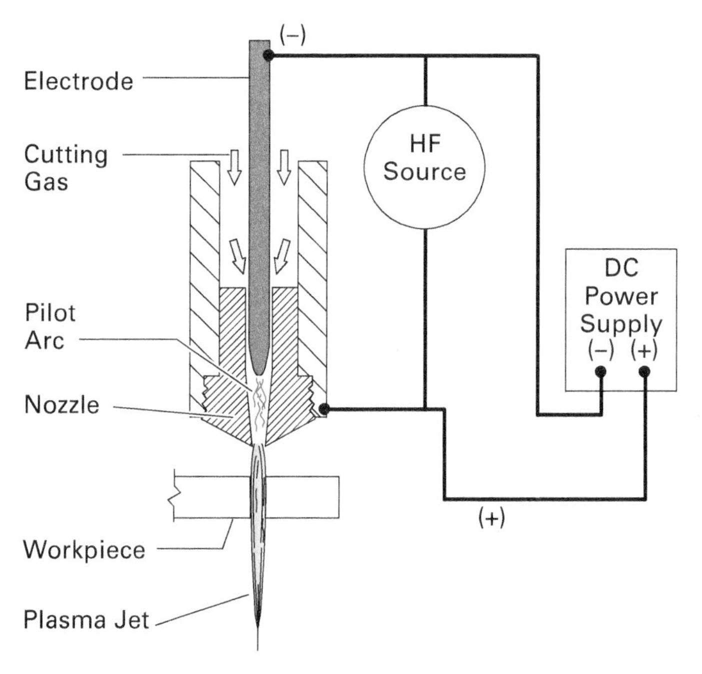

Figure 4 Non-transferred arc cutter diagram.

1.1.3 PAC Operations

Plasma arc cutters can produce a variety of cuts. When equipment manufacturers specify a PAC unit's maximum cutting thickness, they usually specify the kind of cut(s) possible at that limit. Always read specifications and documentation carefully.

A regular (clean) cut has smooth edges with little dross. Most PAC units can't produce a clean cut when operating near their maximum cutting thickness. When a project's specifications require a clean cut, choose a PAC unit with a larger maximum cutting thickness than the material you're cutting.

A **severance cut** simply divides the metal into pieces. The edges will probably be rough, irregular, and have significant dross. PAC units operating near their maximum cutting thickness usually produce severance cuts.

Piercing penetrates the metal, producing a small hole. Welders often pierce a plate before cutting a larger hole or slot. The PAC unit may not be able to pierce as thick a material as it can cut.

Gouging is cutting a groove in the metal's surface. Welders often gouge metal to prepare it for welding or to remove cracks. Gouging requires a different torch tip than regular cutting operations.

Severance cut: A cut that divides metal but leaves behind rough edges with significant dross.

Piercing: To penetrate a metal plate with a cutting torch.

Gouging: Cutting a groove into a surface.

1.2.0 Safety Practices

Plasma arc cutters generate enormous heat with an electric current. Using them safely requires following specific procedures. This section summarizes good practices but isn't a complete PAC safety training. Always complete all training that your workplace requires. Follow all workplace safety guidelines and wear appropriate PPE.

NOTE

Be sure that you've completed NCCER Module 29101, *Welding Safety*, before continuing.

Unlike oxyfuel cutting equipment, PAC machines use non-flammable gases like air or nitrogen for cutting. They do use high voltages, generate very high temperatures, and emit UV radiation. Always practice the following safety guidelines when using a PAC machine:

- PAC torches deliver voltages as high as 400V. Disconnect all energy sources before disassembling a PAC torch or replacing its consumable components.
- The plasma jet can easily cut through your gloves and cause severe injuries. Be extremely careful when it's running.

WARNING!

Energized PAC equipment is extremely dangerous. Always disconnect the power supply and all other energy sources before servicing or otherwise working with the equipment. Never place your hands under the workpiece near the cutting jet. The torch can injure you even though you're wearing PPE.

1.2.1 Protective Clothing and Equipment

Welding and cutting tasks are dangerous. Unless you wear all required PPE, you're at risk of serious injury. The following guidelines summarize PPE for plasma arc cutting:

- Always wear safety glasses plus a full face shield, helmet, or hood. The glasses, face shield, or helmet/hood lens must have the proper shade value for PAC work (*Figure 5*). Never view the cutting arc directly or indirectly without using the proper lens.
- Wear protective leather and/or flame-retardant clothing along with welding gloves that will protect you from flying sparks and molten metal.
- Wear high-top safety shoes or boots. Make sure the pant leg covers the tongue and lace area. If necessary, wear leather spats or chaps for protection. Boots without laces or holes are better for welding and cutting.

PAC Hazards

PAC generates UV radiation, molten metal spatter, and noise. Good PPE protects you from many of these. Industrial plasma arc cutters, such as those used in CNC machines, have features to manage hazards and control noise.

- Wear a 100 percent cotton cap with no mesh material in its construction. The bill should point to the rear or the side with the most exposed ear. If you must wear a hard hat, use one with a face shield.

WARNING!

Never wear a cap with a button on top. The conductive metal under the fabric is a safety hazard.

- Wear earplugs to protect your ear canals from sparks. Since the torch is loud, always wear hearing protection.

Arc Cutting Processes

Process	Electrode	Amperage	Lens Shade Numbers (2 3 4 5 6 7 8 9 10 11 12 13 14)
Plasma Arc Cutting (PAC)		< 20A	4
		20A – 40A	5
		40A – 60A	6
		60A – 80A	8
		80A – 300A	8 9
		300A – 400A	9 10 11 12
		400A – 800A	10 11 12 13 14

Lens shade values are based on OSHA minimum values and ANSI/AWS suggested values. Choose a lens shade that's too dark to see the weld zone. Without going below the required minimum value, reduce the shade value until it's visible.

Figure 5 Plasma Arc Cutting – Guide to Lens Shade Numbers.

1.2.2 Fire/Explosion Prevention

PAC produces extremely high temperatures to cut metal. Welding or cutting a vessel or container that once contained combustible, flammable, or explosive materials is hazardous. Residues can catch fire or explode. Before welding or cutting vessels, check whether they contained explosive, hazardous, or flammable materials. These include petroleum products, citrus products, or chemicals that release toxic fumes when heated.

American Welding Society (AWS) F4.1, Safe Practices for the Preparation of Containers and Piping for Welding and Cutting and *ANSI/AWS Z49.1* describe safe practices for these situations. Begin by cleaning the container to remove any residue. Steam cleaning, washing with detergent, or flushing with water are possible methods. Sometimes you must combine these to get good results.

WARNING!

Clean containers only in well-ventilated areas. Vapors can accumulate in a confined space during cleaning, causing explosions or toxic substance exposure.

NOTE

A *lower explosive limit (LEL)* gas detector can check for flammable or explosive gases in a container or its surroundings. Gas monitoring equipment checks for specific substances, so be sure to use a suitable instrument. Test equipment must be checked and calibrated regularly. Always follow the manufacturer's guidelines.

After cleaning the container, you must formally confirm that it's safe for welding or cutting. *American Welding Society (AWS) F4.1, Safe Practices for the Preparation of Containers and Piping for Welding and Cutting* outlines the proper procedure. The following three paragraphs summarize the process.

Immediately before work begins, a qualified person must check the container with an appropriate test instrument and document that it's safe for welding or cutting. Tests should check for relevant hazards (flammability, toxicity, etc.). During work, repeated tests must confirm that the container and its surroundings remain safe.

Alternatively, fill the container with an inert material ("inerting") to drive out any hazardous vapors. Water, sand, or an inert gas like argon meets this requirement. Water must fill the container to within 3" (~7.5 cm) or less of the work location. The container must also have a vent above the water so air can escape as it expands (*Figure 6*). Sand must completely fill the container.

When using an inert gas, a qualified person must supervise, confirming that the correct amount of gas keeps the container safe throughout the work. Using an inert gas also requires additional safety procedures to avoid accidental suffocation.

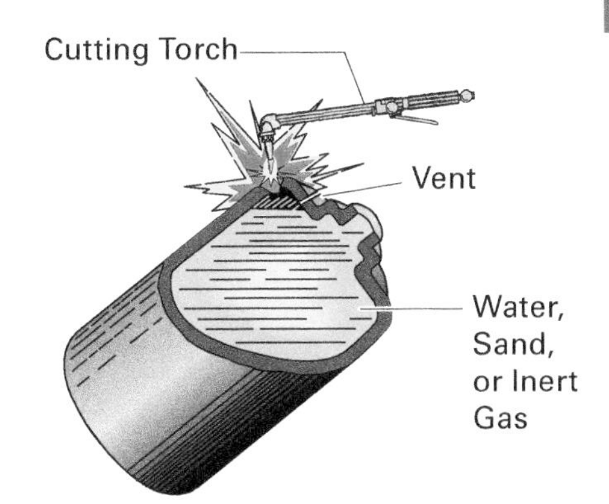

Note: *ANSI Z49.1* and AWS standards should be followed.

Figure 6 "Inerting" a container.

WARNING!

Never weld or cut drums, barrels, tanks, vessels, or other containers until they have been emptied and cleaned thoroughly. Residues, such as detergents, solvents, greases, tars, or corrosive/caustic materials, can produce flammable, toxic, or explosive vapors when heated. Never assume that a container is clean and safe until a qualified person has checked it with a suitable test instrument. Never weld or cut in places with explosive vapors, dust, or combustible products in the air.

WARNING!

Using PAC to cut aluminum on a water table or underwater introduces extra hazards. The metal dust chemically reacts with the water to generate explosive hydrogen gas. This gas can build up under the workpiece or in the environment. Before cutting aluminum around water, you must follow appropriate safety protocols to manage the hydrogen gas and prevent an explosion. *Consult AWS C5.2, Recommended Practices for Plasma Arc Cutting and Gouging*, for more information.

The following general fire- and explosion-prevention guidelines apply to PAC work:

- Never carry matches or gas-filled lighters in your pockets. Sparks can ignite the matches or cause the lighter to explode.
- Always comply with all site and employer requirements for hot work permits and fire watches.
- Never use oxygen in place of compressed air to blow off the workpiece or your clothing. Never release large amounts of oxygen. Oxygen makes fires start readily and burn intensely. Oxygen trapped in fabric makes it ignite easily if a spark falls on it. Keep oxygen away from oil, grease, and other petroleum products.
- Remove all flammable materials from the work area or cover them with a fire-resistant blanket.
- Before welding, heating, or cutting, confirm that an appropriate fire extinguisher is available. It must have a valid inspection tag and be in good condition. All workers in the area must know how to use it.
- Prevent fires by maintaining a neat and clean work area. Confirm that metal scrap and slag are cold before disposing of them.

1.2.3 Work Area Ventilation

Vapors and fumes fill the air around their sources. Welding and cutting fumes can be hazardous. Welders often work above the area from which fumes come. Good ventilation and source extraction equipment help remove the vapors and protect the welder. The following lists good ventilation practices to use when welding and cutting:

- Always weld or cut in a well-ventilated area. Operations involving hazardous materials, such as metal coatings and toxic metals, produce dangerous fumes. You must wear a suitable respirator when working with them. Follow your workplace's safety protocols and confirm that your respirator is in good condition.

WARNING!

Cadmium, mercury, lead, zinc, chromium, and beryllium produce toxic fumes when heated. When cutting or welding these materials, wear appropriate respiratory PPE. Supplied-air respirators (SARs) are best for long-term work.

- Never weld or cut in a confined space without a confined space permit. Prepare for working in the confined space by following all safety protocols.
- Set up a confined space ventilation system before beginning work.
- Never ventilate with oxygen or breathe pure oxygen while working. Both actions are very dangerous.

WARNING!

When working around other craftworkers, always be aware of what they're doing. Take precautions to keep your work from endangering them.

1.2.4 PAC-Specific Safety

Transferred arc PAC passes an electric current through the workpiece. If it arcs or flows through certain components, it could cause damage. Carefully positioning the workpiece clamp can prevent this problem. Sometimes, you should remove sensitive components before beginning work.

Never allow the cutting current to pass through bearings, seals, valves, or lubricated parts. When cutting assembled equipment, position the workpiece clamp to avoid these components. Check for possible gaps where the current could arc. Reposition the clamp if necessary.

The cutting current can easily damage electrical and electronic components. If necessary, remove them or have an electrician isolate them. If the equipment contains a battery, disconnect the lead connected to the equipment frame (the *ground lead*). Cutting current can cause a battery to explode or catch fire. Never cut near a battery. Whenever possible, remove it entirely.

WARNING!

Cutting current entering a battery can have serious consequences. Lead acid batteries can explode and shower the work area with corrosive sulfuric acid. Lithium batteries can catch fire and burn vigorously or even explode.

1.0.0 Section Review

1. The plasma arc cutter type used *most* frequently is the _____.
   a. electrical welding cutter
   b. non-transferred arc cutter
   c. oxyfuel pilot cutter
   d. transferred arc cutter

2. When operating plasma arc cutting equipment, wear a hat made of pure _____.
   a. polyester
   b. nylon
   c. cotton
   d. rayon

2.0.0 Plasma Arc Cutting Equipment

Objective

Summarize plasma arc cutting equipment.

a. Describe plasma arc cutting power units.
b. Describe plasma arc cutting gases and gas control devices.
c. Describe plasma arc cutting torches and accessories.

Performance Tasks

There are no Performance Tasks in this section.

PAC equipment styles and features vary by manufacturer. The following sections outline typical equipment. Before using an unfamiliar machine, however, always consult the manufacturer's documentation. Using a PAC machine incorrectly can injure you, damage the workpiece, or damage the machine itself.

2.1.0 PAC Power Supplies

Plasma arc cutters generate the cutting arc with direct current (DC). Their power supplies convert normal alternating current (AC) to the required DC voltage. In most cases, the power supply's negative lead (–) connects to the torch electrode. The positive lead (+) connects to the workpiece. This system is called *direct-current electrode-negative* (DCEN).

2.1.1 Power Supply Sizes

Power supplies come in many sizes and configurations. Their cutting ability depends on how much electric current they can deliver to the torch (the **amperage**). Small units can cut sheet metal and thin plate. These plug into **single-phase** 120VAC or 240VAC outlets. Most have a **duty cycle** of around 35 percent.

Some commercial-grade power supplies operate on single-phase power, but most require **three-phase** power. These can cut materials up to 1¾" (~4.5 cm). Some will sever metal up to 2¼" (~6 cm) thick and pierce plates as thick as 1" (~2.5 cm). They usually have duty cycles of 50 percent to 60 percent. When cutting thick material, however, they may have to run with a reduced duty cycle.

Large industrial power supplies can operate at 100 percent duty cycle (continuous operation). They require 480VAC three-phase power and draw very large currents. Some can cut metal up to 3" (~8 cm) thick. Many are water-cooled, requiring dedicated plumbing.

Portable units and smaller commercial units have simple control panels with the following controls and indicators:

- Power On/Off switch
- Power On / Unit Ready indicator
- Gas On/Off switch and flow adjustment
- Output current (amperage) adjustment
- Trouble/Fault indicator(s)

NOTE

Duty cycle refers to the percentage of time that a PAC unit can cut in a 10-minute period. It must cool off the rest of the time, or it will overheat. For example, if a unit has a duty cycle of 40 percent, it can cut for a total of 4 minutes out of 10. It must cool off for 6 minutes out of 10. Cutting and cooling periods can take place all at once or be distributed throughout the 10-minute period.

Amperage: The size of an electric current, measured in amperes or *amps* (A).

Single-phase: An alternating current service that delivers a single sine wave to the powered device.

Duty cycle: The percentage of time a welding or cutting machine can operate without overheating within a 10-minute period.

Three-phase: An alternating current service that delivers three overlapping sine waves to the powered device.

Figure 7 shows a commercial-grade PAC unit's control panel.

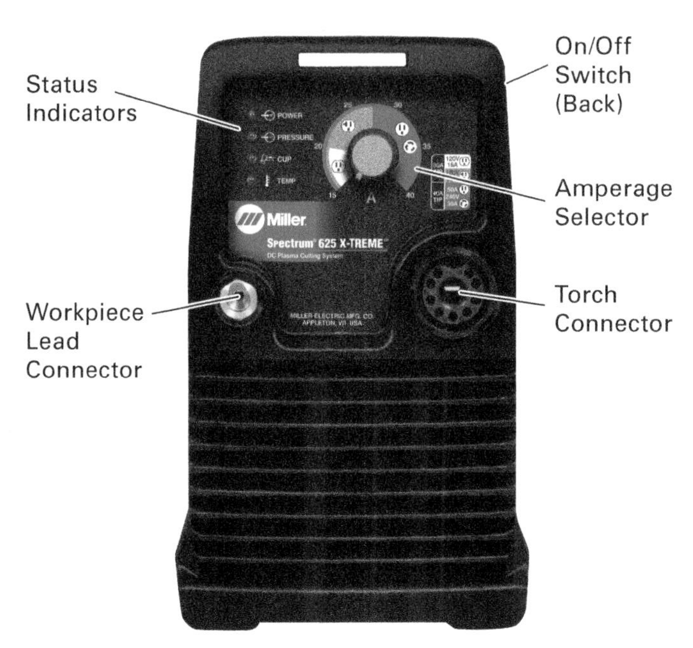

Figure 7 PAC power supply control panel.
Source: © Miller Electric Mfg. LLC

Industrial units have more complex control panels. They have the same controls as simpler units but may also include meters for monitoring voltage and current. Some have sophisticated gas selection and flow management controls too. Power supplies in CNC cutting systems communicate with the machine's industrial control computer.

PAC Power Supply Ratings

Manufacturers rate their product's cutting capacity in various ways. Some use amperage or duty cycle. Others list the maximum severance cut that the machine can produce or the maximum thickness that it can pierce. Be aware, however, that some machines can't cut at their maximum thickness without severely reducing the duty cycle. When purchasing or replacing equipment, read the manufacturer's specifications very carefully. Be sure that you understand their implications and limitations.

Auto-Voltage PAC Machines

Newer PAC power supplies monitor their input voltage and frequency, adjusting themselves as required to operate correctly. This feature lets the PAC power supply work in countries with different electrical power standards. It also helps the power supply operate more reliably and stably by adjusting it to accommodate changes in the local electrical power.

2.1.2 Workpiece Clamp

The workpiece clamp (*Figure 8*) electrically links the workpiece to the power supply. The contact area must be clean, bare metal for good current flow. Clamps come in many sizes and styles for different applications. They're rated by the amount of current they can safely carry. Always replace a clamp with one that's rated at or above the PAC power supply's output amperage.

2.2.0 PAC Gas Delivery Systems

All PAC machines require one or more gases to run. One is the cutting gas, but some units require other gases for shielding or cooling. A shielding gas protects the cut metal from corrosion and moisture. A cooling gas flows through the torch to prevent it from overheating. The gas delivery system supplies each gas at the correct pressure and flow rate. Simple PAC machines have very basic gas delivery systems that just turn the gas on and off. More complex systems can select between gases and adjust their flow rates and pressures.

2.2.1 PAC Gases

The cutting gas determines which metals a PAC unit can cut. They also affect the cut's quality. Common gases include air, nitrogen, oxygen, argon, hydrogen, and carbon dioxide. Some are free (air) or inexpensive (nitrogen), while others are more costly (argon).

WARNING!

When using any gas other than air for PAC work, be especially careful. Hydrogen is flammable and explosive. Oxygen supports and encourages combustion. Nitrogen, argon, and carbon dioxide can displace oxygen-containing air from their surroundings. Follow all safety protocols for working with gases.

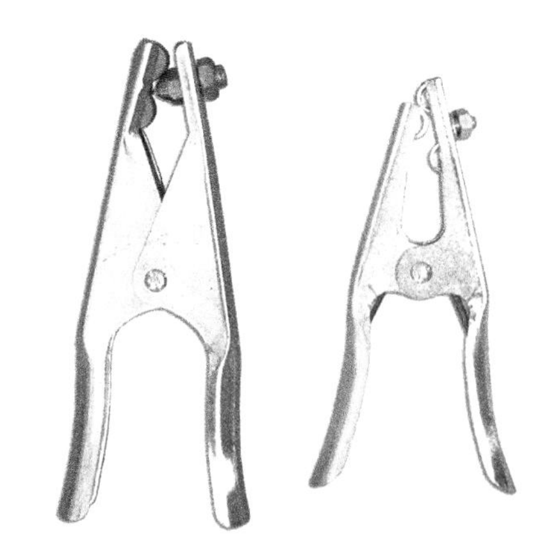

Figure 8 Workpiece clamps.

Simple PAC units use clean compressed air to cut carbon steel, stainless steel, and aluminum up to about 3⁄16" (~5 mm) thick. Heavy-duty units that use air as the cutting gas can cut carbon steel and stainless steel up to about 1¾" (~4.5 cm) thick. They can usually sever somewhat thicker materials.

More sophisticated PAC units can switch from air to other gases supplied from cylinders. Dual-flow units use one gas for cutting and another for cooling or shielding. *Table 1* summarizes combinations of cutting and shielding gases with several common metals. Always confirm that a PAC unit works with a particular gas before using it. Supplying an inappropriate gas could damage the machine.

TABLE 1 Recommended Cutting/Shielding Gas Combinations

Material	Air/Air	O_2/Air	N_2/CO_2	N_2/Air	H35/N_2
Carbon steel	Most economical Good cut quality Good speed Good gouging Good weldability	Best cut quality Maximum cut speed Best weldability	Some dross Long electrode life	Not recommended	Best gouging Long electrode life Some dross
Stainless steel	Most economical Good speed Some dross	Not recommended	Good cut quality Good gouging Minimal dross Long electrode life	Long electrode life Lowest shield gas cost	Best cut quality Best gouging Minimal dross Long electrode life Cuts thicker material
Aluminum	Most economical Good speed Some dross	Not recommended	Good cut quality Good gouging Minimal dross Long electrode life	Not recommended	Best cut quality Best gouging Minimal dross Long electrode life Cuts thicker material

O_2 = Oxygen
N_2 = Nitrogen
CO_2 = Carbon Dioxide
H35 = Mixture of 35% Hydrogen and 65% Oxygen

2.2.2 PAC Gas Controls

Many PAC machines use air for cutting and cooling. Small units often contain their own compressor to supply it (*Figure 9*). Larger units, particularly those in factories, use the facility's compressed air supply or an external compressor. Regardless of the source, compressed air must be very clean and dry, as well as oil-free. Placing a filter, separator, and dryer in front of the PAC unit's air inlet accomplishes this. Good-quality air gives better cuts. It also extends the PAC machine's lifespan, as well as that of its consumables.

Compressed air must enter the PAC unit at the correct pressure. If it's too high, it can damage the equipment. Many compressed air distribution systems supply air at too great a pressure. A *regulator* in front of the PAC machine's

Figure 9 Light-duty PAC unit with built-in compressor.
Source: Photo courtesy of Hypertherm Inc.

inlet reduces the pressure to the required value (*Figure 10*). For clean cuts, the machine must also receive sufficient air volume (*flow rate*). Manufacturers specify the required flow rate in cubic feet per hour (cfh) or standard cubic feet per hour (scfh).

PAC machines that use gases other than air receive them through hoses connected to gas cylinders. These hold the gases at pressures between 1,500 psig and 2,000 psig (~10,300 kPaG and 13,800 kPaG). A regulator on the cylinder reduces the pressure to whatever the machine requires. Dual-flow machines have separate cylinders, regulators, and hoses for each gas (*Figure 11*).

Figure 10 Pressure regulator.
Source: © Miller Electric Mfg. LLC

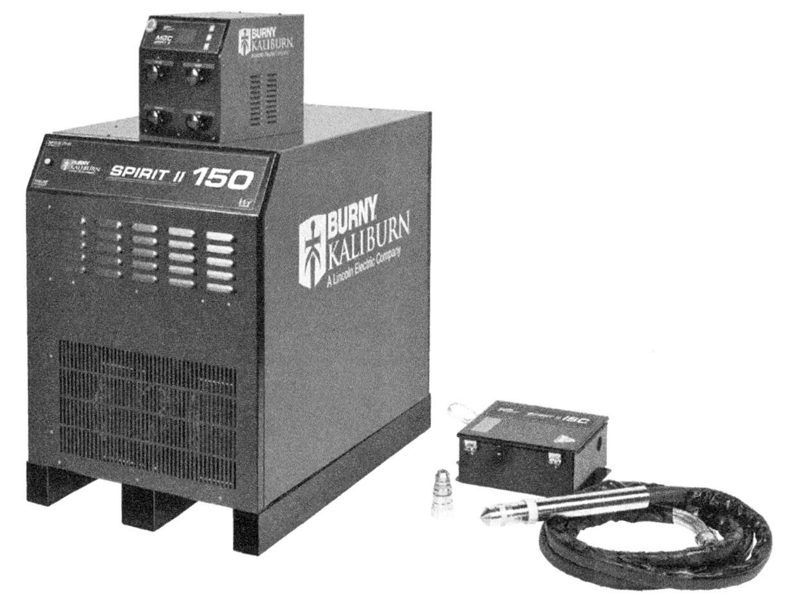

Figure 11 Dual-flow PAC machine.
Source: The Lincoln Electric Company, Cleveland, OH, USA

Regulator Selection

To select a PAC machine regulator, consult the manufacturer's documentation to determine the correct operating pressure. Also be sure that the regulator is designed for the gas that you plan to use. Regulators work most reliably when the target pressure is in the middle of their operating range. For example, if the target pressure is 50 psig, a regulator with a range of 0 psig to 100 psig is ideal.

Interlocks

Running a PAC unit without a good cutting gas supply can damage the torch. For this reason, many systems include *interlocks* that won't allow the machine to operate unless everything is working properly. A gas supply interlock monitors the cutting gas pressure and won't let you start the torch if it's too low. The interlock will also shut down the torch if the cutting gas pressure drops too low during cutting.

The power supply contains a temperature interlock that won't allow it to operate if it's too hot. Sophisticated PAC machines contain other interlocks that protect them or add safety features protecting the operator.

Valves on the cylinders and inside the PAC unit manage gas flow. An electrically controlled **solenoid valve** controls gas flow to the torch itself. A pushbutton on the torch signals the solenoid to open or close as the worker requires. Each gas has its own valves and solenoids.

Solenoid valve: An electrically controlled valve that manages gas or liquid flow.

Going Green

Fumes and Gases

In the past, most companies just released waste gases and fumes from welding and cutting operations directly into the atmosphere. While convenient, this practice was polluting and wasteful. Today, many companies capture fumes and waste gases rather than releasing them into the surrounding air. Some systems even recycle the more expensive gases so they can be reused.

2.3.0 PAC Torches

Transferred arc torches are the style commonly found in workplaces that cut metals. They start up by generating plasma with a pilot arc between the nozzle and electrode. When the worker moves the nozzle close to the workpiece, the plasma jet completes an electrical circuit between it and the torch electrode. This "transfers" the arc to the workpiece (*Figure 12*). Control circuits in the torch shut off the pilot arc and increase the voltage. This heats the plasma to the cutting temperature.

PAC torches come in many styles. Some are handheld, while others attach to mechanized cutting machines. Handheld torches range from those used for light-duty cutting to those meant for heavier work. Some include separate orifices for shielding or cooling gases. The largest torches cool themselves with circulating water. A button or lever on the torch starts the pilot arc. Some PAC machines also have foot pedal controls.

Mechanized PAC torches fit into automated carriers. Most will work with carriers designed for oxyfuel cutting. The carrier must move faster, however, since plasma arc cutting melts the metal more rapidly. Many mechanized torches are water-cooled. Some use *water injection* to make cleaner cuts with almost no dross. Water swirls around the plasma jet, compressing the gas into a tighter stream.

Several torch components erode each time you start the arc. You must replace these consumable parts regularly. *Figure 13* shows a disassembled torch. The retaining cup, nozzle, and electrode require replacement most frequently. Always keep spare parts handy.

Most torches accept a variety of shrouds, heat shields, nozzles, and electrodes. These adapt the torch to different metals, cut styles, and thicknesses. Craftworkers often swap nozzles to change the plasma jet width, which affects the kerf.

Figure 12 A transferred plasma cutting arc.
Source: ESAB Welding & Cutting Products

NOTE

It's a smart idea to replace the nozzle and electrode at the same time.

Reducing Interference

Many PAC torches start the pilot arc with a DC current that pulses on and off rapidly (high frequency start). This method generates a lot of electromagnetic noise that can interfere with nearby electronics. Newer PAC torches start the pilot arc with other methods that generate less interference.

Capacitor-start systems charge up a capacitor and discharge it between the electrode and nozzle to start the arc. Another method shorts the electrode against the nozzle. When the cutting gas starts flowing, it pushes the electrode away from the nozzle, which creates the arc.

Going Green

Cooling Water

Welders often allow cooling water to flow into the ground or down the drain. To better protect the environment, collect the cooling water for reuse or recycling.

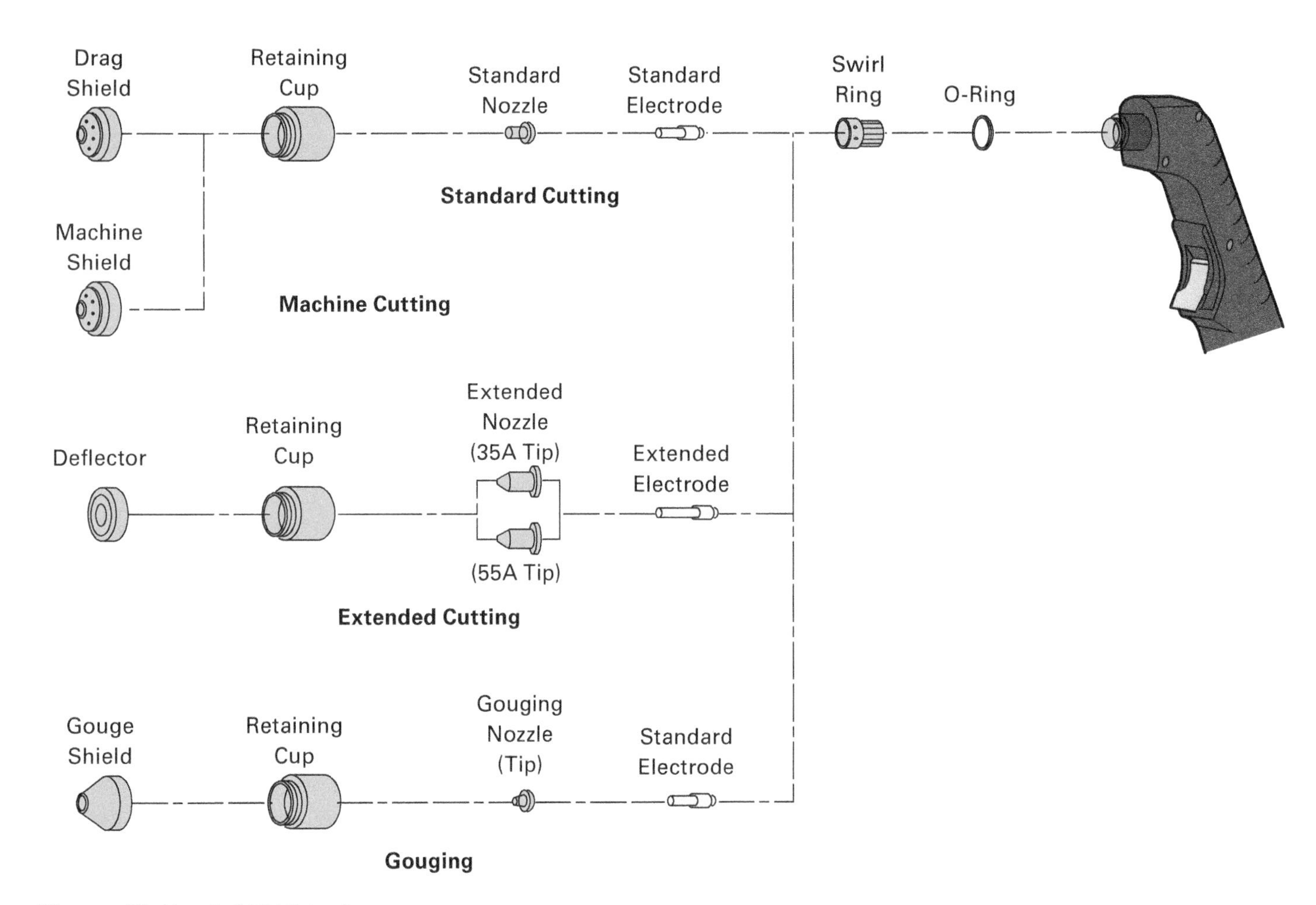

Figure 13 Handheld PAC torches.
Source: The Lincoln Electric Company, Cleveland, OH, USA

When guiding the torch across the workpiece, the craftworker must keep the tip the right distance from the surface. For most work, a *standoff distance* between 1⁄16" and 1⁄4" (~2 mm and 6 mm) works well. Using a *standoff guide* makes precise cutting easier since it supports the torch at the correct height. For low-amperage work, **drag cutting** is simpler. The worker fits the torch with a *drag tip* or *drag shield* and rests the torch directly on the workpiece. *Figure 14* shows accessories that support each method.

Drag cutting: A cutting method in which the torch tip or shield rests on the workpiece.

(A) Drag Shield

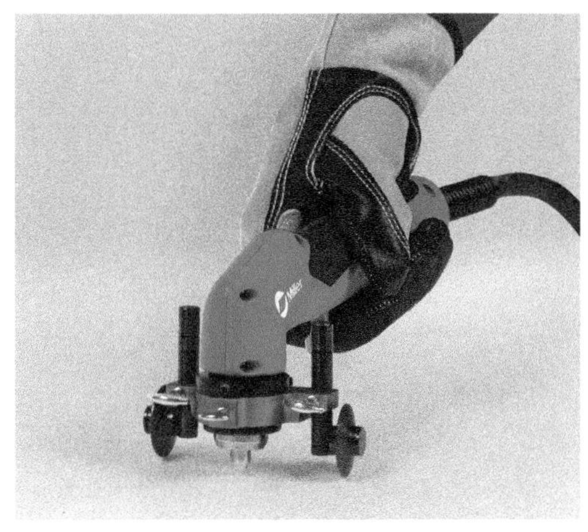

(B) Standoff Roller Guide

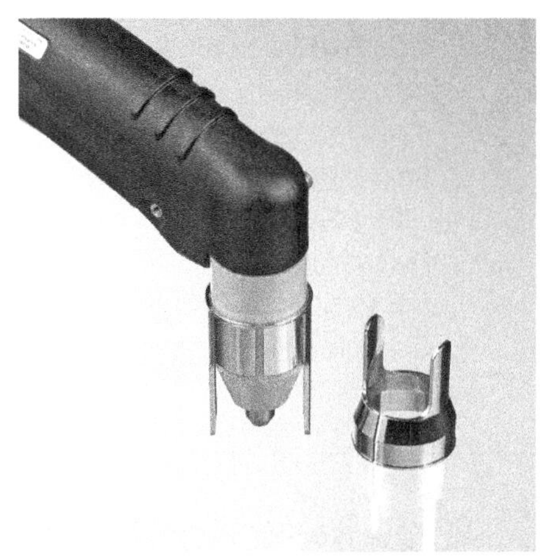

(C) Standoff Post Guide

Figure 14 Tools to manage standoff distance.
Sources: © Miller Electric Mfg. LLC (14A–14B); Victor Technologies (14C)

Plasma Torch Cutting Guides

Many manufacturers furnish accessories that make it easier to cut certain shapes. A circle-cutting guide kit also works as a freehand standoff roller guide.

Source: The Eastwood Company

2.0.0 Section Review

1. A light-duty PAC power supply has a duty cycle of around _____.
   a. 35 percent
   b. 50 percent
   c. 75 percent
   d. 90 percent

2. The *most* common cutting and cooling gas for simple PAC machines is _____.
   a. nitrogen
   b. argon
   c. carbon dioxide
   d. air

3. Torch parts that erode every time the arc starts are called _____.
   a. consumables
   b. standoff guides
   c. high frequency
   d. sacrificial

3.0.0 Using Plasma Arc Cutting Equipment

Performance Tasks

1. Set up plasma arc cutting equipment.
2. Select amperage, gas pressure, and flow rate appropriate to the metal type and thickness.
3. Square-cut metal with plasma arc cutting equipment.
4. Bevel-cut metal with plasma arc cutting equipment.
5. Pierce and cut slots in metal with plasma arc cutting equipment.
6. Dismantle and store plasma arc cutting equipment.

Objective

Outline setting up, safely operating, and maintaining plasma arc cutting equipment.

a. Explain how to set up a plasma arc cutting project.
b. Explain how to operate plasma arc cutting equipment safely.
c. Explain how to maintain plasma arc cutting equipment.

Plasma arc cutters are wonderful tools. They're fast, clean, and efficient. But they're also hazardous and produce poor cuts when used carelessly. The following sections explore setup, cutting, and equipment care. As always, safety remains a high priority when cutting with a PAC machine.

3.1.0 Setting Up

PAC equipment usually comes as a complete system, so setup is relatively simple. Always confirm the following factors as you prepare for a cutting job:

- The power supply can handle the material and its thickness.
- The power supply has a high enough duty cycle to complete the job in a reasonable time.
- The local electrical supply is appropriate for the PAC power supply (phase, voltage, and current).
- The torch nozzle is correct for the material and cut style.
- The required gas or gases are available.
- Spare consumables are available.

3.1.1 Preparing the Work Area

Locate the PAC unit close enough to the workpiece that you can maneuver easily. Confirm that hoses and cables reach without creating hazards. When possible, place the workpiece at a comfortable height.

Plasma arc cutting can spray molten metal up to 40' (~12 m) away. If you can't remove combustible materials within this radius, cover them with fireproof blankets or place shields around the work area. If others are working in the area, place welding curtains around the work area to control glare and UV. Follow all company protocols for hot work permits and fire watches.

Plasma arc cutting generates smoke and fumes. Remember, some metals and many coatings produce toxic fumes when heated. Check the work area's ventilation. Bring in a source extraction unit if necessary. You may also need respiratory PPE. Follow all company safety protocols.

WARNING!

Stainless steel contains chromium, which can turn into toxic Cr(VI), also called *hexavalent chromium*, when heated. Always wear appropriate respiratory PPE when cutting stainless steel with a PAC machine.

Assemble all appropriate clothing and PPE. Get the right eye protection for PAC work. Wear ear protection that keeps sparks out of the ear canals. PAC is noisy, so be sure the PPE protects your hearing too.

3.1.2 Selecting the Cutting Amperage

The material, its thickness, the desired cutting speed, and the cutting gas all affect the correct cutting amperage. Light-gauge sheet steel may require as little as 7A, while thick plate might need 250A (*Figure 15*). Follow the manufacturer's recommendations since each PAC unit is different. Choosing the wrong amperage will produce a poor cut or even damage the torch.

(A) Thin Material

(B) Thick Material

Figure 15 Material type and thickness affect the cutting amperage.
Sources: ESAB Welding & Cutting Products (15A); Victor Technologies (15B)

3.1.3 Selecting Gases and Connecting the Gas Supply

Gas choice depends on the material and its thickness. Dual-flow PAC machines may use different combinations, depending on the application. Always follow the manufacturer's guidelines. Using the wrong gases will produce a poor cut. Never use a gas unless the manufacturer specifically approves it for the PAC unit.

Some gases, like oxygen and hydrogen, are dangerous because they support combustion or are flammable. Others, like nitrogen and argon, aren't toxic, but they can displace oxygen-containing air from the work area. Cylinders contain gases at high pressures, making them hazardous. Handle all gases and their supporting equipment carefully. Follow all safety protocols.

If you're using air as the only gas, setup is simple. A small PAC machine that contains its own compressor is ready to go. If the machine requires an external air source, connect it with the following steps:

Step 1 If you're using the facility's compressed air supply, confirm that the air passes through a filter, separator, dryer, and regulator before it enters the PAC unit. A standalone compressor should have similar conditioning equipment.

Step 2 Connect the PAC machine's hose to the regulator outlet.

Step 3 Start the compressor (if applicable). Adjust the regulator so the output pressure is correct for the machine. If you're uncertain about the correct pressure, consult the manufacturer's documentation.

Gas Flow Rates

PAC gas flow rates vary with equipment design, gas type, and metal thickness. Thin materials like 0.1" (~2.5 mm) carbon steel, stainless steel, or aluminum might require 15 cfh of nitrogen. Thick materials like 3" (~7.5 cm) carbon steel, on the other hand, might require 62 cfh of argon. Always follow the manufacturer's recommendations for the best cut.

Step 4 Open the valve between the air supply and the PAC unit.

When using gas cylinders with a PAC machine, connect them with the following steps:

Step 1 Confirm that the gas is correct for the application. Secure the cylinder near the PAC machine in an upright position.

Step 2 Remove the valve's safety cap.

WARNING!

When using oxygen, keep cloth away from the fittings. Oil or grease can explode if mixed with oxygen. For this reason, never bring dirty waste rags near oxygen.

Step 3 With eye protection in place, stand to one side of the cylinder valve outlet. Crack open the valve to blow out any dirt or debris. Close the valve immediately.

WARNING!

To avoid injury, never stand in front of the cylinder or regulator outlet when opening the valve. Confirm that nearby workers aren't standing in front of it either.

CAUTION

Never tighten the regulator's connections with pliers or a pipe wrench. Use an adjustable wrench, an open-end wrench, or a torch wrench. Connections are soft brass or bronze. Tighten them carefully, but never overtighten.

Step 4 Connect the regulator to the cylinder valve and tighten with the proper wrench. Don't overtighten. Connect the PAC unit's hose to the regulator outlet.

Step 5 Turn the regulator's adjustment knob or screw fully counterclockwise, so the regulator is closed. Open the cylinder valve very slowly at first, and then open fully. Wait for the inlet pressure gauge to rise and stop.

Step 6 Adjust the regulator so the outlet pressure gauge shows the correct value for the PAC machine. If you're uncertain about the correct pressure, consult the manufacturer's documentation.

Step 7 Open the valve between the gas supply and the PAC unit.

3.2.0 Cutting

PAC equipment isn't hard to operate. Getting clean cuts, however, requires some skill and attention to detail. The manufacturer provides specific guidelines for different materials and thicknesses. These include setting the amperage, as well as adjusting the gas pressure and flow rate. Consult the documentation for the proper torch standoff distance and cutting speed too.

The following steps outline the general PAC process:

CAUTION

If cutting on machinery or other assembled equipment, position the workpiece clamp so the cutting current will not pass through seals, bearings, or components that arcing or heat could damage. Have an electrician isolate any nearby electrical or electronic components from the PAC circuit.

Step 1 Prepare the work area. Assemble the necessary protective clothing and PPE.

Step 2 Locate the electrical supply's primary disconnect so you can use it in an emergency.

Step 3 Confirm that the PAC power supply can handle the cutting task.

Step 4 Start the compressor or open the cylinder valves, as applicable. Confirm adequate pressure and adjust the regulator(s) if necessary.

Step 5 Connect the workpiece clamp to the workpiece. It must attach firmly over bright metal. If necessary, clean the metal with emery paper first.

Step 6 Plug the PAC power unit into the electrical supply (if required). Energize the PAC power supply and switch the unit on.

Step 7 Follow the manufacturer's guidelines to set the amperage, gas pressure, and gas flow to values appropriate to the material and thickness.

WARNING!

Plasma arc cutting generates very high temperatures. Anything nearby that's flammable or explosive could ignite. Confirm that you've followed all safety protocols and have a fire extinguisher within reach.

Step 8 Put on your PPE.

Step 9 Turn on the gas.

Step 10 Hold the torch the recommended standoff distance above the cut point. Lower your mask or flip the tinted lens into position. Press the torch start button or pedal.

Step 11 As soon as the cutting arc starts, move the torch along the cut path at the recommended speed. Keep the nozzle perpendicular (90 degrees) to the surface.

Step 12 When you've completed the cut, release the button or pedal.

NOTE

Holding the torch perpendicular to the work surface is called a *0-degree work angle.*

3.2.1 Cutting Square Edges

To cut square edges with PAC, hold the torch at the 0-degree work angle and move it smoothly with no side-to-side motion. Clamp a metal straightedge or scrap angle iron to the workpiece and use it to guide the torch.

When cutting thick workpieces, the cut will taper slightly as it narrows with depth. This produces cuts whose faces aren't quite square. To compensate for this taper, lean the torch slightly towards the scrap side. The cut will be square on one side and tapered on the scrap side (*Figure 16*).

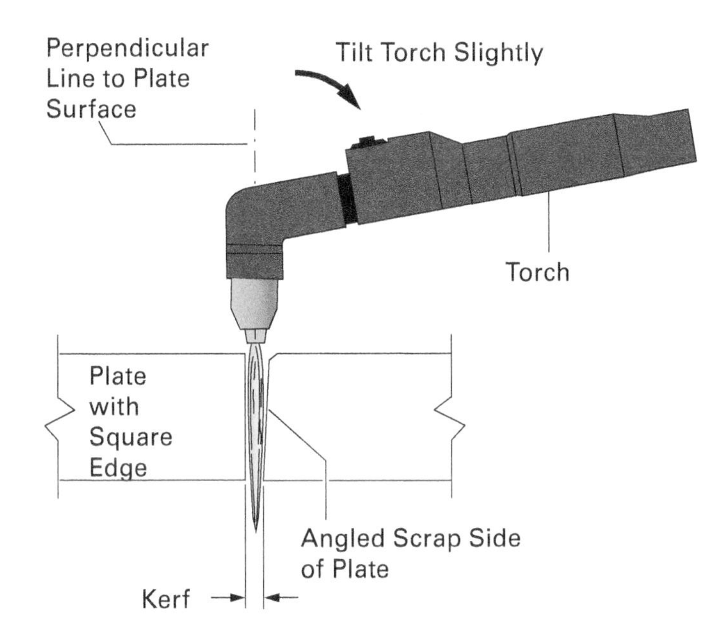

Figure 16 Compensating for taper.

Starting the Torch on a Thick Workpiece

When cutting very thick material by hand, use the same technique that you use with oxyfuel cutting. Start at the edge and tilt the torch slightly so the plasma jet begins cutting at the top edge only. Then, tilt the torch to the 0-degree work angle and start moving along the cutting path.

3.2.2 Cutting Beveled Edges

To produce a bevel (angle) cut, use the same technique. Lean the torch at the correct angle to produce the bevel. Place an angle iron scrap parallel to the cut line. Shim it at the correct angle and clamp it in place. Use it to guide the torch (*Figure 17*).

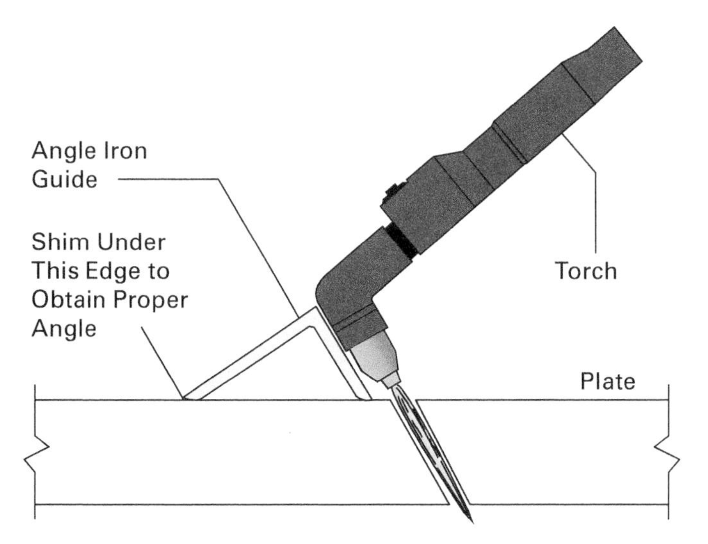

Figure 17 Cutting a bevel.

3.2.3 Piercing and Cutting Slots

Plasma arc cutting can pierce metal and cut slots or other shapes. Wear full body protection, along with head, eye, and ear protection since this task generates metal spatter.

To pierce very thin metal, hold the torch directly over the workpiece at a 0-degree work angle, and press the torch start button. As soon as the jet passes completely through the workpiece, move the torch head in a pattern to produce the desired hole shape.

When piercing thicker materials, rotate the torch to a 10-degree work angle. This will prevent metal from blowing back into the torch and damaging the nozzle. When the jet passes through the workpiece, rotate the torch back to the 0-degree work angle and complete the cut.

NOTE

To produce complex patterns, guide the torch with a template.

3.3.0 Caring for PAC Equipment

Plasma arc cutting equipment is relatively simple and easy to maintain. Keeping it in good condition requires basic maintenance, replacing consumables, and properly storing it when you're finished cutting.

3.3.1 Storing PAC Equipment

When you're done with the machine, perform the following steps to store it:

Step 1 Switch off the PAC power supply.

Step 2 Switch off the compressor or close all gas supply valves, as applicable. Bleed pressure from the lines and disconnect them from the PAC unit.

Step 3 If using gas cylinders, close the cylinder valves, remove the regulators, and replace the cylinder safety caps (*Figure 18*).

WARNING!

Keep safety caps with their cylinders. Always replace the safety cap over the tank valve immediately after removing the regulator assembly. This protects the valve if the cylinder should fall over. A broken valve could turn the cylinder into a small rocket, possibly causing injury or death. Never release a tank from its support or move it without the safety cap in place.

Step 4 Unplug and coil the power cord, as well as all gas hoses.

Step 5 Coil the torch cable and the workpiece lead.

Step 6 Return the equipment to storage.

Step 7 Clean up the work area. Sweep up any slag or debris. Handle hot debris carefully to avoid burns or starting a fire.

Figure 19 shows a properly stored PAC machine.

Figure 18 Cylinder safety cap.

3.3.2 Basic Maintenance

Maintaining PAC equipment is simple. Mostly, you need to check the torch, cables, and hoses for damage. You should also replace consumables. The following procedure outlines inspection and maintenance:

- Check the shield for signs of wear. It should be clean and free from metal debris.
- Unscrew the shield and check the inside surface for wear. If the holes are blocked, try to clean them with a torch tip cleaner. If you can't clean the shield, replace it.
- Inspect the retaining cup for damage and replace if necessary.
- Examine the nozzle for wear or damage. If the orifice is worn or oval-shaped, replace the nozzle.
- Remove the electrode and check the tip for pitting. If pits are more than $\frac{1}{16}$" (~2 mm) deep, replace the electrode.
- Inspect the swirl ring. It should be clean and its holes free from clogs. If the swirl ring is damaged, replace it.
- Check the O-ring. Replace it if it's distorted or degraded. Lubricate the O-ring with an appropriate lubricant before installing it.

Figure 19 Properly stored PAC unit.
Source: The Lincoln Electric Company, Cleveland, OH, USA

Figure 20 Inline air filter.
Source: Motor Guard Corporation

- Check the torch cable for burns, abrasions, or other damage. Examine the connector for damage, deposits, or corrosion. If you can't correct any damage that you discover, replace the cable or the entire torch.
- Check the workpiece cable for burns, abrasions, or missing insulation. If you can't correct any damage that you discover, replace the cable.
- Check the gas supply hoses for burns or abrasions. Confirm that their fittings aren't corroded or damaged. If you can't correct any damage that you discover, replace the hose.

Many PAC units have an inline air filter that cleans the air before it enters the machine (*Figure 20*). Check the filter element regularly and replace it if it's dirty. Clogged filter elements reduce airflow.

If a compressed air source's separator and dryer aren't doing their job, moisture can get into the equipment. If this happens, the torch will hiss or sputter. Check the PAC unit or supply's air filter and separator bowl. Drain the bowl and replace the filter element. Check the dryer too. If it has a replaceable drying agent (a *desiccant*), it may be saturated with moisture. Replace it, if necessary.

Many PAC units clean their cooling air with a filter. Check it regularly and replace as necessary. A clogged air filter restricts airflow and can cause the power supply to overheat. If a power supply keeps shutting down, check the filter and air intake.

3.0.0 Section Review

1. To remove, attach, or tighten a pressure regulator on a compressor or gas cylinder, you should use _____.
   a. no tools since these connections are to be hand tightened only
   b. a correctly sized wrench with an added extension for maximum leverage
   c. a properly sized pipe wrench adjusted to the regulator fitting's size
   d. an adjustable wrench, an open-end wrench, or a torch wrench

2. To pierce thin metal with a PAC torch, you should _____.
   a. hold the torch directly over the point at a 0-degree work angle
   b. start from the bottom, piercing the metal from below
   c. hold the torch directly over the point at a 10-degree work angle
   d. hold the torch directly over the point at a 45-degree work angle

3. If the PAC equipment hisses or sputters when cutting, you should _____.
   a. attach a vacuum pump to the air filter / moisture separator assembly
   b. replace the filter element and the air filter / moisture separator assembly
   c. drain the air filter / moisture separator assembly and replace the filter element
   d. flush the air filter / moisture separator assembly with argon for 12 hours

Module 29103 Review Questions

1. Plasma arc cutting generates a jet of plasma that can be as hot as ____.
   a. 10,000°F (~5,500°C)
   b. 20,000°F (~11,000°C)
   c. 25,000°F (~14,000°C)
   d. 30,000°F (~17,000°C)

2. What are the two PAC machine types?
   a. Transferred arc and non-transferred arc
   b. Transferred arc and phased arc
   c. Phased arc and non-transferred arc
   d. Oxyfuel pilot arc and non-transferred arc

3. In the more common of the two PAC machine types, the pilot arc ____.
   a. cuts the material
   b. creates the plasma at startup
   c. establishes contact with the metal
   d. maintains the plasma's temperature

4. With the less common of the two PAC machine types, the cut material is ____.
   a. above the arc circuit
   b. dependent on the arc circuit
   c. electrically connected to the arc circuit
   d. not electrically connected to the arc circuit

5. Which of the following statements about PAC safety is *true*?
   a. Caps should be made of 100 percent polyester or nylon.
   b. You can safely look at the cutting arc without tinted lenses for short periods.
   c. Boots without laces or holes are the best choice for PAC work.
   d. PAC keeps heat focused on the cut, so nearby flammable materials can't catch fire.

6. *Never* allow a PAC current to pass through a ____.
   a. workpiece
   b. cable
   c. casting
   d. bearing

7. Light-duty PAC equipment typically requires a ____.
   a. 120VAC single-phase power service
   b. 240VAC three-phase power service
   c. 277VAC single-phase power service
   d. 480VAC three-phase power service

8. Large industrial PAC systems usually ____.
   a. have a duty cycle between 50 percent and 60 percent
   b. require dedicated plumbing for water cooling
   c. operate on 120VAC single-phase power
   d. can only cut carbon steel and stainless steel

9. The workpiece clamp is rated by the ____.
   a. voltage that it can safely carry
   b. material that it's meant to pierce
   c. current that it can safely carry
   d. gas volume that it can carry

10. Cutting and shielding gases supplied in cylinders usually have a pressure between ____.
    a. 200 psig and 400 psig (~1,400 kPaG and 2,800 kPaG)
    b. 800 psig and 1,000 psig (~5,500 kPaG and 6,900 kPaG)
    c. 1,000 psig and 1,500 psig (~5,500 kPaG and 10,300 kPaG)
    d. 1,500 psig and 2,000 psig (~10,300 kPaG and 13,800 kPaG)

11. Mechanized PAC torches are designed to be ____.
    a. handheld
    b. mounted in a vise
    c. used at a slow speed
    d. mounted on an automated carrier

12. When cutting materials that release toxic fumes, you *must* _____.
    a. use a small fan near the cutting area
    b. wear an appropriate respirator
    c. work near an open window
    d. attach an airline to the torch

13. Cutting sheet steel requires as little as _____.
    a. 7A
    b. 9A
    c. 15A
    d. 20A

14. To cut square edges in *thin* materials, hold the torch at the _____.
    a. 75-degree work angle
    b. 60-degree work angle
    c. 45-degree work angle
    d. 0-degree work angle

15. Replace the torch electrode if it has pits deeper than _____.
    a. 1⁄32" (~1 mm)
    b. 1⁄16" (~2 mm)
    c. 1⁄8" (~3 mm)
    d. 3⁄16" (~5 mm)

Answers to odd-numbered Module Review Questions are found in *Appendix A*.

Cornerstone of Craftsmanship

Curtis Casey
Welding Department Director
Mesa Community College

How did you choose a career in the industry?

I joined the US Navy and received welding training in plate welding, high-pressure pipe welding, and nuclear components welding. For six years, I was stationed on submarine tenders, which performed maintenance on nuclear submarines.

Who inspired you to enter the industry?

My stepfather was a career sailor in the US Navy. His stories piqued my interest, so I joined directly out of high school as a hull maintenance technician for ships.

What types of training have you been through?

In the Navy, I received training in the metal trades to be a hull maintenance technician, learning to weld on common and exotic materials. After the military, I went to college, earning an A.A.S. Degree in Welding Technology while I worked as a welding inspector in the nuclear power industry. I started teaching welding inspection part time while attending college and found that I loved to teach my craft to others.

How important is education and training in construction?

Education is an ongoing process of lifelong learning. Education and training are the tools we fit ourselves with to be able to create and mold our careers. In welding, many people receive lab training and master the skills of welding, but the true welder is one who has taken the time to focus on the theory of the craft. Basic metallurgy, electricity fundamentals, consumable usage, equipment maintenance, welding symbols, etc., are all invaluable subjects to a person whose goal is to advance in the industry.

What kinds of work have you done in your career?

I started my adult working life as a welder, then worked as a welding inspector, a field welding engineer, a welding business owner, a welding instructor, welding department chair/director, welding curriculum developer, and welding consultant. The common denominator is "Welding." I still maintain American Welding Society Certified Welding Inspector (CWI) and Certified Welding Educator (CWE) credentials.

Tell us about your current job.

I am currently employed as the Welding Program Director at Mesa Community College in Mesa, Arizona, while also working as a welding consultant for my company Gillett Consulting LLC.

What has been your favorite project?

As cliché as it sounds, my favorite project is student development. I receive more personal satisfaction and reward knowing that many of my alumni go on to create very successful welding careers for themselves.

What do you enjoy most about your job?

My passion truly is teaching my craft to inquiring minds, watching the light bulbs go on as they understand an abstract concept of electricity or metallurgy or when they are able to "read the weld puddle" to make consistent quality weld beads.

What factors have contributed most to your success?

Work ethics! I learned at an early age to be responsible for my own successes and failures. I pitched newspapers on a paper route from the time I was 11 years old. That taught me to be dependable, consistent, and exposed me to the business side of dealing with customers.

Would you suggest construction as a career to others? Why?

Yes, because there are dozens of welding processes in the industry, a career in welding will never get boring and will give a person satisfaction seeing an idea or planned project come to fruition. Opportunities for change and advancement are limitless if you are willing to work for them.

What advice would you give to those new to the field?

When you are new and working with a mentor, pay close attention to the process and anticipate the next step. If you do your best to make the job easier by prepping material, cleaning areas, or organizing tools, your supervisors will notice your value as a team player.

Interesting career-related fact or accomplishment?

I have had the opportunity to teach over 1,000 people the craft that has given me my career. Many former students return to share their success in the industry, which reinforces my career choice as a teacher.

How do you define craftsmanship?

Developing a good balance between technical and artistic skill sets, then mastering them through practice. When you make a difficult job look easy because of your skills, then you're a craftsperson!

Answers to Section Review Questions

Answer	Section Reference	Objective
Section 1.0.0		
1. d	1.1.0	1a
2. c	1.2.1	1b
Section 2.0.0		
1. a	2.1.1	2a
2. d	2.2.1	2b
3. a	2.3.0	2c
Section 3.0.0		
1. d	3.1.3	3a
2. a	3.2.3	3b
3. c	3.3.2	3c

User Update

Did you find an error? Submit a correction by visiting **https://www.nccer.org/olf** or by scanning the QR code using your mobile device.

Air-Carbon Arc Cutting and Gouging

Source: The Lincoln Electric Company, Cleveland, OH, USA

Objectives

Successful completion of this module prepares you to do the following:

1. Describe air-carbon arc cutting and its safety procedures.
   a. Summarize air-carbon arc cutting.
   b. List air-carbon arc cutting safety procedures.
2. Summarize air-carbon arc cutting equipment.
   a. Describe air-carbon arc power units.
   b. Describe air-carbon arc torches.
   c. Describe air-carbon arc electrodes.
3. Outline setting up, safely operating, and maintaining air-carbon arc equipment.
   a. Explain how to set up an air-carbon arc cutting project.
   b. Explain how to gouge and wash metals with air-carbon arc cutting equipment.
   c. Explain how to maintain air-carbon arc cutting equipment.

Performance Tasks

Under supervision, you should be able to do the following:

1. Set up air-carbon arc cutting equipment.
2. Select and install air-carbon arc cutting electrodes.
3. Gouge metal with air-carbon arc cutting equipment.
4. Wash metal with air-carbon arc cutting equipment.
5. Dismantle and store air-carbon arc cutting equipment.

Overview

Welders cut metals using several different technologies. Air-carbon arc cutting uses an electric current to melt metal and an air jet to blast it away. The result is a clean cut that's free from contamination. This module introduces the method, its equipment, procedures, and safety practices.

NOTE

This module uses US standard and metric units in up to three different ways. This note explains how to interpret them.

Exact Conversions

Exact metric equivalents of US standard units appear in parentheses after the US standard unit. For example: "Measure 18" (45.7 cm) from the end and make a mark."

Approximate Conversions

In some cases, exact metric conversions would be inappropriate or even absurd. In these situations, an approximate metric value appears in parentheses with the ~ symbol in front of the number. For example: "Grip the tool about 3" (~8 cm) from the end."

Parallel but not Equal Values

Certain scenarios include US standard and metric values that are parallel but not equal. In these situations, a slash (/) surrounded by spaces separates the US standard and metric values. For example: "Place the point on the steel rule's 1" / 1 cm mark."

Digital Resources for Welding

Scan this code using the camera on your phone or mobile device to view the digital resources related to this craft.

Industry Recognized Credentials

If you are training through an NCCER-accredited sponsor, you may be eligible for credentials from NCCER's Registry. The ID number for this module is 29104. Note that this module may have been used in other NCCER curricula and may apply to other level completions. Contact NCCER's Registry at 1.888.622.3720 or go to **www.nccer.org** for more information.

You can also show off your industry-recognized credentials online with NCCER's digital badges. Transform your knowledge, skills, and achievements into badges that you can share across social media platforms, send to your network, and add to your resume. For more information, visit **www.nccer.org**.

1.0.0 Air-Carbon Arc Cutting

Performance Tasks

There are no Performance Tasks in this section.

Objective

Describe air-carbon arc cutting and its safety procedures.

a. Summarize air-carbon arc cutting.

b. List air-carbon arc cutting safety procedures.

Air-carbon arc cutting (A-CAC, CAC-A, or CAC) cuts, gouges, and washes metals with an electric arc between the workpiece and an electrode. An electric current flowing between the electrode and the workpiece heats and melts the base metal. An air jet from the electrode holder then blasts away the molten metal. Nothing contaminates or alters the surrounding base metal, so the cut is very clean.

1.1.0 The A-CAC Process

A-CAC works with any material that conducts electricity. It's popular for cutting steels, cast iron, aluminum, and copper. Welders use it to gouge weld grooves, back-gouge welds, gouge out defective welds, and prepare cracks for welding. They also wash and bevel with A-CAC. *Figure 1* shows A-CAC **gouging** and **washing**.

A-CAC uses a **carbon-graphite electrode**. The mixture of soft carbon and hard graphite conducts electricity but doesn't melt and blend with the cut as a standard welding electrode often does. The electrode erodes, however, so the welder regularly advances it during cutting. Eventually, it's replaced with a fresh one.

A-CAC uses a regular welding machine as its power supply. The electric current, however, is usually much higher, perhaps as much as 1,000A. The torch head includes special orifices (holes) that direct compressed air under the electrode and into the cut. As the craftworker moves the torch along the base metal's surface, it produces a groove or bevel.

The cut's depth and width depend on the following factors:

- Electrode size
- Arc current (amperage)
- Electrode angle
- Cutting speed and movement
- Air pressure

Gouging: Cutting a groove into a surface.

Washing: Cutting off bolts, rivets, and other projections from a metal surface.

Carbon-graphite electrode: A rod-shaped electrical cutting component made from soft carbon and hard graphite.

A Different Kind of Cut

Unlike oxyfuel cutting methods, A-CAC doesn't depend on oxidation. For this reason, it can cut metals that oxyfuel methods can't. Welders treat each cutting method that they learn as another tool to help them solve problems.

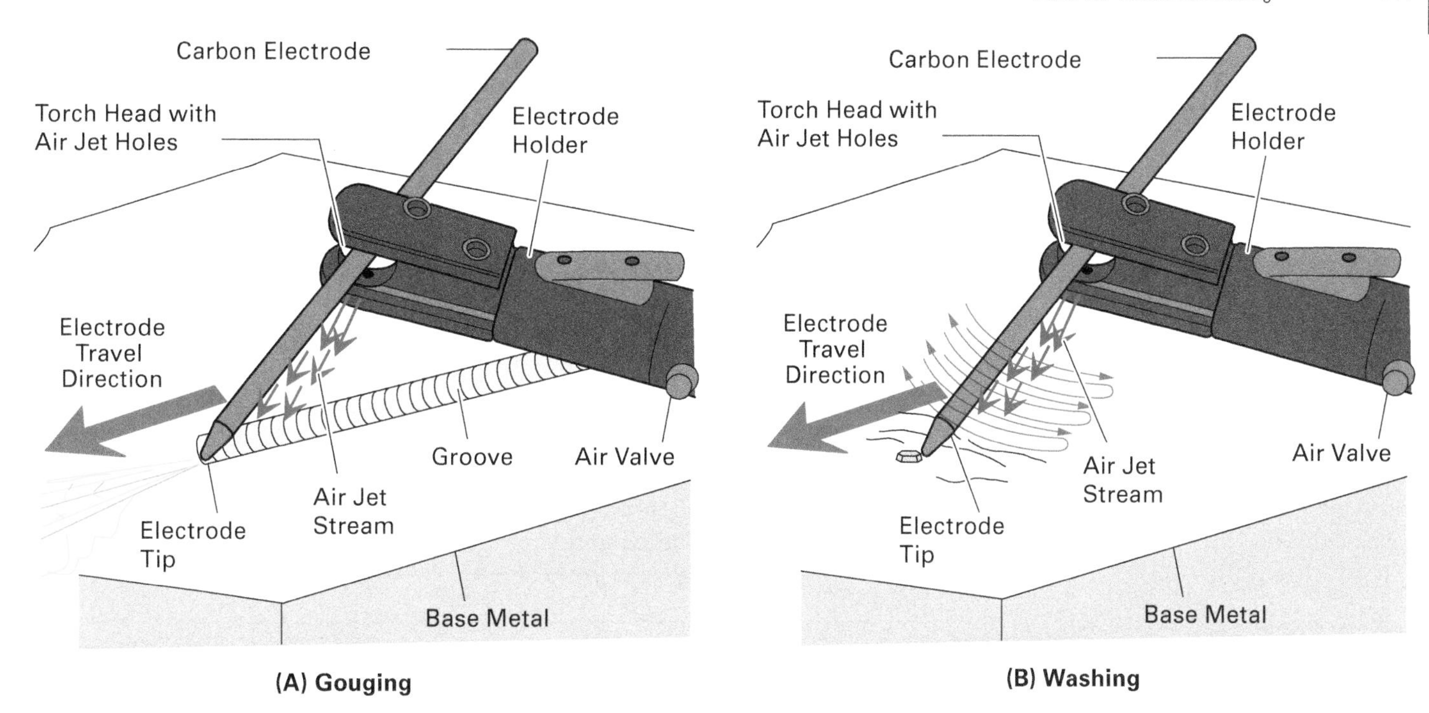

Figure 1 Gouging and washing with Air-Carbon Arc Cutting.

The A-CAC process generates temperatures close to that of some oxyfuel processes. Temperatures over 5,500°F (~3,000°C) are typical. While not as hot as plasma arc cutting (PAC), it's still an aggressive and dangerous process.

WARNING!

Molten material can travel as far as 40' (~12 m) from the workpiece. Take precautions to protect nearby workers and to prevent fires or other damage.

1.2.0 Safety Practices

A-CAC generates significant heat with an electric current. Using this process safely requires following specific procedures. This section summarizes good practices but isn't a complete A-CAC safety training. Always complete all training that your workplace requires. Follow all workplace safety guidelines and wear appropriate PPE too.

Unlike oxyfuel cutting equipment, A-CAC torches use air rather than flammable or explosive gases. While they output relatively low voltages (usually less than 80V), they still can shock you. A-CAC machines can deliver very large electric currents, so they can do a lot of damage and heat metal to very high temperatures. The arc emits UV radiation. Wet A-CAC electrodes can explode during cutting. Treat A-CAC equipment with respect.

NOTE

Be sure that you've completed NCCER Module 29101, *Welding Safety*, before continuing.

WARNING!

Energized A-CAC equipment is dangerous. Always disconnect the power supply and all other energy sources before servicing or otherwise working with the equipment. Never place your hands near the electrode when you're cutting. Always cut away from yourself to avoid injury from molten metal.

A-CAC Hazards

A-CAC generates UV radiation, molten metal spatter, and noise. Good PPE protects you from many of these.

1.2.1 Protective Clothing and Equipment

Welding and cutting tasks are dangerous. Unless you wear all required PPE, you're at risk of serious injury. The following guidelines summarize PPE for A-CAC:

- Always wear safety glasses, plus a helmet or hood. The glasses or helmet/hood lens must have the proper shade value for A-CAC work (*Figure 2*). Never view the cutting arc directly or indirectly without using the proper lens.
- Wear protective leather and/or flame-retardant clothing along with welding gloves that will protect you from flying sparks and molten metal.
- Wear high-top safety shoes or boots. Make sure the pant leg covers the tongue and lace area. If necessary, wear leather spats or chaps for protection. Boots without laces or holes are better for welding and cutting.
- Wear a 100 percent cotton cap with no mesh material in its construction. The bill should point to the rear or the side with the most exposed ear.

WARNING!

Never wear a cap with a button on top. The conductive metal under the fabric is a safety hazard.

- Wear earplugs to protect your ear canals from sparks. Since the torch is loud, always wear hearing protection.

WARNING!

Because of the arc's high temperature and intensity, choose a lens shade that's darker than the normal shade you'd use for welding similar material. Depending on the current level, a number 10 to 14 lens shade is best for A-CAC work.

1.2.2 Fire/Explosion Prevention

A-CAC produces high temperatures to cut metal. Welding or cutting a vessel or container that once contained combustible, flammable, or explosive materials is hazardous. Residues can catch fire or explode. Before welding or cutting vessels, check whether they contained explosive, hazardous, or flammable materials. These include petroleum products, citrus products, or chemicals that release toxic fumes when heated.

American Welding Society (AWS) F4.1, Safe Practices for the Preparation of Containers and Piping for Welding and Cutting and *ANSI/AWS Z49.1* describe safe practices for these situations. Begin by cleaning the container to remove any residue. Steam cleaning, washing with detergent, or flushing with water are possible methods. Sometimes you must combine these to get good results.

WARNING!

Clean containers only in well-ventilated areas. Vapors can accumulate in a confined space during cleaning, causing explosions or toxic substance exposure.

Arc Cutting Processes

Process	Electrode	Amperage	Lens Shade Numbers 2 3 4 5 6 7 8 9 10 11 12 13 14
Air-Carbon Arc Cutting (A-CAC)	Light	< 500A	10 11 12
	Heavy	500A – 1,000A	11 12 13 14

Lens shade values are based on OSHA minimum values and ANSI/AWS suggested values. Choose a lens shade that's too dark to see the weld zone. Without going below the required minimum value, reduce the shade value until it's visible.

Figure 2 Air-Carbon Arc Cutting – Guide to Lens Shade Numbers.

After cleaning the container, you must formally confirm that it's safe for welding or cutting. *American Welding Society (AWS) F4.1, Safe Practices for the Preparation of Containers and Piping for Welding and Cutting* outlines the proper procedure. The following three paragraphs summarize the process.

Immediately before work begins, a qualified person must check the container with an appropriate test instrument and document that it's safe for welding or cutting. Tests should check for relevant hazards (flammability, toxicity, etc.). During work, repeated tests must confirm that the container and its surroundings remain safe.

Alternatively, fill the container with an inert material ("inerting") to drive out any hazardous vapors. Water, sand, or an inert gas like argon meets this requirement. Water must fill the container to within 3" (~7.5 cm) or less of the work location. The container must also have a vent above the water so air can escape as it expands (*Figure 3*). Sand must completely fill the container.

When using an inert gas, a qualified person must supervise, confirming that the correct amount of gas keeps the container safe throughout the work. Using an inert gas also requires additional safety procedures to avoid accidental suffocation.

NOTE

A *lower explosive limit* (*LEL*) gas detector can check for flammable or explosive gases in a container or its surroundings. Gas monitoring equipment checks for specific substances, so be sure to use a suitable instrument. Test equipment must be checked and calibrated regularly. Always follow the manufacturer's guidelines.

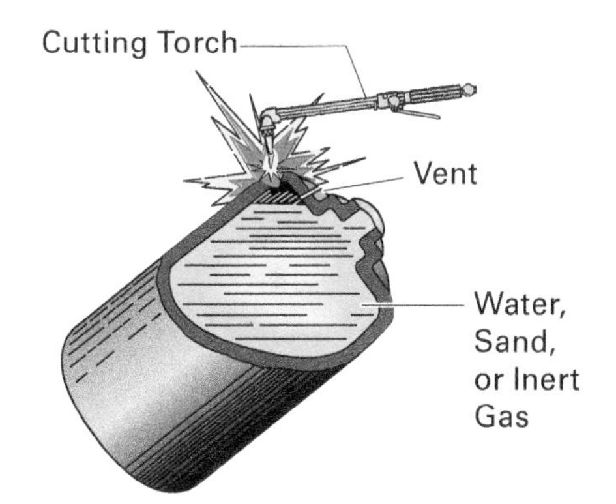

Note: *ANSI Z49.1* and AWS standards should be followed.

Figure 3 "Inerting" a container.

WARNING!

Never weld or cut drums, barrels, tanks, vessels, or other containers until they have been emptied and cleaned thoroughly. Residues, such as detergents, solvents, greases, tars, or corrosive/caustic materials, can produce flammable, toxic, or explosive vapors when heated. Never assume that a container is clean and safe until a qualified person has checked it with a suitable test instrument. Never weld or cut in places with explosive vapors, dust, or combustible products in the air.

The following general fire- and explosion-prevention guidelines apply to A-CAC work:

- Never carry matches or gas-filled lighters in your pockets. Sparks can ignite the matches or cause the lighter to explode.
- Always comply with all site and employer requirements for hot work permits and fire watches.
- Never use oxygen in place of compressed air to blow off the workpiece or your clothing. Never release large amounts of oxygen. Oxygen makes fires start readily and burn intensely. Oxygen trapped in fabric makes it ignite easily if a spark falls on it. Keep oxygen away from oil, grease, and other petroleum products.
- Remove all flammable materials from the work area or cover them with a fire-resistant blanket.
- Before welding, heating, or cutting, confirm that an appropriate fire extinguisher is available. It must have a valid inspection tag and be in good condition. All workers in the area must know how to use it.
- Prevent fires by maintaining a neat and clean work area. Confirm that metal scrap and slag are cold before disposing of them.

1.2.3 Work Area Ventilation

Vapors and fumes fill the air around their sources. Welding and cutting fumes can be hazardous. Welders often work above the area from which fumes come. Good ventilation and source extraction equipment help remove the vapors and protect the welder. The following lists good ventilation practices to use when welding and cutting:

- Always weld or cut in a well-ventilated area. Operations involving hazardous materials, such as metal coatings and toxic metals, produce dangerous fumes. You must wear a suitable respirator when working with them. Follow your workplace's safety protocols and confirm that your respirator is in good condition.

WARNING!

Cadmium, mercury, lead, zinc, chromium, and beryllium produce toxic fumes when heated. When welding or cutting these materials, wear appropriate respiratory PPE. Supplied-air respirators (SARs) are best for long-term work.

- Never weld or cut in a confined space without a confined space permit. Prepare for working in the confined space by following all safety protocols.
- Set up a confined space ventilation system before beginning work.
- Never ventilate with oxygen or breathe pure oxygen while working. Both actions are very dangerous.

WARNING!

When working around other craftworkers, always be aware of what they're doing. Take precautions to keep your work from endangering them.

1.0.0 Section Review

1. A-CAC's primary application is _____.
   a. gouging welding grooves, back-gouging welds, and gouging out defective welds
   b. controlling base metal temperature during cutting by cooling it with compressed air
   c. introducing additional metal alloys from the melting electrode into the base metal
   d. cutting or gouging metal with a very low electric current

2. One way to prevent explosions during A-CAC work is to _____.
   a. cover the floor under the workpiece with sand
   b. make sure that the gouge width equals the electrode diameter
   c. check that the electrode is dry before starting
   d. not use inverters or electronically controlled welding machines

2.0.0 Air-Carbon Arc Cutting Equipment

Performance Tasks

There are no Performance Tasks in this section.

Objective

Summarize air-carbon arc cutting equipment.

a. Describe air-carbon arc power units.
b. Describe air-carbon arc torches.
c. Describe air-carbon arc electrodes.

A-CAC equipment is almost identical to the equipment used for shielded metal arc welding (SMAW). The torch is different, however, since it connects to an air supply. Most significantly, A-CAC electrodes are quite different since they don't melt as SMAW electrodes do. *Figure 4* shows a complete A-CAC system.

2.1.0 A-CAC Power Supplies

SMAW (stick) welding machines work well as A-CAC power supplies, provided they can deliver enough current (amperage). A-CAC requires very large electric currents, often several hundred amperes or more. Smaller welding machines won't be able to meet this requirement.

SMAW welding machines deliver either alternating current (AC) or direct current (DC). They have constant current (CC) outputs. That is, they vary their output voltage to maintain a fixed current flow to the torch. An AC machine requires special AC-rated electrodes to do A-CAC work.

Multipurpose welding machines can also do A-CAC if they have a SMAW mode or can be switched to constant current DC operation. *Figure 5* shows welding machines suitable for A-CAC.

NOTE

Avoid constant voltage (CV) welding machines. While they can do A-CAC work, they're not well-suited to it. Many have duty cycle limitations that prevent them from being a good choice as well.

2.1.1 Current and Polarity Requirements

Light-duty gouging and cutting requires 200A or less. Medium-duty cutting requires up to 600A. Heavy-duty cutting may require as much as 2,000A. Always confirm that the power supply can deliver what the job requires.

Besides the correct amperage, DC welding machines must have the correct **polarity** for A-CAC work. Polarity refers to whether the electrode is electrically positive (+) or negative (–). It determines the direction that current flows between the workpiece and the electrode. Some machines have a switch that sets polarity. Others require you to swap the leads between the machine's + and – terminals.

Polarity: A terminal's electrical charge, either positive or negative, which determines whether current flows into or away from it.

A DC welding machine that makes the electrode positive and the workpiece negative is *direct current electrode positive* (*DCEP*). One that makes the electrode negative and the workpiece positive is *direct current electrode negative* (*DCEN*). AC machines do not have polarity since the current changes direction many times per second.

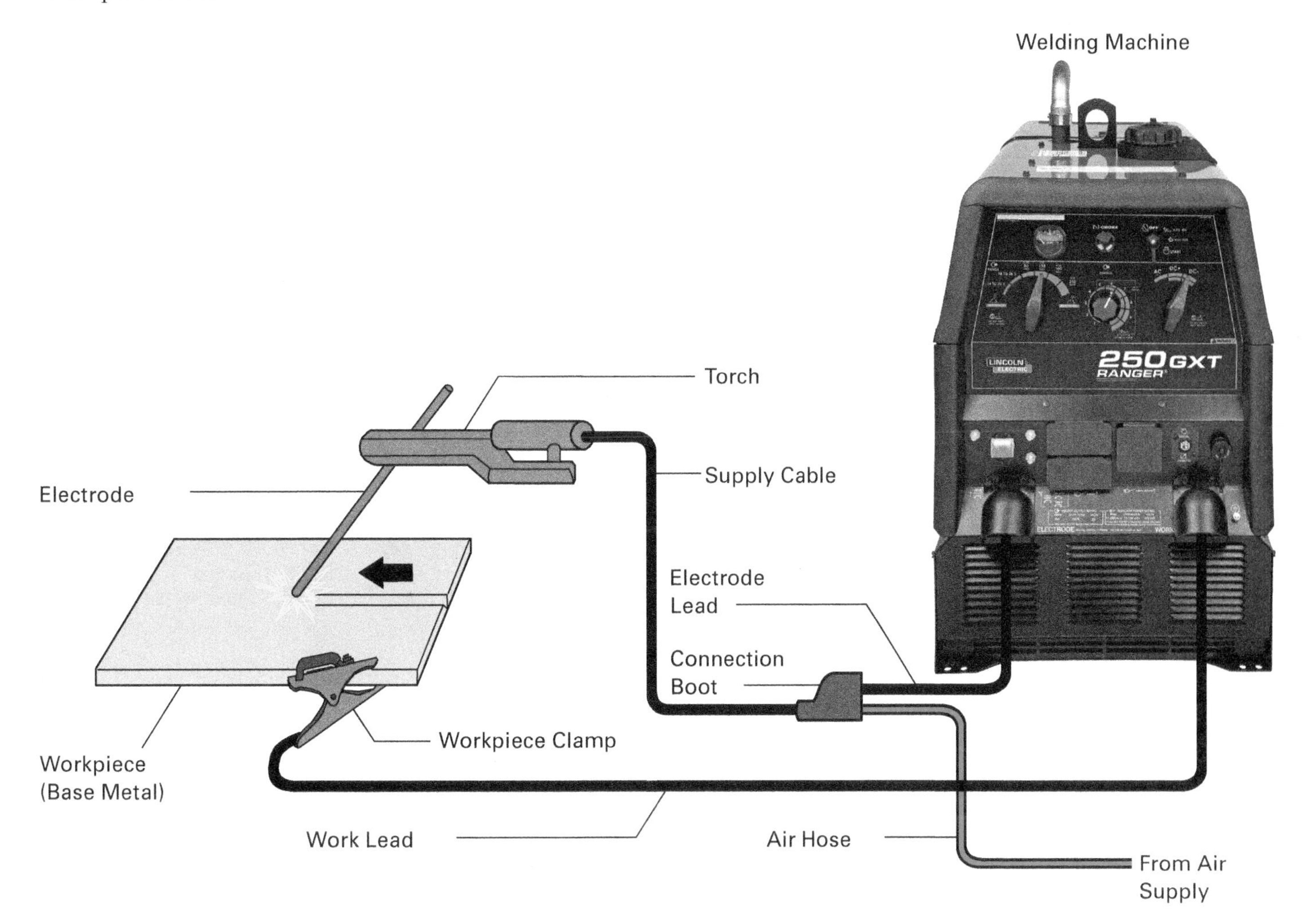

Figure 4 A-CAC system.
Source: The Lincoln Electric Company, Cleveland, OH, USA

(A) Medium-Duty/Heavy-Duty Multiprocess

(B) Medium-Duty SMAW

Figure 5 Welding machines for A-CAC.

Sources: © Miller Electric Mfg. LLC (8A); The Lincoln Electric Company, Cleveland, OH, USA (8B)

2.2.0 A-CAC Torches

While an A-CAC power supply is just a regular welding machine, an A-CAC torch is special. It not only holds the electrode, but it also directs an air jet onto the workpiece at the correct angle. Choosing the wrong torch configuration will produce poor results. *Figure 6* shows an A-CAC torch and its components.

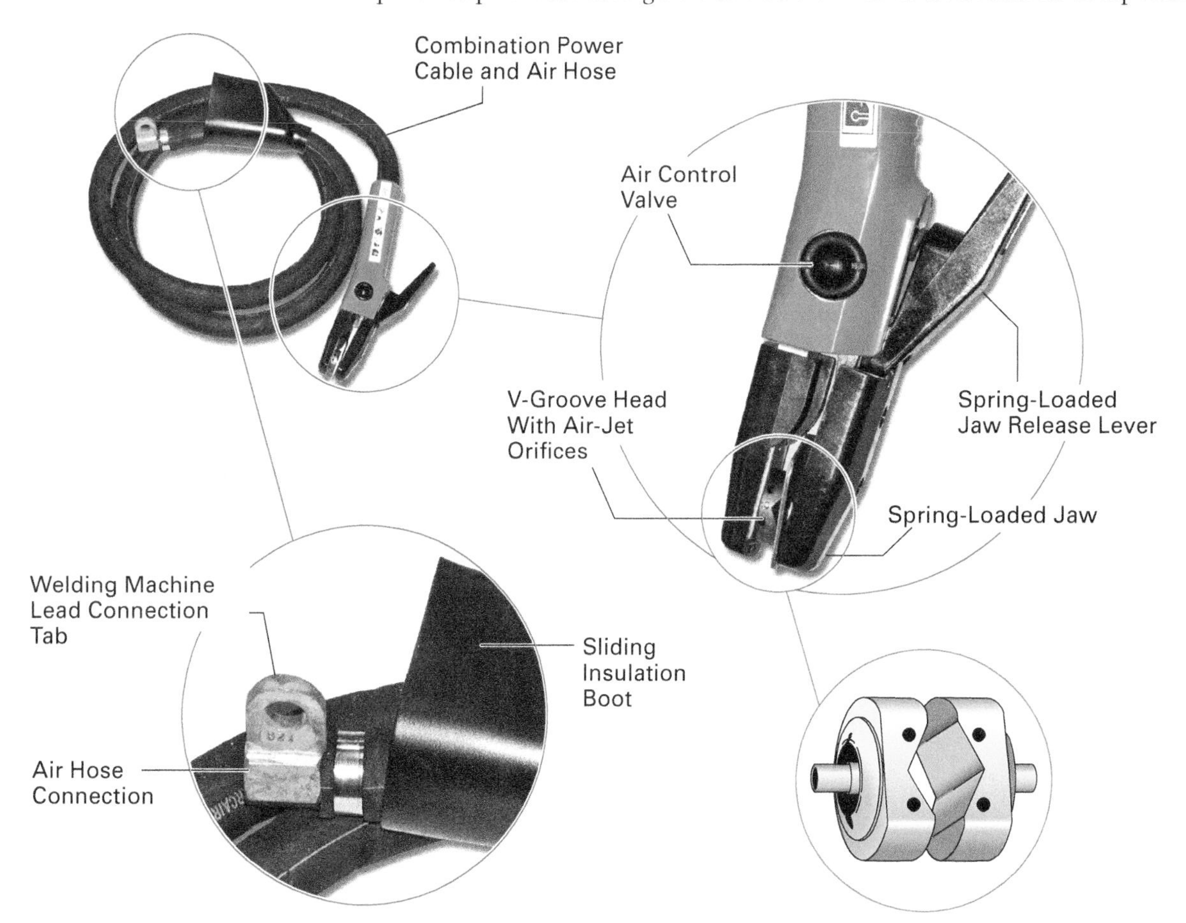

Figure 6 A-CAC torch.

2.2.1 Torches

An A-CAC torch's spring-loaded jaws grip the electrode between two metal "heads." One has a V-shaped groove in which the electrode sits. The other can be flat or V-shaped. V-shaped heads transfer more current to the electrode than flat ones do. Usually, the heads swivel so you can change the electrode's position or angle. To insert, remove, or adjust the electrode, squeeze the handle to open the jaws. The electrode tip should be no more than 6" (~15 cm) from the head or it could overheat.

Torch heads come in different sizes and styles for light, medium, and heavy work. A light-duty head might hold a 1⁄8" (3.2 mm) electrode, while a heavy-duty one can handle electrodes as large as 1" (25 mm). Orifices in one or both heads direct air towards the cut.

Welders swap heads for different jobs. When cutting or gouging, the orifices should be below the electrode. This position directs the air jet to blow metal out from the bottom of the cut. When washing, the orifices should direct the air to the electrode's side. This prevents the arc from cutting into the surface. *Figure 7* shows different head styles.

A short cable several feet long delivers both air and electrical power to the torch. One end attaches to the torch, while the other has two separate connectors. One is for the air supply. Some torches use a *quick-connect* fitting, while others have a screw fitting. The other is for the electrical supply lead and is thick metal. A rubber *boot* slides over both, preventing the electrical connection from touching bare metal and arcing. *Figure 8* shows these features.

A light-duty torch's supply cable is an air hose and a thick wire bundled inside a single jacket. Heavy-duty torches have a **concentric cable** instead. This is a special hollow cable made from heavy wires jacketed by insulation. Air flows through the cable's core, while electrical power flows through the wires. The air cools the wires, which can get quite hot. *Figure 9* shows both supply cable styles.

When the supply cable enters the torch, electrical power goes to the jaws and heads. Air passes through channels to the head orifices. A button- or lever-operated valve turns the air on and off.

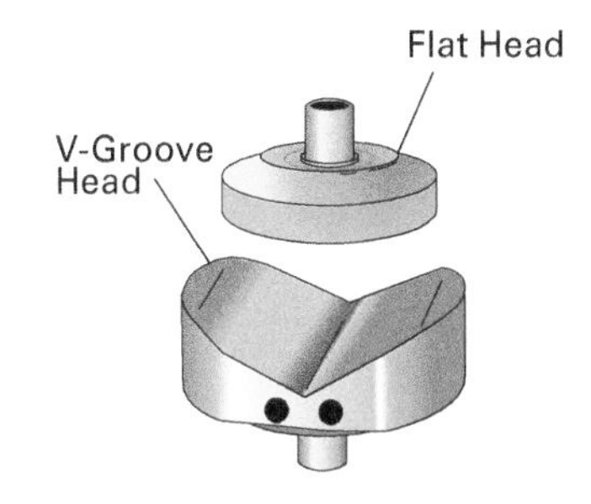

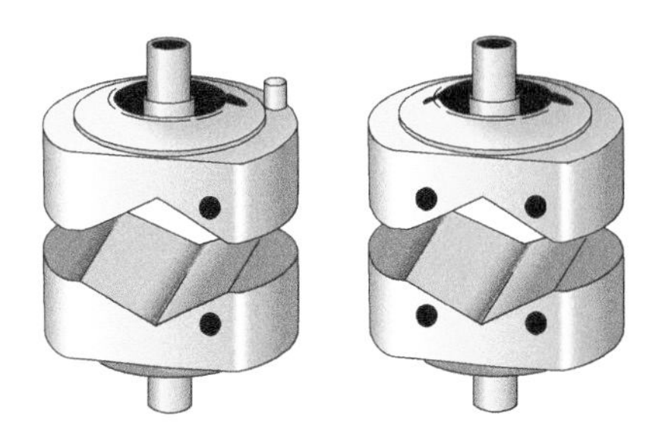

Figure 7 Torch head styles.

Concentric cable: A special torch cable that carries electricity through its outer shell and compressed air through its hollow core.

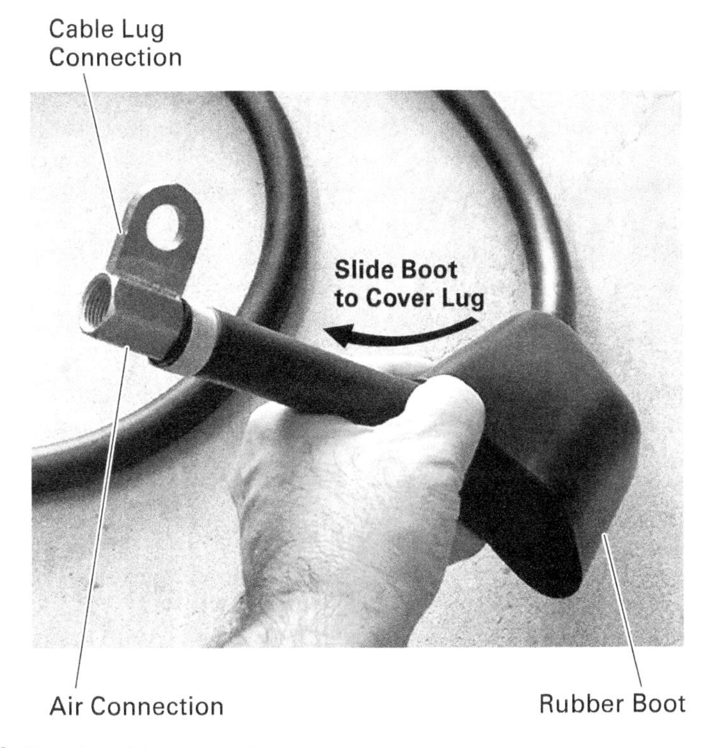

Figure 8 Supply cable connections.

Foundry Torch

Foundry torches are extra-heavy-duty torches meant to carry very large currents (1,000A or more). They have large, durable heads. These torches work well for defect removal, pad washing, and as general-purpose tools.

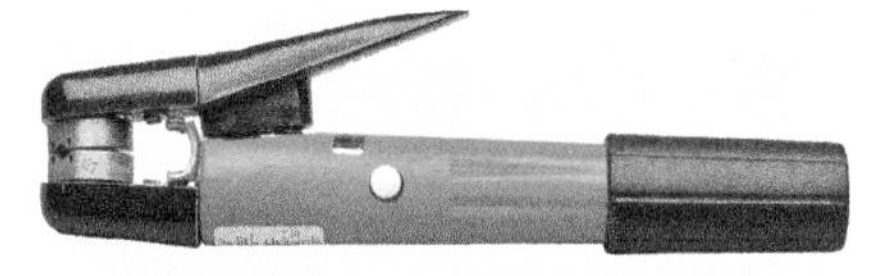

Source: Victor Technologies

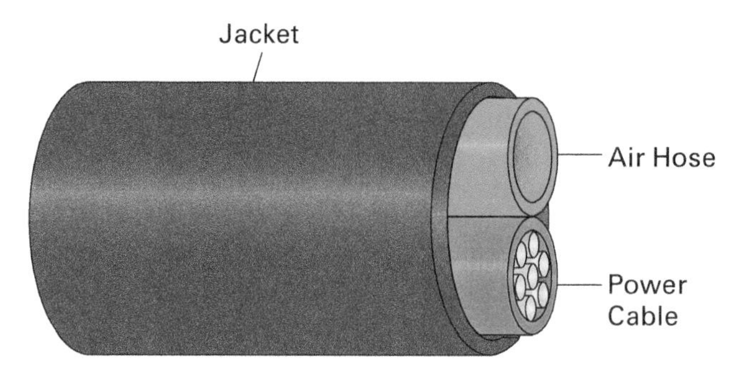

Low-Current Applications

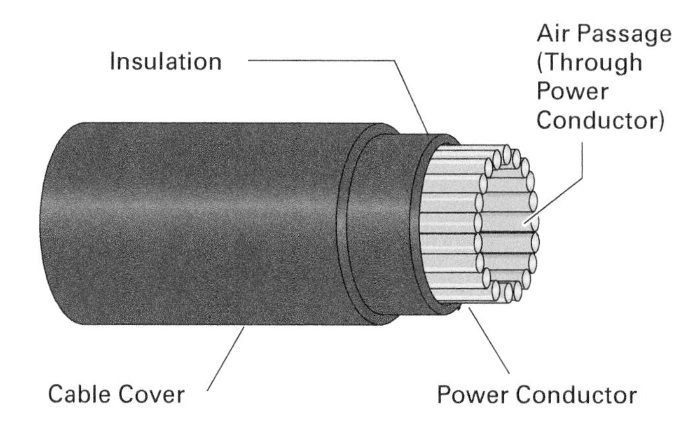

High-Current Applications

Figure 9 Supply cable styles.

2.2.2 Torch Air Supply

A-CAC torches require a steady flow of clean air at the correct pressure. Pressures of 40 psig to 100 psig (~275 kPaG to 690 kPaG) are typical. The craftworker adjusts the air pressure and flow to blow the molten metal cleanly away from the cut. Insufficient airflow or weak pressure makes a poor-quality cut.

Manufacturers specify a torch's airflow requirements in cubic feet per minute (cfm) or standard cubic feet per minute (scfm). Manufacturers in metric-speaking countries may use liters per minute (lpm or L/min) instead. Cutting airflow ranges from 8 cfm (~225 lpm) for light work to 50 cfm (~1,400 lpm) for heavy work.

Air can come from a standalone compressor, a tank, or a facility's compressed air supply. A small compressor or a tank is portable and works fine with smaller electrodes. For bigger cutting jobs, a medium-sized compressor or facility air supply is better. Air should be filtered and regulated.

2.3.0 A-CAC Electrodes

A-CAC electrodes contain two kinds of carbon—soft carbon and hard graphite—plus a binder. The manufacturer bakes the mixture into long, thin rods in one of several shapes. Some electrodes have an outer coating or small amounts of metal added to them. The electrode's size, shape, and materials determine its suitability for different A-CAC tasks and equipment.

WARNING!

Keep all A-CAC electrodes dry. A wet electrode can explode!

2.3.1 Electrode Types

A-CAC electrode types fall under the following three categories:

Plain — These electrodes have no coating or additives. They're meant for DC operation (usually DCEP). They tend to run hot and **oxidize** quickly. The section between the torch heads and the tip gets thinner as you work. Plain electrodes can't carry as much current as other types, but they give a very clean cut with no contamination.

Oxidize: To chemically combine with oxygen either quickly or slowly.

Copper-clad — These electrodes have a copper coating (*Figure 10*). They're meant for DC operation (usually DCEP). The coating helps them carry larger currents, run cooler, and not get thinner as you work. The vaporized copper does contaminate the base metal slightly, so they aren't appropriate for some jobs.

AC — These electrodes work with AC welding machines or DC machines operating in DCEN mode. They too have a copper coating. Plain and regular copper-clad electrodes produce an erratic arc when powered by an AC machine. AC electrodes contain small amounts of *rare earth metals* that overcome this problem. Like regular copper-clad electrodes, they also slightly contaminate the base metal.

Did You Know?

Not Really Rare

Rare earth metals aren't truly rare. They're abundant but are never found in large quantities in one place. Even though many industries use them, most people have never heard of them. Cerium is one of the rare earth metals used in AC electrodes.

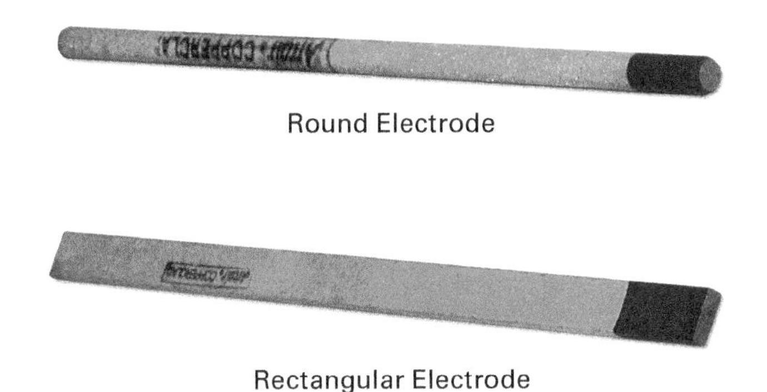

Figure 10 Copper-clad electrodes.

2.3.2 Electrode Shapes

Electrodes come in several shapes, each of which produces a different cut. The following are common shapes:

Round — These electrodes are cylindrical. The tip tapers to a point during cutting. This shape works well for many common cutting and gouging tasks.

Round-jointed — These electrodes are also round, but they have joints that link them together. Automatic-feed equipment uses them for uninterrupted cutting operations.

Special-shape — These electrodes produce grooves and cuts with specific shapes. Flat electrodes perform close-tolerance metal removal. Welders use them to remove weld buildups, shape dies, remove welded lugs, and bevel edges. Half-round electrodes cut wide grooves. *Figure 11* shows common special-shape electrode profiles.

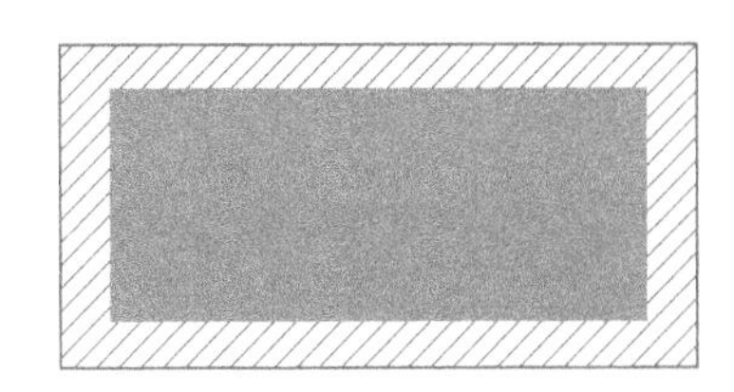

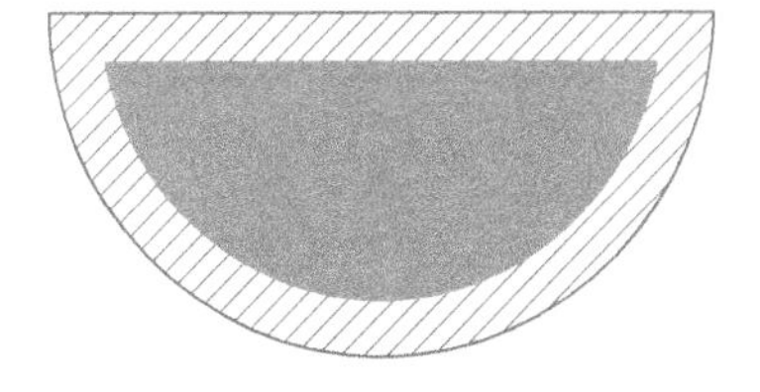

Figure 11 Special-shape electrode profiles.

2.3.3 Electrode Sizes and Amperages

Manufacturers specify round electrodes by diameter and flat ones by their width. They range from ⅛" to 1" (3.2 mm to 25 mm). Not all types and shapes come in all sizes. Electrodes for hand torches are usually 12" (~30 cm) long. Jointed electrodes for automatic feed machines are longer.

Every electrode has a rating that specifies how much current it can safely carry. Large electrodes can carry bigger currents than small electrodes. A copper-clad electrode can carry more current than the same size plain electrode. Copper-clad AC electrodes are an exception. They carry the same current as the equivalent plain electrode.

Electrode Availability

Local welding supply stores usually stock round and rectangular copper-clad DC electrodes. Most don't stock other styles or AC electrodes. Be prepared to order these when needed.

Electrode Consumption

When making a gouge equal to the electrode diameter, you'll get about 8" (~20 cm) of groove for every 1" (~2.5 cm) of electrode consumed. Actual consumption depends on angle, cutting speed, current, and material.

Always consult the electrode manufacturer for current ratings, as these vary from one brand to another. *Table 1* shows typical current ratings for plain electrodes and copper-clad AC electrodes. Copper-clad DC electrodes can carry about 10 percent more current.

TABLE 1 Typical Electrode Current Ratings

Electrode Size		Minimum current	Maximum Current
1⁄8"	3.2 mm	30A	60A
5⁄32"	4.0 mm	90A	150A
3⁄16"	4.8 mm	150A	200A
1⁄4"	6.4 mm	200A	400A
5⁄16"	8.0 mm	250A	450A
3⁄8"	9.5 mm	350A	600A
1⁄2"	13 mm	600A	1,000A
5⁄8"	15 mm	800A	1,200A
3⁄4"	19 mm	1,200A	1,600A
1"	25 mm	1,800A	2,200A

2.0.0 Section Review

1. What kind of welding machine works well for A-CAC?
   a. GMAW
   b. SMAW
   c. FCAW
   d. PAC

2. When installing the electrode in an A-CAC torch, the tip should be no farther from the heads than _____.
   a. 3" (~8 cm)
   b. 6" (~15 cm)
   c. 8" (~20 cm)
   d. 12" (~30 cm)

3. What *must* the manufacturer add to copper-clad A-CAC electrodes if they're to be used with an AC welding machine?
   a. Cobalt
   b. Nickel
   c. Silica mixed with clay
   d. A rare earth metal

3.0.0 Using Air-Carbon Arc Cutting Equipment

Performance Tasks

1. Set up air-carbon arc cutting equipment.
2. Select and install air-carbon arc cutting electrodes.
3. Gouge metal with air-carbon arc cutting equipment.
4. Wash metal with air-carbon arc cutting equipment.
5. Dismantle and store air-carbon arc cutting equipment.

Objective

Outline setting up, safely operating, and maintaining air-carbon arc equipment.

a. Explain how to set up an air-carbon arc cutting project.
b. Explain how to gouge and wash metals with air-carbon arc cutting equipment.
c. Explain how to maintain air-carbon arc cutting equipment.

A-CAC quickly removes material, so it's ideal for reducing preparation time or grinding work. A-CAC equipment is simple and powerful. The process is fast, clean, and efficient too. But A-CAC equipment is also hazardous and produces poor cuts when used carelessly. The following sections explore setup, cutting, and equipment care. As always, make safety a high priority when cutting with an A-CAC machine.

3.1.0 Setting Up

A-CAC equipment is simple and has just a few components, so setup is relatively easy. Always confirm the following factors as you prepare for a cutting job:

- The welding machine can deliver enough current for the job.
- The local electrical supply is appropriate for the machine (phase, voltage, and current).
- The torch, its heads, and the electrode are correct for the material and cut style.
- A suitable compressed air source is available.
- Spare consumables are available.

3.1.1 Preparing the Work Area

Place the A-CAC machine close enough to the workpiece so you can maneuver easily. Confirm that the cable reaches without creating hazards. When possible, place the workpiece at a comfortable height.

A-CAC can spray molten metal up to 40' (~12 m) away. If you can't remove combustible materials within this radius, cover them with fireproof blankets or place shields around the work area. If others are working in the area, place welding curtains around the work area to control glare and UV. Follow all company protocols for hot work permits and fire watches.

A-CAC generates smoke and fumes. Remember, some metals and many coatings produce toxic fumes when heated. Check the work area's ventilation. Bring in a source extraction unit if necessary. You may also need respiratory PPE. Follow all company safety protocols.

WARNING!

Stainless steel contains chromium, which can turn into toxic Cr(VI), also called *hexavalent chromium*, when heated. Always wear appropriate respiratory PPE when cutting stainless steel with an A-CAC machine.

Assemble all appropriate clothing and PPE. Get the right eye protection for A-CAC work. Wear ear protection that keeps sparks out of the ear canals. A-CAC is noisy, so be sure your PPE protects your hearing too.

3.1.2 Selecting the Electrode

The electrode size and shape determine the cut style and width. Electrode angle, travel speed, side-to-side movement, air pressure, and cutting current control the groove's depth and contour. With copper-clad electrodes and up to ½" (13 mm) thick carbon steel, the groove will be about ⅛" (~3 mm) wider than the electrode.

Select the correct electrode with these factors in mind. *Table 2* lists example groove sizes, along with cutting recommendations (electrode size, current, and travel speed) for copper-clad electrodes.

TABLE 2 Groove Size Cutting Recommendations

Groove Width (in, mm)	Groove Depth (in, mm)	Electrode Size (in, mm)	Current (DC Amperes)	Travel Speed (in/min)
1/4, 6.4	1/16, 1.6	3/16, 4.8	200	82
5/16, 7.9	1/8, 3.2	1/4, 6.4	300	51
5/16, 7.9	1/4, 6.4	1/4, 6.4	300	29
5/16, 7.9	3/8, 9.5	1/4, 6.4	300	15
3/8, 9.5	1/8, 3.2	5/16, 7.9	320	65
3/8, 9.5	1/4, 6.4	5/16, 7.9	420	31
3/8, 9.5	1/2, 12.7	5/16, 7.9	540	27
7/16, 11.1	1/8, 3.2	3/8, 9.5	560	82
7/16, 11.1	1/4, 6.4	3/8, 9.5	560	75
7/16, 11.1	1/2, 12.7	3/8, 9.5	560	15
7/16, 11.1	11/16, 17.5	3/8, 9.5	560	12
9/16, 14.3	1/4, 6.4	1/2, 12.7	1,200	22
9/16, 14.3	3/8, 9.5	1/2, 12.7	1,200	21
9/16, 14.3	1/2, 12.7	1/2, 12.7	1,200	18
9/16, 14.3	3/4, 19.1	1/2, 12.7	1,200	12
13/16, 20.6	1/4, 6.4	5/8, 15.9	1,300	29
13/16, 20.6	1/2, 12.7	5/8, 15.9	1,300	14
13/16, 20.6	3/4, 19.1	5/8, 15.9	1,300	11

Select an electrode with the following steps:

Step 1 Identify the base metal and cut specifications.

NOTE

If the cut specifications require that the base metal not be contaminated, you must use a plain electrode. Otherwise, a copper-clad electrode is usually a better choice.

WARNING!

If the base metal contains toxic metals, like cadmium, or metals that can become toxic when heated, like chromium, be sure to wear appropriate respiratory PPE. If uncertain, always wear respiratory PPE.

NOTE

Always have extra electrodes available to avoid wasting time if you should run out before finishing.

Step 2 Identify the welding machine current type (AC or DC).

Step 3 Select an electrode type compatible with the base metal and the welding machine.

Step 4 Select an electrode size and shape appropriate to the cutting task.

Step 5 Estimate and collect the number of electrodes required for the job.

3.1.3 Setting up the Machine

The following steps outline setting up the A-CAC machine:

Step 1 Install the correct torch heads. They should be the right size for the electrode and suitable for the cutting job. Washing, for example, requires different heads than gouging.

Step 2 Slide back the protective boot from the torch's supply cable end.

Step 3 Attach the air supply connector to the compressed air source.

Step 4 Attach the electrical supply connector to the welding machine's electrode lead.

Step 5 Slide the protective boot over the torch supply cable end.

WARNING!

Never operate the welding machine without the boot covering the supply cable connections. If the electrical connection touches bare metal, it could arc and heat the metal to a high temperature.

Step 6 If the welding machine has multiple operating modes, select SMAW mode or constant current DC mode.

Step 7 If using a DC welding machine, set the polarity switch to the correct position (usually DCEP). If the machine doesn't have a polarity switch, connect the leads to the correct terminals.

3.2.0 Gouging and Washing

A-CAC equipment isn't hard to operate. Getting clean cuts, however, requires good technique. Before beginning, evaluate the task and work out the best way to perform it. Be sure that you know the right electrode position for each operation. Inspect the surface and deal with any problems first.

3.2.1 Gouging

Gouging is cutting a groove into a surface, usually to prepare it for welding. The electrode size determines the minimum groove width. Special-shape electrodes will produce special groove shapes. You can increase the groove width by using a larger electrode or weaving the electrode tip slightly from side to side. *Figure 12* shows A-CAC gouging.

The groove's depth and contour depend on the electrode travel angle, side-to-side movement, and travel speed. For a shallow groove, advance the electrode rapidly. For a deeper groove, go more slowly or increase the cutting current. As you'll discover, relatively small changes in your technique will greatly alter the outcome.

The following steps outline A-CAC gouging:

Step 1 Prepare the work area. Assemble the necessary protective clothing and PPE. Set up the A-CAC equipment.

Step 2 Locate the electrical supply's primary disconnect so you can use it in an emergency.

Step 3 Confirm that the welding machine can handle the gouging task.

Step 4 Start the compressor or open the compressed air supply valve. Confirm adequate pressure. Adjust the regulator if necessary.

Step 5 Connect the workpiece lead's clamp to the workpiece. It must attach firmly over bright metal. If necessary, clean the metal with emery paper first.

Figure 12 A-CAC gouging.
Source: The Lincoln Electric Company, Cleveland, OH, USA

CAUTION

If cutting on machinery or other assembled equipment, position the workpiece clamp so the cutting current will not pass through seals, bearings, or components that arcing or heat could damage. Have an electrician isolate any nearby electrical or electronic components from the A-CAC circuit.

Step 6 Plug the welding machine into the electrical supply (if required). Energize the welding machine and switch the unit on.

Step 7 Select the proper electrode size and shape for the groove.

Step 8 Select the correct current (amperage) for the electrode and cutting type.

WARNING!

A-CAC generates very high temperatures. Anything nearby that's flammable or explosive could ignite. Confirm that you've followed all safety protocols and have a fire extinguisher within reach.

Step 9 Put on your PPE.

Step 10 Open the torch's jaws. Position the electrode between the heads. It should protrude about 6" (~15 cm) for most metals (*Figure 13*). For aluminum, 3" (~8 cm) is better.

WARNING!

Confirm that the electrode is dry before using it. A wet electrode can explode!

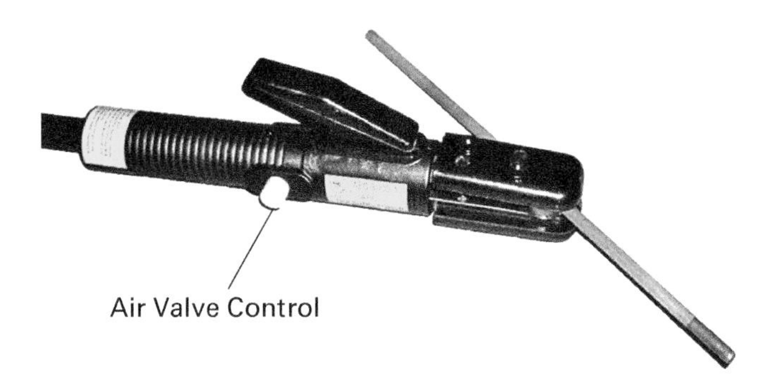

Figure 13 Installed electrode.

Step 11 Briefly operate the torch air valve to confirm that air flows freely from the head orifices.

Step 12 Adjust the electrode and torch position so the electrode forms a 35- to 45-degree angle with the workpiece (*Figure 14*). The head orifices should be positioned to blow air below the electrode into the groove.

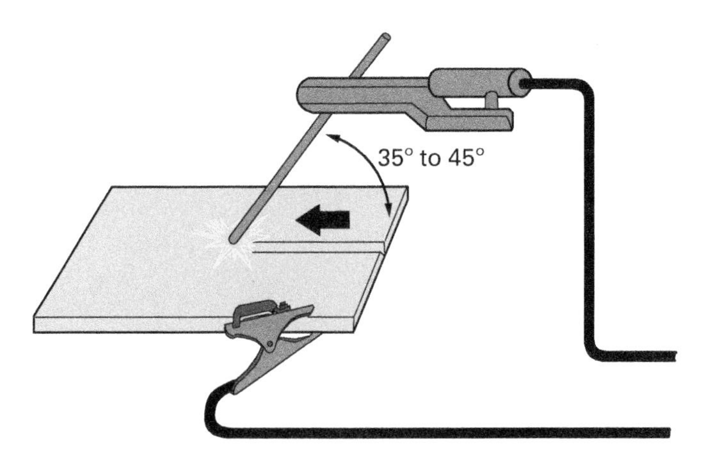

Figure 14 Proper angle for gouging.

Step 13 Lower your mask or flip the tinted lens into position. Open the torch air valve.

Step 14 Strike an arc by touching the electrode to the workpiece. Don't pull back after establishing the arc. Maintain a short arc and begin the operation.

Step 15 Push the torch away from you to cut the groove. Move at a suitable speed for the groove depth.

Step 16 When finished, move the torch away from the workpiece and shut off the torch air valve.

3.2.2 Washing

Washing is cutting objects or projections—bolt heads, old welds, or unwanted materials—from a surface, leaving it smooth. Weaving the electrode from side to side across the projection cuts it away. Washing works best with special heads in the torch. These have orifices on only one side of the electrode. This configuration keeps the torch from gouging the surface. Cutting, however, only works in one direction. *Figure 15* shows A-CAC washing.

Figure 15 A-CAC washing.
Source: The Lincoln Electric Company, Cleveland, OH, USA

The following steps outline A-CAC washing:

Step 1 Prepare the work area. Assemble the necessary protective clothing and PPE. Set up the A-CAC equipment.

Step 2 Locate the electrical supply's primary disconnect so you can use it in an emergency.

Step 3 Confirm that the welding machine can handle the cutting task.

Step 4 Start the compressor or open the compressed air supply valve. Confirm adequate pressure and adjust the regulator if necessary.

Step 5 Connect the workpiece lead's clamp to the workpiece. It must attach firmly over bright metal. If necessary, clean the metal with emery paper first.

CAUTION

If cutting on machinery or other assembled equipment, position the workpiece clamp so the cutting current will not pass through seals, bearings, or components that arcing or heat could damage. Have an electrician isolate any nearby electrical or electronic components from the A-CAC circuit.

Step 6 Plug the welding machine into the electrical supply (if required). Energize the welding machine and switch the unit on.

Step 7 Select the proper electrode size and shape for washing.

Step 8 Select the correct current (amperage) for the electrode.

WARNING!

A-CAC generates very high temperatures. Anything nearby that's flammable or explosive could ignite. Confirm that you've followed all safety protocols and have a fire extinguisher within reach.

Step 9 Put on your PPE.

Step 10 Open the torch jaws. Position the electrode between the heads. It should protrude about 6" (~15 cm) for most metals (*Figure 13*). For aluminum, 3" (~8 cm) is better.

WARNING!

Confirm that the electrode is dry before using it. A wet electrode can explode!

Step 11 Briefly operate the torch air valve to confirm that air flows freely from the head orifices.

Step 12 Adjust the electrode and torch position so the electrode forms a 35- to 45-degree angle with the workpiece (*Figure 14*). The head orifices should be positioned to blow air to one side of the electrode.

Vertical Washing and Gouging

Positioning the workpiece vertically uses gravity, as well as the air jets, to eject metal from the cut. Always be especially careful of your feet when gouging or washing vertically. Wearing spats over your shoes is a good idea.

Step 13 Lower your mask or flip the tinted lens into position. Open the torch air valve.

Step 14 Strike an arc at the edge of the projection you're removing. As the arc melts the metal, swing it gradually across the projection. Continue the process until the surface is flush.

Step 15 When finished, move the torch away from the workpiece and shut off the torch air valve.

3.2.3 Inspecting A-CAC Work

After gouging or cutting, inspect the cut surfaces. They should have smooth edges that are free from notches. Notches can trap slag during welding. Grooves and surfaces should also be free from dross or carbon deposits. During welding, these combine with the base metal to produce hard, brittle zones or other weld defects. After A-CAC work and before welding, always clean grooves and surfaces.

NOTE

NCCER Module 29106, *Weld Quality*, covers this topic in much greater depth. Always follow the job specifications and your company's welding quality standards.

3.3.0 Caring for A-CAC Equipment

A-CAC equipment is relatively simple and easy to maintain. Keeping it in good condition requires properly storing it when you're finished cutting. You must also perform basic maintenance as required.

3.3.1 Storing A-CAC Equipment

When you're done with the machine, perform the following steps to store it:

Step 1 Switch off the welding machine. Unplug or de-energize it.

Step 2 Switch off the compressor or close the air supply valve, as applicable. Bleed pressure from the line and disconnect it from the torch supply cable.

Step 3 Disconnect the electrode lead from the torch supply cable. Slide the boot over the supply cable end.

Step 4 Coil all cables neatly and store them.

Step 5 Return the welding machine to storage.

Step 6 Clean up the work area. Sweep up any slag or debris. Handle hot debris carefully to avoid burns or starting a fire.

Step 7 Store unused electrodes properly so they stay clean and dry.

3.3.2 Basic Maintenance

A-CAC equipment maintenance is relatively basic. Mostly, you need to check the torch, as well as its hoses and cables, for damage. The following procedure outlines torch inspection and maintenance:

- Check the torch heads for damage, excessive deposits, or pitting. Discard and replace ones that are badly damaged. Clean off minor deposits and pits with emery paper.
- Check the torch itself. Confirm that the jaws open and close properly. Examine the air valve and make sure that it operates smoothly.
- Check the torch supply cable for burns, abrasions, or missing insulation. Examine the supply cable connectors (air and electrical) for damage, deposits, or corrosion. Confirm that the boot is in good condition. If you can't correct any damage that you discover, replace the cable or the entire torch.
- Check the welding machine's cables for burns, abrasions, or missing insulation. If you can't correct any damage that you discover, replace the cables.
- Check the air supply hose for burns or abrasions. Confirm that its fittings aren't corroded or damaged. If you can't correct any damage that you discover, replace the hose.

3.0.0 Section Review

1. Which factor influences the groove *depth* when gouging with A-CAC?
   a. The electrode size
   b. The electrode material
   c. The electrode angle
   d. The electrode polarity

2. Washing prepares a base metal surface by _____.
   a. introducing additional metal alloys to the surface
   b. creating a narrow groove
   c. heating it up evenly, followed by a slow cooldown
   d. removing objects and projections

3. Maintaining an A-CAC machine mostly involves _____.
   a. testing the welding machine's voltage
   b. checking the torch and cables
   c. cleaning the outlet air filter
   d. removing pits from the electrode

Module 29104 Review Questions

1. An A-CAC torch uses compressed air to blow the _____.
   a. fume plume away from the cut
   b. molten metal away from the cut
   c. contamination away from the cut
   d. metal vapor away from the cut

2. A-CAC melts the base metal with a high-current electric arc between the base metal and the end of a(n) _____.
   a. carbon-graphite electrode
   b. high-frequency pilot arc
   c. plasma jet
   d. idle arc

3. A-CAC work requires eye protection with a lens shade value of _____.
   a. 2 to 5
   b. 5 to 7
   c. 7 to 9
   d. 10 to 14

4. To use a multipurpose welding machine for A-CAC work, switch it to _____.
   a. constant current DC mode
   b. DC constant voltage mode
   c. High-frequency mode
   d. DC open circuit mode

5. Which welding machine would be appropriate for light-duty A-CAC work?
   a. 10A unit
   b. 200A unit
   c. 600A unit
   d. 1,200A unit

6. When cutting and gouging with A-CAC equipment, the torch head orifices should be _____.
   a. to one side of the electrode
   b. above the electrode
   c. inside the electrode
   d. below the electrode

7. What do rare earth metals do for A-CAC electrodes?
   a. They prevent the electrodes from absorbing moisture and exploding.
   b. They make the electrodes last longer when they're supplied by a DC machine.
   c. They make the electrodes stronger, so they can be made longer than 12" (~30 cm).
   d. They make the arc stable when the electrodes are used with an AC machine.

8. The electrode typically used for general-purpose A-CAC cutting and gouging is the _____.
   a. flat electrode
   b. round electrode
   c. half-round electrode
   d. special-shape electrode

9. When gouging with an A-CAC torch, which factor determines the *narrowest* groove that you can cut?
   a. The torch size
   b. The electrode size
   c. The cutting current
   d. The electrode angle

10. Connect the A-CAC torch's compressed air supply to _____.
   a. the welding machine's shielding gas port
   b. the connector on the supply cable
   c. the welding machine's positive (+) lead
   d. the welding machine's negative (–) lead

11. One way to make a groove deeper when gouging with an A-CAC torch is to _____.
   a. increase the cutting current
   b. switch to a copper-clad electrode
   c. increase the travel speed
   d. decrease the air supply pressure

12. When washing a projection from a metal plate with an A-CAC torch, hold the electrode at a _____.
   a. 30-degree angle to the surface
   b. 35- to 45-degree angle to the surface
   c. 50- to 60-degree angle to the surface
   d. 90-degree angle to the surface

13. Washing heads that have orifices on one side of the electrode limit cutting to _____.
   a. one direction only
   b. two directions only
   c. vertical orientation cutting only
   d. horizontal orientation cutting only

14. If you leave dross or carbon deposits in an A-CAC groove and then create a weld, the finished weld will be _____.
   a. hard and brittle with defects
   b. oxidized and/or corroded
   c. stronger than necessary
   d. acceptable for most applications

15. If you discover that an A-CAC torch's heads are badly pitted, you should _____.
   a. grind them smooth with a wire wheel
   b. use a bead blaster to remove the pitting
   c. sand them with emery paper
   d. replace them

Answers to odd-numbered Module Review Questions are found in *Appendix A*.

Cornerstone of Craftsmanship

Brian Dennis

Corporate Quality Manager

United Group Services

How did you choose a career in the industry?

I knew at a young age I was not going to be the "college type." I opted to go to trade school my junior and senior years of high school. My father recommended I take welding since his brother was a welder and made a good living doing it.

Who inspired you to enter the industry?

My father.

What types of training have you been through?

Two years of welding trade school, multiple safety and OSHA courses, AWS/CWI, CPWI+, Radiographic Interpretation, ASME seminars, and various AWS seminars.

How important is education and training in construction?

The construction industry is everchanging. Being educated in the latest technologies and processes is crucial in being able to keep up and stay ahead.

What kinds of work have you done in your career?

I have worked in many aspects of the industry as a pipe/pressure vessel and structural welder in the late 90's and 2000's. Some examples would be power plants, steel mills, pharmaceutical, paper mills, and food and beverage processing.

Tell us about your current job.

For the past twelve years, I have been the Corporate Quality Manager for United Group Services. My typical job duties include welder qualification, weld procedure qualification, ASME/NBIC code work, project audits, and quality management in general.

What has been your favorite project?

I always enjoyed shop fabrication.

What do you enjoy most about your job?

In my current position, I enjoy the fact that I am able to travel and am not stuck in the same place for weeks at a time.

What factors have contributed most to your success?

Wanting to stand out because of my work ethic and welding abilities.

Would you suggest construction as a career to others? Why?

Yes, in these current times a skilled construction worker can make as much (and a lot of time more) as a collage graduate and not be overwhelmed with student debt.

What advice would you give to those new to the field?

Pick a trade you enjoy and stick with it. Learn as much about it as you can.

Interesting career-related fact or accomplishment?

I was the youngest person promoted to Corporate Quality Manager in my company's history.

How do you define craftsmanship?

Taking the time to do it right the first time and taking pride in what you do, knowing not everyone can do these things.

Answers to Section Review Questions

Answer	Section Reference	Objective
Section 1.0.0		
1. a	1.1.0	1a
2. c	1.2.0	1b
Section 2.0.0		
1. b	2.1.0	2a
2. b	2.2.1	2b
3. d	2.3.1	2c
Section 3.0.0		
1. c	3.1.2; 3.2.1	3a; 3b
2. d	3.2.2	3b
3. b	3.3.2	3c

Base Metal Preparation

Objectives

Successful completion of this module prepares you to do the following:

1. Summarize metal properties and safely preparing metals for welding.
   a. List safety procedures for preparing metals.
   b. List the types and properties of carbon, alloy, and stainless steels.
   c. Outline cleaning metals correctly and safely.
2. Outline joint and weld types.
   a. Summarize loads and the ways they affect welded joints.
   b. Summarize joint and weld types, along with their qualities.
   c. Describe a Welding Procedure Specification (WPS).
3. Outline preparing joints for welding.
   a. Describe mechanically preparing joints for welding.
   b. Describe thermally preparing joints for welding.

Performance Tasks

Under supervision, you should be able to do the following:

1. Mechanically create a $22\frac{1}{2}$-degree bevel ($\pm2\frac{1}{2}$ degrees) on the edge of a $\frac{1}{4}$" to $\frac{3}{4}$" (~6 mm to 20 mm) thick carbon steel plate.
2. Thermally create a $22\frac{1}{2}$-degree bevel ($\pm2\frac{1}{2}$ degrees) on the edge of a $\frac{1}{4}$" to $\frac{3}{4}$" (~6 mm to 20 mm) thick carbon steel plate.

Overview

Before welders can join metals together, they must properly prepare them. Preparation includes cutting, shaping, and cleaning the metals. Welders can do these tasks correctly only when they understand those metals' properties. This module introduces metals, preparation techniques, and safety. It also introduces standard weld joints and the codes that govern them.

NOTE

This module uses US standard and metric units in up to three different ways. This note explains how to interpret them.

Exact Conversions

Exact metric equivalents of US standard units appear in parentheses after the US standard unit. For example: "Measure 18" (45.7 cm) from the end and make a mark."

Approximate Conversions

In some cases, exact metric conversions would be inappropriate or even absurd. In these situations, an approximate metric value appears in parentheses with the ~ symbol in front of the number. For example: "Grip the tool about 3" (~8 cm) from the end."

Parallel but not Equal Values

Certain scenarios include US standard and metric values that are parallel but not equal. In these situations, a slash (/) surrounded by spaces separates the US standard and metric values. For example: "Place the point on the steel rule's 1" / 1 cm mark."

Digital Resources for Welding

Scan this code using the camera on your phone or mobile device to view the digital resources related to this craft.

Industry Recognized Credentials

If you are training through an NCCER-accredited sponsor, you may be eligible for credentials from NCCER's Registry. The ID number for this module is 29105. Note that this module may have been used in other NCCER curricula and may apply to other level completions. Contact NCCER's Registry at 1.888.622.3720 or go to **www.nccer.org** for more information.

You can also show off your industry-recognized credentials online with NCCER's digital badges. Transform your knowledge, skills, and achievements into badges that you can share across social media platforms, send to your network, and add to your resume. For more information, visit **www.nccer.org**.

1.0.0 Basic Welding Safety and Cleaning

Performance Tasks

There are no Performance Tasks in this section.

Objective

Summarize metal properties and safely preparing metals for welding.

a. List safety procedures for preparing metals.
b. List the types and properties of carbon, alloy, and stainless steels.
c. Outline cleaning metals correctly and safely.

Base metal: A metal to be welded, cut, or brazed.

Metallurgy: The science of metals, their chemistry, and their mechanical properties.

Excellent welds don't just happen. They require the welder to prepare the **base metal** carefully and correctly before welding it. To prepare metals, welders must learn some basic **metallurgy**. Understanding a metal's properties helps craftworkers produce welds that meet or exceed code requirements. As with any welding process, however, safety remains a top priority.

1.1.0 Safety Practices

Welders prepare base metals in several ways. They can use hand or powered tools to cut, shape, and clean them (*Figure 1*). Or they can use thermal (heat) processes to do the work (*Figure 2*). Trainee welders learn these skills early since they're vital to all welding work.

Figure 1 Grinding metal.
Source: The Lincoln Electric Company, Cleveland, OH, USA

Figure 2 Cutting metal with an oxyfuel torch.
Source: © Miller Electric Mfg. LLC

All tools have their specific hazards. Cutting, grinding, and cleaning metal produces dust, fragments, and hot sparks. These can burn skin or injure eyes. Rotating tools like cutting discs and wire brushes can fling sharp objects and particles around the room. Wearing the right PPE protects the craftworker from these (*Figure 3*). Welding screens around the work area protect others working nearby.

Preparing metal can start fires, so think ahead and take precautions. Welding screens and blankets protect combustible materials that you can't move. Always have a suitable fire extinguisher ready. Follow all company fire watch protocols.

Think about electrical safety when using powered tools. Powered equipment has unique hazards. Disconnect all tools from the power supply before working on them. Be particularly careful with energized welding and cutting machines.

When possible, prepare metal in the shop where you have safe workspaces and good clamping devices. If you must work in the field, carefully secure parts before working on them. If you must work at height, wear fall prevention and fall arrest equipment. Inspect this equipment every time you use it.

Figure 3 Worker with PPE preparing metal.
Source: The Lincoln Electric Company, Cleveland, OH, USA

NOTE

Be sure that you've completed NCCER Module 29101, *Welding Safety*, before continuing.

This section summarizes good practices but isn't a complete safety training. Always complete all training that your workplace requires. Follow all workplace safety guidelines and wear appropriate PPE too.

1.1.1 Protective Clothing and Equipment

Welding and cutting tasks are dangerous. Unless you wear all required PPE, you're at risk of serious injury. The following guidelines summarize PPE for this work:

- For cutting, always wear close-fitting goggles or safety glasses, plus a full face shield, helmet, or hood. The goggles, glasses, face shield, or helmet/hood lens must have the proper shade value for the type of cutting work (*Figure 4*). Never cut without using the proper lens.
- For welding, always wear safety glasses, plus a helmet or hood. The glasses or helmet/hood lens must have the proper shade value for the welding process (*Figure 4*). Never view the cutting arc directly or indirectly without using the proper lens.

WARNING!

Don't open your welding helmet/hood lens and expose your eyes when grinding or cleaning a weld. Either use a helmet or hood with an auto-darkening lens or a clear lens under a flip-up shaded lens. Alternatively, remove the helmet or hood and use a full face shield.

- Wear protective leather and/or flame-retardant clothing along with welding gloves that will protect you from flying sparks and molten metal.
- Wear high-top safety shoes or boots. Make sure the pant leg covers the tongue and lace area. If necessary, wear leather spats or chaps for protection. Boots without laces or holes are better for welding and cutting.
- Wear a 100 percent cotton cap with no mesh material in its construction. The bill should point to the rear or the side with the most exposed ear. If you must wear a hard hat, use one with a face shield.

WARNING!

Never wear a cap with a button on top. The conductive metal under the fabric is a safety hazard.

- Wear earplugs to protect your ear canals from sparks. Since the torch makes a continuous noise, always wear hearing protection.

1.1.2 Fire/Explosion Prevention

Metal cutting equipment, especially oxyfuel torches, causes many welding environment fires. Operators must be properly trained in oxyfuel equipment before using it. Even tools like grinders produce sparks that can start fires. Never use a cutting tool without proper training.

When possible, do cutting operations in a dedicated welding shop space. This area should have a concrete floor, protective drapes, and appropriate fire extinguishers. When you must cut outside the shop, remove all flammables from the work area. If this isn't possible, set up protective screens and cover flammables with welding blankets (*Figure 5*). Most companies require posting at least one fire watch with a fire extinguisher during hot work operations.

Thermal cutting equipment produces extremely high temperatures to cut metal. Welding or cutting a vessel or container that once contained combustible, flammable, or explosive materials is hazardous. Residues can catch fire or explode. Before welding or cutting vessels, check whether they contained explosive, hazardous, or flammable materials. These include petroleum products, citrus products, or chemicals that release toxic fumes when heated.

Arc Welding Processes

Process	Electrode	Amperage	Lens Shade Numbers
			2 3 4 5 6 7 8 9 10 11 12 13 14
Shielded Metal Arc Welding (SMAW)	< 3/32" (2.4 mm)	< 60A	7
	3/32" – 5/32" (2.4 mm – 4.0 mm)	60A – 160A	8 9 10
	5/32" – 1/4" (4.0 mm – 6.4 mm)	160A – 250A	10 11 12
	> 1/4" (6.4 mm)	250A – 550A	11 12 13 14
Gas Metal Arc Welding (GMAW) and Flux Cored Arc Welding (FCAW)		< 60A	7
		60A – 160A	10 11
		160A – 250A	10 11 12
		250A – 500A	10 11 12 13 14
Gas Tungsten Arc Welding (GTAW)		< 50A	8 9 10
		50A – 150A	8 9 10 11 12
		150A – 500A	10 11 12 13 14
Plasma Arc Welding (PAW)		< 20A	6 7 8
		20A – 100A	8 9 10
		100A – 400A	10 11 12
		400A – 800A	11 12 13 14
Carbon Arc Welding (CAW)			14

Arc Cutting Processes

Process	Electrode	Amperage	Lens Shade Numbers
			2 3 4 5 6 7 8 9 10 11 12 13 14
Plasma Arc Cutting (PAC)		< 20A	4
		20A – 40A	5
		40A – 60A	6
		60A – 80A	8
		80A – 300A	8 9
		300A – 400A	9 10 11 12
		400A – 800A	10 11 12 13 14
Air-Carbon Arc Cutting (A-CAC)	Light	< 500A	10 11 12
	Heavy	500A – 1,000A	11 12 13 14

Gas Processes

Process	Type	Plate Thickness	Lens Shade Numbers
			2 3 4 5 6 7 8 9 10 11 12 13 14
Oxyfuel Welding (OFW)	Light	< 1/8" (3 mm)	4 5
	Medium	1/8" – 1/2" (3 mm – 13 mm)	5 6
	Heavy	> 1/2" (13 mm)	6 7 8
Oxyfuel Cutting (OC)	Light	< 1" (25 mm)	3 4
	Medium	1" – 6" (25 mm – 150 mm)	4 5
	Heavy	> 6" (150 mm)	5 6
Torch Brazing (TB)			3 4
Torch Soldering (TS)			2

Lens shade values are based on OSHA minimum values and ANSI/AWS suggested values. Choose a lens shade that's too dark to see the weld zone. Without going below the required minimum value, reduce the shade value until it's visible. If possible, use a lens that absorbs yellow light when working with fuel gas processes.

Figure 4 Guide to lens shade numbers.

Figure 5 Welding blanket.

American Welding Society (AWS) F4.1, Safe Practices for the Preparation of Containers and Piping for Welding and Cutting, and *ANSI/AWS Z49.1* describe safe practices for these situations. Begin by cleaning the container to remove any residue. Steam cleaning, washing with detergent, or flushing with water are possible methods. Sometimes you must combine these to get good results.

WARNING!

Clean containers only in well-ventilated areas. Vapors can accumulate in a confined space during cleaning, causing explosions or toxic substance exposure.

After cleaning the container, you must formally confirm that it's safe for welding or cutting. *American Welding Society (AWS) F4.1, Safe Practices for the Preparation of Containers and Piping for Welding and Cutting,* outlines the proper procedure. The following three paragraphs summarize the process.

NOTE

A *lower explosive limit (LEL)* gas detector can check for flammable or explosive gases in a container or its surroundings. Gas monitoring equipment checks for specific substances, so be sure to use a suitable instrument. Test equipment must be checked and calibrated regularly. Always follow the manufacturer's guidelines.

Immediately before work begins, a qualified person must check the container with an appropriate test instrument and document that it's safe for welding or cutting. Tests should check for relevant hazards (flammability, toxicity, etc.). During work, repeated tests must confirm that the container and its surroundings remain safe.

Alternatively, fill the container with an inert material ("inerting") to drive out any hazardous vapors. Water, sand, or an inert gas like argon meets this requirement. Water must fill the container to within 3" (~7.5 cm) or less of the work location. The container must also have a vent above the water so air can escape as it expands (*Figure 6*). Sand must completely fill the container.

When using an inert gas, a qualified person must supervise, confirming that the correct amount of gas keeps the container safe throughout the work. Using an inert gas also requires additional safety procedures to avoid accidental suffocation.

WARNING!

Never weld or cut drums, barrels, tanks, vessels, or other containers until they have been emptied and cleaned thoroughly. Residues, such as detergents, solvents, greases, tars, or corrosive/caustic materials, can produce flammable, toxic, or explosive vapors when heated. Never assume that a container is clean and safe until a qualified person has checked it with a suitable test instrument. Never weld or cut in places with explosive vapors, dust, or combustible products in the air.

The following general fire- and explosion-prevention guidelines apply to welding and cutting work:

- Never carry matches or gas-filled lighters in your pockets. Sparks can ignite the matches or cause the lighter to explode.
- Always comply with all site and employer requirements for hot work permits and fire watches.
- Never use oxygen in place of compressed air to blow off the workpiece or your clothing. Never release large amounts of oxygen. Oxygen makes fires start readily and burn intensely. Oxygen trapped in fabric makes it ignite easily if a spark falls on it. Keep oxygen away from oil, grease, and other petroleum products.
- Remove all flammable materials from the work area or cover them with a fire-resistant blanket.
- Before welding, heating, or cutting, confirm that an appropriate fire extinguisher is available. It must have a valid inspection tag and be in good condition. All workers in the area must know how to use it.

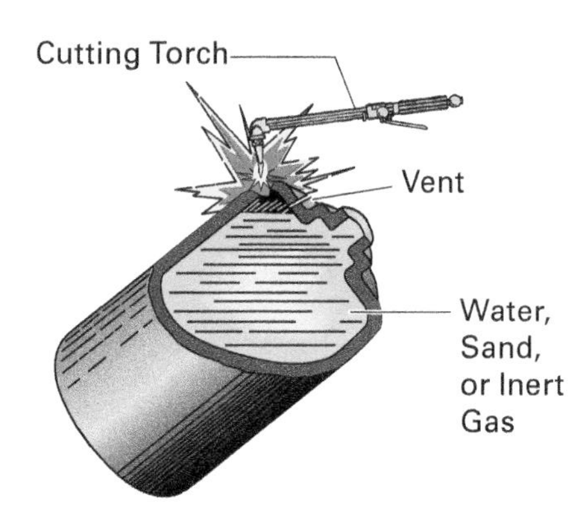

Note: *ANSI Z49.1* and AWS standards should be followed.

Figure 6 "Inerting" a container.

- Never release a large amount of fuel gas into the work environment. Some gases (propane and propylene) accumulate on the floor, while others (acetylene and natural gas) accumulate near the ceiling. Either can ignite far from the release point. Acetylene will explode at lower concentrations than other fuel gases.
- Prevent fires by maintaining a neat and clean work area. Confirm that metal scrap and slag are cold before disposing of them.

1.1.3 Work Area Ventilation

Vapors and fumes fill the air around their sources. Welding and cutting fumes can be hazardous. Welders often work above the area from which fumes come. Good ventilation and source extraction equipment help remove the vapors and protect the welder. The following lists good ventilation practices to use when welding and cutting:

- Always weld or cut in a well-ventilated area. Operations involving hazardous materials, such as metal coatings and toxic metals, produce dangerous fumes. You must wear a suitable respirator when working with them. Follow your workplace's safety protocols and confirm that your respirator is in good condition.
- Mechanically cutting and grinding metals produces metal dust. All metal dust is hazardous, but toxic metal dust is especially dangerous. Know the metals you're working with and wear appropriate respiratory PPE.

WARNING!

Cadmium, mercury, lead, zinc, chromium, and beryllium produce toxic fumes when heated. When cutting these materials, wear appropriate respiratory PPE. Supplied-air respirators (SARs) are best for long-term work.

- Never weld or cut in a confined space without a confined space permit. Prepare for working in the confined space by following all safety protocols.
- Set up a confined space ventilation system before beginning work.
- Never ventilate with oxygen or breathe pure oxygen while working. Both actions are very dangerous.
- Grinding wheels containing aluminum oxide, silicon carbide, or other abrasives are a respiratory hazard. Consult their SDS before starting any grinding activities. Wear appropriate respiratory PPE.

WARNING!

When working around other craftworkers, always be aware of what they're doing. Take precautions to keep your work from endangering them.

Going Green

Welding Water and Fumes

Many companies are taking measures to protect the environment from welding byproducts. Rather than letting wastewater run onto the ground or down a storm drain, collect it for proper processing. Use source extraction equipment that filters the air rather than simply venting welding fumes into the environment.

1.2.0 Properties of Steel

Of the 118 chemical elements, about 95 are metals. Industry uses around 70 of these. Metals share certain common qualities. They're shiny when polished. They can carry an electric current and transfer heat efficiently. Most are **malleable** and **ductile**. Some, like iron, aluminum, and titanium, are quite strong.

Malleable: Metals that can be shaped by hammering or pressure and not crack.

Ductile: Metals that can be significantly stretched without breaking.

1.2.1 Basic Metallurgy

Over several thousand years, humans have learned to work with metals. Combining several metals or mixing metals with nonmetals produces an *alloy*. Alloys are better than a single, pure metal. They're usually stronger, harder, or resist corrosion better. Most industrial metals are alloys.

Tempering: Heating and then slowly cooling metal to improve its strength or toughness.

Quenching: Rapidly cooling a metal by plunging it into a liquid.

Annealing: Heating and then gradually cooling a metal to relieve its internal stresses.

Ferrous metals: Metals containing iron.

Nonferrous metals: Metals that don't contain significant amounts of iron.

Wrought: Metal shaped by beating with a hammer.

NOTE

The term *metallurgical properties* refers to a metal's chemical and mechanical properties. These depend on the metal's composition and manufacturing process. Post-manufacturing processes like heat treatments affect metallurgical properties too. Welders can intentionally or unintentionally change a base metal's metallurgical properties through the welding process.

Casting: A method of making metal objects by pouring molten metal into a mold.

Heat treating metals alters their mechanical properties. Heat treatment processes involve heating the metal to a specific temperature. It's then cooled, either quickly or slowly. **Tempering** metals makes them tougher. Heating and then **quenching** a metal makes it harder but less ductile and more brittle. Tempering after quenching reduces this undesirable brittleness. **Annealing** a metal reduces stress inside its structure. It's more ductile and malleable afterwards.

Metallurgists divide metals into two broad categories: **ferrous metals** and **nonferrous metals**. Ferrous metals contain mostly iron. They include cast iron, **wrought** iron, malleable iron, ductile iron, and all steels. Steel is an alloy of iron and carbon (and often other metals too). Ferrous metals are the backbone of modern industry and construction.

Nonferrous metals contain little or no iron. Copper, aluminum, magnesium, and titanium play major roles in industry. Other nonferrous metals like nickel, chromium, molybdenum, and manganese usually form alloys with metals like iron or aluminum. Some nonferrous metals like zinc and beryllium act as protective coatings on other metals.

1.2.2 Steels

In general, steel is more useful than pure iron, because it's harder and stronger. All steels contain carbon. Besides carbon, most steels contain small amounts of other metals. Chromium, molybdenum, manganese, nickel, and cobalt are common. Some steels contain other nonmetals like phosphorus, silicon, and sulfur.

Industry uses steels for countless applications. It's formed by **casting**, forging, rolling, machining, stamping, and extruding. Since steels are the most popular construction metals, welders need to understand them thoroughly. Each steel has its own requirements for giving quality welds.

Welding codes are very specific about metals and the issues surrounding them. For example, welders must choose electrodes that are compatible with the base metals. Using the wrong combination can produce weak or defective welds. Understanding steels is the first step towards understanding how to produce good welds. Metallurgists divide steels into three main categories.

Carbon Steels

These steels are the most popular. They contain 1.7 percent or less carbon. As its carbon content increases, a steel becomes harder but also less ductile and malleable. Carbon steels divide into the following types:

Low-carbon — 0.10–0.15 percent carbon, 0.25–1.50 percent manganese
Mild-carbon — 0.15–0.30 percent carbon, 0.60–0.70 percent manganese
Medium-carbon — 0.30–0.50 percent carbon, 0.60–1.65 percent manganese
High-carbon — 0.50–1.00 percent carbon, 0.30–1.00 percent manganese

Many general-purpose steels fall into the mild-carbon group. They machine easily and work well in numerous applications. High-carbon steels, on the other hand, harden well. They're common in edged tools and springs. But they're also harder to machine.

Alloy Steels

These steels, often called *high-strength, low-alloy (HSLA)*, contain more than trace amounts of other materials. Each alloy steel has a specific advantage, such as greater toughness, hardness, or corrosion resistance. Metallurgists have produced thousands of alloys, each ideally suited to a particular application.

Stainless Steels

These steels are alloys containing at least 11 percent chromium. They usually contain other metals, like nickel, too. Their main advantage is superior corrosion resistance. Unlike carbon steels, they don't rust. For this reason, they're extremely popular and important. There are over 150 different stainless steels.

Carbon and Alloy Steels

There are currently around 3,500 different grades of steel, each with unique properties. Most were developed in the past 50 years. In 2020, the world produced about two billion tons of steel. Much of it was carbon and alloy steel.

Source: Photo Courtesy of U.S. Army

1.2.3 AISI/SAE Steel Classification System

Organizing and classifying steels is complicated. Manufacturers have their own trade names and classification schemes. Standards organizations, such as the American Iron and Steel Institute (AISI), the Society of Automotive Engineers (SAE), the American Society of Mechanical Engineers (ASME), and ASTM International, have their own systems.

For example, the SAE and ASTM International manage the Unified Numbering System (UNS). The SAE also manages the popular AISI/SAE system, which has been around since the 1930s. It identifies most steels with a four-digit number. An identifier may include an optional letter prefix. *Figure 7* shows what each symbol represents.

The letter prefix, if present, identifies the process that created the steel. Letters have the following meanings:

A — Open-hearth steel
B — Acid-Bessemer carbon steel
C — Basic open-hearth carbon steel

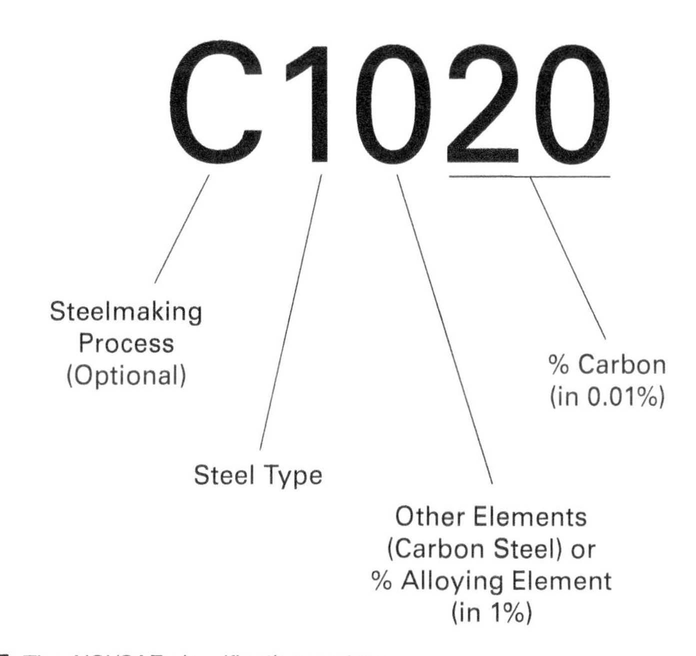

Figure 7 The AISI/SAE classification system.

D — Acid open-hearth carbon steel
E — Electric furnace steel

The first digit identifies the steel type. Carbon steel identifiers, for example, all begin with 1. Numbers have the following meanings:

1 — Carbon steels
2 — Nickel steels
3 — Nickel-chromium steels
4 — Molybdenum steels
5 — Chromium steels
6 — Chromium-vanadium steels
7 — Tungsten steels
8 — Nickel-chromium-molybdenum steels
9 — Silicon-manganese steels

For carbon steels, the second digit identifies other elements added to the steel. Numbers have the following meanings:

0 — Plain carbon steel with a maximum of 1.00 percent manganese
1 — Resulfurized carbon steel
2 — Resulfurized and rephosphorized carbon steel
3 — Carbon steel with up to 1.75 percent manganese
5 — Plain carbon steel with a maximum of 1.00–1.65 percent manganese

For other steel types, the second digit identifies the approximate percentage of the main alloying element. As an example, nickel steel with a 3 as the second digit contains about 3.5 percent nickel.

The third and fourth digits indicate approximately how much carbon the steel contains in hundredths of a percent.

For example, AISI/SAE C1020 is a regular carbon steel (1). It has no significant alloying elements (0). It contains 0.20 percent carbon ($20 \times {}^1\!/_{100} = 0.20$ percent). The C prefix indicates that it was produced by the basic open-hearth process.

As another example, AISI/SAE E2512 is a nickel steel (2). The 5 indicates that it contains 5 percent nickel. It contains 0.12 percent carbon ($12 \times {}^1\!/_{100} = 0.12$ percent). The E prefix indicates that it was produced by the electric furnace process.

NOTE

Some AISI/SAE identifiers have *five* digits instead of four. These identify steels containing more than 1 percent carbon. The first two digits have the usual meaning. The last three digits indicate the carbon content in hundredths of a percent. For example, if the last three digits are 100, the steel contains 1 percent carbon ($100 \times {}^1\!/_{100} = 1$ percent).

1.2.4 Stainless Steels

Stainless steels are steel alloys containing significant amounts of chromium. They come in many types, but all have an important quality—they don't rust. Instead, they form a transparent layer that protects the metal underneath.

Metallurgists form the many different stainless steel alloys by adding other metals. These include nickel, manganese, molybdenum, and silicon. The alloying elements enhance desirable properties. For example, some stainless steels resist heat exceptionally well.

Stainless steels do have several disadvantages. They don't conduct heat as well as mild steels. They expand and contract with temperature changes more than mild steel does. These qualities can lead to **distortion** during welding. Stainless steels don't conduct electricity as well as mild steels do either, so they get hotter during welding.

Distortion: Changes in a welded assembly's shape caused by its parts changing size as the weld joint is heated and cooled.

Metallurgists classify stainless steels by the way their atoms form crystals. Both manufacturing and heat treatment processes affect crystallization. **Austenitic** stainless steel isn't magnetic when it's annealed. Heat treatment won't harden it either. **Cold working**, on the other hand, will increase its hardness. Austenitic stainless steels resist oxidation and heat very well. They also perform effectively at low temperatures. Because of their useful mechanical properties, austenitic stainless steels are the most common type. Welders encounter them frequently.

NOTE

Austenitic stainless steels may become magnetized after welding or cold working. Heat treating usually removes this magnetization.

Ferritic stainless steels are magnetic. Heat treatment won't harden them either, but cold working will. These steels have decent oxidation and heat resistance. Their mechanical properties are generally good. They don't have as wide a temperature range, however, as austenitic stainless steels do. They're also harder to weld. Ferritic stainless steels contain a significant percentage chromium but few other elements. They contain less than 0.20 percent carbon.

Martensitic stainless steels are magnetic. Heat treatments, such as quenching and tempering, will harden them. This quality makes them good for edged tools. They work well in mild environments and can tolerate weak acids. They can't resist highly corrosive chemicals. Martensitic stainless steels are harder to weld. They contain relatively low percentages of chromium but high amounts of carbon. Some contain nickel to improve their corrosion resistance and toughness.

Austenitic: A nonmagnetic type of stainless steel whose internal crystal structure prevents it from being hardened by heat treatment.

Cold working: Methods of shaping metal without applying heat.

Ferritic: A magnetic type of stainless steel whose internal crystal structure prevents it from being hardened by heat treatment.

Martensitic: A magnetic type of stainless steel whose internal crystal structure allows it to be hardened by heat treatment.

1.2.5 Stainless Steel Classification Systems

Several different systems classify stainless steels. The AISI/SAE system is popular and well understood. It identifies each alloy with a three-digit number. The first digit indicates the general stainless steel category:

100 series — General-purpose austenitic stainless steels
200 series — Austenitic chromium-nickel-manganese stainless steels
300 series — Austenitic chromium-nickel stainless steels
400 series — Ferritic and martensitic chromium stainless steels
500 series — Heat-resistant chromium stainless steels
900 series — Austenitic chromium-molybdenum stainless steels

Unlike AISI/SAE identifiers for carbon and HSLA steels, however, the second and third digits are just identifiers and don't have specific meanings.

For example, 304 stainless steel is one of the most popular varieties. It's an austenitic stainless steel containing 16–24 percent chromium. It may also contain up to 35 percent nickel. It contains relatively small amounts of carbon.

Some stainless steels come in both low- and high-carbon forms. An L or an H after the number identifies these. For example, 304L is a low-carbon form of 304 stainless steel. Low-carbon stainless steels are useful in pipe fittings that can't be annealed after welding. High-carbon stainless steels are good in extreme temperature conditions.

The Unified Numbering System (UNS) offers another way to identify stainless steels. Identifiers begin with the letter S ("stainless steel"), followed by a five-digit number. The first three digits match the AISI/SAE number. The remaining two digits identify alloy variants. For example, S30400 is the UNS equivalent of 304 stainless steel.

NOTE

Historically, stainless steels have had other classification systems. One type used number pairs to identify the percentage of chromium and nickel in the metal. Examples include $^{1}/_{8}$, $^{18}/_{10}$, and $^{25}/_{20}$. Today, most people use the AISI/SAE system or the UNS.

NOTE

The UNS classifies other metals too. The initial letter identifies the metal type. For example, an A indicates aluminum. The five-digit number identifies the specific alloy and its variants.

1.3.0 Cleaning Base Metals

All base metals require cleaning before welding. Even though a metal may look clean, it's probably not. Manufacturing can leave an oily film on the surface. Handling and shipping often contaminate metals too. Equipment in service may have protective coatings, like paint. Most working equipment will be oily or greasy. Heating contaminated metals produces fumes that may be toxic. Surface contaminants can also harm the weld by creating defects.

Many metals *oxidize* when they combine with oxygen from the air. Moisture speeds up oxidation. Surface **oxides** can be white or even transparent. Just because a metal looks shiny doesn't mean that it's oxide-free. Aluminum and stainless steels, for example, form invisible oxide layers. These are beneficial because they protect the surface from further oxidation.

Oxides: A scale or film on metal surfaces formed by oxygen in the air combining with the metal.

A more familiar oxide, *rust*, forms on carbon steel. This highly visible oxidation is often destructive. Some steels form rust in large flakes that break away, exposing the steel to further rusting (*Figure 8*). Manufacturers sometimes add chromium and copper to carbon steel to reduce this problem. These **weathering steels** still rust, but the rust layer is very fine. It bonds tightly to the surface and protects it from further rusting. Weathering steels work well for outdoor applications.

Weathering steels: Steel alloys that form a dense oxide layer on their surface, which prevents further oxidation.

Codes require welders to clean metal before thermally cutting or welding it. Mechanical and/or chemical processes remove contaminants, oxide films, rust, and scale. The prepared surface must also be free from paint, oil, and grease. These could harm the weld or base metal, creating defects.

Porosity: Gas pockets, or voids in weld metal.

Piping porosity: Porosity approximately perpendicular to the weld face that is deeper than it is wide. Sometimes called *wormholes*.

Porosity is a common example (*Figure 9*). Gases from the contaminants dissolve in the weld pool and form voids (holes) when the weld solidifies. Voids from **piping porosity** (wormholes) come from gas bubbles floating upwards through successive weld layers. Porosity weakens the weld. It may or may not be visible at the surface.

Properly handling and storing metals reduces contamination. Never handle metals with oily gloves. Store different metals separately. Keep aluminum indoors. Carbon steel resting on alloy steels can contaminate them. Even handling some metals with carbon steel equipment like chains and forklifts can be a problem. Marking metals with chlorine-based markers will contaminate them.

Figure 8 Destructive rusting.
Source: iStock@JJSINA

1.3.1 Mechanical Cleaning

Mechanical cleaning methods remove contaminants by scraping or abrading them off. These methods utilize both hand and powered tools. Since mechanical cleaning produces dust and flying debris, always wear suitable PPE.

Flexible or rigid scrapers remove dirt, grease, and paint from component surfaces. Wire brushes, both hand and powered, remove paint and surface oxidation. These may have carbon steel, stainless steel, or brass bristles. Files remove deeper surface oxidation. When possible, use fine files rather than coarse ones. Clean their teeth regularly so they don't scratch surfaces (*Figure 10*).

WARNING!

Never use a file without a handle. The sharp tang can easily pierce your hand.

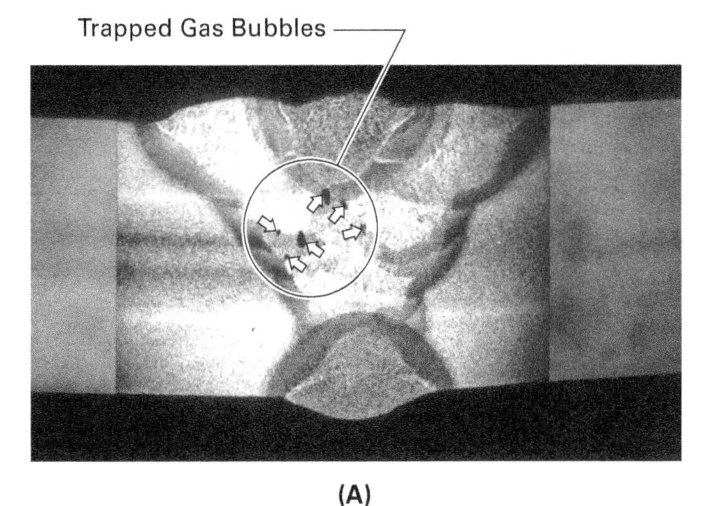

(A)

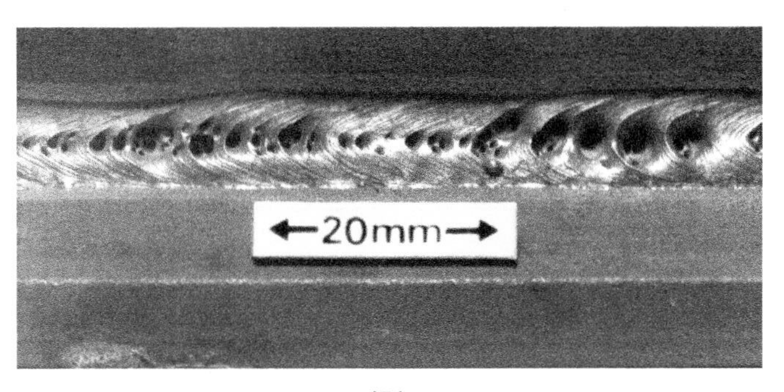

(B)

Figure 9 Porosity.
Source: Image copyright TWI Ltd, reproduced by permission

Figure 10 File and cleaning tools.

Remove tight oxidation with an electric or pneumatic grinder. Angle grinders clean large surfaces quickly. Die grinders and small angle grinders clean weld grooves and bevel edges. Die grinders also work well for cleaning the insides of pipes. Grinders often have specialized attachments. These include grinding discs, cutoff wheels, wire brushes, rotary files, and flapper wheels (*Figure 11*).

WARNING!

Use welding screens to contain the sparks produced by grinding tools. Always wear appropriate PPE.

Grinding discs and cutoff wheels are metal-specific. Use the correct one for the metal you're cleaning. For example, grind aluminum or stainless steel with aluminum oxide discs. When grinding, be careful not to reduce the base metal below the minimum thickness.

WARNING!

Grinding discs and cutoff wheels can shatter without warning. The tool will fling the sharp fragments perpendicularly to the shaft. Never stand in this position or let anyone else do so. Always wear a full face mask and gloves when grinding or cutting.

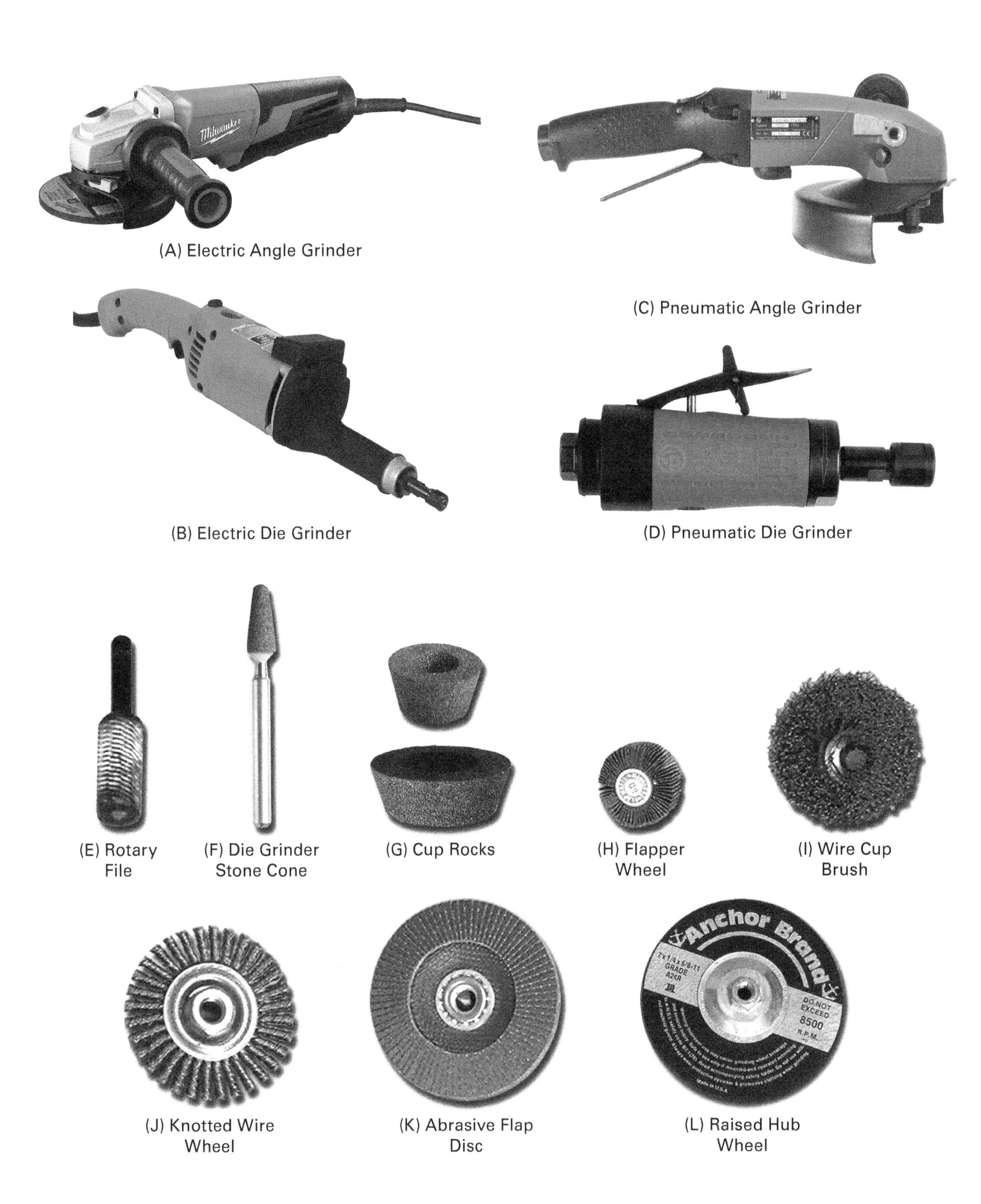

Figure 11 Handheld grinders and attachments.
Sources: Photos courtesy of Milwaukee Tool (11A–11B); Courtesy of Chicago Pneumatic Tools (11C–11D)

Weld slag chippers and needle scalers are pneumatic tools (*Figure 12*). They clean surfaces and remove slag from welds. They're also handy for removing paint and hardened dirt. They don't remove corrosion very well. A weld slag chipper has a single chisel, while a needle scaler has 18–20 blunt steel needles. Most weld slag chippers can convert to needle scalers with an attachment.

Sandblasting is excellent for removing major corrosion. It's a messy process, however, so work in a suitable place and try to confine the abrasive. It's not suitable for cleaning aluminum. Always wear appropriate PPE and respiratory protection.

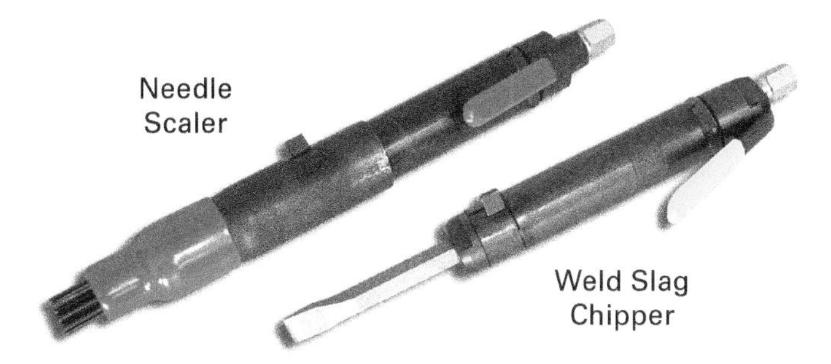

Figure 12 Pneumatic weld slag chipper and needle scaler.

1.3.2 Avoiding Contamination

When cleaning, be very careful not to contaminate the weld zone. Contaminants will produce defective welds. The tool itself can contaminate the metal. For example, a wire brush with carbon steel bristles will contaminate stainless steel or aluminum. Always use stainless steel bristles on these metals. If possible, use bristles made from low-carbon stainless steel.

Tools can transfer contaminants too. For example, using a stainless steel brush on carbon steel contaminates the bristles. If you later use the brush on stainless steel, you'll contaminate the metal with the carbon steel dust. Similarly, using a file, grinding disc, or cutoff wheel on carbon steel and then on stainless steel will contaminate the stainless steel. Using a contaminated file card or brush on a file can transfer contaminants to its teeth.

NOTE

To prevent cross-contamination, use separate sets of tools for each metal type. Label each by metal name to prevent confusion.

Even metal workbenches and supports, like pipe jacks, can contaminate metals. For example, placing stainless steel parts on a carbon steel workbench will contaminate them. Cover incompatible surfaces to prevent these problems.

1.3.3 Chemical Cleaning

Sometimes, mechanical cleaning isn't enough. Surface films, oils, and other residues may remain to contaminate the weld. Following mechanical cleaning with a chemical cleaner readies the metal for welding.

Always use compatible cleaners. Read the product instructions and SDS before using a cleaner. Many are flammable or toxic. Some give off dangerous fumes when used around welding processes that generate ultraviolet light. Whenever possible, use the safest and least aggressive chemical possible.

WARNING!

Handle all chemicals carefully. Consult the SDS for the proper PPE. It will always include eye protection and gloves. Many chemicals require additional skin and respiratory protection. Treat all chemical contacts or injuries seriously. Follow the SDS's guidelines when managing chemical accidents and emergencies.

Acetone is a popular cleaning solvent. It's colorless, has a sharp odor, and is highly flammable. It dissolves resins and epoxies. It also melts many plastics. Acetone removes oils and greases. It's a respiratory irritant and can injure eyes. Prolonged exposure may cause liver damage.

Methyl ethyl ketone (MEK or butanone) is another popular solvent. It's colorless, has a sweet, sharp odor, and is highly flammable. Like acetone, it dissolves many surface contaminants. It too is a respiratory irritant and can injure eyes. Its long-term effects are uncertain, so always read a current SDS before using it.

Some locations, such as Southern California, have banned MEK as an environmental pollutant. Other states restrict its use. Check state and local regulations before using it.

WARNING!

Acetone and MEK are both extremely flammable. Acetone has an autoignition temperature of 869°F (465°C). MEK's auto ignition temperature is 759°F (403°C). Keep these solvents away from any ignition source, including hot plates.

Never use a solvent not rated for welding applications. For example, brake cleaning fluids are *not* appropriate. Many contain chlorinated hydrocarbons. Exposing these chemicals to the UV light from a welding arc generates highly toxic phosgene gas. This gas can damage lung tissue. OSHA regulations specify that chlorinated solvents may not be used within 200' (60 m) of an exposed or unshielded welding arc. Any surface cleaned with a chlorinated solvent must be totally dry before welding.

If the surface is very oily or greasy, clean it chemically before mechanically cleaning it. This prevents unnecessarily contaminating tools with oil or grease. A second chemical cleaning may be necessary after the mechanical cleaning. If surfaces are not especially oily or greasy, mechanically clean them first.

To chemically clean a surface, apply the cleaner with a brush or spray. Give it time to dissolve the surface contaminants. Finally, wipe it off with a clean cloth. Remember, some contaminants can give off vapors when they react with the solvent. When in doubt, wear suitable respiratory protection.

WARNING!

Used cleaning rags are a fire hazard. Dispose of them in appropriate containers.

1.0.0 Section Review

1. Why does using a grinding wheel require respiratory PPE?
   a. Aluminum oxide grinding wheels give off phosgene during grinding.
   b. Silicon carbide grinding wheels can generate flammable acetylene gas.
   c. Grinding wheels produce potentially hazardous dust.
   d. Ultraviolet light from grinding generates irritating ozone gas.

2. Approximately how much carbon does AISI/SAE A1030 steel contain?
   a. 0.10 percent
   b. 0.30 percent
   c. 1.0 percent
   d. 3.0 percent

3. It's acceptable to clean stainless steel with a stainless steel wire brush that's also been used to clean carbon steel.
   a. True
   b. False

2.0.0 Joint and Weld Types

Objective

Outline joint and weld types.

a. Summarize loads and the ways they affect welded joints.
b. Summarize joint and weld types, along with their qualities.
c. Describe a Welding Procedure Specification (WPS).

Performance Tasks

There are no Performance Tasks in this section.

Almost all welding tasks involve producing *joints*—the places where welds unite base metal components. The quality of the finished **weldment** depends on the quality of its joints. Joint quality, in turn, depends on *joint preparation*. This process includes machining the joint surfaces to the correct shape and positioning them properly.

The weld begins at the **joint root**—the place where the weldment components are closest together. The first weld pass, the *root pass*, is critical. It unites the components and affects the overall weld quality. Problems at the **weld root** can compromise the finished weld. After completing the root pass, the welder may produce additional passes. How many and how they're positioned depends on the joint type.

Trainee welders learn to produce many weld joint types. Each works well for specific situations. Joints must be both strong and safe when they experience the forces that the application places on them. The following sections explore joint types and the welds that create them.

Weldment: An assembly made from parts fastened together by welded joints.

Joint root: The place in the joint where the weld members are closest together.

Weld root: The point where the back of the weld extends farthest into the joint.

Basic Weld Anatomy

Trainee welders must learn their way around a weld so they can produce good ones. You'll learn more about weld features and terminology in NCCER Module 29106, *Weld Quality*. For now, however, recognizing the following four weld features is enough.

The *joint root* and *weld root* refer to the same general area, but there's a crucial difference. The joint root is the place where the weldment components come closest together. The weld root is the point where the back of the weld extends farthest into the joint. It's deeper in the base metal because the welding process melts the base metal so it can combine with the filler metal.

The joint root and weld root are deep inside the weld. The *weld face*, however, is the surface that you see. The places where it merges with the base metal surface are the *weld toes*. When welders and welding inspectors assess a finished joint, they'll always examine the weld face and weld toes since problems often appear there.

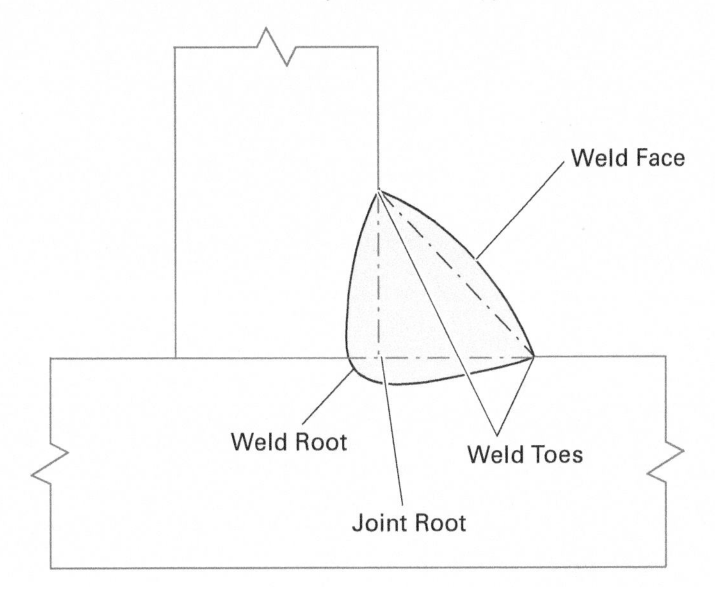

NOTE

The melted base metal or metals make up the weld itself. If the welding process uses a filler metal, it combines with these. The resulting solidified mixture, the *weld metal*, forms the finished weld.

2.1.0 Joints and Loads

Loads: Forces applied to a material or structure.

Every application generates forces that act on welded joints. These **loads** stress the joint, pulling, pushing, bending, or twisting it. When properly prepared, welded components behave as a single unit. Poorly prepared joints or badly made welds will fail under load. For these reasons, welders must understand the forces that challenge their work.

2.1.1 Load Types

Every force has a size and a direction in which it acts. Forces can work together or act against each other. Loads are usually combinations of forces. *Figure 13* shows the loads that can stress welded joints. Each letter F represents a force. The arrow next to it shows its direction. The following descriptions explain how each load stresses a welded joint. Never forget that welded joints may experience multiple loads simultaneously.

Tensile Loads

Ultimate tensile strength: The amount of pulling force a material can tolerate before breaking. Often shortened to *tensile strength*.

These loads try to pull the joint apart. As *Figure 13* shows, two forces (F_1 and F_2) pull in opposite directions. If they're powerful enough, either the joint or the base metal itself will break. All metals have an **ultimate tensile strength**. It's the amount of pulling force that the metal can tolerate before breaking. Welded joints have an ultimate tensile strength too.

Welding processes heat and then cool the metal. These changes can alter its tensile strength. Ideally, a well-prepared joint will be stronger than the base metal itself. In many cases, the base metal will break under tension before the joint does. Improperly welding the metal can weaken it or produce a poor joint that fails easily.

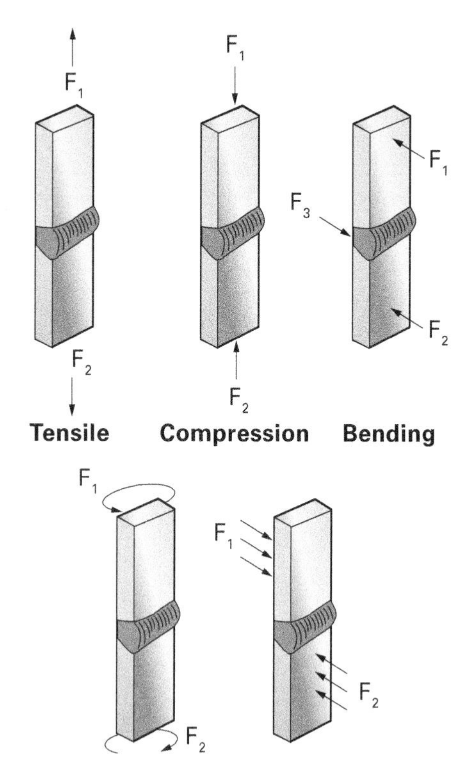

Figure 13 Forces acting on welded joints.

Compression Loads

A compression load squeezes the joint. As *Figure 13* shows, two forces (F_1 and F_2) push against the joint. These forces can distort the base metal or the joint. Many materials, however, can carry large compression loads.

Bending Loads

These loads change a structure's shape, usually forcing it into a curve. As *Figure 13* shows, two forces (F_1 and F_2) push against the structure perpendicularly to the joint. A third force (F_3) pushes back, also perpendicularly to the joint. This force combination can bend the structure at the joint. If the joint is weak, it will fail.

Torsional Loads

These loads twist the structure. As *Figure 13* shows, two forces (F_1 and F_2) rotate in opposite directions. The metal itself may twist and distort. Since the joint is between the two twisting forces, it may break apart.

Shearing Loads

Shearing loads are parallel forces that push against the structure in opposite directions. As *Figure 13* shows, two forces (F_1 and F_2) push perpendicularly against the structure. Since they're parallel, however, they don't push against each other. Instead, they work like a pair of scissors with the joint in the middle. If the joint is weak, it will fail.

2.2.0 Joint and Weld Types

When welders join metals, they can choose from several joint types. Each joint handles loads differently. Each also requires a particular base metal preparation. Joints fall into five basic types: butt, lap, T, edge, and corner. Some types have variants that handle specialized situations. *Figure 14* shows the five joint types and their variants.

Welders produce these joints with a variety of welds. Many of these have descriptive names, such as *V-groove* or *fillet*. Often, the **Welding Procedure Specification (WPS)** designates the joint and weld types. Sometimes, welders must make their own choices. The following sections introduce weld types and the joints they can produce.

Welding Procedure Specification (WPS): A document specifying all essential procedural details related to project-specific welds.

2.2.1 Surfacing Welds

Surfacing is the one weld type that doesn't join metal. Instead, it builds up the base metal, making it thicker. Welders create surfacing welds when the base metal is worn or too thin. In the tool and die industry, welders use this technique to rebuild expensive equipment. Sometimes, they use this process to strengthen base metal. Surfacing welds can improve corrosion resistance, wear resistance, or heat resistance too.

Surfacing: Welding, brazing, or thermally spraying a layer of metal onto a surface to increase its thickness.

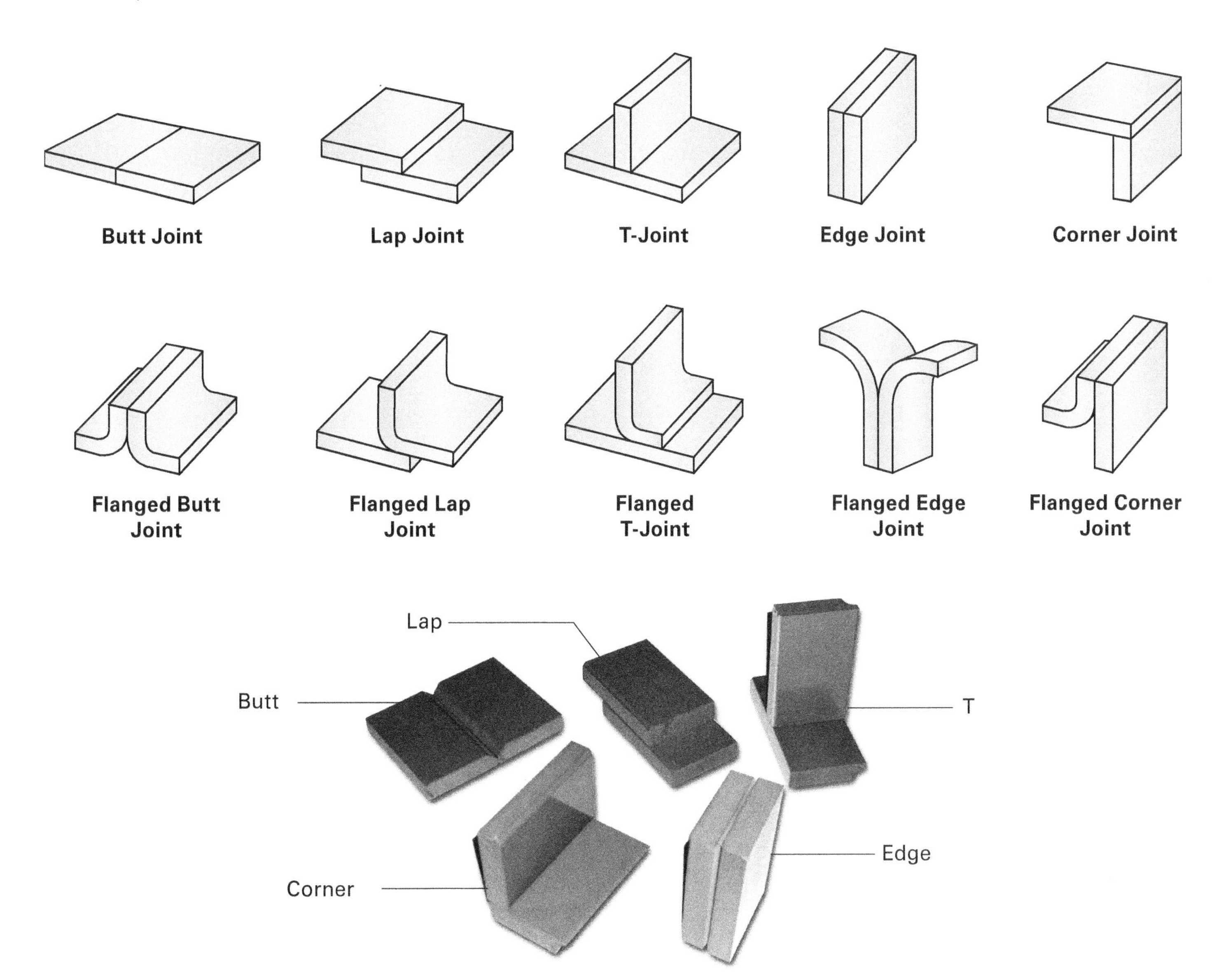

Figure 14 Five basic joint types and variants.

Root opening: The distance between the two base metal parts at the weld root.

Welders may use a surfacing technique called *buttering* to join two incompatible metals. The buttering weld metal is compatible with the metals that it joins. Buttering welds also bring a butt joint's **root opening** into tolerance when it's too large. Similarly, welders can use buttering welds to improve fit-up.

A surfacing weld, also called a *pad*, is a series of parallel weld beads (*Figure 15*). Before laying them, clean the base metal to remove contaminants, oxidation, and dirt. When using an oxyfuel welding process, preheat the surface to prevent warping. After laying the first bead layer, chip, scrape, and clean it before laying the next.

Figure 15 Surfacing weld on an inclined vertical surface.

Rebuilding with Surfacing Welds

Equipment operating in the marine environment can corrode and wear rapidly. Gearbox shafts are a common example. Sometimes, maintenance personnel will decide to rebuild rather than replace them. A welder builds up the worn shaft with surfacing welds. A machinist then turns the shaft back to its correct dimensions and balances it, so it will run at high speed without vibration.

2.2.2 Plug and Slot Welds

Plug and slot welds join parts when their edges can't be welded. Sometimes, a lap joint needs extra strength to meet requirements. A plug or slot weld can provide it. These welds leave a finished surface and don't add extra thickness. Plug welds are round, while slot welds are elongated. The weld metal partially or completely fills the hole or slot. *Figure 16* shows lap joints produced with plug and slot welds. Usually, the welder overfills the hole or slot and then grinds off the excess.

Preparation involves cleaning contaminants, oxidation, and dirt from the base metal. The welder then drills or cuts the hole/slot. It too requires cleaning before welding.

Figure 16 Plug and slot welds.

2.2.3 Fillet Welds

Fillet welds are usually triangular (*Figure 17*). Some have faces that curve out (convex) or curve in (concave). Fillet welds produce lap, T, and corner joints. Unlike most welds, they require little preparation other than thoroughly cleaning the base metal.

After positioning the base metals, the welder applies one or more weld passes (beads) where they come together. Fillet welds may or may not have a root opening. If the weld requires more than one pass, the welder must chip, scrape, and clean the bead before applying the next. Dross, slag, or oxides will cause defects.

When producing outside corner welds, the welder can choose between two styles. The *half-lap joint* is easier to assemble and requires less weld metal. It's also less likely to burn through at the corner. It often requires a second weld on the inside. Dimensions may require adjustment to accommodate the half-lap.

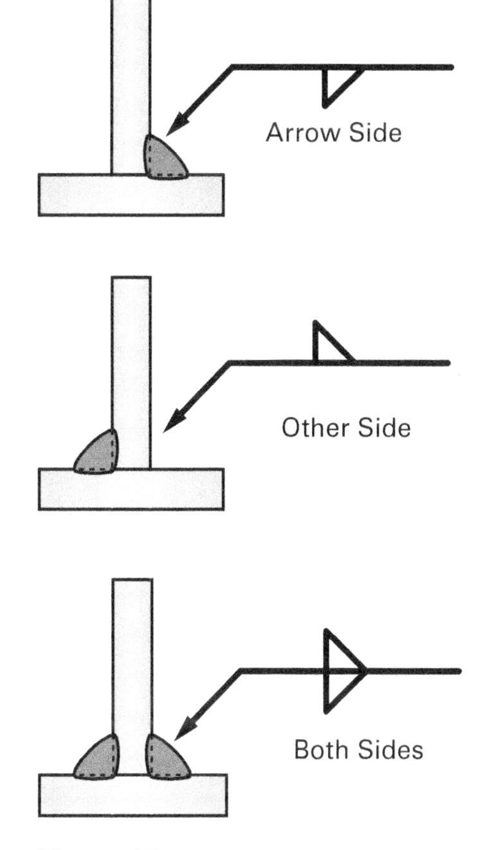

Figure 17 Fillet welds and drawing symbols.

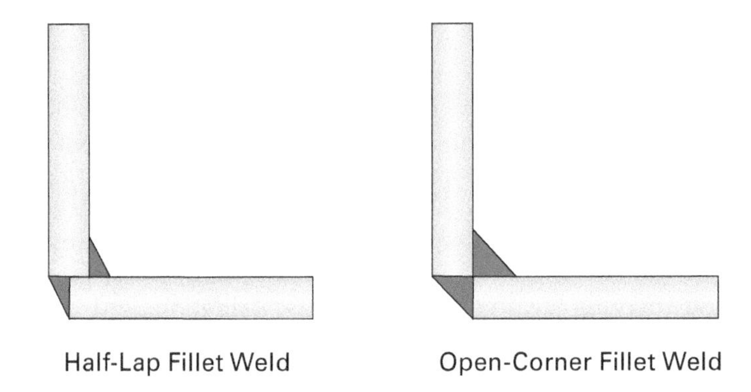

Figure 18 Half-lap and open-corner fillet welds.

The alternative, the *open-corner joint*, is harder to assemble. The plates can't support each other. It's also easy to burn through the joint. *Figure 18* shows both types.

2.2.4 Groove Welds

Fillet welds are simple and require little preparation. Welders produce them quickly and easily. Sometimes, however, a different weld type works better. *Groove welds* are welds that have a space (the groove) between the base metal parts. It holds the weld metal and unites the parts. Some groove welds can produce all five joint types.

Most groove welds require significant preparation beyond cleaning. Usually, the welder must cut or machine the groove in one or both parts. Grooves come in many shapes and styles. The finished weld may go all the way through the base metal—*complete joint penetration (CJP)*. Alternatively, it may penetrate only partway through—*partial joint penetration (PJP)*. CJP groove welds are stronger than PJP ones. The following sections explore the many groove weld styles.

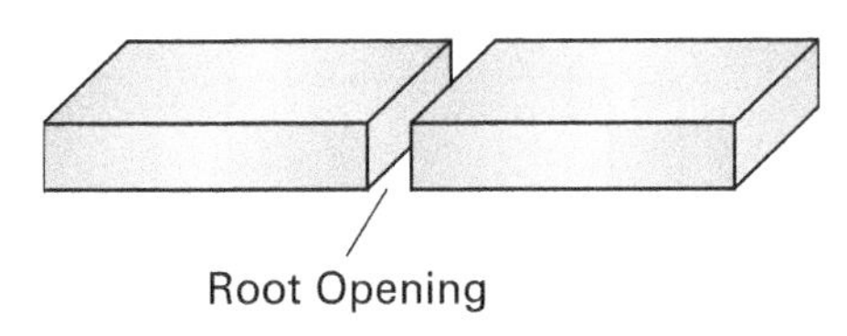

Figure 19 Preparation for a square-groove butt weld.

2.2.5 Square-Groove Welds

Square-groove welds are the simplest groove weld. They can produce butt, T, edge, and corner joints. The welder must set the base metal parts to produce the correct root opening (*Figure 19*).

Welding codes place restrictions on CJP square-groove welds. For example, when producing them with SMAW under *AWS D1.1*, the welder must use a root opening equal to half the base metal's thickness. The maximum base metal thickness is $^1/_4$" (~6 mm), and the root opening tolerance is $+^1/_{16}$" and $-^1/_8$". The welder must weld from both sides and gouge before making the second weld. The GMAW and FCAW processes have the same requirements, except that the base metal may be up to $^3/_8$" (~10 mm) thick.

NOTE

Never forget that other codes may have different guidelines or tolerances. Also remember to follow the WPS and any jobsite quality specifications.

2.2.6 Single-Bevel Groove Welds

A single-bevel groove weld is like a square-groove weld but with a crucial difference. One edge is beveled (angled). This change produces a stronger weld since the groove has more surface area. Single-bevel groove welds can produce all five joint types. The welder prepares a single-bevel groove weld by cutting the bevel. Often, the bevel will have a specified root opening at the bottom (*Figure 20*). It's usually relatively small.

Figure 20 Single-bevel corner and butt welds.

2.2.7 V-Groove Welds

V-groove welds, as their name suggests, have a bevel on each edge. Together, they form a V-shaped groove (*Figure 21*). V-groove welds are popular and form butt, edge, and corner joints. They're common in both plate and pipe welds.

V-groove bevels usually have an angle between $22^1/_2$ and $37^1/_2$ degrees. In many cases, the bevel doesn't extend completely through the base metal. Instead, there's a small flat section at the bottom called the **root face** or *land*. The root opening is the gap between the two root faces.

Root face: The flattened area at the end of a groove weld's bevel. Also called the *land*.

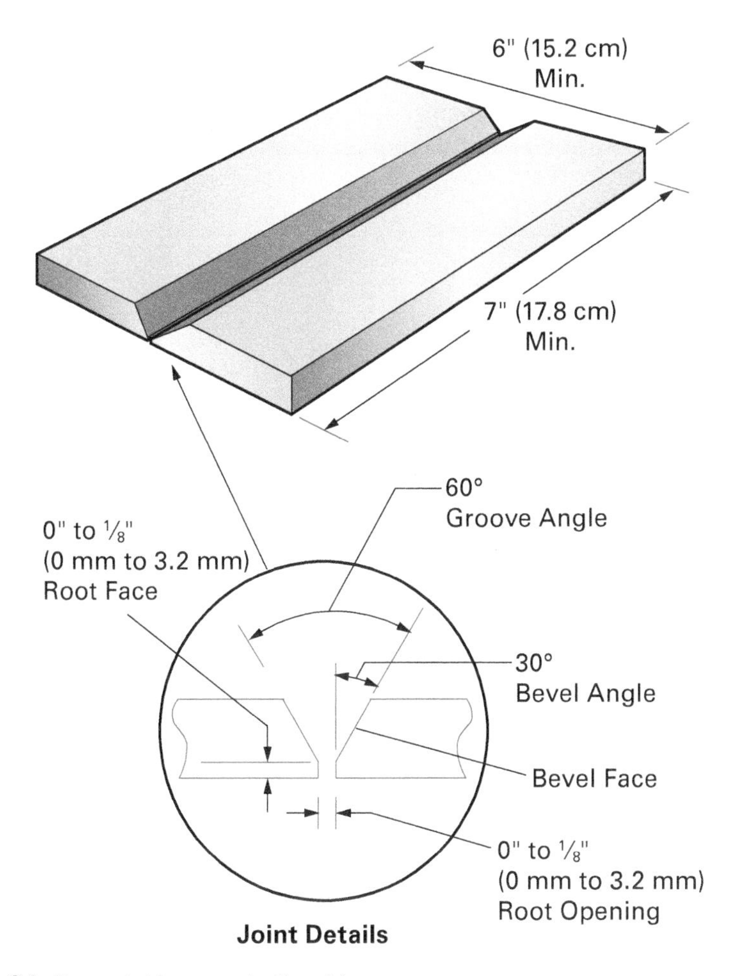

Figure 21 Example V-groove butt weld.

Some V-groove welds have two grooves, one on each side of the base metal. This arrangement produces an hourglass-shaped weld. These are *double V-groove welds* (*Figure 22*). In general, double V-groove welds are better than single ones. They require less weld metal. Welding from each side reduces distortion because the heat-induced forces counteract each other.

2.2.8 J-Groove and U-Groove Welds

J-groove welds are like single-bevel groove welds except that the bevel is J-shaped rather than flat. They require less weld metal than bevel and V-groove welds but are harder to prepare. This weld style works for all five joint types.

U-groove welds are simply two J-groove welds facing each other. They share many qualities with J-groove welds. They form only butt, edge, and corner joints. *Figure 23* shows both weld types.

2.2.9 The Weld Root and Root Preparation

Groove welds can have several root types. An *open-root groove weld* has a gap between the two base metal parts. The weld completely penetrates the joint. Alternatively, welders can place a **backing** that bridges the gap between the base metal parts. This transforms the joint into a *groove weld with backing*. Finally, some groove welds don't have a gap between the base metal parts. These are *partial joint penetration groove welds* since the weld doesn't go all the way through.

Root preparation is the process welders use to prepare the root. It includes selecting the groove angle, the root face size, and the root opening. The *groove angle* is the total included angle between the two base metal faces. If a V-groove weld has $22^1/_2$-degree bevels, the groove angle is twice this value—45 degrees.

Backing: A weldable or non-weldable material placed behind a weld's root opening.

NOTE

A root face isn't necessary with a groove weld backed by a metal strip. The grooves taper to a knife edge that meets the backing.

NOTE

The optimal joint preparation requires the smallest amount of weld metal, yet it provides enough room for the electrode/torch to reach the root opening. Its design also helps the welder manage the heat properly.

Melt-through: Complete joint penetration.

The root preparation significantly controls the heat during welding. Proper root preparation ensures sufficient penetration without **melt-through**. In general, decreasing the groove angle requires increasing the root opening to compensate. *Figure 24* shows good and bad root preparations and their consequences.

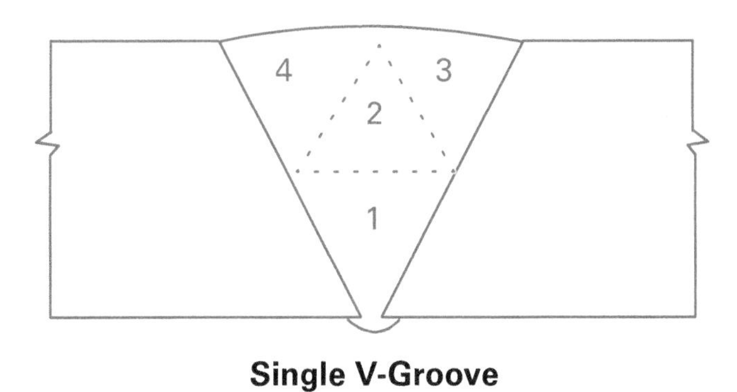

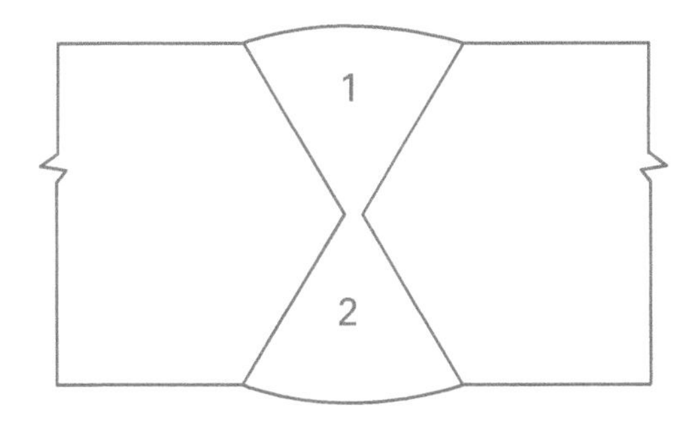

Figure 22 Double and single V-groove welds.

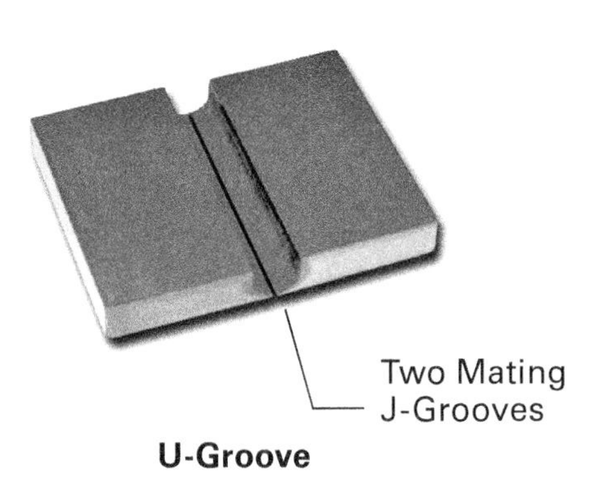

Figure 23 J-groove and U-groove welds.

For open-root welds on plate, the groove angle should be 60 degrees. The root opening and root face should be no more than 1/8" (3.2 mm). With V-groove welds, the 60-degree groove angle comes from the two 30-degree bevel angles. For open-root welds on pipe, the groove angle is 60 or 75 degrees, depending on the WPS or jobsite quality specifications. *Figure 25* shows groove angles for open-root welds on plate and pipe.

For groove welds with backing, the backing type determines the root preparation. Backing can be a metal strip identical to the base metal. It can also be flux-coated tape, fiberglass tape, ceramic tape, or even just a gas. The WPS will specify when to use backing and what type to use.

When using a backing strip identical to the base metal, the groove angle should be 45 degrees. The root opening will usually be 1/4" (6.4 mm). Codes may differ, however, so consult the applicable code, WPS, or jobsite quality specifications.

NOTE

The following guidelines are general and conform to some common welding codes. Other codes, however, have different guidelines. Specialized welds may have different requirements. Always follow the applicable code or specification when preparing the joint.

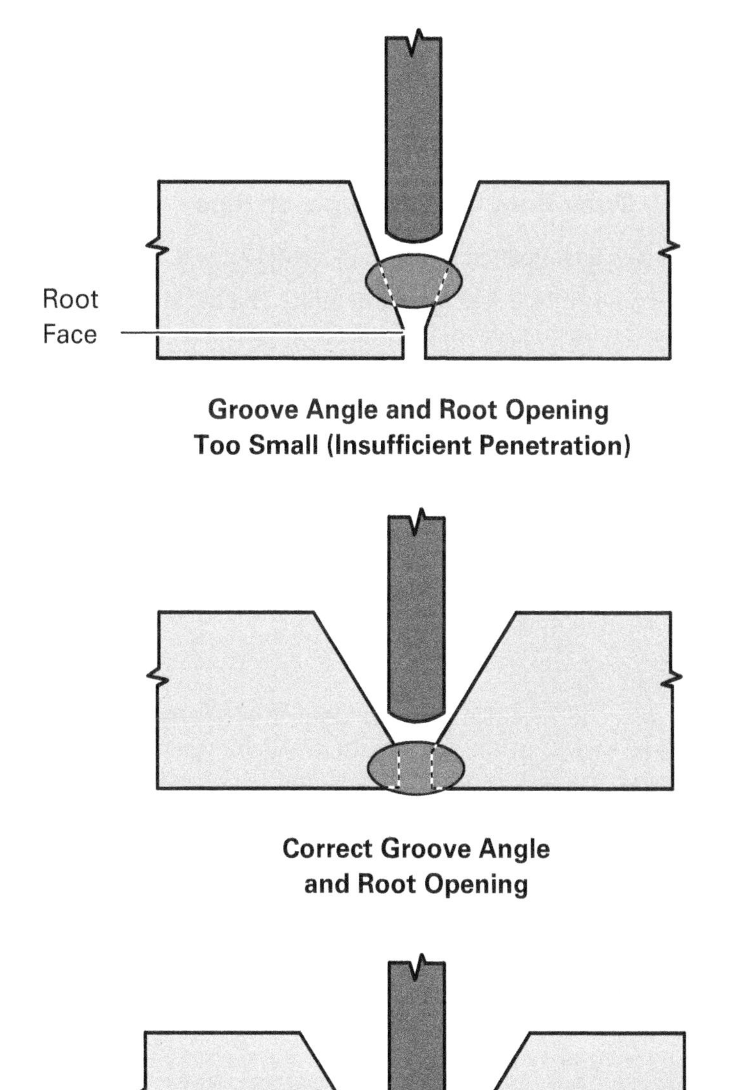

Figure 24 Good and bad root preparations.

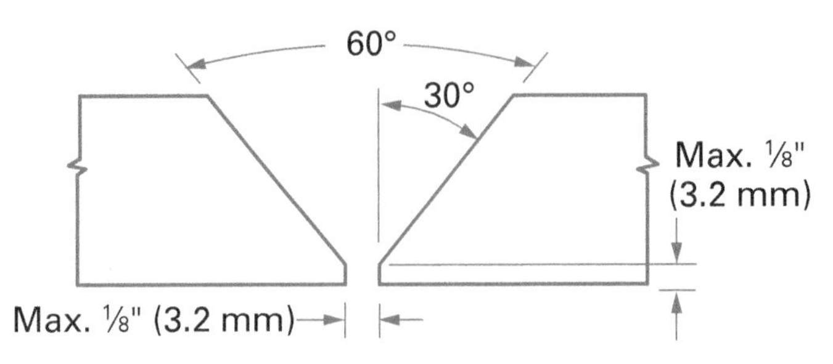

Open-Root Groove Angle on Plate

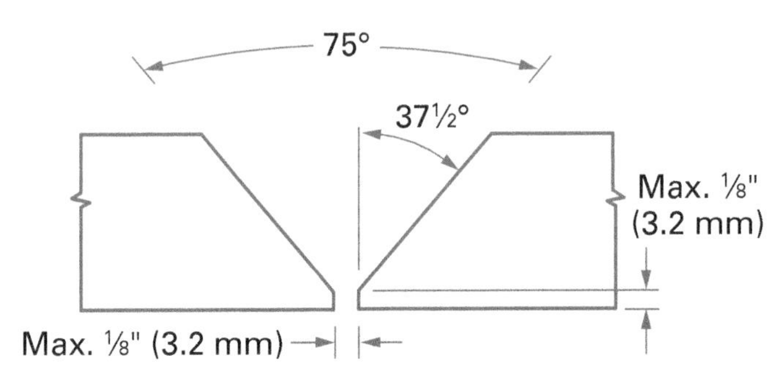

Open-Root Groove Angle on Pipe

Figure 25 Example open-root groove weld root preparations.

Feather: Tapered.

There will be no root face, just a **feather** edge. The backing strip must be sufficiently thick and wide to prevent burn-through at the root. It must also be able to absorb and dissipate the root pass welding heat. Center it over the groove and secure it tightly into place with several small welds (tacks).

When using a tape backing, the groove angle should be 60 degrees. The maximum root opening and root face should be $^{3}/_{16}$" (4.8 mm). Clean the base metal surface so the tape adheres tightly. Apply the tape centered over the groove.

When using gas as a backing, the groove angle should be 60 degrees. The root opening and root face should be no more than $^{1}/_{8}$" (3.2 mm). *Figure 26* summarizes each root preparation.

NOTE

Always check the WPS or the jobsite specifications for backing requirements.

2.2.10 Welding Position

It's easier and faster to produce fillet welds in the flat or horizontal position. Groove welds are easiest in the flat position. Whenever possible, position the weldment so you can work in the flat position. If the weldment has multiple welds, create as many as you can on the bench. Then tack the assembly together and complete the remaining welds.

When welding pipe, weld as many fittings as possible. Then tack the pipe into position and complete the remaining welds. Good planning ensures that the most difficult welds are in the best possible position.

Out-Of-Position Welding

Out-of-position welding refers to welding positions other than flat or horizontal fillet. Welding out-of-position generally requires smaller electrodes with more passes to fill the joint. Since it's more difficult, weld defects are more likely.

Backing Gas Hazards

Welders often use nitrogen, argon, or carbon dioxide as a backing gas. These gases aren't toxic, but they don't support life. They can displace the oxygen-containing air from a confined space. Since all three gases are colorless and odorless, craftworkers won't know they're present. Nitrogen is lighter than air, so it accumulates near the ceiling in still air. Argon and carbon dioxide do the opposite, accumulating near the floor in still air.

Before using a backing gas, follow all required precautions against asphyxiation. Confirm adequate ventilation, especially if working in a confined space. In many cases, forced ventilation is the only way to ensure safe oxygen levels. Sometimes, a supplied air respirator (SAR) is required.

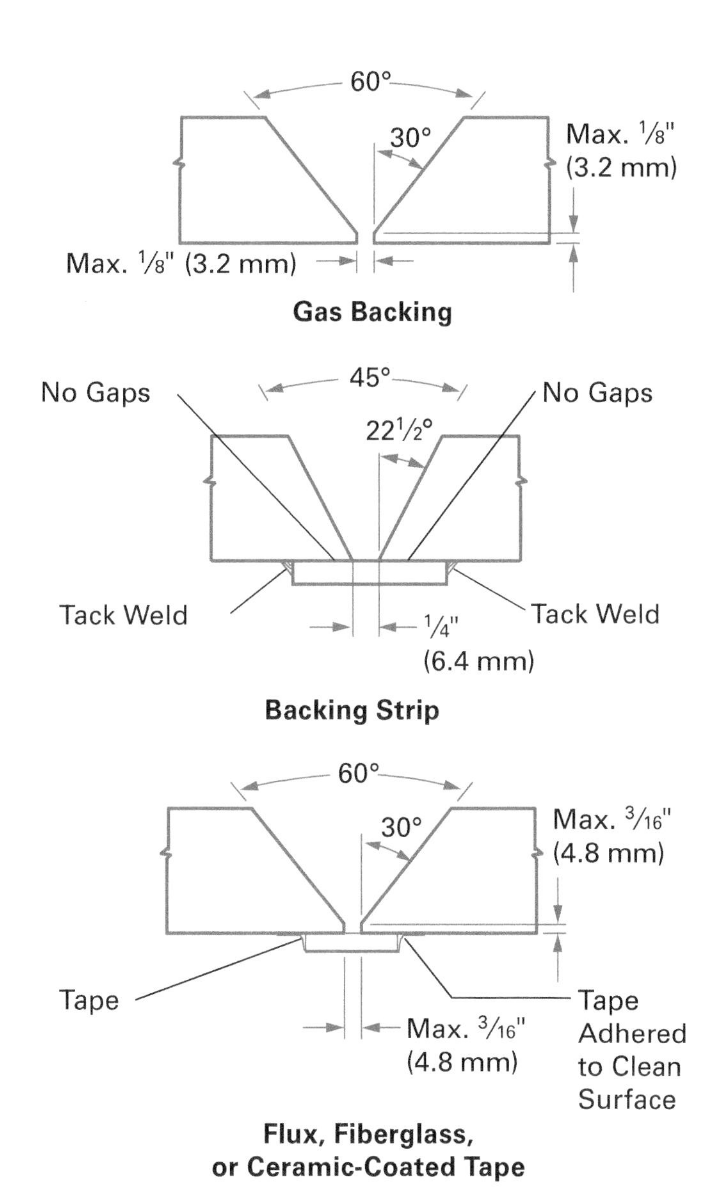

Figure 26 Example root preparations for groove welds with backing.

Groove Angles

If the groove angle is larger than necessary, the weld will require extra metal. This will increase the weld's cost and completion time. There may also be more distortion. If the groove angle is too small, it may cause weld defects.

2.3.0 Codes and Welding Procedure Specifications

Codes, standards, and specifications can govern almost every aspect of welding work. A welding code defines a product or process' minimum requirements. Standards establish the details that implement the code in actual work. Following a code's standards ensures that the work meets the code's requirements. Code-driven work may exceed the code's requirements, but it must never fall below them. Clients usually specify a particular welding code in their contracts.

In code-driven projects, each weld procedure has an associated WPS (*Figure 27*). The WPS specifies the joint type, weld type, and any required preparatory steps. All WPSs go through a qualifying process that confirms they work properly and meet code requirements.

Not all projects are code driven, but all welds should follow industry standards. Blueprints and welding symbols often cite a WPS. If you're unsure whether a weld requires a WPS, ask your supervisor.

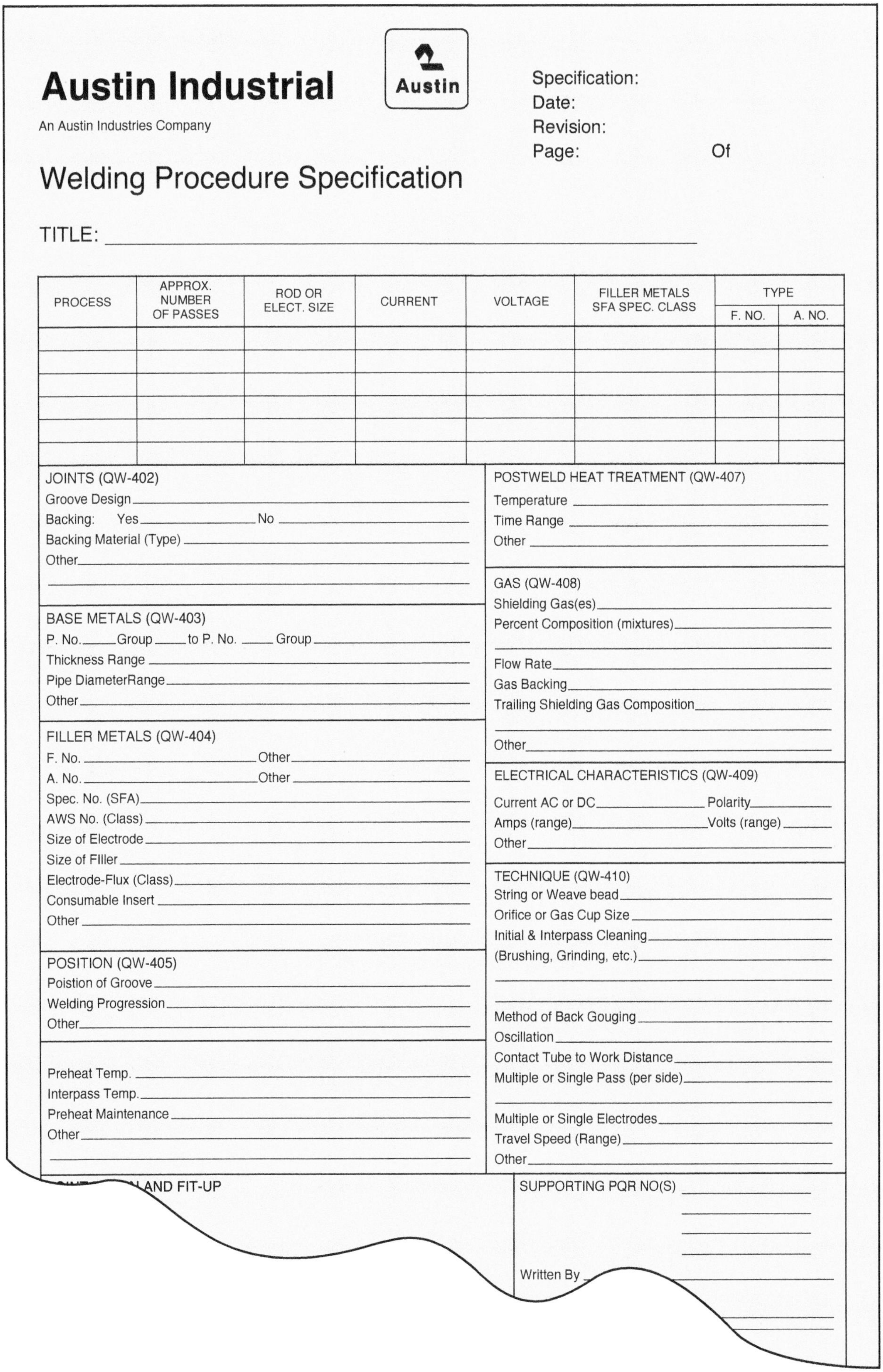

Austin Industrial

Austin

An Austin Industries Company

Specification:
Date:
Revision:
Page: Of

Welding Procedure Specification

TITLE: ______________________________

PROCESS	APPROX. NUMBER OF PASSES	ROD OR ELECT. SIZE	CURRENT	VOLTAGE	FILLER METALS SFA SPEC. CLASS	TYPE	
						F. NO.	A. NO.

JOINTS (QW-402)
Groove Design ____
Backing: Yes ____ No ____
Backing Material (Type) ____
Other ____

BASE METALS (QW-403)
P. No. ____ Group ____ to P. No. ____ Group ____
Thickness Range ____
Pipe DiameterRange ____
Other ____

FILLER METALS (QW-404)
F. No. ____ Other ____
A. No. ____ Other ____
Spec. No. (SFA) ____
AWS No. (Class) ____
Size of Electrode ____
Size of FIller ____
Electrode-Flux (Class) ____
Consumable Insert ____
Other ____

POSITION (QW-405)
Poistion of Groove ____
Welding Progression ____
Other ____

Preheat Temp. ____
Interpass Temp. ____
Preheat Maintenance ____
Other ____

...AND FIT-UP

POSTWELD HEAT TREATMENT (QW-407)
Temperature ____
Time Range ____
Other ____

GAS (QW-408)
Shielding Gas(es) ____
Percent Composition (mixtures) ____
Flow Rate ____
Gas Backing ____
Trailing Shielding Gas Composition ____
Other ____

ELECTRICAL CHARACTERISTICS (QW-409)
Current AC or DC ____ Polarity ____
Amps (range) ____ Volts (range) ____
Other ____

TECHNIQUE (QW-410)
String or Weave bead ____
Orifice or Gas Cup Size ____
Initial & Interpass Cleaning ____
(Brushing, Grinding, etc.) ____
Method of Back Gouging ____
Oscillation ____
Contact Tube to Work Distance ____
Multiple or Single Pass (per side) ____
Multiple or Single Electrodes ____
Travel Speed (Range) ____
Other ____

SUPPORTING PQR NO(S) ____

Written By ____

Figure 27 Example WPS.

2.0.0 Section Review

1. When a weld joint experiences forces that try to pull it apart, it's under a _____.
   a. tensile load
   b. torsional load
   c. shearing load
   d. compression load

2. When creating a multilayer surfacing weld, _____.
   a. chip, scrape, and clean the previous layer before welding the next
   b. don't worry about cleaning each layer since the welding heat removes any impurities
   c. carefully heat the last bead before welding the next layer
   d. gouge each layer before welding the next one

3. If a weld requires a WPS, it will be identified _____.
   a. within the company's procedural handbook
   b. on blueprints or in welding symbols
   c. in the electrode's SDS
   d. on the base metal

3.0.0 Welding Joint Preparation

Objective

Outline preparing joints for welding.

a. Describe mechanically preparing joints for welding.
b. Describe thermally preparing joints for welding.

Performance Tasks

1. Mechanically create a $22^1/_2$-degree bevel ($\pm 2^1/_2$ degrees) on the edge of a $^1/_4$" to $^3/_4$" (~6 mm to 20 mm) thick carbon steel plate.
2. Thermally create a $22^1/_2$-degree bevel ($\pm 2^1/_2$ degrees) on the edge of a $^1/_4$" to $^3/_4$" (~6 mm to 20 mm) thick carbon steel plate.

Welders have many choices when they prepare a joint for welding. They can use hand or power tools to mechanically cut or shape it. Alternatively, they can use a thermal (heat) process. Their selection depends on convenience, the available tools, and the base metal type. Of course, codes may play a role as well.

Before preparing a joint, check whether a WPS or the jobsite quality specifications cover it. In this case, the specifications will govern the preparation. Remember, these standards are *not* optional. If you're uncertain about joint preparation, check with your supervisor.

3.1.0 Mechanical Joint Preparation

Mechanical joint preparation methods work best on alloy steel, stainless steel, and nonferrous metals. They're slower than thermal methods. They're usually more precise, however, not leaving oxides behind as thermal methods do. Since they don't heat the base metal nearly as much, they rarely cause warping or distortion.

WARNING!

When using joint preparation tools, always follow the manufacturer's guidelines and safety procedures. These machines can do a lot of damage and can easily injure you. Always wear the correct PPE.

3.1.1 Grinders

Handheld electric or pneumatic grinders are popular joint preparation tools. Welders use them in the shop or field to cut and bevel pipe or plate. *Figure 28* shows a craftworker using a handheld angle grinder to prepare a pipe elbow.

WARNING!

Grinding discs and cutoff wheels are material specific. Always use the right disc or wheel for the base metal. Don't use a contaminated tool on the wrong base metal. Remember that discs and wheels can shatter without warning. Wear the correct PPE and stay out of the path of flung fragments. Confirm that nearby workers aren't in the path either. Verify that the disc or wheel has a speed rating appropriate to the grinder's maximum speed.

3.1.2 Pipe Beveling Machines

Many industries use welded piping. All pipes and fittings require preparation before welding. Fittings usually come properly beveled from the factory. The welder or fitter just needs to clean them before welding. Pipes require cutting, squaring, and beveling. Mechanical or thermal pipe cutting machines handle these tasks.

Mechanical pipe cutting and beveling machines are electrically, hydraulically, or pneumatically powered (*Figure 29*). Some work like a lathe. Various adjustable cutting attachments produce whatever joint preparation the job requires. These machines come in sizes for both small and large pipes.

Figure 28 Preparing a weld joint with a grinder.

Special mechanical cutoff machines prepare pipe in the field (*Figure 30*). They clamp onto the pipe with a ring or chain. A blade then cuts the pipe to the right length.

3.1.3 Nibblers

Nibblers prepare the edges of plates and pipes (*Figure 31*). The cutting tool moves up and down, biting off a chip with each stroke. An adjustable guide sets the bevel angle. Nibblers work only on edges. To create an internal cut, drill a pilot hole for the nibbler first.

Figure 29 Pipe beveling machine.
Source: Courtesy of Tri Tool Inc.

Figure 30 Pipe cutoff machine.
Source: Courtesy of Tri Tool Inc.

Figure 31 Nibbler.
Source: TRUMPF Inc.

3.2.0 Thermal Joint Preparation

Thermal joint preparation methods use heat to cut and shape the base metal. Oxyfuel cutting and plasma arc cutting (PAC) are popular methods. Oxyfuel works best on carbon steel. PAC works on many metals. While air-carbon arc cutting (A-CAC) works too, it produces poorer results. Carbon deposits from the electrodes can contaminate the weld joints. Most welders use A-CAC only for gouging seams and cracks.

NOTE

NCCER Module 29102, *Oxyfuel Cutting*, introduces oxyfuel cutting processes. NCCER Module 29103, *Plasma Arc Cutting*, introduces PAC. NCCER Module 29104, *Air-Carbon Arc Cutting*, introduces A-CAC. Consult these modules for more information.

WARNING!

When using thermal cutting tools, completely follow the manufacturer's guidelines and safety procedures. Always wear the correct PPE. Be particularly careful with oxyfuel equipment since it starts many jobsite fires. Follow all fire watch protocols and have a fire extinguisher ready.

3.2.1 Torches and Cutting Machines

Handheld oxyfuel and PAC torches are useful for many preparation tasks, particularly smaller ones. For large, repetitive, or precision jobs, however, welders may prefer mechanized cutting machines. These hold the torch in a fixture and move it in a prescribed path.

A *track burner* cuts plate with a motorized carriage that runs on a guiding track (*Figure 32*). An adjustable fixture holds the torch at the correct position and angle. After selecting the motor's speed and direction, the welder simply turns on the machine and lets it run down the track, cutting as it goes.

Figure 32 Track burner.

Thermal pipe cutting and beveling machines clamp onto the pipe with a ring or chain (*Figure 33*). A fixture holds the torch at the correct position and angle. Some have a hand crank that the welder turns to drive the fixture around the pipe. Others are powered and operate at a preset speed. Some pipe cutting machines can even compensate for out-of-round pipes.

A special version of this machine mounts inside large pipes with powerful magnets (*Figure 34*). It also works outside pipe and can cut plate.

NOTE

When preparing the joint with a thermal process, preheating the base metal may be necessary. If so, the WPS should specify this step.

3.2.2 Final Joint Preparation

Thermal preparation processes do have one disadvantage. They leave oxides and dross on the cut surfaces. Air-carbon arc cutting leaves carbon deposits. Any of these can cause weld defects, including porosity, hard spots, and cracking.

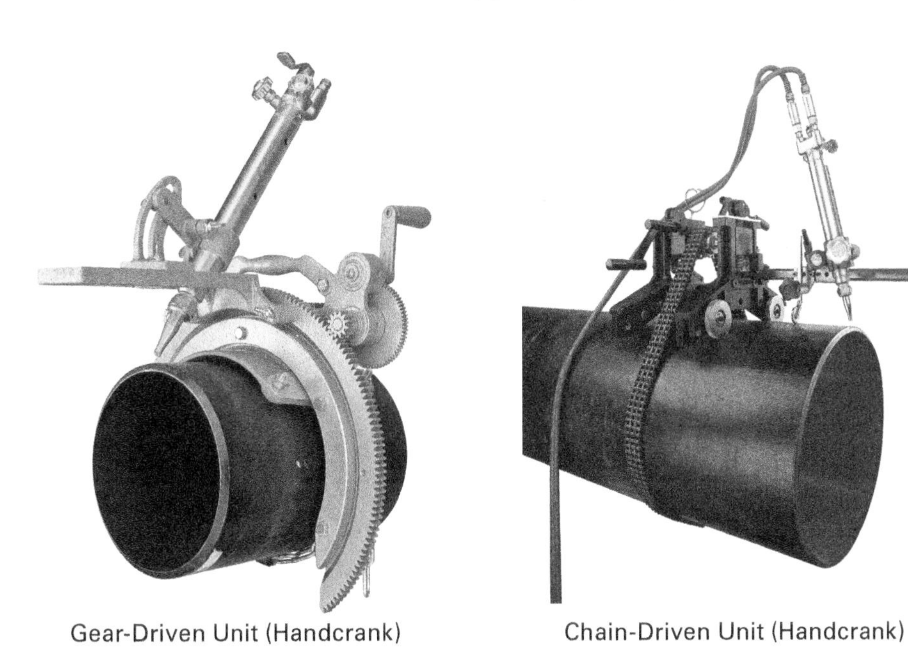

Figure 33 Pipe cutting and beveling equipment.
Source: Courtesy of Mathey Dearman

Figure 34 Inside pipe cutting machine.
Source: Courtesy of Mathey Dearman

Automatic Pipe Bevelers

Automatic pipe bevelers and cutters save both time and cutting gases. Once properly set up, they use the same settings over and over. They don't require resetting the preheat flame either and will shut off the cutting gases when they're finished.

After thermally preparing the joint, mechanically clean it by chipping, scraping, and brushing. A grinder can speed up larger cleaning jobs. If necessary, use a chemical cleaner to remove oils or other contaminants.

3.0.0 Section Review

1. Which of the following is an advantage of mechanical joint preparation methods over thermal methods?
   a. They're more precise and heat the base metal less.
   b. They're faster and require less work.
   c. Mechanically prepared joints require less weld metal.
   d. Mechanically prepared joints require fewer weld passes.

2. Which thermal preparation process is a poor choice because it can leave deposits in the base metal?
   a. Air-carbon arc cutting
   b. Oxyfuel cutting
   c. Plasma arc cutting
   d. Pipe-end prep lathe

Module 29105 Review Questions

1. When cutting metal, _____.
   a. have your eyes examined every six months
   b. wear ordinary sunglasses to protect against bright sparks
   c. don't look straight at the work area
   d. wear goggles and a face shield with a properly tinted lens

2. Mild-carbon steel contains _____.
   a. 0.10–0.15 percent carbon
   b. 0.15–0.30 percent carbon
   c. 0.30–0.50 percent carbon
   d. 0.50–1.00 percent carbon

3. What kind of steel is 310 steel?
   a. Stainless steel
   b. Weathering steel
   c. HSLA steel
   d. Carbon steel

4. Adding copper to carbon steel produces _____.
   a. spring steel
   b. weathering steel
   c. mild steel
   d. stainless steel

5. A pneumatic tool that removes surface oxidation with several blunt rods is a _____.
   a. needle scaler
   b. weld slag chipper
   c. needle-nose chipper
   d. tattoo gun scaler

6. A torsional load applies a _____.
   a. pulling force
   b. twisting force
   c. pushing force
   d. bending force

7. The five basic joint types are butt, lap, corner, T, and _____.
   a. chamfer
   b. square
   c. bevel
   d. edge

8. Welders can produce lap joints, T-joints, and corner joints with _____.
   a. surface welds
   b. butt welds
   c. fillet welds
   d. spot welds

9. U-groove welds require _____.
   a. less preparation time but use more weld metal than V-groove welds
   b. more preparation time but use less weld metal than V-groove welds
   c. more preparation time and use more weld metal than V-groove welds
   d. less preparation time and the same weld metal as V-groove welds

10. Which of the following is *not* a common backing material?
    a. Flux-coated tape
    b. A gas
    c. Ceramic tape
    d. Asbestos card

11. When using a backing strip identical to the base metal, the root face should be _____.
    a. a feather edge
    b. $\frac{1}{16}$" (1.6 mm)
    c. $\frac{1}{8}$" (3.2 mm)
    d. $\frac{1}{4}$" (6.4 mm)

12. Whenever possible, produce groove welds in the _____.
    a. vertical position
    b. overhead position
    c. flat position
    d. horizontal position

13. In code-driven projects, each welding procedure will have a(n) _____.
    a. PQR
    b. WPS
    c. AISI
    d. SAE

14. Which of the following is *not* true of grinding discs and cutoff wheels?
    a. They can't cross-contaminate metals.
    b. They're material specific.
    c. They can shatter without warning.
    d. They have a speed rating.

15. Which of the following thermal cutting processes is better for gouging than for joint preparation?
    a. Plasma arc cutting
    b. SMAW cutting
    c. Oxyfuel cutting
    d. Air-carbon arc cutting

Answers to odd-numbered Module Review Questions are found in *Appendix A*.

Cornerstone of Craftsmanship

Ashley Applegate
Director of Training
Kentucky Welding Institute

How did you choose a career in the industry?

Growing up, I saw family members who were welders, and they were very successful. Also, skilled training was less time-consuming than college, so I ended up choosing that route.

Who inspired you to enter the industry?

My father and two high school instructors were the inspiration to join the industry. My dad had the ability to repair our own things when it came to agriculture on the farm, so we didn't have to depend on someone else. My two instructors were very adamant about welding being something you could do to build a great career moving forward and to make a good income for you and your family.

What types of training have you been through?

I went through two years of a tech program in a vocational technical school. I went to community college for my associate degree. I earned the credentials for OSHA Instructor, MSHA Instructor, and then NCCER Master Trainer and Administrator. Lately, I have completed AWS and CWI training. That's what it takes to be Director of Training at my current institution. I was recently doing more training with AWS.

How important is education and training in construction?

If you have a need you can fill with your education, you can do that and make sure you continue growing. Education has the utmost importance. Education builds your skill level and gives you the confidence that you need to perform the skill; it also creates a safer workforce. In welding, carpentry, or any other craft, safety comes into play every single day. The environment takes its toll on craftsmen and the project. Training gets you ready to have the skills to build something that outlasts the environmental challenges thrown at it.

What kinds of work have you done in your career?

Just about everything. I had my own Production and Fabrication shop for 20 years. We built whatever the customer wanted—boilers, truck beds, etc.—and did maintenance. We performed production work for steel mills and for different facilities around country. Almost anything welding related, we did. We serviced the northeastern region of Kentucky as well as the southeastern region of Ohio. We shipped products across the US and even serviced five more countries around the world.

Tell us about your current job.

As Director of Training for KWI, I oversee all the other instructors. I make sure they have what they need to succeed, from welding rods to the most up-to-date curriculum. I make sure the trainees are getting enough time and that they can get the skills the industry is looking for. Trainees are at the forefront of our curriculum, and I want to make sure the industry is getting the employees that they want and that better suit the current market.

What has been your favorite project?

Absolutely founding, building, and establishing Kentucky Welding Institute and seeing the success of the trainees.

What do you enjoy most about your job?

The success of the trainees. Seeing someone who struggles but works hard and then they get promoted, or pass that test, or get that first paycheck, that is the most rewarding thing in this job. I enjoy being part of the process of changing someone's life for the better.

What factors have contributed most to your success?

Solid work ethic. That is number one in all the work you do. I grew up on a farm, so no matter the challenge, the crop had to be taken care of and the animals had to get fed. If you apply those principles to any career, you will be successful.

Would you suggest construction as a career to others? Why?

Yes. Skilled trades are in high demand. By year 2024, we will need 300,000 craftsmen to work in the field. When you look anywhere with job listings, there is an endless list of employers looking for craftsmen.

What advice would you give to those new to the field?

Be ready to earn it. Don't expect success in the job or promotion to happen overnight. Be ready to pursue what you want, put in the time, learn from your superiors, and be humble and courteous to coworkers. Work for your success and go out there and earn it.

Interesting career-related fact or accomplishment?

It is a great accomplishment to see trainees from KWI being successful in the workforce, at home, and then coming full circle to KWI as instructors. It is rewarding to see someone who a few years ago was struggling as a trainee, and now is helping the next generation. It has been a huge accomplishment of mine to see that.

How do you define craftsmanship?

Craftsmanship is taking pride in your work and doing it until it's right. Quality craftsmanship is the best form of advertisement. If you are doing a great job over and over in your field, you are going to keep those life-changing opportunities and high-paying jobs rolling in.

Answers to Section Review Questions

Answer	Section Reference	Objective
Section 1.0.0		
1. c	1.1.3	1a
2. b	1.2.3	1b
3. b	1.3.2	1c
Section 2.0.0		
1. a	2.1.1	2a
2. a	2.2.1	2b
3. b	2.3.0	2c
Section 3.0.0		
1. a	3.1.0	3a
2. a	3.2.0	3b

User Update

Did you find an error? Submit a correction by visiting **https://www.nccer.org/olf** or by scanning the QR code using your mobile device.

Source: iStock@Berkut_34

Objectives

Successful completion of this module prepares you to do the following:

1. Summarize welding codes, code standards organizations, and welding qualifications.
   a. Identify welding codes and the key welding code standards organizations.
   b. Define and describe welding qualifications.
2. Identify weld discontinuities and their causes.
   a. Describe acceptable and unacceptable weld profiles.
   b. Describe common discontinuities and their causes.
3. Summarize nondestructive and destructive weld tests.
   a. Describe visual weld inspection methods.
   b. Describe nondestructive weld testing methods.
   c. Describe destructive weld testing methods.
4. Summarize welder performance qualification testing.
   a. Describe welder performance qualifications.
   b. List AWS, ASME, and API welder qualification testing requirements.
   c. Describe a welder performance qualification test.

Performance Task

Under supervision, you should be able to do the following:

1. Visually inspect (VT) a fillet and/or groove weld and complete an inspection report.

Overview

All welders must make high-quality welds that consistently pass inspection. Learning to recognize weld imperfections is crucial. It is even more important, however, to learn what causes imperfections. To produce good welds, welders must avoid making these mistakes. Only then can they look at finished welds and know they are truly good.

NOTE

This module uses US standard and metric units in up to three different ways. This note explains how to interpret them.

Exact Conversions

Exact metric equivalents of US standard units appear in parentheses after the US standard unit. For example: "Measure 18" (45.7 cm) from the end and make a mark."

Approximate Conversions

In some cases, exact metric conversions would be inappropriate or even absurd. In these situations, an approximate metric value appears in parentheses with the ~ symbol in front of the number. For example: "Grip the tool about 3" (~8 cm) from the end."

Parallel but not Equal Values

Certain scenarios include US standard and metric values that are parallel but not equal. In these situations, a slash (/) surrounded by spaces separates the US standard and metric values. For example: "Place the point on the steel rule's 1" / 1 cm mark."

Digital Resources for Welding

Scan this code using the camera on your phone or mobile device to view the digital resources related to this craft.

Industry Recognized Credentials

If you are training through an NCCER-accredited sponsor, you may be eligible for credentials from NCCER's Registry. The ID number for this module is 29106. Note that this module may have been used in other NCCER curricula and may apply to other level completions. Contact NCCER's Registry at 1.888.622.3720 or go to **www.nccer.org** for more information.

You can also show off your industry-recognized credentials online with NCCER's digital badges. Transform your knowledge, skills, and achievements into badges that you can share across social media platforms, send to your network, and add to your resume. For more information, visit **www.nccer.org**.

1.0.0 Welding Codes and Qualifications

Performance Tasks

There are no Performance Tasks in this section.

Objective

Summarize welding codes, code standards organizations, and welding qualifications.

a. Identify welding codes and the key welding code standards organizations.
b. Define and describe welding qualifications.

Successful construction depends on welds doing their job reliably. Faulty welds not only cost money but also can injure or kill. Welders rely on their skills to produce excellent welds. They also rely on detailed information that helps them make the decisions required to produce quality welds.

1.1.0 Welding Codes

Codes: Documents establishing the minimum requirements of products or processes.

Standards: Documents that implement code requirements.

Specifications: Detailed documents defining the work performed and the materials used in specific products or processes.

Addendum: Supplementary information that corrects or revises a document.

Codes, **standards**, and **specifications** can govern almost every aspect of a welding project. A welding code defines the minimum requirements of a product or process. Standards establish the details that implement the code in actual work. Following a code's standards ensures that the work meets the code's requirements. Code-driven work may exceed the code's requirements but must never fall below them. Since welding codes help ensure safety, they're often integrated into a country's, state's, or region's laws.

Welding codes are very detailed and specific. They address materials, fabrication, inspection, testing, service limitations, and qualifications. *Standards organizations* develop welding codes and their associated standards. Made up of representatives from across the industry, they meet regularly to produce, update, and revise the documents that welders follow when working on code-driven projects.

Some organizations issue updates and revisions quite frequently. They identify these by attaching the revision year to the code's identifier—for example, *AWS D1.1:2020*. Other organizations don't revise their codes as often. Instead, they issue **addendum** pages to modify relevant sections in the existing code. Since each organization handles changes differently, you'll need to read the code's introductory material to learn how it identifies changes. As you use codes, you'll become more familiar with how they work.

Specifications are detailed documents that define how to produce a product or perform a process. They're like standards in that they specify materials, procedures, and quality requirements. Unlike standards, however, they're created by individual companies and apply to a specific project only. Specifications frequently cite codes and their associated standards.

When developing a project's contract, clients specify the codes and standards that they wish to govern the work. They also lay out specifications since these will be more detailed and job-specific than the codes and standards. Specifications frequently cite specific parts of applicable codes. When citing codes, clients must include the year, so welders will know which code revision to follow.

In the United States, several standards organizations create and maintain welding codes. Other countries may follow different standards organizations. Depending on the industry in which you work, you may encounter one or more of these. The following sections introduce the major welding standards organizations. You should become familiar with these, as well as the codes that they maintain. The longer you work in the welding field, the more comfortable you'll become reading and following codes.

1.1.1 American Society of Mechanical Engineers (ASME)

The ASME publishes two welding codes that you should recognize. Both apply to applications involving pressurized liquids or gases.

ASME Boiler and Pressure Vessel Code (BPVC)

Power generation and heating equipment, as well as certain specialized machines, heat water in enclosed vessels called *boilers*. This code covers various aspects of boiler construction and safety. Welders mostly reference the following sections:

Section II, Materials — This section pairs base and filler metals in acceptable combinations. These specifications often match and have the same number designation as the comparable AWS specifications.

Section V, Nondestructive Examination — This section covers methods and standards for **nondestructive testing (NDT)**.

Section IX, Welding, Brazing, and Fusing Qualifications — This section covers qualifications for those who weld and braze boilers and pressure vessels. It also covers applicable welding and **brazing** procedures. Other codes and standards often cite this section in their own welding qualifications.

Nondestructive testing (NDT): Testing methods that don't change or damage the test specimen.

Brazing: Joining metal by melting a filler metal at a temperature above 450°C (~840°F) over unmelted base metal.

ASME B31, Code for Pressure Piping

This code covers pipes that carry pressurized liquids and gases. As you can imagine, it applies to numerous industries and applications. The code includes many sections, each covering a particular piping system. All include qualifications for welding procedures and welders. The ASME has retired some sections that other codes now cover. The following are the active ones that welders work with frequently:

B31.1, Power Piping

B31.3, Process Piping

B31.4, Pipeline Transportation Systems for Liquids and Slurries

B31.5, Refrigeration Piping and Heat Transfer Components

B31.8, Gas Transmission and Distribution Piping Systems

B31.9, Building Services Piping

B31.12, Hydrogen Piping and Pipelines

1.1.2 American Welding Society (AWS)

The AWS publishes numerous codes, standards, specifications, recommended practices, and guides. Welders most frequently reference *AWS D1.1/D1.1M, Structural Welding Code—Steel*. It covers welding and qualification requirements for welded structures made from carbon and low-alloy steels. It doesn't apply to pressure vessels, pressurized piping, or base metals less than $\frac{1}{8}$" (3.2 mm) thick.

Welders also reference the following related codes:

AWS D1.2/D1.2M, Structural Welding Code—Aluminum

AWS D1.3/D1.3M, Structural Welding Code—Sheet Steel

AWS D1.5M/D1.5, Bridge Welding Code

AWS D1.6/D1.6M, Structural Welding Code—Stainless Steel

1.1.3 American Petroleum Institute (API)

As its name suggests, the API publishes codes and standards for the petroleum industry. *API 1104, Welding of Pipelines and Related Facilities* covers both arc and oxyfuel welding. It applies to welding for petroleum piping, as well as pumping, transmission, and distribution systems. The standard outlines methods and procedures for making welds. It also covers weld quality testing.

1.1.4 American National Standards Institute (ANSI)

ANSI is a private organization that doesn't create standards. Instead, it adopts and endorses other organizations' standards. By placing its mark of approval on those standards, it elevates their importance and exposure. ANSI standards cover a huge range of topics and industries. Many welding codes have become ANSI standards.

1.1.5 Other Standards Organizations

Many other national and international organizations publish welding codes, standards, and specifications. Some governments have official standards organizations. The following are a few that you may encounter:

International Organization for Standardization (ISO) — This international body is extremely influential and important. Its members come from the standards organizations of 165 countries. It develops and maintains thousands of standards, including nearly 300 that apply to welding.

ASTM International — Another international standards organization, ASTM International, develops and maintains over 12,000 standards applying to many industries, including welding. Many of these deal with materials and testing procedures.

American Society for Nondestructive Testing (ASNT) — This organization publishes standards associated with nondestructive testing. It also provides educational materials and forums for NDT professionals. *ANSI/ASNT CP-189, ASNT Standard for Qualification and Certification of Nondestructive Testing Personnel* and *ANSI/ASNT CP-105, ASNT Standard Topical Outlines for Qualification of Nondestructive Testing Personnel* are relevant to welders involved in NDT.

1.1.6 Maritime Welding Guides and Specifications

With its unique environment and challenges, the maritime world has its own welding codes and standards. The US Navy has codes too, although these differ significantly from commercial marine codes. Welders working in the marine environment should familiarize themselves with the applicable codes. The following AWS marine codes are typical examples:

AWS D3.5, Guide for Steel Hull Welding (Historical) — This guide covers welding for steel hulls and related structures. It includes information on materials, equipment, processes, techniques, qualifications, and safety.

AWS D3.6M, Underwater Welding Code — This specification covers the requirements for welding underwater structures or components.

AWS D3.7, Guide for Aluminum Hull Welding (Historical) — This guide covers welding for aluminum hulls and related structures. It includes information on materials, equipment, processes, techniques, qualifications, and safety.

NOTE

Codes marked "historical" are no longer being updated under that title. In some cases, the standards organization has merged them into another code. If so, they'll be maintained and updated under that code's identifier. For reference purposes, standards organizations usually make older versions available through their online library or bookstore.

1.1.7 Quality Workmanship and Inspections

Codes and standards help welders consistently produce quality work. Welding inspectors work with welders to confirm that weldments meet those standards. Ideally, inspectors should examine every weld, but this is rarely practical. In most settings, inspectors check a specific percentage of the total welds. These samples represent the whole job. Based on them, the inspector decides if the work meets the job's standards and specifications.

Individual welders contribute to the quality control process in two ways. First, they must meticulously follow codes, standards, and specifications. They must also hold to high personal standards, producing the best welds possible. Second, they must support the quality control process by cooperating with welding inspectors and watching for problems.

If quality problems arise, report them through the proper channels. Occasionally, however, you may have to bypass the normal channels. For example, if your instructions require you to do something unsafe, you should try to resolve the problem with your supervisor. Similarly, speak to your supervisor if your instructions require you to do work for which you aren't certified. If you can't resolve the situation at that level, speak to the general foreman, site superintendent, project manager, or safety officer.

Did You Know?

PHMSA

In 2004, the US Department of Transportation (DOT) established the Pipeline and Hazardous Materials Safety Administration (PHMSA). It develops and enforces regulations governing the millions of miles of pipelines in the United States. PHMSA manages pipeline inspection to ensure safety for people, property, and the environment.

Welding Inspections

Welding joins metals in many critical applications, including structures, vehicles, machinery, and pipelines. Weld failures in these could cause major damage, injuries, or fatalities. For this reason, welding inspectors continually inspect welders' work to make sure that it meets quality standards. Expect more frequent inspections during critical projects or those with significant hazards.

The AWS publishes standards covering weld inspections. *Certified Welding Inspectors (CWIs)* know these thoroughly and use them daily:

AWS B1.10M/B1.10, Guide for the Nondestructive Inspection of Welds
AWS B1.11M/B1.11, Guide for the Visual Examination of Welds
AWS B5.1, Specification for the Qualification of Welding Inspectors
AWS QC1, Standard for AWS Certification of Welding Inspectors
AWS WIT-T, Welding Inspection Technology

1.2.0 Welding Qualifications

All welding codes include standards for *qualifications*. Think of qualifications like specific certifications or formal approvals. For example, welders can obtain performance and operator qualifications. These confirm that they can do specific welding tasks or set up automated equipment to do it. Welding procedures must also obtain qualifications. These confirm that the procedure works correctly and produces the desired result. The following sections introduce different qualifications.

NOTE

This information is general. Check with your supervisor if you're uncertain about project qualifications.

1.2.1 Welding Procedure Qualifications

Welding procedures are documents outlining all the information associated with a welding task. This includes materials, methods, processes, electrode types, techniques, and other essential details. Before welders can start using a welding procedure, however, it must be qualified.

Example WPS (Single-Process)
WELDING PROCEDURE SPECIFICATION (WPS)

RED Inc.		**2010**	**0**	12/01/2020
Company Name		WPS No.	Rev. No.	Date
J. Jones	12/01/2015	**231**	No	
Authorized by	Date	Supporting PQR(s)	CVN Report	

BASE METALS	Specification	Type or Grade	AWS Group No.
Base Material	ASTM A131	A	I
Welded To	ASTM A131	A	I
Backing Material	ASTM A131	A	I
Other			

BASE METAL THICKNESS	As-Welded	With PWHT
CJP Groove Welds	3/4–1-1/2 in	–
CJP Groove w/CVN	–	–
PJP Groove Welds	–	–
Fillet Welds	–	–
DIAMETER	–	–

JOINT DETAILS	
Groove Type	Single V Groove Butt Joint
Groove Angle	35° included
Root Opening	1/4 in
Root Face	–
Backgouging	None
Method	–

POSTWELD HEAT TREATMENT	
Temperature	None
Time at Temperature	–
Other	–

JOINT DETAILS (Sketch)

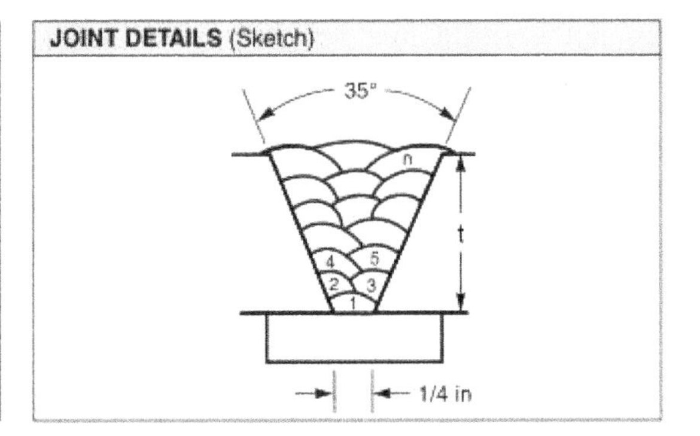

PROCEDURE								
Weld Layer(s)	All							
Weld Pass(es)	All							
Process	**FCAW**							
Type *(Semiautomatic, Mechanized, etc.)*	Semiauto							
Position	OH							
Vertical Progression	–							
Filler Metal (AWS Spec.)	A5.20							
AWS Classification	E71T-1C							
Diameter	0.045 in							
Manufacturer/Trade Name	–							
Shielding Gas (Composition)	100% CO_2							
Flow Rate	45–55 cfh							
Nozzle Size	#4							
Preheat Temperature	60° min.							
Interpass Temperature	60°–350°							
Electrical Characteristics	—							
Current Type & Polarity	DCEP							
Transfer Mode	–							
Power Source Type (*cc, cv, etc.*)	CV							
Amps	180–220							
Volts	25–26							
Wire Feed Speed	(Amps)							
Travel Speed	8–12 ipm							
Maximum Heat Input	–							
Technique	—							
Stringer or Weave	Stringer							
Multi or Single Pass (per side)	Multipass							
Oscillation (*Mechanized, Automatic*)	–							
Number of Electrodes	1							
Contact Tube to Work Dist.	1/2–1 in							
Peening	None							
Interpass Cleaning	Wire Brush							
Other	–							

Form J-2 (See http://go.aws.org/D1forms)

415

Figure 1 An example WPS.
Source: Permission granted by the American Welding Society

Welding procedure qualifications are limiting instructions that explain how to do a welding operation. These instructions appear in a document called a *welding procedure specification (WPS)*. *Figure 1* shows an example WPS.

The WPS defines and documents in detail the *variables* related to project-specific welds. **Essential variables** are items in the WPS that can't be changed without requalifying the entire procedure. If the procedure includes testing for **notch toughness**, additional essential variables enter the picture. These are **supplemental essential variables**.

Essential variables vary by code and the welding process. The following are common examples:

- Base metal type
- Filler metal classification
- Material thickness
- Joint design
- Welding process
- Current type
- Preheat and postheat treatments

Non-essential variables don't affect the WPS's qualification status. They may be changed within the limits established by the code. The following are common examples:

- Amperage
- Travel speed
- Shielding gas flow (if applicable)
- Electrode and filler wire size
- Rod travel angle

Qualifying a WPS involves creating **weld coupons** by following the WPS. A technician then tests the coupons by methods that the code specifies. Tests confirm the weld's physical, mechanical, and metallurgical properties. The technician documents the results with a **procedure qualification record (PQR)**. The PQR includes information about the specimens, the testing methods, and the test results. If all tests confirm that the weld meets the code requirements, the WPS becomes qualified. Most codes require a WPS to have a matching PQR. *Figure 2* shows a PQR example.

1.2.2 Welder Performance Qualifications

A *welder performance qualification* test confirms that a welder can correctly perform a specific welding procedure. Welders must take and pass the tests that qualify them for a particular job's procedures. Later sections examine performance qualifications in more detail. *Figure 3* shows an example performance qualification.

1.2.3 Welding Machine Operator Qualifications

Welding operators who set up, operate, and monitor automatic welding machines must be qualified too. They take qualification tests that confirm their abilities with a particular machine and procedure. Not all welding codes contain qualifications for automatic machine welding.

Welding procedure qualifications: Demonstrating through testing that welds made by following a welding procedure specification (WPS) meet prescribed standards.

Essential variables: Elements in a welding procedure specification (WPS) that can't be changed without requalifying the WPS.

Notch toughness: A material's ability to resist breaking at a point of concentrated stress.

Supplemental essential variables: Variables affecting a weld's toughness that are relevant when the welding procedure specification (WPS) requires toughness testing.

CAUTION

Don't change any essential or supplemental essential variable without discussing it with your supervisor.

Non-essential variables: Elements in a welding procedure specification (WPS) that can be changed without requalifying the WPS.

Weld coupons: Metal pieces that welders use for practice or to produce test welds for qualifications.

Procedure qualification record (PQR): A document recording the test results required to qualify a welding procedure specification (WPS).

FORM QW-483 SUGGESTED FORMAT FOR PROCEDURE QUALIFICATION RECORDS (PQR)
(See QW-200.2, Section IX, ASME Boiler and Pressure Vessel Code)
Record Actual Variables Used to Weld Test Coupon

Organization Name ______
Procedure Qualification Record No. ______ Date ______
WPS No. ______
Welding Process(es) ______
Types (Manual, Automatic, Semi-Automatic) ______

JOINTS (QW-402)

Groove Design of Test Coupon
(For combination qualifications, the deposited weld metal thickness shall be recorded for each filler metal and process used.)

BASE METALS (QW-403)
Material Spec. ______
Type or Grade, or UNS Number ______
P-No. ____ Group No. ____ to P-No. ____ Group No. ____
Thickness of Test Coupon ______
Diameter of Test Coupon ______
Maximum Pass Thickness ______
Other

FILLER METALS (QW-404)	1	2
SFA Specification		
AWS Classification		
Filler Metal F-No.		
Weld Metal Analysis A-No.		
Size of Filler Metal		
Filler Metal Product Form		
Supplemental Filler Metal		
Electrode Flux Classification		
Flux Type		
Flux Trade Name		
Weld Metal Thickness		
Other		

POSITION (QW-405)
Position(s) ______
Weld Progression (Uphill, Downhill) ______
Other

PREHEAT (QW-406)
Preheat Temperature ______
Interpass Temperature ______
Other

POSTWELD HEAT TREATMENT (QW-407)
Temperature ______
Time ______
Other

GAS (QW-408)

	Gas(es)	Percent Composition (Mixture)	Flow Rate
Shielding			
Trailing			
Backing			
Other			

ELECTRICAL CHARACTERISTICS (QW-409)
Current ______
Polarity ______
Amps. ______ Volts ______
Waveform Control ______
Power or Energy ______
Arc Time ______
Weld Bead Length ______
Tungsten Electrode Size ______
Mode of Metal Transfer for GMAW (FCAW) ______
Heat Input ______
Other

TECHNIQUE (QW-410)
Travel Speed ______
String or Weave Bead ______
Oscillation ______
Multipass or Single Pass (Per Side) ______
Single or Multiple Electrodes ______
Other

(07/17)

Figure 2A An example PQR (1 of 2).
Source: Reprinted from ASME 2021 edition, BPVC, Section IX by permission of The American Society of Mechanical Engineers.

FORM QW-483 (Back)

PQR No. ____________

Tensile Test (QW-150)

Specimen No.	Width	Thickness	Area	Ultimate Total Load	Ultimate Unit Stress, (psi or MPa)	Type of Failure and Location

Guided-Bend Tests (QW-160)

Type and Figure No.	Result

Toughness Tests (QW-170)

Specimen No.	Notch Location	Specimen Size	Test Temperature		Toughness Values		Drop Weight Break (Y/N)
				ft-lb or J	% Shear	Mils (in.) or mm	

Comments ____________

Fillet-Weld Test (QW-180)

Result — Satisfactory: Yes ______ No ______ Penetration into Parent Metal: Yes ______ No ______

Macro — Results ____________

Other Tests

Type of Test ____________

Deposit Analysis ____________

Other

....

Welder's Name ____________ Clock No. ______ Stamp No. ______

Tests Conducted by ____________ Laboratory Test No. ______

We certify that the statements in this record are correct and that the test welds were prepared, welded, and tested in accordance with the requirements of Section IX of the ASME Boiler and Pressure Vessel Code.

Organization ____________

Date ____________ Certified by ____________

(Detail of record of tests are illustrative only and may be modified to conform to the type and number of tests required by the Code.)

(07/17)

Figure 2B An example PQR (2 of 2).

Source: Reprinted from ASME 2021 edition, BPVC, Section IX by permission of The American Society of Mechanical Engineers.

Example Welder Qualification (Single-Process)
WELDER, WELDING OPERATOR, OR TACK WELDER PERFORMANCE QUALIFICATION TEST RECORD

Name	Z. W. Elder	OPTIONAL PHOTO ID	Test Date	12/12/2020	Rev.
ID Number	00-001-ZWE		Record No.	WPQ-001	0
Stamp No.	ZWE-1		Std. Test No.	ST-001	0
Company	RED Inc.		WPS No.	WPS-001	0
Division	–		Qualified To	AWS D1.1	

BASE METALS	Specification	Type or Grade	AWS Group No.	Size (NPS)	Schedule	Thickness	Diameter
Base Material	ASTM A36	UNS K02600	I	–	–	3/8 in	–
Welded To	ASTM A36	UNS K02600	I	–	–	3/8 in	–

VARIABLES	Actual Values	RANGE QUALIFIED
Type of Weld Joint	Plate – Groove (Fig. 6.20) with Backing	Groove, Fillet, Plug, and Slot Welds (T-, Y-, K-Groove PJP only)
Base Metal	Group I to Group I	Any AWS D1.1 Qualified Base Metal

	Groove	Fillet	Groove	Fillet
Plate Thickness	3/8 in	–	1/8 in – 3/4 in	1/8 in min.
Pipe/Tube Thickness	–	–	1/8 in – 3/4 in	Unlimited
Pipe Diameter	–	–	24 in min.	Unlimited

Welding Process	**GMAW**	**GMAW**
Type *(Manual, Semiautomatic, Mechanized, Automatic)*	Semiautomatic	Semiautomatic, Mechanized, Automatic
Backing	With	With (incl. Backgouging and Backwelding)
Filler Metal (AWS Spec.)	A5.18	A5.xx
AWS Classification	ER70S-6	All
F-Number	–	–
Position	2G, 3G, and 4G	
Groove – Plate and Pipe ≥ 24 in		All
Groove – Pipe < 24 in		–
Fillet – Plate and Pipe ≥ 24 in		All
Fillet – Pipe < 24 in		All
Progression	Vertical Up	Vertical Up
GMAW Transfer Mode	Globular	Spray, Pulsed, Globular
Single or Multiple Electrodes	Single	Single
Gas/Flux Type	A5.32 SG-C	A5.xx Approved

TEST RESULTS

Type of Test	Acceptance Criteria	Results	Remarks
Visual Examination per 6.10.1	6.10.1	Acceptable	–
Each Position: 1 Root Bend per 6.10.3.1 and Fig. 6.8	6.10.3.3	Acceptable	–
Each Position: 1 Face Bend per 6.10.3.1 and Fig. 6.8	6.10.3.3	Acceptable	3G: Small (<1/16 in) Opening

CERTIFICATION

Test Conducted by	
Laboratory	Welding Forms Lab
Test Number	Fictitious Test XYZ
File Number	Welding Forms/Sample-WPQ-for-GMAW.pdf

We, the undersigned, certify that the statements in this record are correct and that the test welds were prepared, welded, and tested in accordance with the requirements of Clause 6 of AWS D1.1/D1.1M (_______) *Structural Welding Code—Steel.*
(year)

Manufacturer or Contractor **Red Inc.** Authorized by **E. M. Ployee (Q.C. Mgr.)**

Date **12/12/2020**

Form J-4 (See http://go.aws.org/D1forms)

418

Figure 3 Welder performance qualification example.
Source: Permission granted by the American Welding Society

1.0.0 Section Review

1. Which of the following often become integrated into laws?
   a. Specifications
   b. Codes
   c. Standards
   d. WPSs

2. A welder who successfully passes a performance qualification test may perform a specific welding procedure.
   a. True
   b. False

2.0.0 Weld Discontinuities and Their Causes

Objective

Identify weld discontinuities and their causes.

a. Describe acceptable and unacceptable weld profiles.

b. Describe common discontinuities and their causes.

Performance Tasks

There are no Performance Tasks in this section.

Codes and standards define what produces a weld that will serve its intended function for the weldment's expected life. Some welds, however, contain problems that keep them from meeting the minimum code requirements. These welds are unacceptable and require replacement.

A weld **defect** is always unacceptable. Welding inspectors automatically reject them. On the other hand, a weld **discontinuity** may or may not be unacceptable. The AWS defines a discontinuity as an interruption in the weldment's typical structure. This could be insufficient **homogeneity** in the weldment's mechanical, metallurgical, or physical properties. Sometimes, a weld may have one or more discontinuities and yet still be acceptable. On the other hand, a single large discontinuity or a certain combination of discontinuities can make the weld defective.

Defect: A discontinuity or other imperfection that makes a weld unacceptable according to code.

Discontinuity: A change or break in a weld's shape or structure.

Homogeneity: Having a uniform structure or composition.

Ideally, welds shouldn't have any discontinuities, but many will have them. Welders should be able to identify them and understand how they affect the weld. Simple visual inspection can reveal some. Others, however, aren't visible and require tests to detect them.

When evaluating a weld, note a discontinuity's type, size, and location. Any of these can turn the discontinuity into a defect. For example, discontinuities at stress points can expand, making them riskier than those in less-crucial places. Surface or near-surface discontinuities may be more harmful than similar internal discontinuities.

2.1.0 Weld Profiles

Welders refer to a weld's cross-sectional view as its *profile* (*Figure 4*). One or more weld beads (passes) create the profile. A weld's profile can affect the joint's performance under load as much as discontinuities can. Both single-pass and multi-pass weld profiles affect the weld's behavior. Unacceptable profiles can also encourage discontinuities to form.

NOTE

The following guidelines are general. Refer to the WPS and jobsite quality specifications for specific requirements. Check with your supervisor if you're uncertain.

2.1.1 Fillet Welds

A fillet weld has a right-triangle profile. It forms T, lap, and corner joints. Welding symbols specify a fillet weld's size and location. A *convex* fillet weld bows outwards like a ball's surface. A *concave* fillet weld bows inwards like the

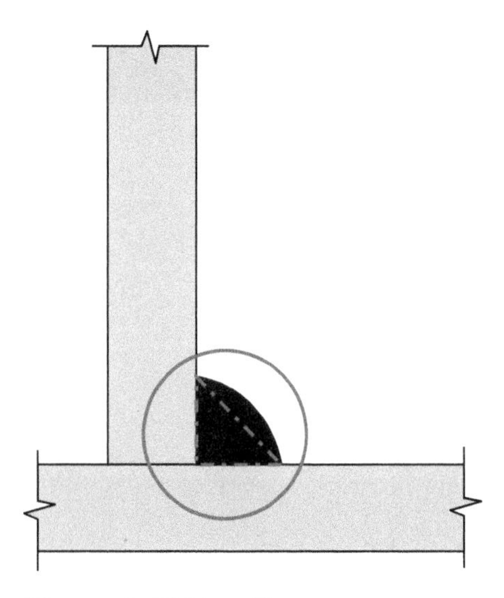

Figure 4 Weld profile.

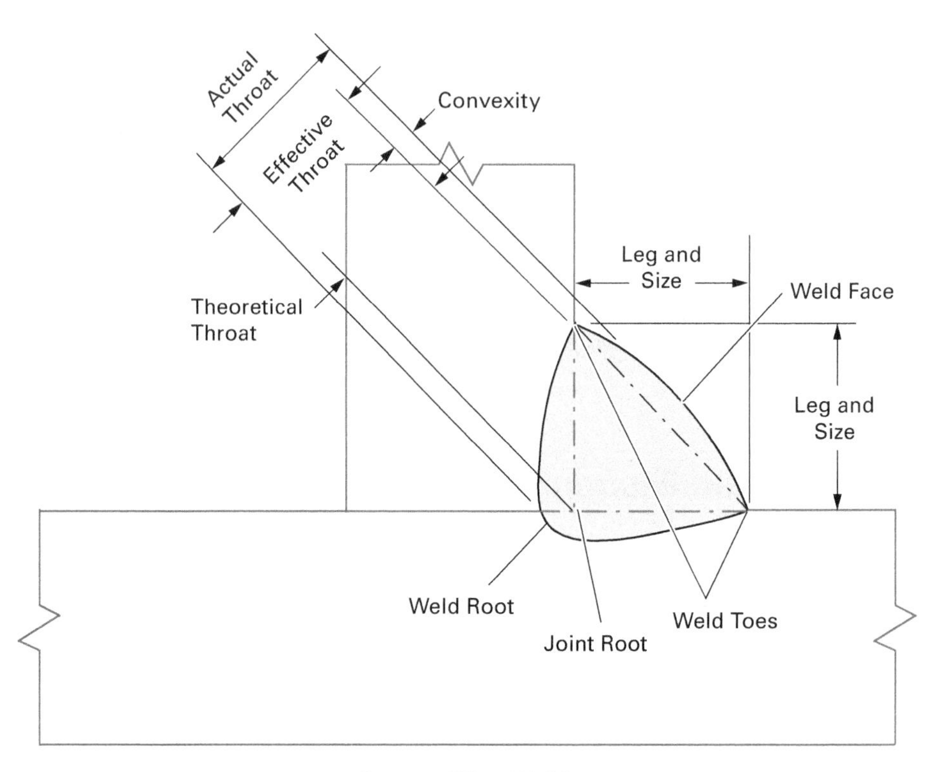

Convex Fillet Weld

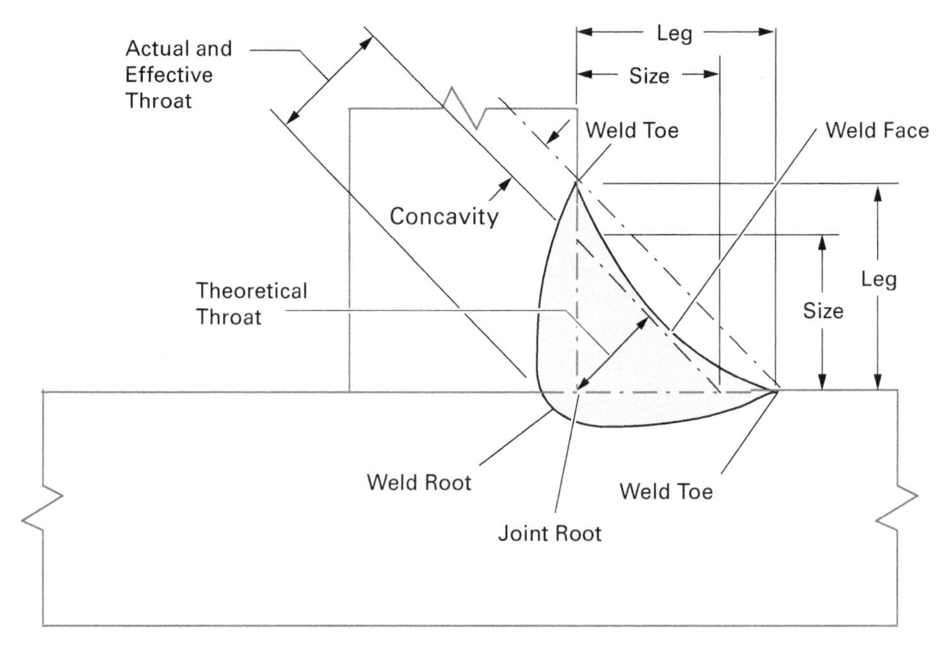

Concave Fillet Weld

Figure 5 Convex and concave fillet welds.

inside of a bowl. Welders describe fillet welds with the following terminology (*Figure 5*):

Weld face — The weld's exposed surface on the side where the welder produced it.

Weld toes — The points at the weld face where the weld and base metals meet.

Weld root — The point where the back of the weld extends farthest into the joint.

Joint root — The place in the joint where the weld members are closest together.

Leg — The distance from the joint root to the weld toe.

Size — The leg lengths of the largest right triangle that can fit within the weld profile.

Actual throat — The shortest distance from the weld root to its face.

Effective throat — The actual throat value minus the weld face's convexity.

Theoretical throat — The perpendicular distance from the hypotenuse of the profile's imaginary right triangle to the joint root.

Welding codes accept fillet welds with equal or unequal legs (*Figure 6*). Codes permit slightly convex or slightly concave faces. Faces must be relatively uniform, although slight non-uniformity is acceptable too. The following rules determine the maximum permissible convexity:

- If the weld size or individual surface bead is ≤ 5⁄16" (8 mm), the maximum convexity is 1⁄16" (1.6 mm).
- If the weld size or individual surface bead is > 5⁄16" (8 mm) and < 1" (25 mm), the maximum convexity is 1⁄8" (3.2 mm).
- If the weld size or individual surface bead is ≥ 1" (25 mm), the maximum convexity is 3⁄16" (4.8 mm).

Any of the following profile discontinuities make a fillet weld unacceptable (*Figure 7*):

- Insufficient throat
- Excessive convexity
- Excessive undercut
- Overlap
- Insufficient leg
- Incomplete fusion

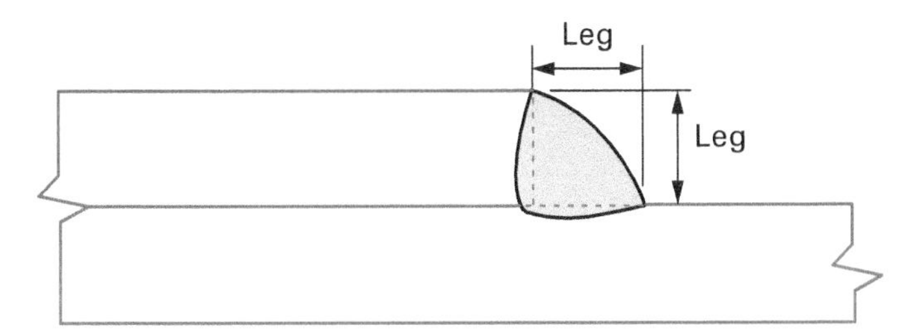

Equal Leg Fillet Weld

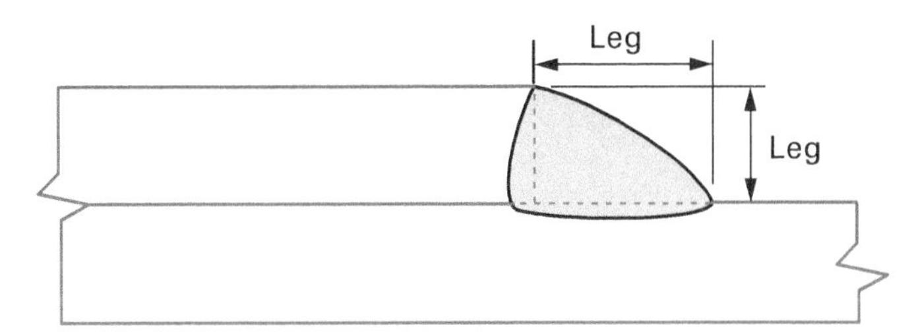

Unequal Leg Fillet Weld

Figure 6 Equal and unequal leg fillet welds.

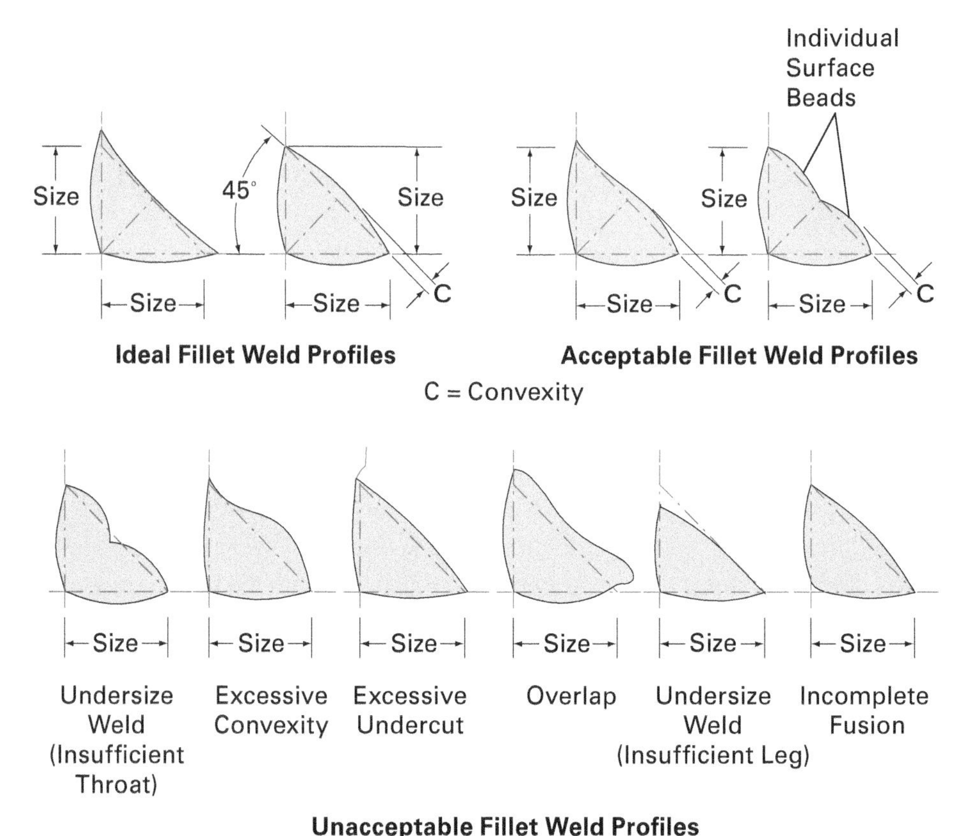

Figure 7 Acceptable and unacceptable fillet weld profiles.

NOTE

NCCER Module 29109, *SMAW – Beads and Fillet Welds*, discusses fillet welds in more detail.

Fillet welds require minimal base metal preparation. But the weld area must be clean and free from dross. Dross from cutting operations will cause discontinuities in the weld. Codes require removing it before welding.

2.1.2 Groove Welds

Groove weld profiles are triangle- or hourglass-shaped but come in many variants. The face should be slightly convex. Reinforcement—weld metal extending above the base metal on a groove weld's face—must not exceed a code-specified amount. Most codes determine acceptable reinforcement from the base metal's thickness. Under *AWS D1.1*, it may fall between 1⁄8" (3.2 mm) and 1⁄4" (6.4 mm) for plates under 2" (~50 mm) thick. For thicker plates, it's limited to 3⁄16" (4.8 mm). Codes regulating pipes also determine the acceptable reinforcement from the base metal's thickness.

NOTE

NCCER Module 29111, *SMAW – Groove Welds with Backing (Plate)*, and NCCER Module 29112, *SMAW – Open-Root Groove Welds (Plate)*, discuss groove welds in more detail.

Reinforcement must transition gradually to the base metal at each toe. The bead width shouldn't exceed the groove width by more than the code-specified amount. Excess reinforcement, insufficient throat, excessive undercut, or overlap are all unacceptable profile discontinuities (*Figure 8*).

2.2.0 Discontinuity Types

Discontinuities vary widely in their appearance and causes. The following sections introduce common examples and describe their causes. Learn to identify them. Even more importantly, learn how to avoid creating them. As you become more skillful and experienced, they'll occur less often.

2.2.1 Porosity

Porous materials contain holes (voids), making them look like a sponge. Gas trapped in the molten weld pool creates voids that remain when the metal cools. Sometimes, the gas works its way to the surface before the weld solidifies, creating surface imperfections. These indicate internal porosity. In many cases, however, you can't spot porosity visually.

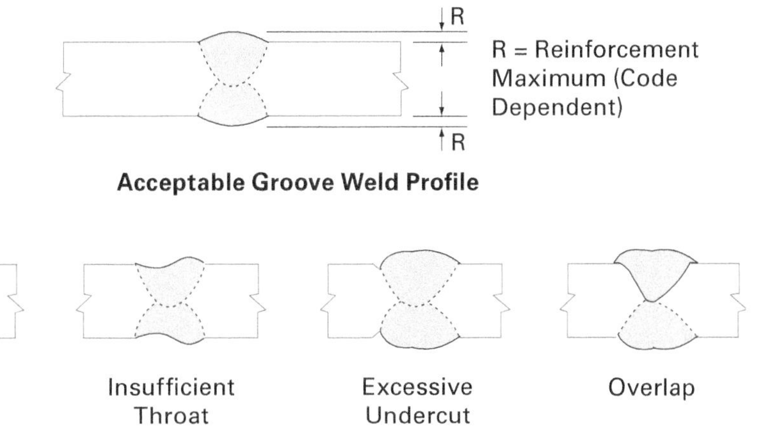

Figure 8 Acceptable and unacceptable groove weld profiles.

Improper welding technique or contamination causes most porosity. If there isn't enough shielding gas to protect the weld, oxygen or moisture can dissolve in the weld pool, producing voids. The welding heat breaks down paint, dirt, oil, and other contaminants, producing hydrogen. If this gas dissolves in the weld pool, it will produce voids.

Excessive porosity affects the joint mechanically, usually making it weak. Some codes permit limited porosity, but it's better to avoid it altogether. Prevent it by properly cleaning the base metal, keeping electrodes dry, and using the proper welding technique. Remove excessive surface porosity by grinding it out.

Figure 9 shows several surface porosity examples.

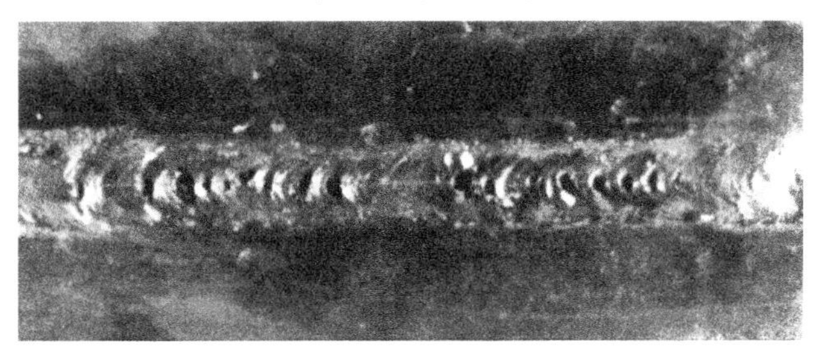

(A) Linear Porosity

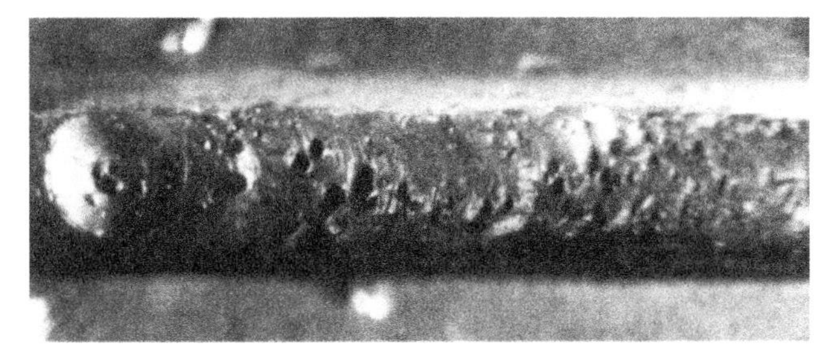

(B) Scattered Surface Porosity

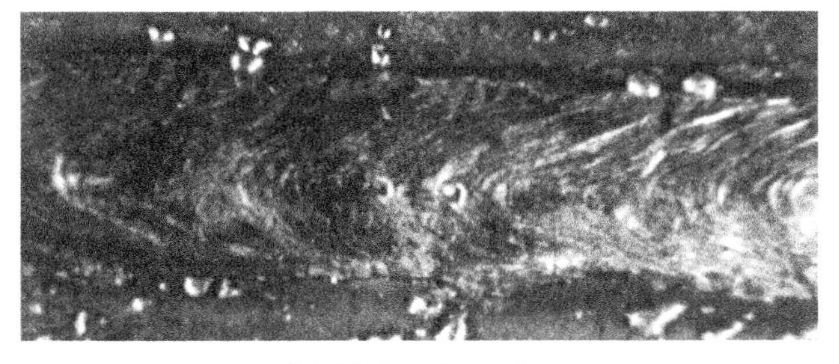

(C) Piping Porosity

Figure 9 Porosity.
Source: Permission granted by the American Welding Society

Linear Porosity

This type forms a line along the place within the weld where the base and weld metals meet. It can also appear at the weld root or the boundary between weld beads.

Scattered Porosity

This type is scattered throughout the weld. It may appear in either single- or multi-pass welds.

Clustered Porosity

This type concentrates in one part of the weld. It's caused by improperly starting or stopping the weld.

Piping Porosity

This type has elongated holes, sometimes called *wormholes*. They extend from the weld root toward the face. They don't always reach the surface, so they may not be visible.

2.2.2 Inclusions

Inclusions: Foreign matter trapped in a weld.

Inclusions are foreign matter trapped in the weld, between weld beads, or between the weld metal and the base metal (*Figure 10*). Some are jagged and irregularly shaped, while others form a continuous line. Inclusions concentrate stresses in one area and reduce the weld's strength.

Inclusions find their way into the weld during out-of-position welding. Poor technique can also introduce them. Slag often becomes an inclusion if the welder manipulates the electrode incorrectly. The arc will blow slag particles into the weld, where they become trapped before they can float to the top. Notches in joint boundaries or between weld passes can trap slag too. Tungsten particles from GTAW processes can become inclusions.

Avoid inclusions by practicing proper technique. Use the right electrode and make sure it's in good condition. Position the work properly to prevent slag from becoming trapped in the weld. Removing slag between weld passes helps too. Prepare the weld surface and base metal before starting work. Grind rough surfaces smooth and remove rust or scale.

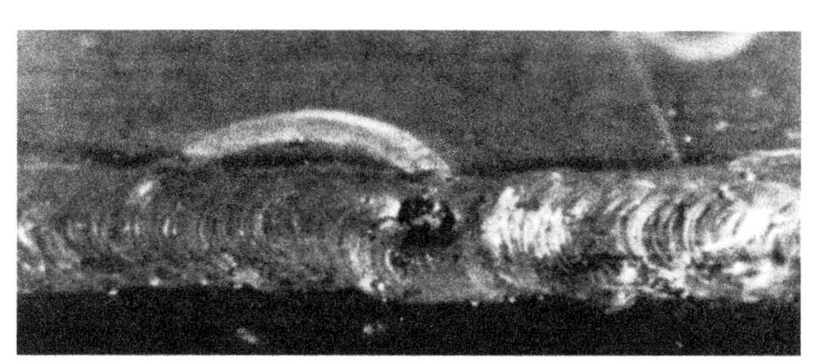

Surface Slag Inclusions

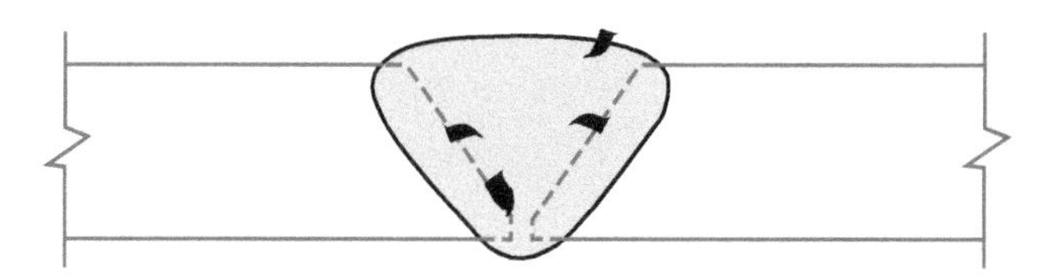

Figure 10 Inclusions.
Source: Permission granted by the American Welding Society

2.2.3 Cracks

Cracks are narrow breaks in the weld metal, the base metal, or the crater formed at the end of a bead (*Figure 11*). They occur when nearby stresses are greater than the metal's strength. Cracks often appear near other discontinuities since these can produce stresses. Cracks can grow (propagate) under load. Some may eventually cause the weld joint to fail. For this reason, welders should treat them as rejectable and repair them.

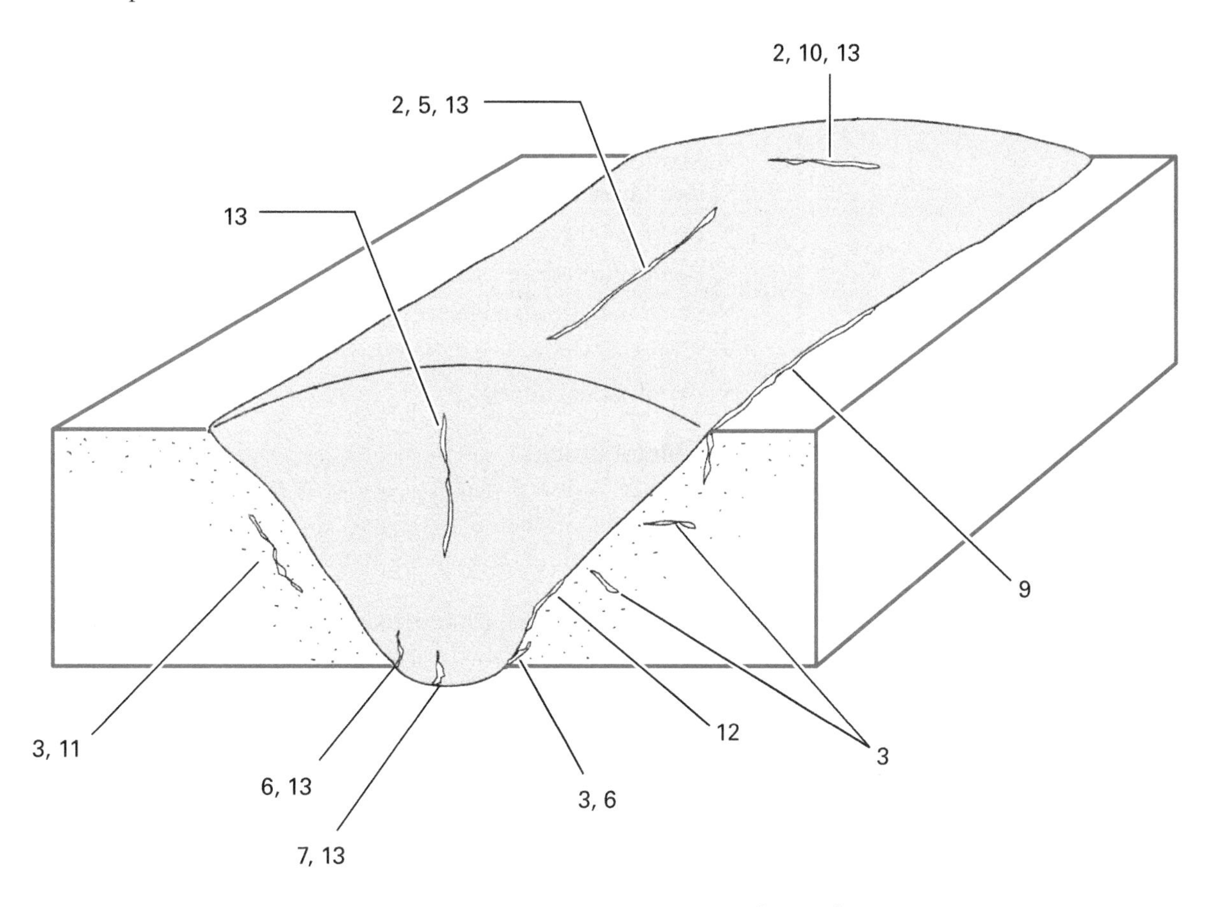

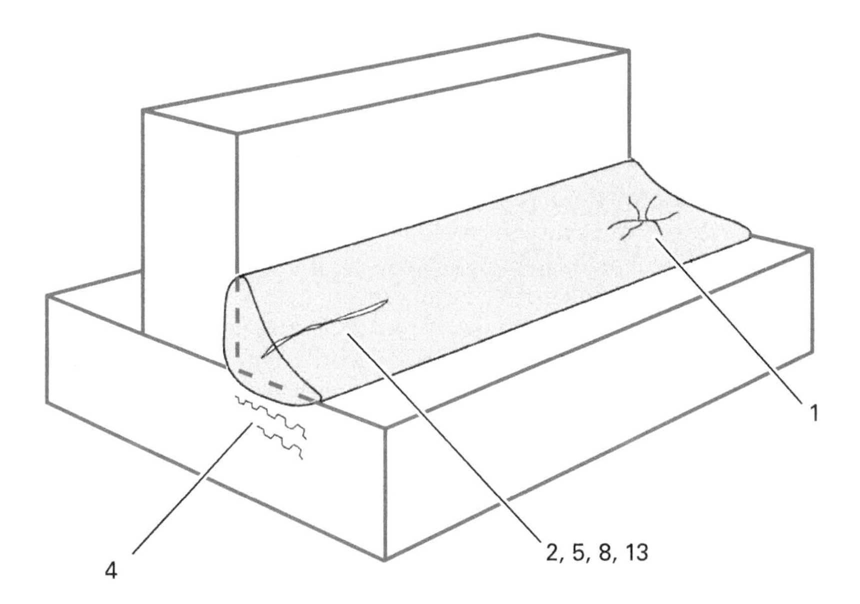

Legend

1. Crater Crack
2. Face Crack
3. Heat-Affected Zone Crack
4. Lamellar Tear
5. Longitudinal Crack
6. Root Crack
7. Root Surface Crack
8. Throat Crack
9. Toe Crack
10. Transverse Crack
11. Underbead Crack
12. Weld Interface Crack
13. Weld Metal Crack

Figure 11 Crack types and locations.
Source: Permission granted by the American Welding Society

Weld Metal Cracks

Transverse cracks run across the weld face and can extend into the base metal. They're more common in highly restrained joints. *Longitudinal cracks* are usually near the weld deposit center. They may start in the crater or first weld layer if the bead is thin. They'll work through the entire weld if you don't deal with them immediately. *Crater cracks* form in the crater if you interrupt the welding operation. They extend to the crater's edge and may start a longitudinal crack. Prevent them by filling craters so they're slightly convex before breaking the arc.

Figure 12 shows various weld metal cracks. Reduce weld metal cracking with one or more of the following strategies:

- Manipulate the electrode or amperage to improve the weld deposit's contour or composition.
- Decrease travel speed to lay a thicker weld that can resist stress better.
- Preheat to reduce thermal stress.
- Use low-hydrogen electrodes.
- Sequence welds to balance shrinkage stresses.
- Avoid rapid cooling.

Base Metal Cracks

Base metal cracking usually occurs within the heat-affected zone. It's more common in **hardenable materials** since these can become **embrittled** after heating and cooling. The cracks appear near the zone's edges and travel into the base metal.

Hardenable materials: Materials that can become harder by heating and then cooling them.

Embrittled: Metal that has become brittle and tends to crack easily.

Underbead cracking: Base metal cracking near the weld but below its surface.

Underbead cracking happens mostly with steel. It occurs under the weld metal, so it's not visible during an inspection. *Toe cracking* happens when the metal cools and puts stress on the heat-affected zone. If it has become embrittled, a crack will start.

Reduce base metal cracking with one or more of the following strategies:

- Preheat to control the cooling rate.
- Control heat input during welding.
- Use the correct electrode.
- Control the welding materials.
- Properly match the filler metal to the base metal.

Hot and Cold Cracks

Hot cracks occur as the weld solidifies. They often appear in metals that don't stretch much before breaking, especially at higher temperatures. *Cold cracks* happen after the weld solidifies. They're usually caused by poor welding technique.

2.2.4 Incomplete Joint Penetration

Incomplete joint penetration happens when the filler metal doesn't penetrate and fuse with part of the weld joint (*Figure 13*). It's always undesirable, especially in single-groove welds whose root experiences tension (pulling) or bending stresses. These stresses can cause the weld to fail.

Inadequate weld heat at the joint root can cause incomplete joint penetration. If the metal above the root melts first, it can prevent heat from reaching the root. Improper joint design is another cause. A root face dimension that is too big, a root opening that is too small, or a groove angle in a V-groove weld that is too small can contribute to this problem. *Figure 14* shows correct and incorrect joint designs.

Even when all other factors are correct, poor arc control can produce incomplete joint penetration. "Long arcing" the electrode is one example. Contributing factors include using too large an electrode, too low an amperage, or too fast a travel speed.

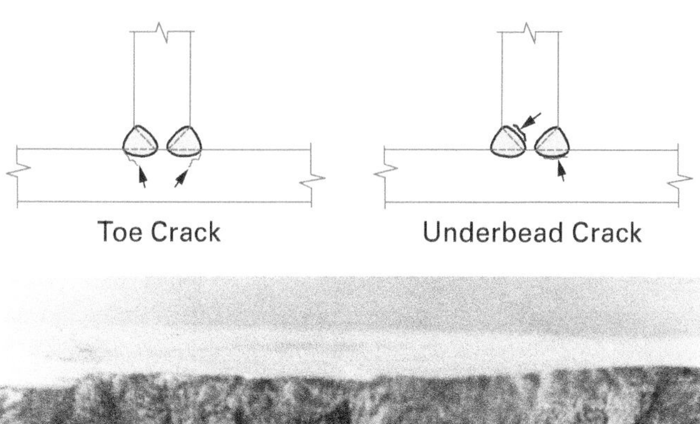

Toe Crack Underbead Crack

Toe Crack

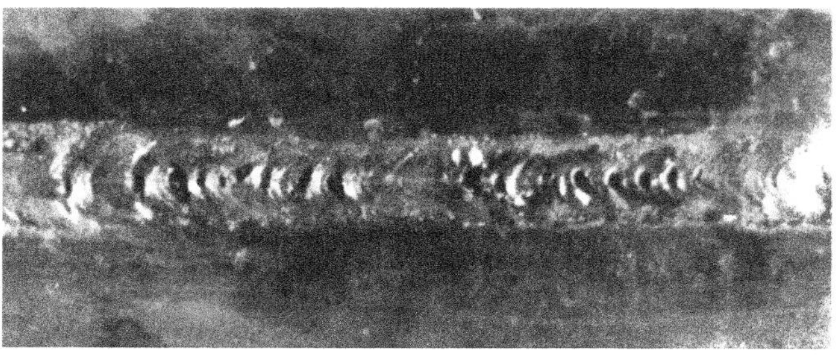

Longitudinal Crack and Linear Porosity

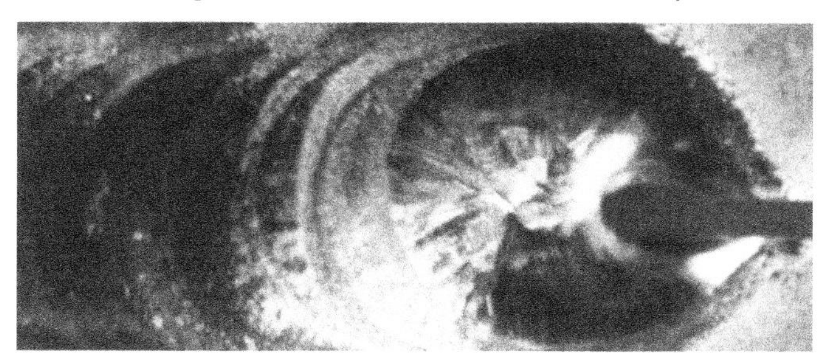

Crater Crack

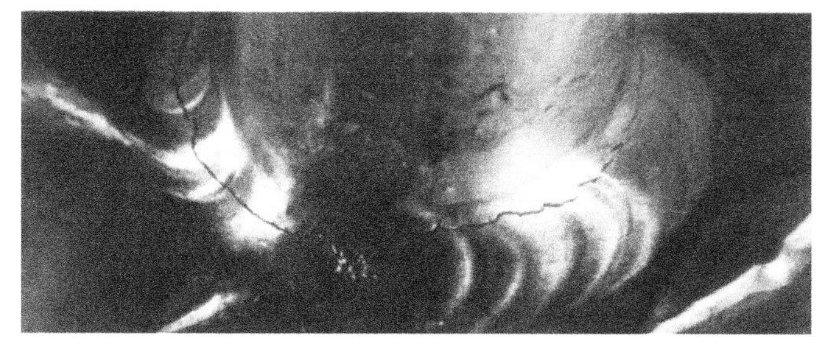

Longitudinal Crack out of Crater Crack

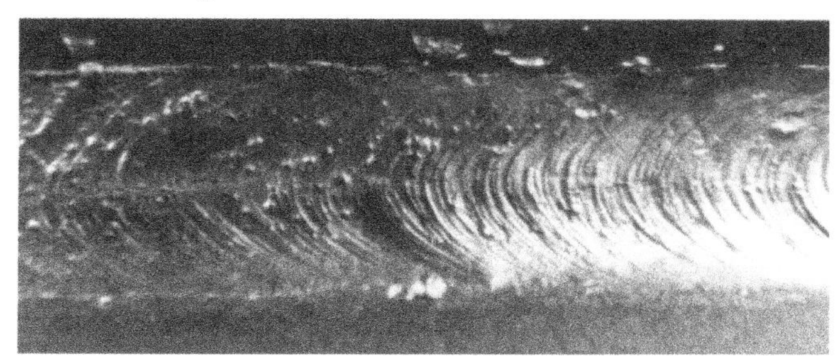

Fillet Weld Throat Crack

Figure 12 Weld metal cracks.
Source: Permission granted by the American Welding Society

Incomplete
Joint Penetration
Incomplete
Joint Penetration

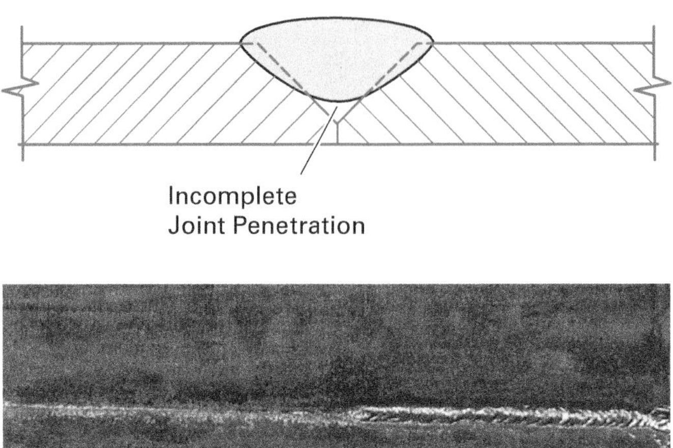

Figure 13 Incomplete joint penetration.

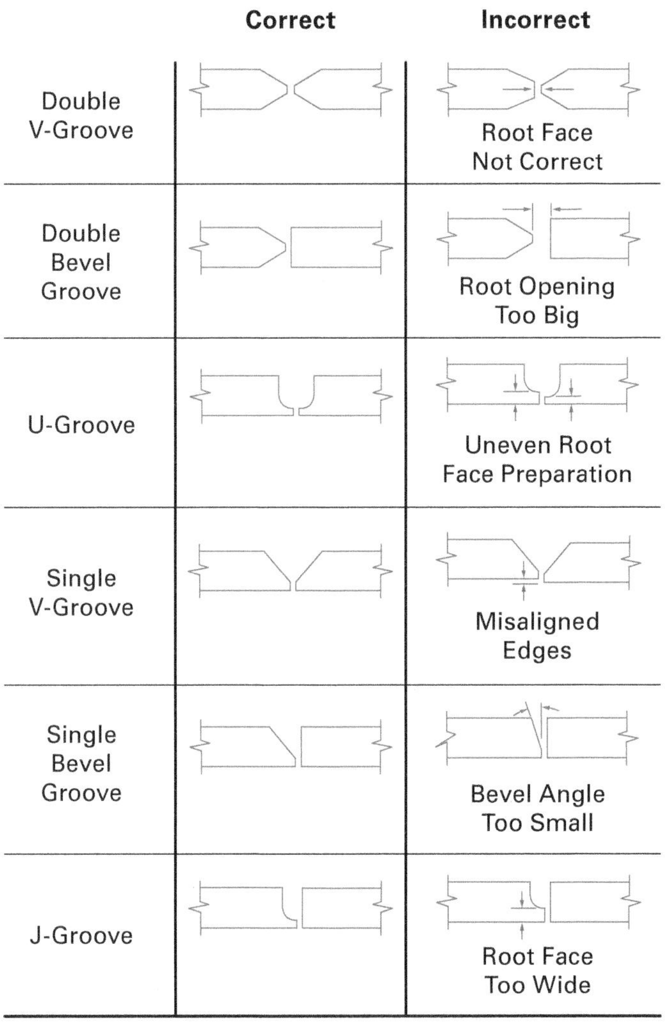

Correct
Incorrect
Double
V-Groove
Root Face
Not Correct
Double
Bevel
Groove
Root Opening
Too Big
U-Groove
Uneven Root
Face Preparation
Single
V-Groove
Misaligned
Edges
Single
Bevel
Groove
Bevel Angle
Too Small
J-Groove
Root Face
Too Wide

Figure 14 Correct and incorrect joint designs.

2.2.5 Incomplete Fusion

This discontinuity occurs when the weld and/or base metals don't fuse (join) together. Some welders confuse it with incomplete joint penetration, but it's different. A weld can have good penetration but poor fusion. Similar to porosity and inclusions, incomplete fusion affects the weld's integrity. Some welders call incomplete fusion *cold lap*.

Incomplete fusion can occur anywhere—including the root—in fillet or groove welds. Often, the weld metal simply rolls over onto the surface (*overlap*). Sometimes, the weld fuses correctly at the root and at the surface but not at the toe. *Figure 15* shows incomplete fusion and overlap. Causes include the following factors:

- Insufficient heat (low amperage, too fast a travel speed, or too close an arc gap)
- Wrong electrode size or type
- Not removing oxide or slag from groove faces or previously deposited beads
- Improper joint design
- Inadequate shielding gas
- Improper electrode angle
- **Arc blow** deflecting the arc

Arc blow: Magnetic forces that deflect the welding arc from its intended path.

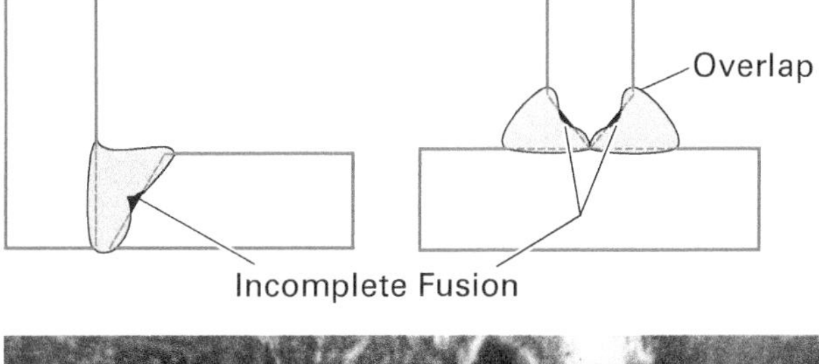

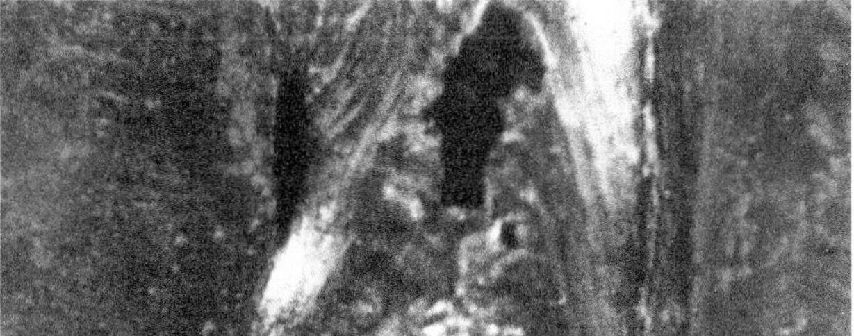

Incomplete Fusion at Weld Face

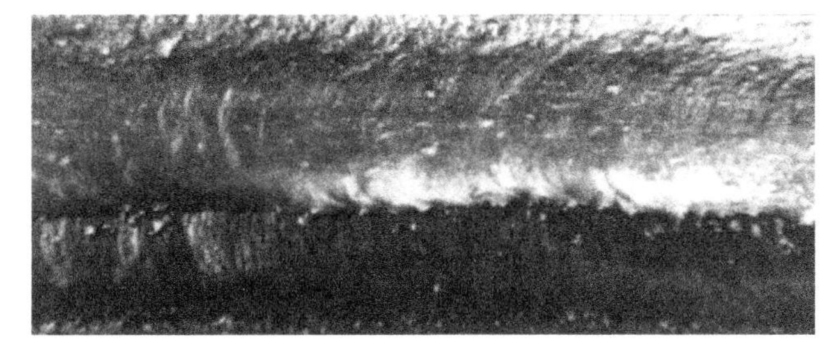

Incomplete Fusion Between Individual Weld Beads

Figure 15 Incomplete fusion and overlap.
Source: Permission granted by the American Welding Society

2.2.6 Undercut

Undercut happens when the welder melts a groove into the base metal at the weld toe. The arc removes more metal from the joint face than it replaces. On multi-pass welds, undercut can occur where a layer meets the groove wall (*Figure 16*).

Most welds have some undercut. Welding inspectors don't treat undercut as a defect if it stays within the specifications' limits and doesn't create sharp or deep notches. Outside these limits, however, it's a defect because it reduces the joint's strength. Causes include the following factors:

- Amperage too high
- Arc gap too long
- Improper electrode angle
- Incorrect travel speed
- Not filling the crater completely

NOTE

Undercut is different from *underfill*. Undefill is caused by using too little filler metal in the weld.

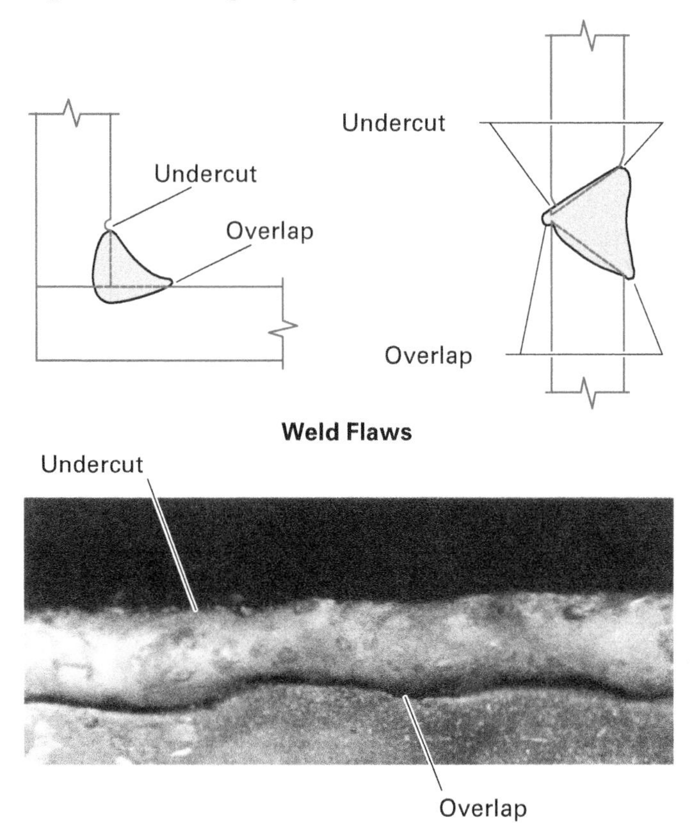

Figure 16 Undercut and overlap.
Source: Permission granted by the American Welding Society

2.2.7 Arc Strikes and Spatter

Arc strikes happen when the welder accidentally starts (strikes) the arc outside the weld zone. This mistake leaves a melted spot on the surface (*Figure 17*). Faulty ground connections sometimes cause arc strikes too.

Arc strikes outside the weld zone are unsightly and suggest sloppy welding. They can also harden the base metal near the strike. The hard spot can start a crack. Welding inspectors normally reject welds with arc strikes outside the weld zone.

Spatter is a spray of fine metal particles on the surface around the weld zone. It makes an ugly weld that's hard to inspect. It can damage coatings too. Sometimes, spatter can cause hard spots just like an arc strike. This too can start cracks. Long arcing, an amperage that is too high, an unstable arc, or improper shielding can cause spatter.

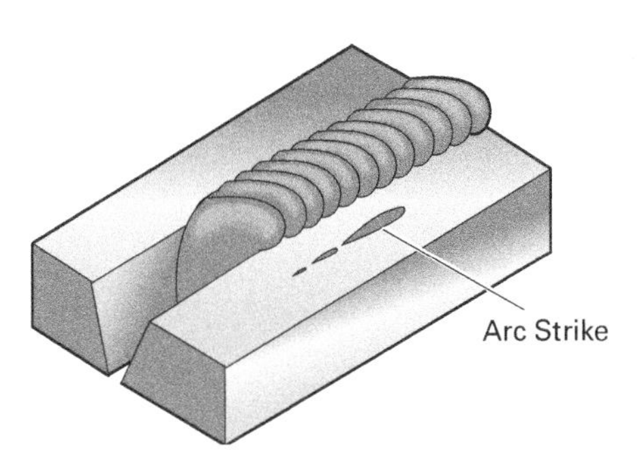

Figure 17 Arc strike.

2.0.0 Section Review

1. A groove weld has reinforcement $^{5}/_{16}$" (8 mm) thick. The weld is defective.
   a. True
   b. False

2. Gas pockets entrapped in the weld pool cause _____.
   a. inclusions
   b. cracks
   c. porosity
   d. undercut

3.0.0 Weld Inspection and Testing

Objective

Summarize nondestructive and destructive weld tests.

a. Describe visual weld inspection methods.
b. Describe nondestructive weld testing methods.
c. Describe destructive weld testing methods.

Performance Task

1. Visually inspect (VT) a fillet and/or groove weld and complete an inspection report.

Some weld defects and discontinuities are obvious. Simple visual inspection reveals them. But many problems are either hidden within the weld or not obvious. These require more advanced methods to locate and identify.

Welding inspectors use many tools to inspect and test welds. Some are simple, while others are sophisticated laboratory instruments. Inspection and testing techniques divide into two categories: nondestructive testing and destructive testing.

Nondestructive testing doesn't alter or damage the test specimen. Methods range from visual inspection techniques to instrument-based tests. Advanced nondestructive testing can look deep within the weld for hidden defects and discontinuities.

Destructive testing, as its name suggests, alters or breaks the test specimen. Most methods use machines that stress welds until they fail. By measuring the applied forces, the machine determines the weld's strength.

Since destructive testing ruins the specimen, it isn't used on production welds. Quality control teams use it for developing and qualifying welding procedures. They also use destructive tests to verify that nondestructive testing methods accurately reveal defects and discontinuities. Once they've confirmed this, welding inspectors can confidently use nondestructive tests to inspect production welds.

Welding inspectors perform some tests in the field. Other tests take place in a material science laboratory equipped with the necessary tools. Lab technicians may do some of these tests rather than welding inspectors. For expensive or exotic tests, companies may send parts to specialist laboratories.

NOTE

Nondestructive testing (NDT) also goes by the names *nondestructive inspection (NDI)* and *nondestructive examination (NDE)*.

3.1.0 Visual Inspection

Visual inspection (VT) is a very basic nondestructive testing method. Welders and welding inspectors examine the weld and base metal surfaces for imperfections. Sometimes they use tools, such as gauges or magnifiers, to assist themselves. Visual inspection is fast and inexpensive. Although it's obviously limited by human visual power, a skillful person can spot more than 75 percent of visible discontinuities, making it extremely useful.

3.1.1 Visual Inspection During Welding

Visual inspection doesn't happen just once. It's an ongoing process during welding. By examining their work at each stage, welders catch many problems before they turn into serious discontinuities.

Before welding, examine the base metal's condition. Measure and prepare it according to the specifications. Correct all problems as you discover them. After assembling the parts, visually check the weld joint for the proper root opening. Consider factors that could affect the outcome such as:

- Proper surface cleaning
- Correct joint preparation and dimensions
- Correct clearance dimensions for backing strips, rings, or consumable inserts
- Correct alignment and fit-up
- Proper welding procedures and machine settings
- Correct preheat temperature (if applicable)
- Good tack weld quality

As you create the required welds, examine each one. Verify that it meets the specifications. Inspect the root pass and each succeeding layer. Check that they follow the correct sequence. Clean welds between passes. If you're going to weld the other side, prepare its root correctly before continuing.

After completing the weld, clean the surface thoroughly. Examine the weld and the surrounding base metal. Visual inspection reveals many defects and discontinuities. Look for cracks, shrinkage cavities, undercuts, incomplete penetration, incomplete fusion, overlap, and crater deficiencies. Deal with problems immediately.

3.1.2 Visual Inspection Tools

Visual inspection includes checking dimensional accuracy. This step confirms that welds are within the limits established by codes and specifications. Checking weld dimensions requires tools called *gauges*. These measure different weld features.

Undercut Gauge

An undercut gauge measures the undercut on the base metal. Many codes allow undercut between 0.01" and 0.031" (0.25 mm and 0.8 mm) deep. An undercut gauge has a pointed end that the welder pushes into the undercut. A scale shows the result. *Figure 18* shows two undercut gauge styles.

NOTE

The bridge cam gauge (Cambridge gauge) in *Figure 18* can measure many other weld features besides undercut. Consult the booklet that accompanies the gauge for more information.

Automatic Weld Size Gauge

This versatile tool checks several weld features (*Figure 19*). It can measure a butt weld's reinforcement. It also measures a fillet weld's leg sizes, as well as its convexity or concavity. After positioning it, the welder slides the pointer into contact with the feature and reads the right scale.

Fillet Weld Blade Gauge Set

This tool is a set of blades that can measure fillet welds between $\frac{1}{8}$" and 1" (3.2 mm and 25 mm). Each blade matches a specific size. The blades have protrusions to measure concave fillets. Cutouts fit around convex fillets. The welder selects the correct blade, places it flat on the base metal, and slides it into contact with the fillet. A correctly sized fillet matches the blade (*Figure 20*).

Versatile Fillet Weld Gauge

This fillet weld gauge measures 11 fillet weld sizes in fractions or decimals. Metric versions are available too.

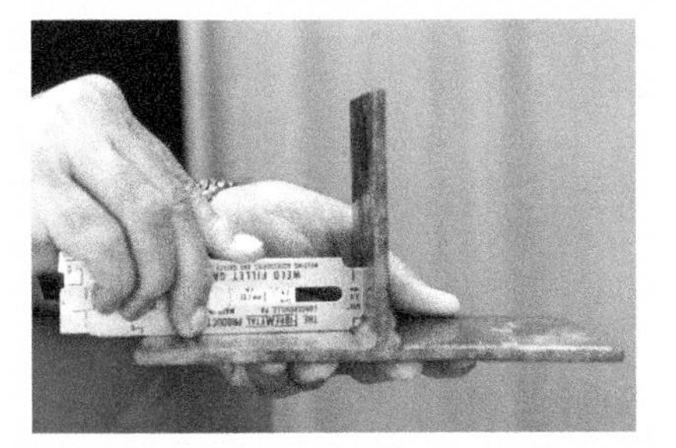

3.1.3 Liquid Penetrant Inspection

Liquid penetrant inspection (PT) is a more sophisticated nondestructive testing method than simple visual inspection. Like visual inspection, it identifies surface problems only. But it can reveal ones that the eye can't see. It's inexpensive and simple. Most liquid penetrant tests come as kits containing three spray cans—a cleaner, a dye, and a developer (*Figure 21*).

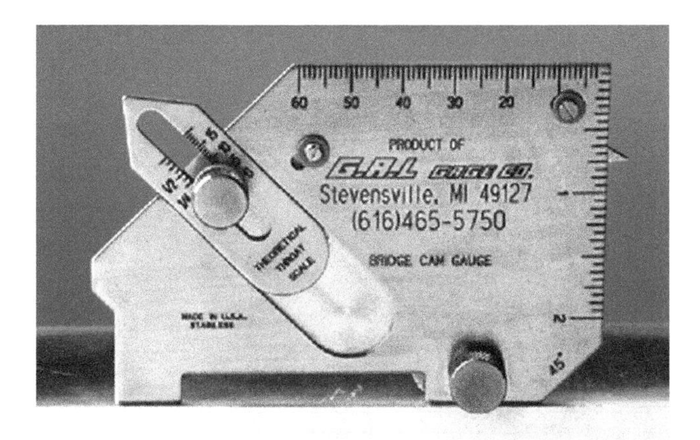

Bridge Cam Gauge

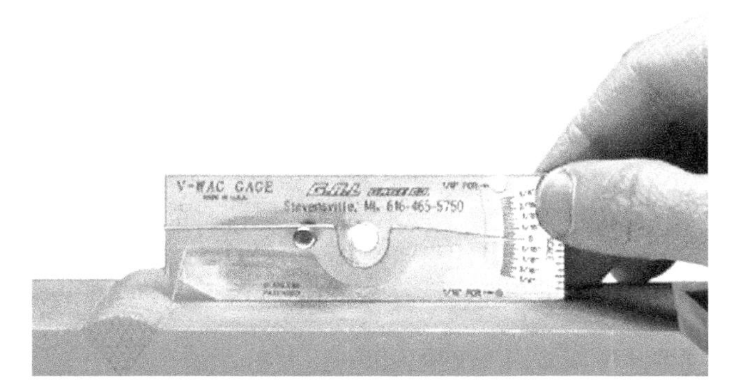

V-WAC Gauge

Figure 18 Undercut gauges.
Source: G.A.L. Gage Co

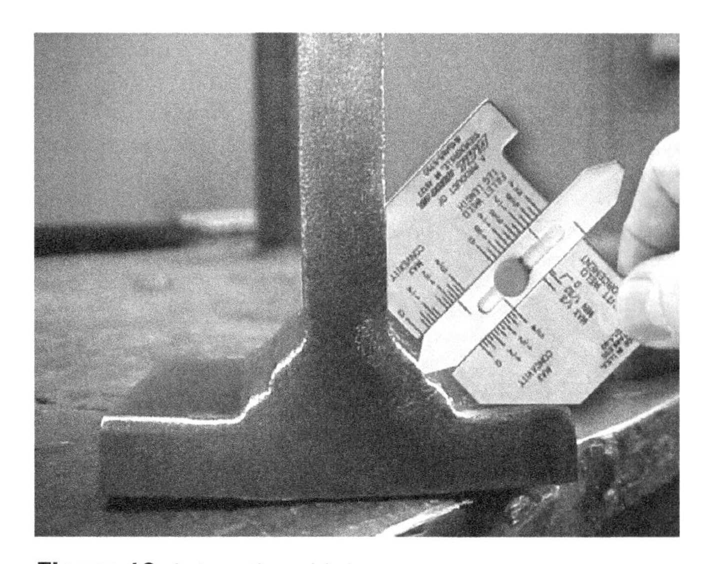

Figure 19 Automatic weld size gauge.
Source: G.A.L. Gage Co

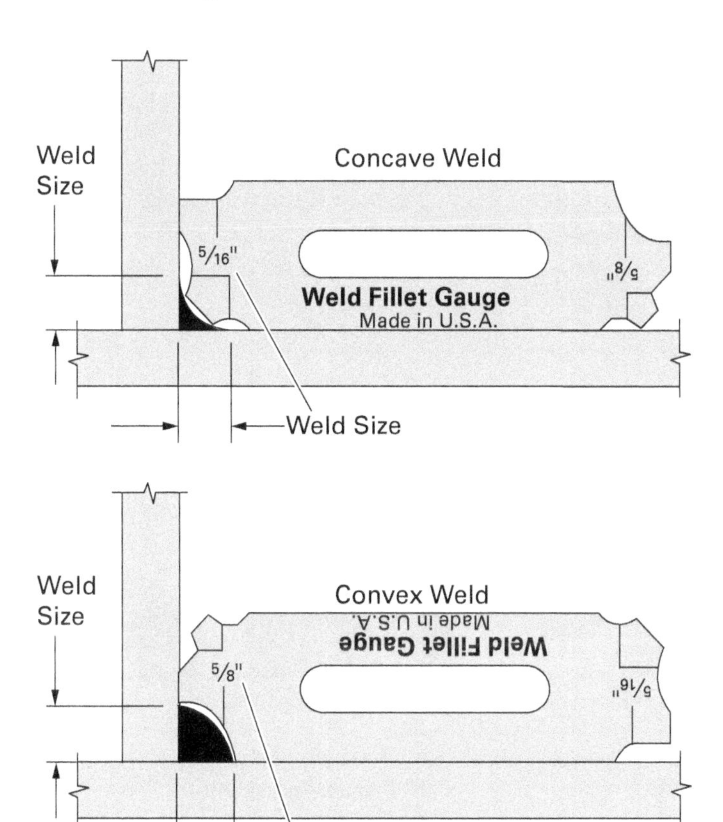

Figure 20 Fillet weld blade gauge.

Figure 21 Liquid penetrant testing.

NOTE

Some liquid penetrant dyes are red. They're easily visible under ordinary light. Others, however, are fluorescent. They're invisible in ordinary light, but they glow brightly under an ultraviolet lamp.

After preparing the metal with the cleaner, the inspector sprays the dye onto the joint and wipes off any excess. Discontinuities and defects draw in the liquid. The inspector then sprays the white developer onto the surface. The developer draws out the dye from any discontinuities, revealing their presence and size. After inspection, the cleaner removes the dye and developer from the surface.

WARNING!

Some liquid penetrants contain chlorine or other hazardous chemicals. Read the label and their SDS before using them. Wear appropriate PPE and take any recommended precautions.

Liquid penetrants readily reveal cracks. Most cracks are irregular in shape. The width of the red dye on the white surface indicates the crack's depth. This test method also reveals surface porosity, oxides, and slag. These show up as circular red spots on the white background.

Liquid penetrant testing works on most metals. It's popular for welds that may have problems like surface cracking. It takes longer than visual inspection, however, and uses chemicals. It doesn't work as well on rough, irregular surfaces or those with weld bead ripples.

3.1.4 Leak Testing

Pipes and pressure vessels require special tests to confirm that they don't leak. Welding inspectors can choose from several methods, depending on the application. Bubble tests involve pressurizing the vessel and immersing it in water or applying a soap solution. Bubbles show any leaks (*Figure 22*). To test an open tank, the inspector fills it with water containing a fluorescent dye. Examining the outside surface with an ultraviolet lamp reveals the leaks as glowing spots.

To find very subtle leaks, pressurizing the vessel with helium instead of air is better. Helium atoms are very small, so they easily pass through tiny weld defects. A gas detector spots the escaping helium, pinpointing the leak (*Figure 22*).

For vessels with only one accessible side, vacuum box testing works well. The vacuum box has a soft rubber seal and a transparent panel. After coating the weld's surface with a soap solution, the inspector places the vacuum box over it. A vacuum pump reduces the pressure inside the box. If the weld leaks, air will percolate through, causing the soap solution to bubble.

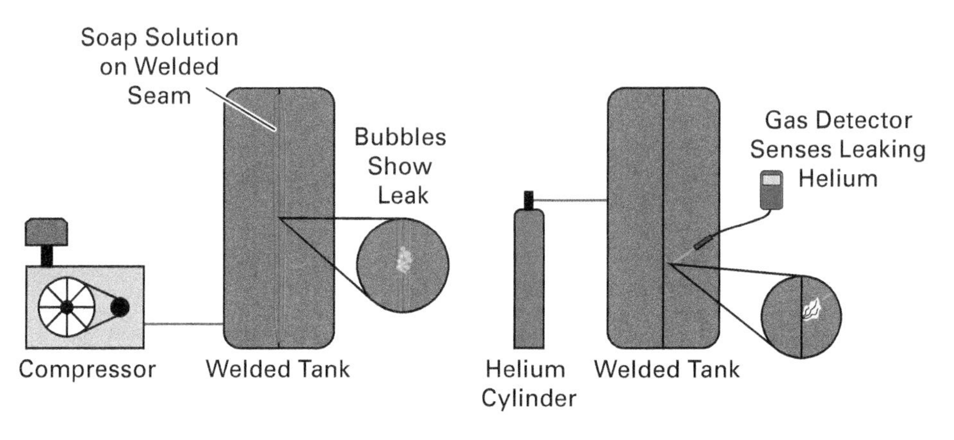

Figure 22 Leak testing.

3.2.0 Sophisticated Nondestructive Testing

Sometimes, simple nondestructive tests like visual inspection or liquid penetrant inspection aren't enough. Defects and discontinuities may be below the surface or deep inside the weld. To find these, as well as very small surface problems, welding inspectors use more sophisticated methods. Some of these require expensive instruments. Inspectors must have special training to interpret the results.

3.2.1 Magnetic Particle Inspection

Magnetic particle inspection (MT) magnetizes the weld area with an electric current (*Figure 23*). Metal particles sprinkled or sprayed onto the surface align with the magnetic field and form patterns. These patterns can reveal surface cracks, incomplete fusion, porosity, and slag inclusion. The method is also useful for inspecting plate edges before welding to reveal surface imperfections.

Figure 23 Magnetic particle inspection (wet).
Source: iStock@Funtay

Magnetic particle inspection is faster than liquid penetrant testing. But it works only with ferrous metals (iron, steel, and some alloys). It can reveal problems at or just below the surface. Defects and discontinuities deep within the weld, however, won't show up. The method also requires an inspector skilled at interpreting the powder's patterns.

For this test to work well, the part must be smooth, clean, oil-free, and dry. Rough surfaces aren't suitable. Remove any slag. Magnetize the part and sprinkle or spray the powder onto the surface. The inspector can then interpret the patterns. Welding codes usually provide criteria for accepting or rejecting the weld based on the results.

3.2.2 Electromagnetic (Eddy Current) Inspection

Welding inspectors have another magnetic test available that works with all metals—electromagnetic testing (ET). Also called *eddy current inspection,* it uses a test instrument to generate a magnetic field around the part. The magnetic field induces (creates) an electric current within the part.

This *eddy current* flows through the metal. Defects and discontinuities alter the flow. These show up as patterns on the instrument's screen. The welding inspector interprets them to identify defects and discontinuities (*Figure 24*).

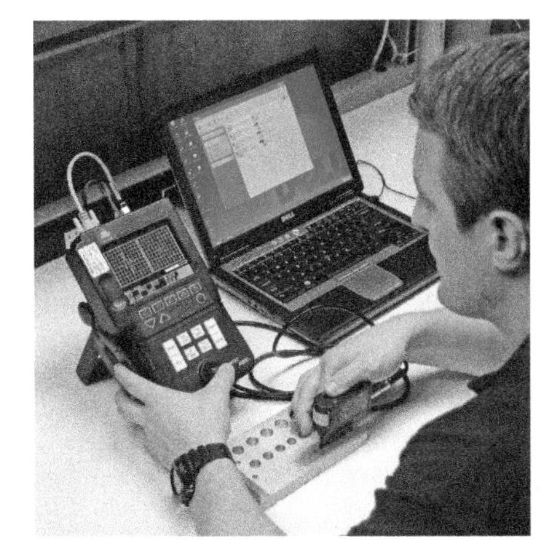

Figure 24 Eddy current testing.
Source: Applied Technical Services

Eddy current testing detects problems only at or near the surface. It can't find defects and discontinuities deep within the part. The method is excellent, however, for testing welded pipes and tubes. It can measure their wall thickness and the material's properties. Eddy current testing detects porosity, pinholes, slag inclusions, cracks, and incomplete fusion.

How well eddy current testing works depends on the welding inspector. The instrument requires careful calibration (setup and adjustment) first. Interpreting the screen requires great skill, particularly when trying to identify problems below the surface.

3.2.3 Radiographic Inspection

Radiographic: Imaging technology that passes X-rays or gamma rays through an object.

Radiographic inspection is a family of techniques related to medical X-rays and CT scans. All look deep into the test specimen, producing either two- or three-dimensional images. Radiographic inspection reveals many defects and discontinuities. These include porosity, undercut, slag inclusions, incomplete fusion, and cracks.

Welding radiography requires expensive equipment and specialist training. All radiographic methods use *ionizing radiation*—either X-rays or gamma rays. These penetrate most materials, including metals. If properly captured after passing through an object, they can form detailed images of it.

WARNING!

Ionizing radiation is dangerous. It can injure or cause long-term health problems, like cancer. Only qualified persons should operate industrial radiographic equipment.

Traditional radiographic inspection (RT) positions the test specimen between the radiation source and a cassette. The cassette holds a sheet of photographic film. During exposure, radiation passes through the specimen and strikes the film. The specimen casts a shadow on the film, revealing its features and interior. After a technician develops the film, the welding inspector can examine and interpret it.

Computed radiography (CR) is like film radiography but with a few advantages. Instead of film, it uses a reusable plate coated with chemicals called *phosphors*. When exposed to the radiation source, the phosphors store the image as energy patterns. Instead of developing the plate, a technician places it in a special scanner that captures the information stored on it. A computer turns that information into an image just like a digital photo.

Digital radiographic inspection (DR), the most modern technology, captures the image with a digital imaging plate. It's an electronic sensor that converts the radiation into signals that a computer turns into an image within seconds. Digital radiography produces more detailed images than older methods. It's also faster and requires less radiation. The images are digital files just like regular digital photos. *Figure 25* shows all three types of radiographic inspection.

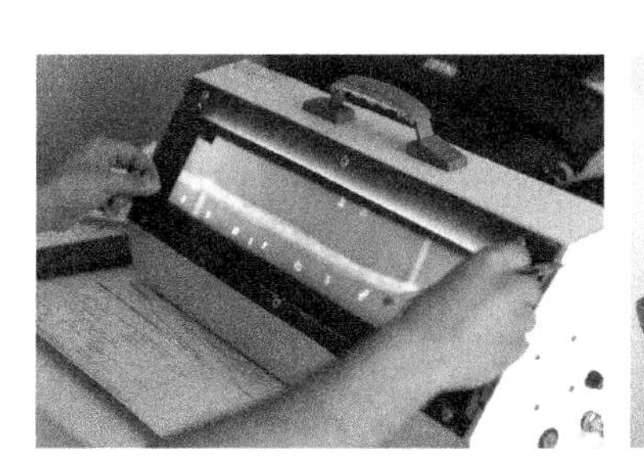

(A) Traditional Radiography

(B) Computed Radiography

(C) Digital Radiography

Figure 25 Radiographic inspection methods.

Sources: iStock@Funtay (25A); Applied Technical Services (25B); Courtesy of Waygate Technologies (25C)

These methods produce two-dimensional images. Sometimes, however, a detailed three-dimensional image is better. *Computed tomography (CT)* is a sophisticated technique that produces 3D representations. The CT scanner uses a flat radiation beam to scan the object one "slice" at a time.

The beam sweeps over the object, or the object rotates within the beam. Meanwhile, a computer captures hundreds of images. It assembles these into a 3D representation of the object. The welding inspector can view the image on a screen, rotating it to any angle.

As with all inspection methods, radiography requires a skilled inspector who can interpret the images. Tiny defects and discontinuities aren't always obvious. Radiography is also expensive and dangerous. Many companies send specimens to specialist companies that perform the radiography and return the results. Some radiography companies have mobile units that travel to the customer site to do radiographic inspections.

3.2.4 Ultrasonic Inspection

Ultrasonic inspection (UT) offers an inexpensive and safe alternative to radiographic inspection. It too can look inside test specimens. An inspector runs a handheld **transducer** probe across the specimen. It emits ultrasonic (very high frequency) soundwaves that pass into the object. The sound travels through the object in a straight line. If it comes to an edge, it bounces and changes direction. Eventually, the sound works its way back to the transducer, which detects the return echo. The ultrasonic testing machine shows the echoes as peaks on a screen. These change as the inspector moves the transducer across the surface in different patterns (*Figure 26*).

Transducer: An instrument that converts energy from one form to another.

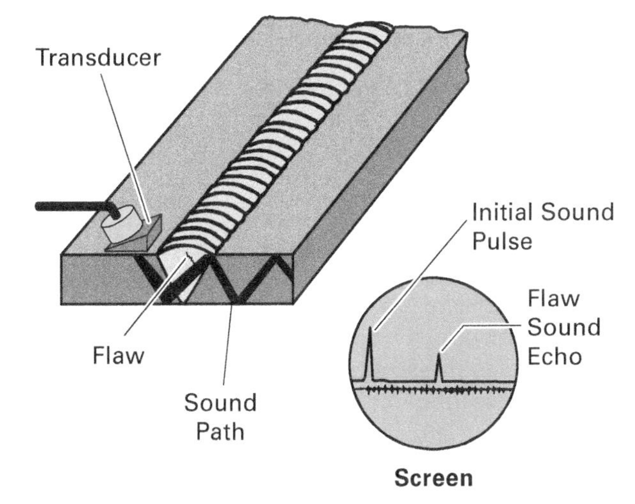

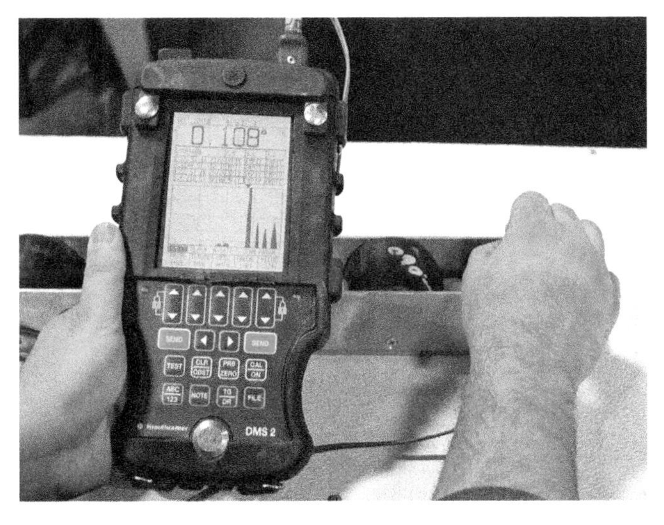

Figure 26 Ultrasonic testing machine.
Source: Applied Technical Services (bottom image)

Anything inside the object that's different from the surrounding material causes the sound waves to change direction and bounce back. Many defects, such as cracks, produce this effect. The inspector identifies what's inside the object by interpreting the peaks and the way they change as the transducer moves (*Figure 27*).

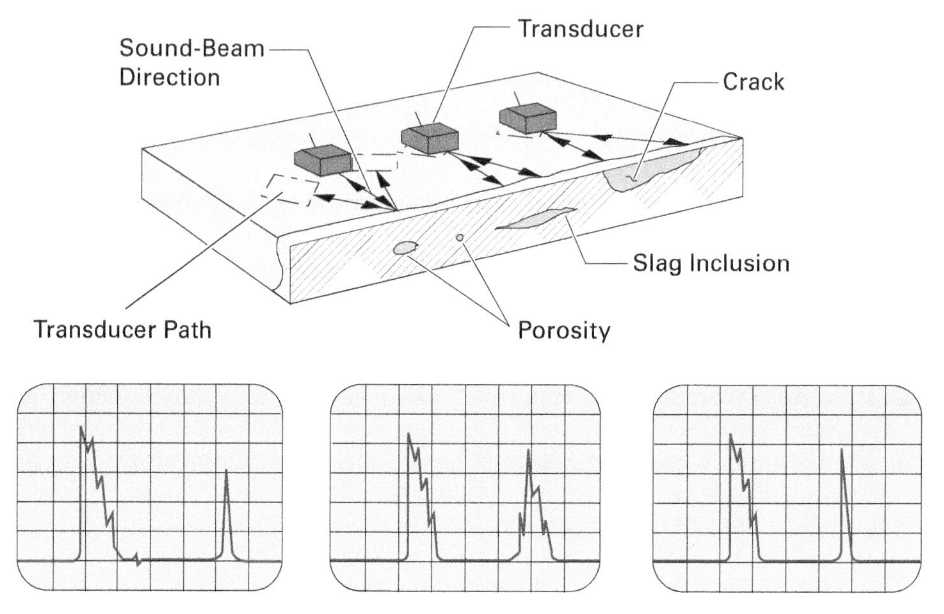

Figure 27 Detecting discontinuities in an object.

Laminations: Separating layers in the base metal.

Ultrasonic testing is very good at identifying small defects. It can find cracks, **laminations**, shrinkage cavities, porosity, slag inclusions, incomplete fusion, and incomplete joint penetration. It's not very good with thin materials, however, since these don't give the sound enough time to bounce back and forth.

Using an ultrasonic testing machine requires significant training and experience. Inspectors must select the best probe for the job. They must calibrate (set up and adjust) the machine properly. The way they move the probe across the surface affects the outcome too. Interpreting the patterns on the screen can be tricky since many look similar. Experienced inspectors, however, can easily identify different defects and discontinuities.

Advanced ultrasonic testing machines offer useful features to make the inspector's life easier (*Figure 28*). These have *phased array (PA)* probes containing several transducers that send sounds through the object at different angles. The machine analyzes this information and produces a detailed picture of the test specimen's interior.

Self-Propelled NDT

This ultrasonic test instrument inspects pipe and tanks. Controlled by a joystick, its magnetic wheels cling to the surface as it moves along.

Source: Applied Technical Services

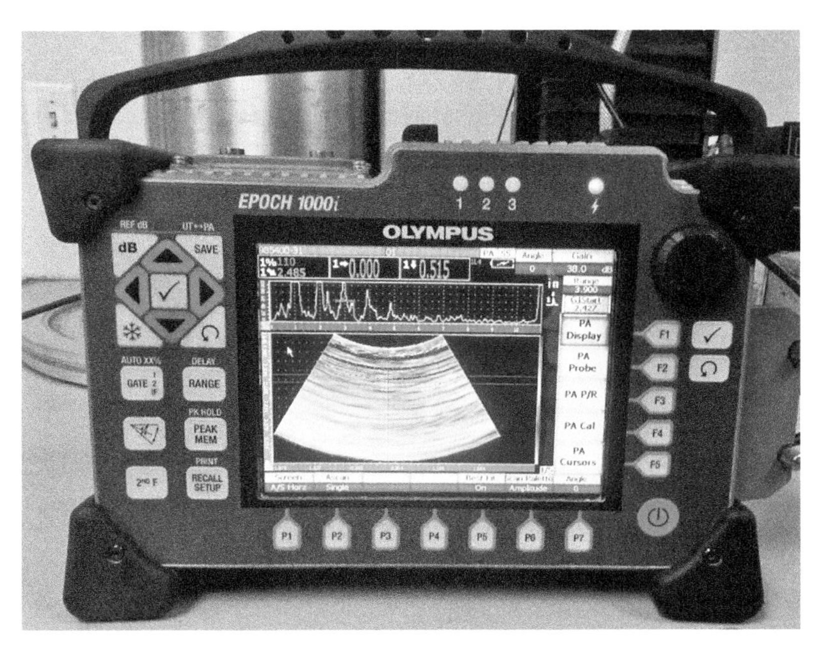

Figure 28 Advanced ultrasonic testing machine.
Source: Olympus NDT Inc.

3.3.0 Destructive Testing

Destructive tests damage or break the test specimen. Technicians perform these tests in a material science lab. Larger companies usually have their own, but smaller companies often use an outside testing service.

Destructive tests use special machines to bend, stretch, strike, or penetrate the test specimen. Some tests do this to determine the weld's **ultimate tensile strength** or **ductility**. Others break the specimen to reveal weld defects and discontinuities. The following are common destructive tests.

Ultimate tensile strength: The amount of pulling force a material can tolerate before breaking. Often shortened to *tensile strength*.

Ductility: A material's ability to be bent, shaped, or stretched without breaking.

3.3.1 Tensile Testing

A material's ultimate tensile strength is the amount of pulling force it can tolerate before breaking. Tensile testing pulls the test specimen with a *universal testing machine (UTM)* until it breaks (*Figure 29*). Sensors monitor the applied force and the amount that the specimen stretches (elongation). A tensile test reveals a weldment's tensile strength and ductility. It can also confirm that a weld is as strong as the base metal.

Figure 29 Tensile testing.
Source: iStock@Funtay

Hardness: A material's ability to resist penetration.

3.3.2 Hardness Testing

Hardness is a material's ability to resist penetration. High hardness can be desirable. But very hard materials can also be brittle and crack easily. Checking a weld's hardness shows its relationship to the base metal. Checking the heat-affected zone shows how the welding process has altered the base metal around the weld.

Lab technicians measure hardness with a hardness testing machine (*Figure 30*). The machine penetrates the test specimen with a pointed or ball-shaped *indenter*. It calculates the hardness value by measuring the applied force and penetration depth. Labs report hardness using one of several hardness scales. The Rockwell and Brinell hardness scales are both popular.

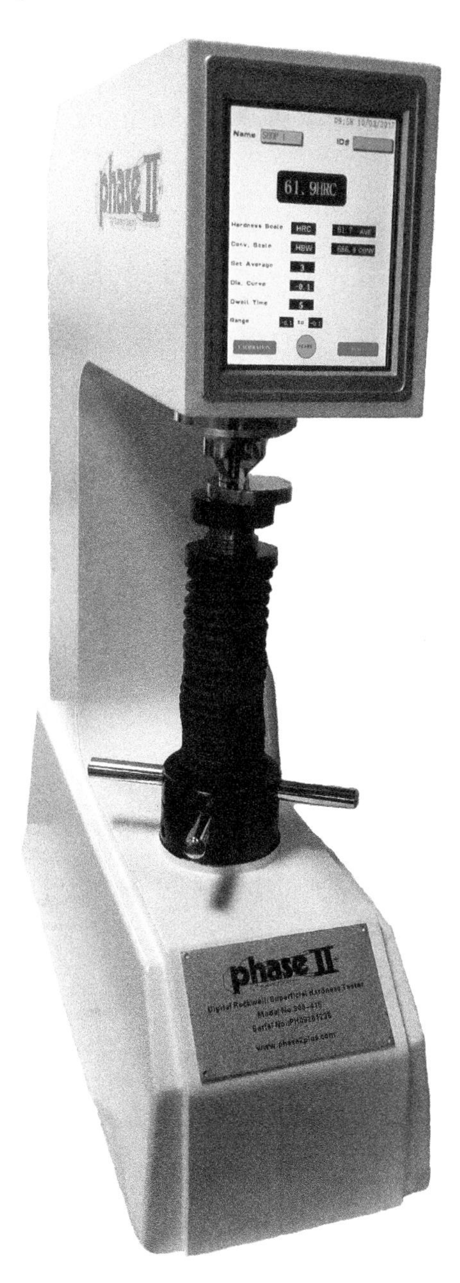

Figure 30 Hardness testing machine.
Source: Brett Gitter

3.3.3 Impact Testing

Impact testing determines a material's notch toughness—its ability to absorb a sudden stress without breaking. To measure a weld's notch toughness, a technician first cuts a V-shaped notch in the test specimen. The notch has a specific shape and depth. Then, the technician places the specimen in an impact testing machine (*Figure 31*).

The machine has a heavy pendulum that swings and strikes the specimen like a hammer, breaking it. The distance that the pendulum bounces back indicates how much energy the specimen absorbed before breaking. This value gives the notch toughness.

Figure 31 Impact testing machine.
Source: Applied Technical Services

NOTE

Welding codes usually provide key details for performing soundness tests. These include fixture types and sizes, as well as test procedures. They also include criteria for evaluating the results.

3.3.4 Soundness Testing

These tests evaluate a weld's quality and reveal defects and discontinuities. All work by bending or breaking the test specimen at the weld point. Most use a universal testing machine configured to apply pressure. Soundness tests are part of welder performance qualifications and welding procedure qualifications. The following are common soundness tests.

Bend Tests

Lab technicians perform bend tests by placing the specimen in a fixture. A universal testing machine applies pressure, bending the specimen 180 degrees around the weld (*Figure 32*). This reveals many weld defects.

Guided-bend tests stress the weld in one of three ways as they bend it into a U-shape (*Figure 33*). Side bends, face bends, and root bends all check for penetration and proper fusion. They also test for porosity and inclusions. Side bends work best for thick materials. Face and root bends are useful for materials up to $^{3}/_{8}$" (~10 mm) thick.

Figure 32 Bend testing.

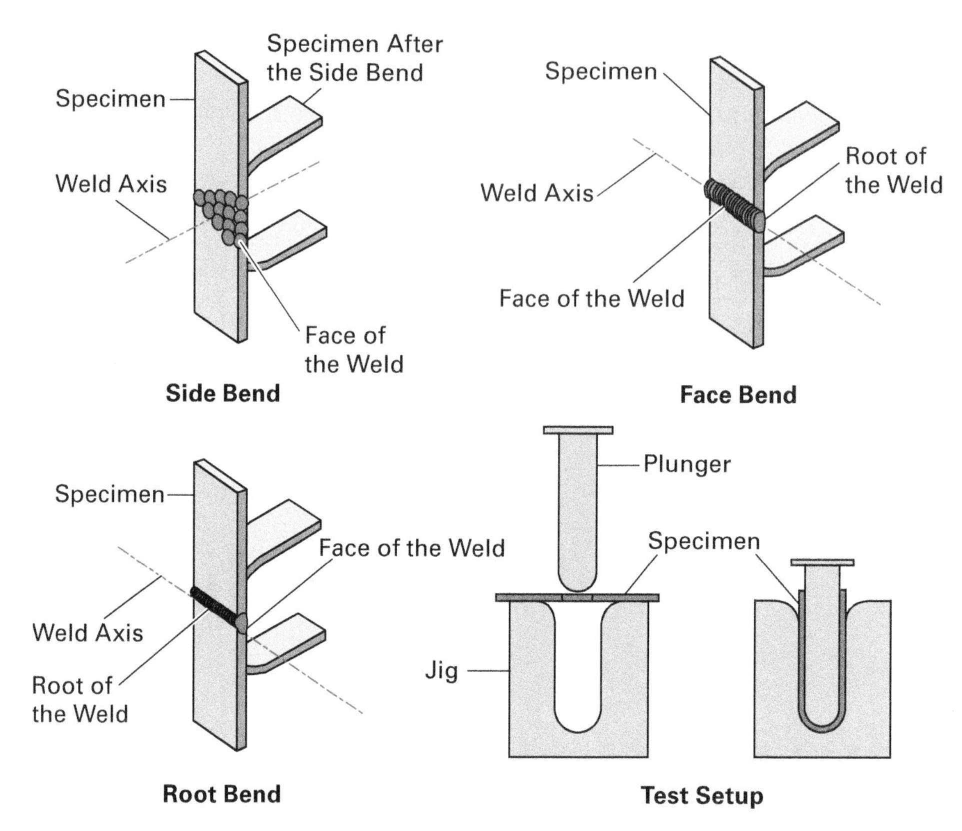

Figure 33 Guided-bend tests.

Welding codes specify criteria for evaluating guided-bend tests. The following example is typical:

1. Single discontinuities on the convex surface exceeding ⅛" (3.2 mm) are rejectable.
2. The total of all discontinuities on the convex surface that measure greater than $^{1}/_{32}$" (0.8 mm) but less than ⅛" (3.2 mm) must be less than ⅜" (9.5 mm).
3. Corner cracks on the convex side are not automatically rejectable unless they develop from inclusions like slag or other visual discontinuities.

Nick Break Test

The pipeline industry checks pipe welds with this test. First, a technician saws two small cuts in the weld. This weakens the weld so it will break at that point. A universal testing machine or hammer stresses the weld until it breaks (*Figure 34*). The technician then examines the weld's interior for defects and discontinuities.

Fillet Weld Break Test

This test examines fillet welds for defects, discontinuities, and fusion problems. First, the welder creates a test specimen by building a T-joint with a fillet weld on one side. A universal testing machine applies pressure to the weld until it breaks (*Figure 35*). A technician then examines the weld for defects and discontinuities.

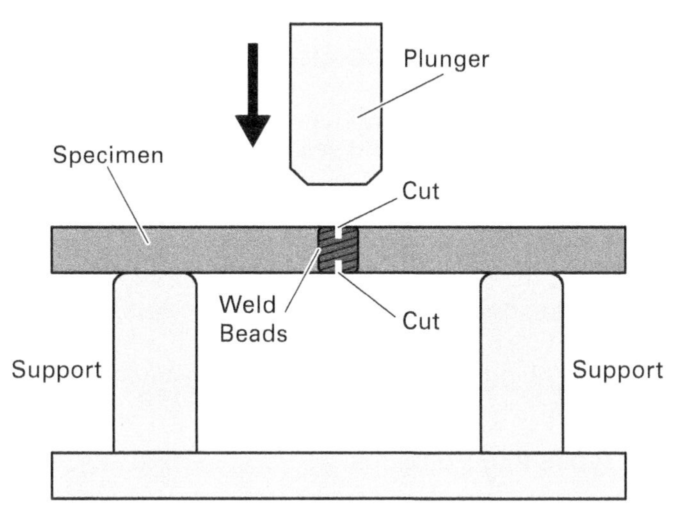

Figure 34 Nick break test.

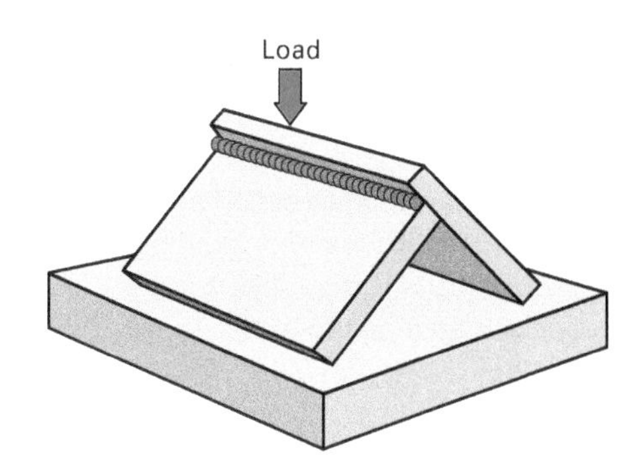

Figure 35 Fillet weld break test.

3.0.0 Section Review

1. Which of the following tools can measure a butt weld's reinforcement?
   a. Automatic weld size gauge
   b. Fillet weld gauge
   c. Liquid penetrants
   d. Magnetic powders

2. Eddy current inspection (ET) can test welds in ferrous metals *only*.
   a. True
   b. False

3. In tensile testing, the test specimen is _____.
   a. bent
   b. pulled
   c. struck
   d. compressed

4.0.0 Welder Performance Qualification Tests

Objective

Summarize welder performance qualification testing.

a. Describe welder performance qualifications.
b. List AWS, ASME, and API welder qualification testing requirements.
c. Describe a welder performance qualification test.

Performance Tasks

There are no Performance Tasks in this section.

The term "qualification" applies to welding procedures and to individual welders. A welding procedure qualification involves producing and testing welds to confirm that a welding procedure works properly. On the other hand, a welder performance qualification confirms that a welder can produce a particular weld type in a particular position. This section explores the second qualification type since welders must achieve them to work on specific jobs.

NOTE

Codes and specifications govern welder performance qualifications. Different codes often have similar requirements and procedures. Always consult the specific code for details. If you're uncertain about the code governing a project, ask your supervisor.

4.1.0 Welder Performance Qualifications

Welders seeking performance qualifications must prove that they can produce a particular weld under specific conditions. They do this by creating test welds that represent various real-world welding scenarios.

A test weld has two essential qualities: type and position. *Type* refers to whether the weld involves plate, pipe, or both. It also indicates whether the weld itself is a fillet weld or a groove weld. *Position* refers to the weld's orientation—flat, horizontal, vertical, or overhead. Each type and position combination has its own unique challenges. For this reason, welders must qualify for each. Sometimes, however, qualifying for one combination automatically qualifies a welder for other types.

NOTE

Some codes include additional requirements in a qualification besides weld type and position. For example, AWS qualifications may specify a particular electrode group.

4.1.1 Welding Types and Positions

All codes use the same system for identifying weld types and positions—a number and a letter. The number identifies the weld's position. The letter identifies the weld's type: F for fillet welds and G for groove welds.

For plate welds, the following positions are possible:

1 – Flat

2 – Horizontal

3 – Vertical

4 – Overhead

For example, a 3F plate weld is a fillet weld in the vertical position. In other words, the fillet itself is in an up-down orientation. A 4G plate weld is a groove weld in the overhead position. The plates are horizontal, but the groove itself is above the welder's head. *Figure 36* summarizes all plate welds.

Pipe weld position identifiers are slightly more complex. Not only do they specify the orientation, but they also specify how the welder must work. For some positions, the welder stands in a fixed location and rotates the pipe as required. In other positions, the welder can't rotate the pipe but must move around it instead. A further complication is that for fillet welds, the AWS and ASME define some positions differently.

For fillet welds on pipes, the following positions are possible:

1 – Flat (rotated)

2 – Horizontal (not rotated / rotated)

3 – Vertical (rotated)

4 – Overhead (not rotated)

5 – Multiple Position (not rotated)

The 1F position can be confusing since the pipe is angled 45 degrees. The fillet weld itself, however, is always flat. The welder stands above it and rotates the pipe, so all welding occurs from above. This is the easiest pipe weld to create. The 1F position is the same for the AWS and the ASME.

The 2F position requires the welder to move around the pipe rather than rotating it. The weld's basic orientation, however, remains fixed—horizontal. It too is relatively easy to produce. The 2F position is the same for the AWS and the ASME.

The AWS and the ASME define the 2FR position differently. The AWS version requires the welder to stand in a fixed location (horizontal/above) and rotate the pipe. The ASME version requires the welder to stand in a fixed location (vertical/the side) and rotate the pipe.

Only the AWS defines a 3F position. It's identical to the ASME's 2FR position. The welder stands in a fixed location (vertical/the side) and rotates the pipe.

The 4F position requires the welder to move around the pipe rather than rotating it. The weld's basic orientation, however, remains fixed—overhead. It's challenging to produce because welding overhead requires more skill. The 4F position is the same for the AWS and the ASME.

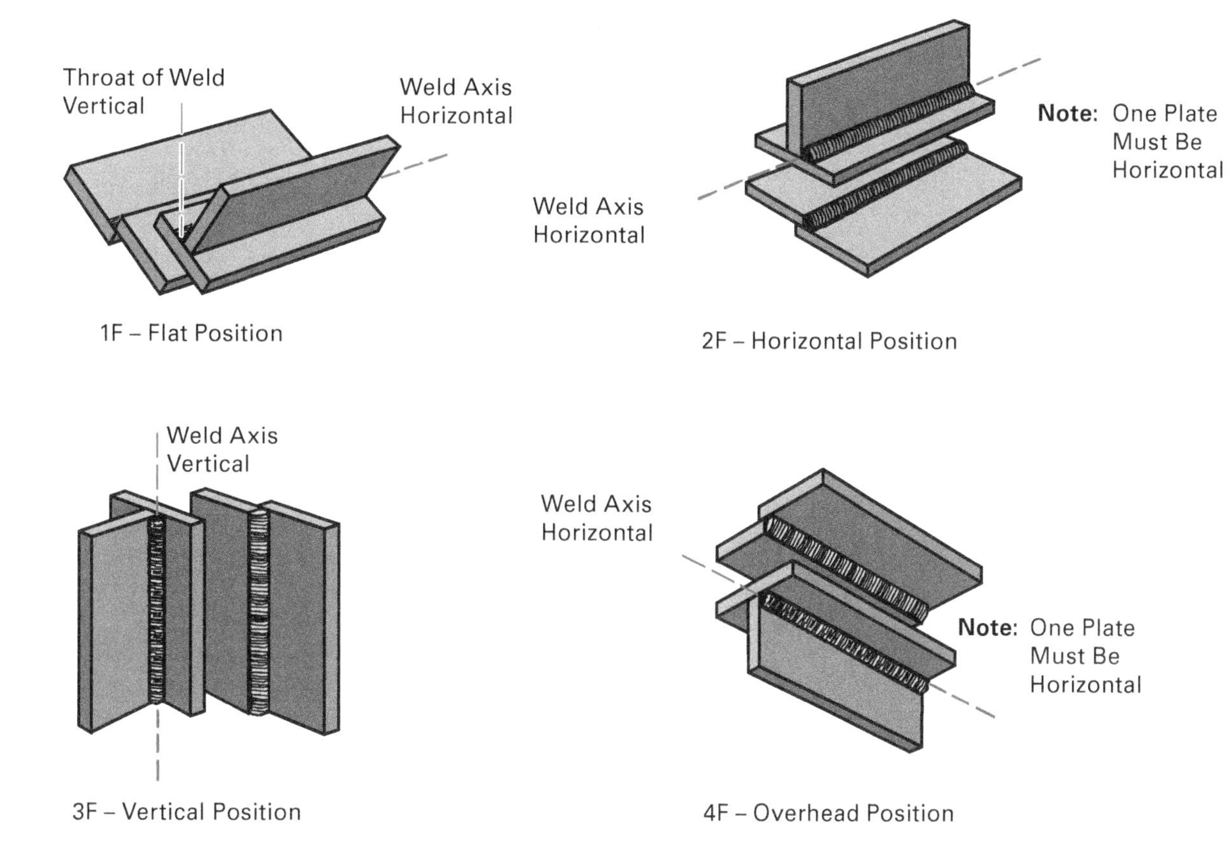

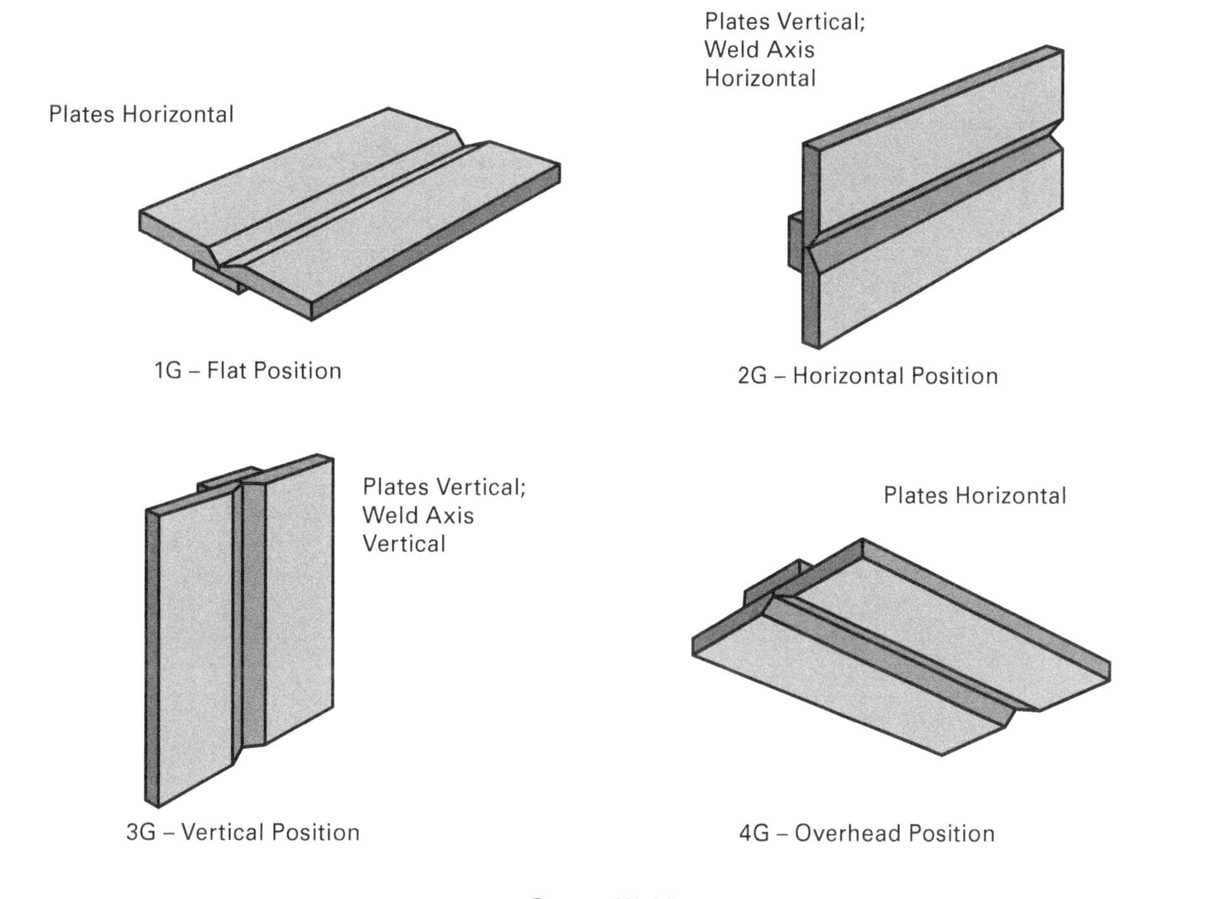

Figure 36 Plate weld types and positions.

The 5F position is the most difficult. The welder can't rotate the pipe but must move to different locations to produce the weld. Unlike the 2F and 4F positions, the weld's basic orientation changes with each location. Essentially, it tests the welder's ability to work in all positions. The 5F position is the same for the AWS and the ASME.

For groove welds on pipes, the following positions are possible:

1 – Flat (rotated)

2 – Horizontal (rotated)

5 – Multiple Position (not rotated)

6 – Angled Multiple Position (not rotated)

While the groove itself is vertical in the 1G position, the welder always stands above it and rotates the pipe. From the welder's perspective, the groove is always flat. Position 2G is similar. The welder stands to the side and rotates the pipe. From the welder's perspective, the groove is always horizontal.

Position 5G is more challenging than 1G or 2G. The welder can't rotate the pipe but must move to different locations to produce the weld. The weld's basic orientation changes with each location. Essentially, it tests the welder's ability to work in all positions.

Position 6G is also very challenging. Like 5G, it requires the welder to move around the pipe rather than rotating it. The pipe is angled, however, which makes producing the weld even more difficult.

Some codes add an extreme challenge by requiring the welder to produce a 6G position weld with a barrier in the way. The barrier, a metal ring placed near the weld, makes it harder to see the weld and maneuver around it. This is position 6GR ("restricted").

Figure 37 summarizes all pipe welds.

With some codes, qualifying for more difficult welds automatically qualifies the welder for easier ones. For example, welders who qualify for pipe welds may automatically qualify for plate welds. Similarly, qualifying for groove welds may qualify the welder for fillet welds. Always consult the specific code for details. Never assume one code's rules match another's.

NOTE

Welders use position identifiers, like 1F or 5G, when referring to qualification tests. When making production welds, they don't use them. Instead, they use the terms "flat," "horizontal," "vertical," and "overhead" to identify position.

4.2.0 Code-Specific Qualifications

The AWS, ASME, and API codes govern many welding projects. Each includes welder performance qualifications that welders can acquire. Besides weld type and position, these may also specify base metal, joint design, electrode type, as well as other elements. The following sections summarize each code.

4.2.1 AWS Structural Steel Code

This code governs welding for building and bridge construction. It includes the usual qualifications for weld types and positions. Qualification for plate welding automatically qualifies a welder for rectangular tubing.

The code also includes the *electrode group* in its qualification requirements. The AWS organizes shielded metal arc welding (SMAW) electrodes into four numbered groups, each beginning with the letter F. *Table 1* shows those used with mild steels.

TABLE 1 F-Number and AWS Electrode Classification

Group	AWS Electrode Classification				
F1 (Fast-Fill)	EXX20	EXX24	EXX27	EXX28	
F2 (Fill-Freeze)	EXX12	EXX13	EXX14		
F3 (Fast-Freeze)	EXX10	EXX11			
F4 (Low-Hydrogen)	EXX15	EXX16	EXX18	EXX28	EXX48

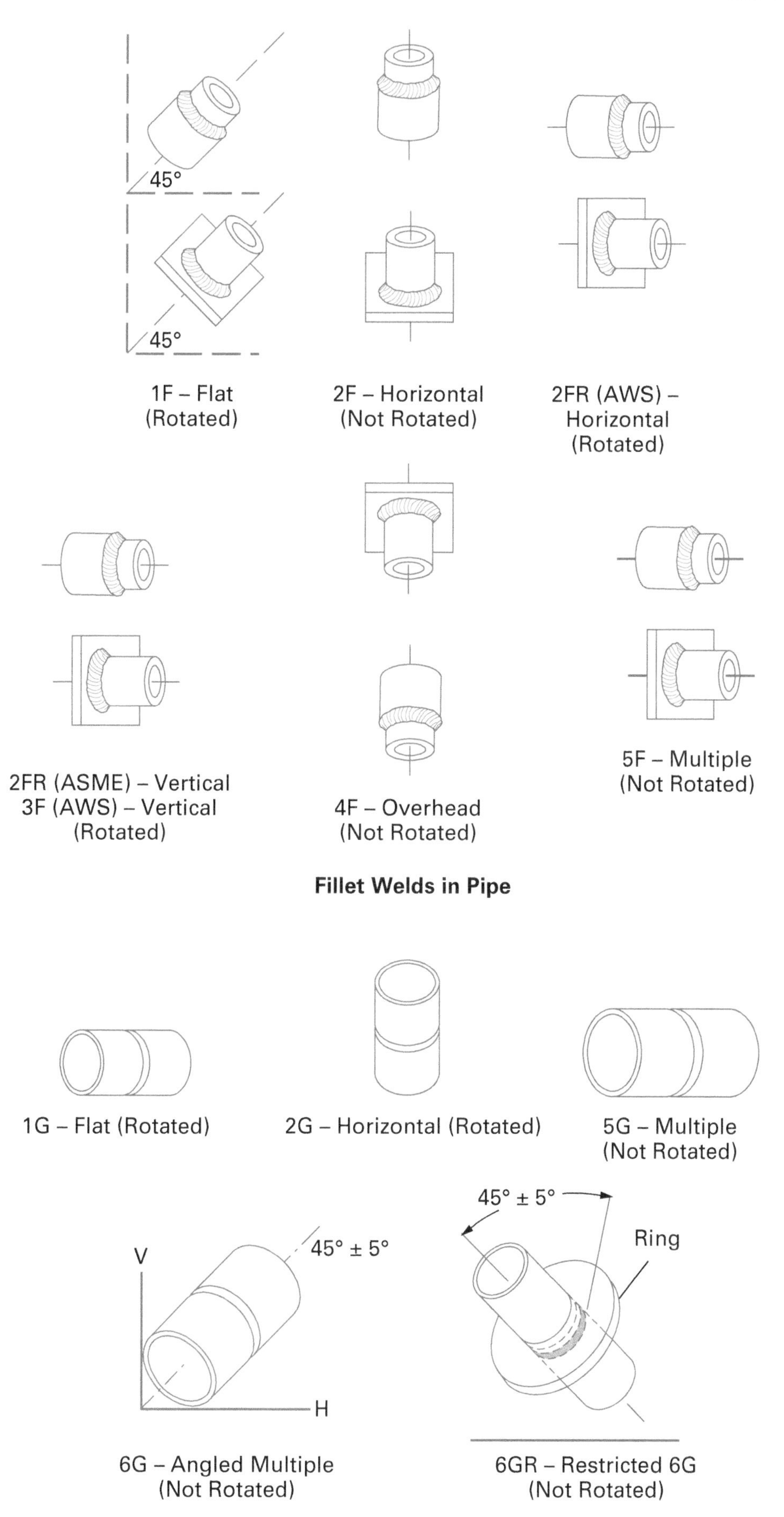

Figure 37 Pipe weld types and positions.

NOTE

NCCER Module 29108, *SMAW Electrodes*, explains electrode classification and group numbers in more detail.

A welder who qualifies in a particular electrode group will automatically qualify for all lower-numbered groups too.

Some AWS qualification tests may also specify material thickness. For example, a welder who successfully welds a test piece may qualify for materials up to twice as thick as the test piece. Other tests, however, are not so demanding and will qualify the welder for any thickness.

A typical AWS welder qualification test might specify a V-groove weld with metal backing in the 3G and 4G positions using an F4 electrode. Passing this test qualifies the welder to weld with F4 or lower electrodes and to make groove and fillet welds in all positions. *Figure 38* shows an example fit-up for an AWS qualification test.

Figure 38 AWS plate test coupon.
Source: Zachry Industrial, Inc.

4.2.2 ASME Code

This code outlines qualifications for welders working on ASME-governed projects. According to Section IX of the *ASME Boiler and Pressure Vessel Code*, qualification on pipe automatically qualifies the welder for plate. Similarly, qualification with groove welds automatically qualifies the welder for fillet welds. The reverse is *not* true for either. It's also possible to qualify for fillet welds only.

An ASME welder qualification test might be to weld pipe in the 6G position using an open root (*Figure 39*). Passing this test qualifies the welder to weld pipe and plate in all positions with both fillet and groove welds.

The ASME code includes electrode groups in its qualifications too. Using only F3 electrodes qualifies the welder only for that level. Using F3 electrodes for the root and F4 electrodes for the filler qualifies the welder to use F3 electrodes. The welder also acquires qualification with F4 electrodes but only with backing.

4.2.3 API Code

This code includes qualifications that require welders to make butt or fillet welds on pipe nipples or pipe nipple segments. They must use qualified procedures during the test. Inspectors check the completed welds visually and by destructive testing or radiographic inspection.

Welders can receive a single qualification by making a single weld in the fixed or rotated position. They can obtain a multiple-weld qualification by making a butt weld in a fixed position on 6" pipe (DN150) with a minimum thickness of 0.250" (6.4 mm). They must then complete a second test by cutting, fitting, and welding a pipe of the same size without a backing strip.

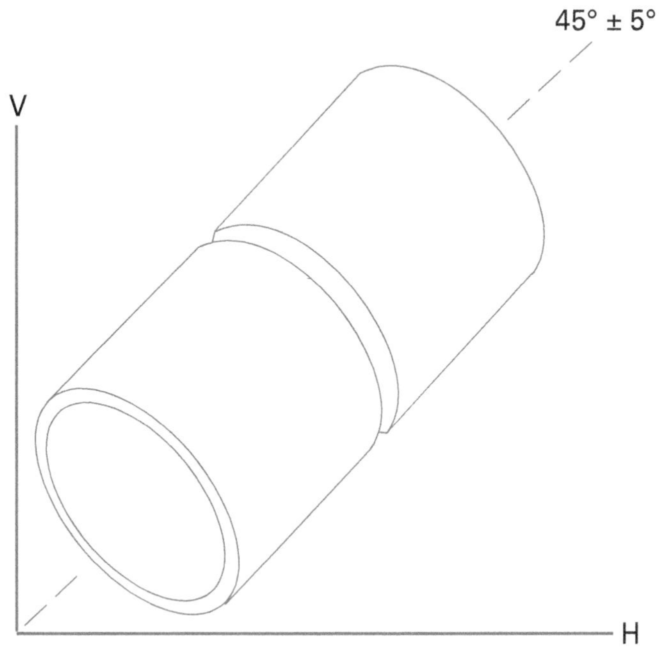

Figure 39 ASME pipe test.

4.3.0 Welder Performance Qualification Tests

Welders take tests to acquire their qualifications. These require producing welds by following a suitable WPS. Welders usually qualify with groove weld tests since these automatically qualify them to make fillet welds. The following sections outline the testing process.

4.3.1 Making the Test Weld

Qualification tests assess the welder's ability to produce specific welds. The welder begins the process by setting up and producing the required weld in the required position. Some qualification tests also specify a particular electrode type. After completing the work, the welder cuts coupons, called *test specimens,* from the weldment.

Creating the test weld is crucial. A poor weld won't survive the qualification tests. Sometimes, however, welders fail for reasons other than their basic ability. Carelessly preparing the weld or the test specimens can lead to failure. Before making the test weld, note the specimen locations. As you create the weld, avoid restarts and other problems in these locations.

4.3.2 Cutting the Test Specimens

After making the weld, cut the test specimens from the plate or pipe by any suitable means. Qualification tests specify the specimen locations within the weldment. *Figure 40* shows examples for plate welds. Limited thickness qualification tests usually use plate $^3/_8$" (~10 mm) thick. Unlimited thickness qualifications use 1" (~25 mm) thick plate. Materials $^3/_8$" (~10 mm) thick require a face bend and a root bend. Thicker materials require side bends instead.

Pipe weld test specimens are similar. After welding short pipe sections in the specified position, cut the specimens from the pipe. *Figure 41* shows the locations and tests specified by some AWS and ASME codes. Notice that pipes with thicker walls require side bends rather than face or root bends. Also notice that specimens vary with position.

NOTE

Even though the locations and tests are the same, don't assume that the AWS and ASME test procedures are identical. Consult the most recent edition of the applicable code. Confirm the specimen locations and dimensions, as well as the test type.

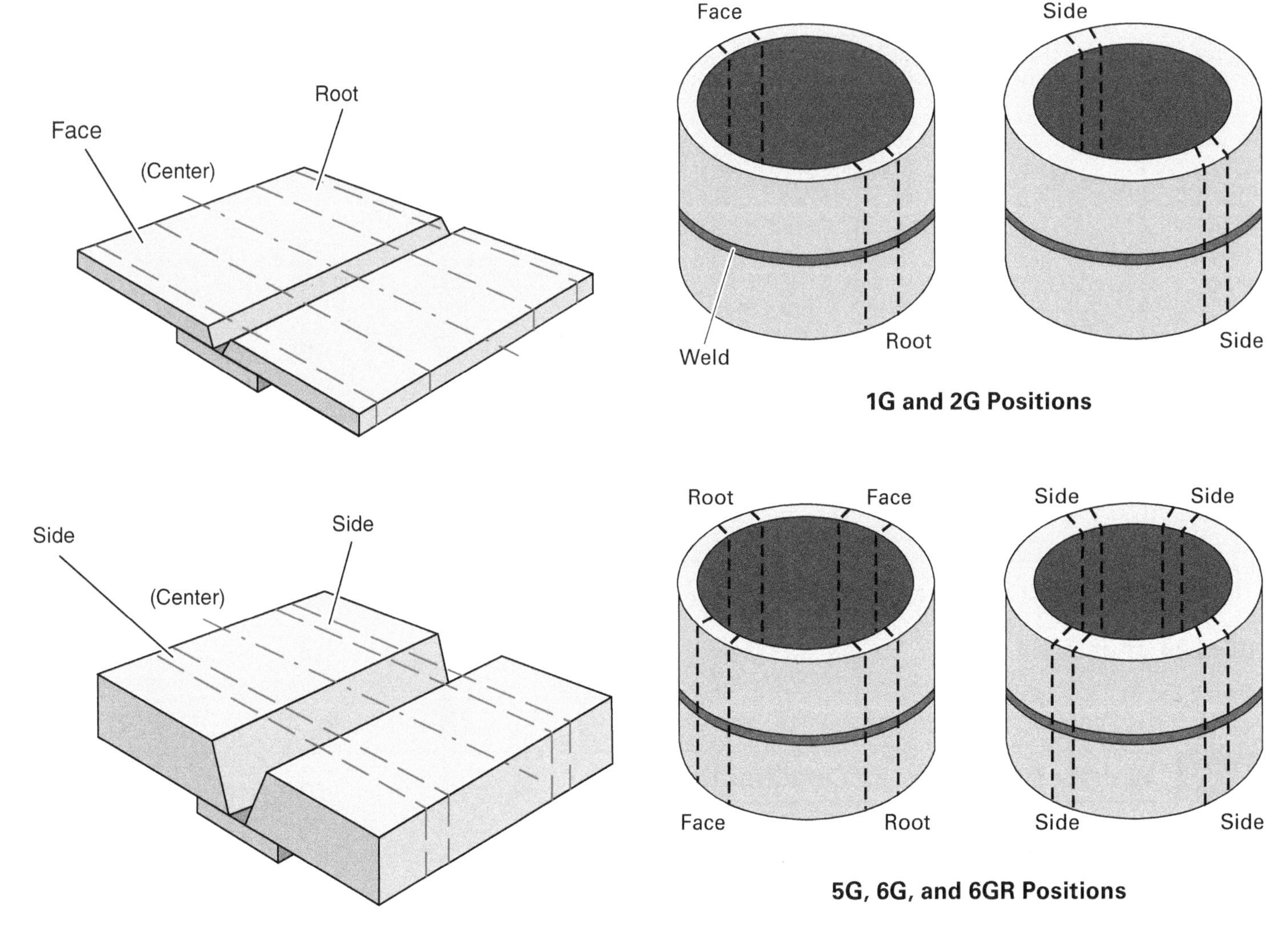

Figure 40 Test specimen locations for plate welds.

Figure 41 Test specimen locations for pipe welds (AWS and ASME).

NOTE

In general, pipes with thicker walls require testing the coupons with side bends. Those with thinner walls require root and face bends. Other tests might be necessary too.

Figure 42 shows the locations and tests specified by some API codes. While the specimen locations are similar to those specified by other codes, the specimen number and test types vary with the pipe size and wall thickness.

The information in this module is general. Always consult the relevant code before preparing and cutting plate or pipe test specimens. Specimen locations and dimensions depend on multiple factors, including thickness, size, and welding position. Codes include detailed drawings that provide guidance. Identify and document each specimen by writing on it with soapstone after you cut it.

4.3.3 Preparing the Specimens

After cutting the specimens, prepare them for testing (*Figure 43*). Poor preparation can cause an otherwise good weld to fail. For example, a slight nick could open during bending, causing failure. To prepare the test specimen, do the following:

- Grind or machine the surface to a smooth finish. All machining marks must run lengthwise. If they run crosswise, they can cause a failure.
- Remove face or root reinforcement from the weldment. This is a requirement. Not doing this can cause a good weld to fail. According to AWS specifications, never grind into the base metal more than $^1/_{32}$" (0.8 mm) or 5 percent of the base material's thickness.
- Round the edges with a file to a smooth $^1/_{16}$" (1.6 mm) radius. *AWS D1.1* permits $^1/_8$" (3.2 mm) radius edges. This practice helps prevent cracks from starting at the corners.

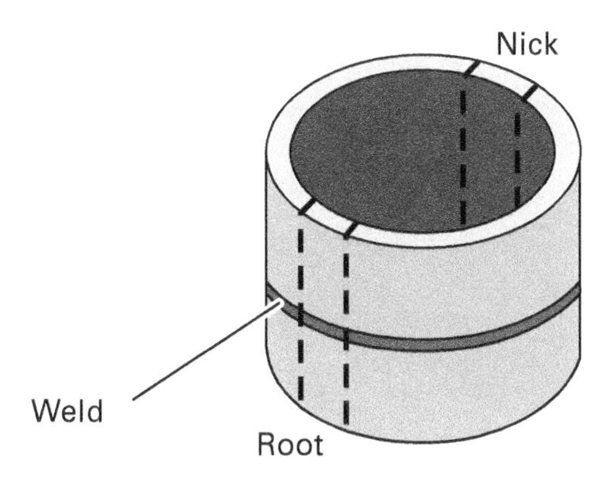

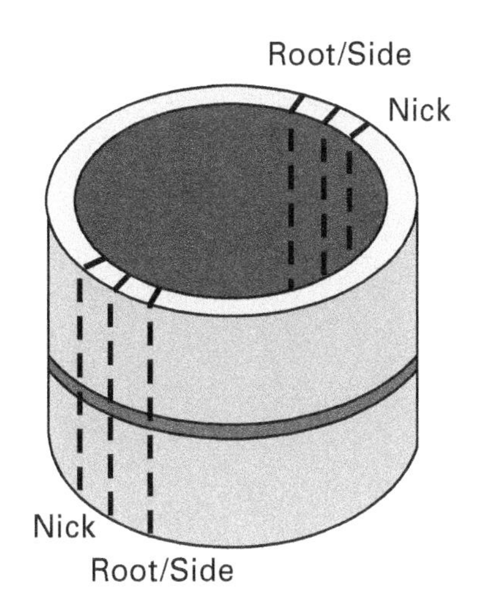

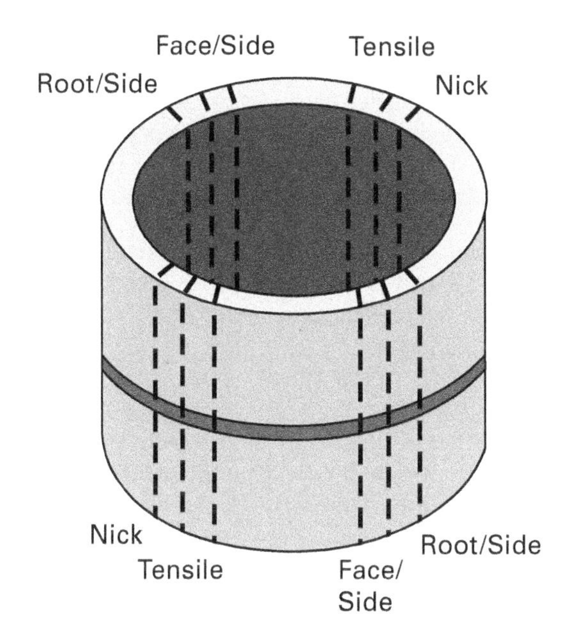

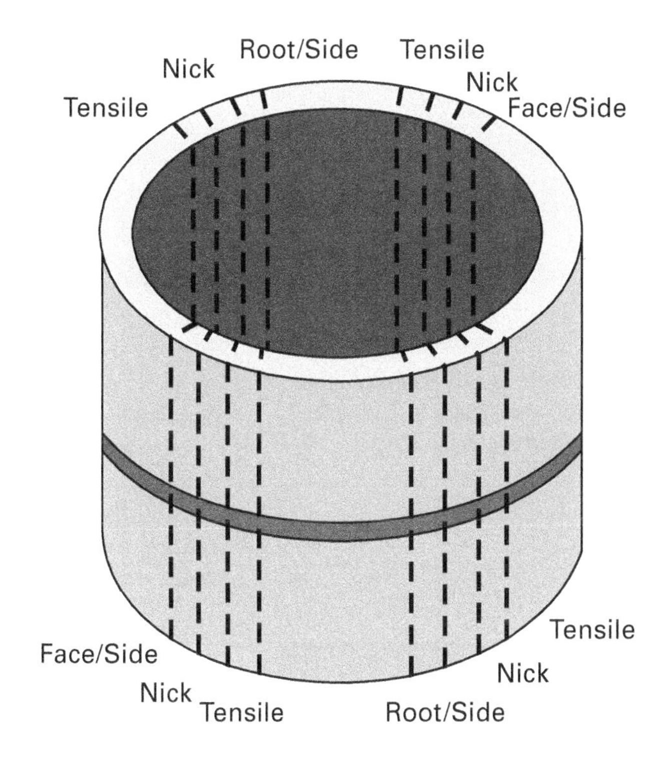

Figure 42 Test specimen locations for pipe welds (API).

- Don't quench (cool in water) specimens after grinding. Quenching can create small surface cracks that might become larger during the bend test.

4.3.4 Bending Test

A lab technician bends the prepared test specimens using the correct fixture and testing machine. This exposes defects and discontinuities. Acceptance criteria vary by code and/or jobsite quality specifications. AWS standards, for example, require that the surface contain no discontinuities exceeding the following dimensions:

- $^1/_8$" (3.2 mm) – Measured in any direction on the surface.
- $^3/_8$" (9.5 mm) – Sum of the greatest dimensions of all discontinuities exceeding $^1/_{32}$" (0.8 mm) but less than or equal to $^1/_8$" (3.2 mm).

NOTE

For some qualification tests, a radiographic inspection replaces the destructive test. This allows the inspector to check the entire weld for small discontinuities anywhere within the weld.

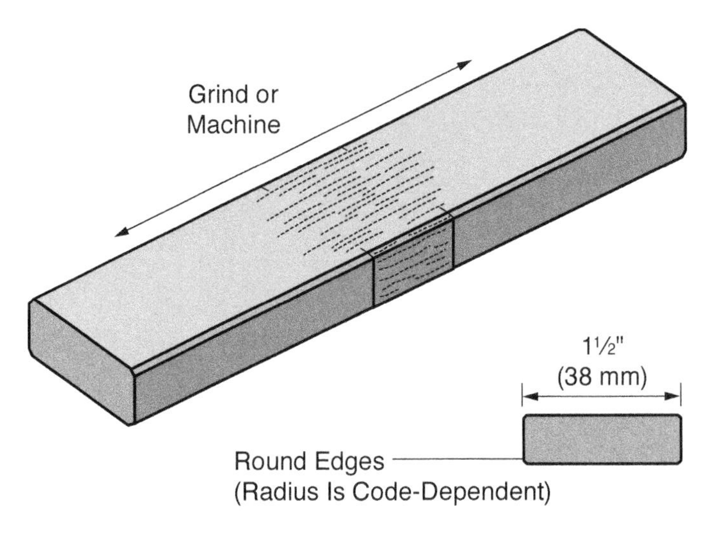

Figure 43 Prepared test specimen.

- ¼" (6.4 mm) – Maximum corner crack, except when the corner crack results from a visible slag inclusion or other fusion-type discontinuities. In those cases, a ⅛" (3.2 mm) maximum applies. If the specimen has corner cracks exceeding ¼" (6.4 mm), with no evidence of slag inclusions or other fusion-type discontinuities, discard it. Test a replacement specimen from the original weldment.

After a welder passes a qualification test, the company updates its records. These list the welder's qualifications and their associated WPSs. The company uses this information in its quality documentation. After qualifying, a welder can start producing the associated weld.

4.3.5 Welder Qualification Limits

NOTE

Retest requirements vary by code and jobsite quality specifications. Consult these if you have any questions.

Usually, welders who fail a qualification test may take an immediate retest. But they must submit *two* test welds for each type that failed. Both must pass for the welder to receive the qualification. Alternatively, the welder may take a complete retest after additional training and practice.

A qualification may expire if the welder hasn't produced the weld type within a specified time. In that case, the welder must requalify for the weld. Consult the applicable code and jobsite quality specifications for guidance. Welders may lose a qualification if their work suggests that they can't reliably produce a particular weld. Welders must also requalify if an essential variable in a WPS changes.

4.0.0 Section Review

1. A weld designated 3G is a _____.
   a. fillet weld in the flat position
   b. groove weld in the flat position
   c. groove weld in the vertical position
   d. fillet weld in the vertical position

2. Which of the following is a *correct* statement about ASME code qualifications?
   a. Qualification testing is done only on pipe.
   b. Qualifying on plate qualifies the welder on pipe sizes less than 24" (61 cm).
   c. Qualification on groove welds also qualifies the welder for fillet welds.
   d. Using F3 electrodes on test welds qualifies the welder for all electrode groups.

3. Once a welder passes a qualification test, the qualification is permanent.
   a. True
   b. False

Module 29106 Review Questions

1. Which standards organization publishes a major code that governs boilers and pressure vessels?
   a. APM
   b. ASME
   c. ASTM
   d. ASNT

2. When working on code-driven structural steel projects, welders frequently reference _____.
   a. *AWS D1.1*
   b. *AWS D1.2*
   c. *ASME BVPC*
   d. *API 1104*

3. If you're working on a project developed outside the United States, from which standards organization would its codes most likely come?
   a. ASME
   b. ANSI
   c. ISO
   d. API

4. If your instructions tell you to perform a weld for which you're not qualified, you should _____.
   a. issue a formal complaint to human resources
   b. report directly to the safety engineer as soon as possible
   c. try to resolve the issue with your immediate supervisor
   d. perform the weld to the best of your ability

5. In a WPS, which of the following is a *non-essential* variable?
   a. Material thickness
   b. Postheat treatment
   c. Travel speed
   d. Welding process

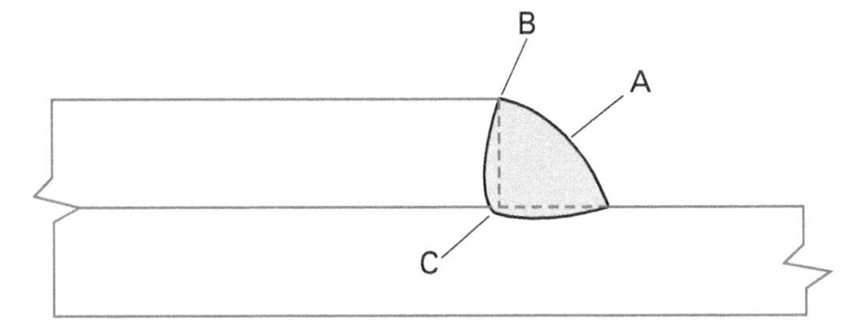

Figure RQ01

6. What part of the fillet weld is labeled A in *Figure RQ01*?
   a. Weld face
   b. Concavity
   c. Weld toe
   d. Actual throat

7. What part of the fillet weld is labeled B in *Figure RQ01*?
   a. Weld face
   b. Weld toe
   c. Weld root
   d. Overlap

8. What part of the fillet weld is labeled C in *Figure RQ01*?
   a. Convexity
   b. Leg
   c. Size
   d. Weld root

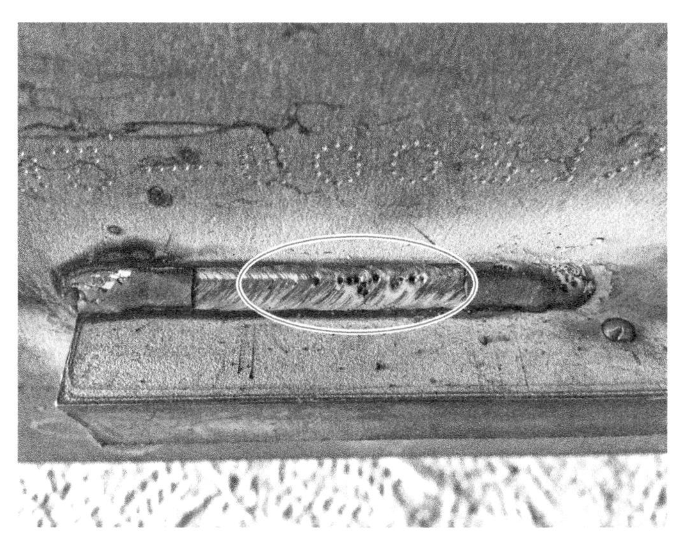

Figure RQ02
Source: iStock@Funtay

9. What kind of weld discontinuity does the circled part of *Figure RQ02* show?
   a. Cracking
   b. Incomplete fusion
   c. Inclusions
   d. Porosity

10. A weld's surface has small bits of slag embedded in it. The weld has a discontinuity called _____.
    a. porosity
    b. inclusions
    c. cracking
    d. spatter

Figure RQ03
Source: iStock@Funtay

11. What kind of weld discontinuity does the circled part of *Figure RQ03* show?
    a. Cracking
    b. Incomplete fusion
    c. Inclusions
    d. Porosity

12. You notice a crack forming next to an arc strike. What probably caused this to happen?
    a. The metal became undercut near the arc strike.
    b. The metal became porous near the arc strike.
    c. The metal became embrittled near the arc strike.
    d. The metal gained tensile strength near the arc strike.

13. To check dimensional accuracy, a welding inspector might use a _____.
    a. liquid penetrant
    b. gauge
    c. magnetic powder
    d. magnifier

Figure RQ04
Source: iStock@Funtay

14. Which nondestructive test does *Figure RQ04* show?
    a. Liquid penetrant inspection (PT)
    b. Eddy current inspection (ET)
    c. Magnetic particle inspection (MT)
    d. Radiographic inspection (RT)

15. Welding inspectors use a vacuum box to check tanks and pipes for _____.
    a. inclusions
    b. porosity
    c. leaks
    d. overlap

16. A welding inspector is testing an aluminum weld. Which of the following tests would *not* work?
    a. Magnetic particle inspection (MT)
    b. Eddy current inspection (ET)
    c. Digital radiographic inspection (DR)
    d. Computed tomography inspection (CT)

17. A nick break test is a type of _____.
    a. tensile test
    b. hardness test
    c. soundness test
    d. nondestructive test

18. What is the weld type and position for a 4G weld?
    a. Groove weld in the flat position
    b. Fillet weld in the vertical position
    c. Fillet weld in the horizontal position
    d. Groove weld in the overhead position

19. Under AWS standards, to qualify for using all electrode groups, which electrode group should a welder use during a qualifying test?
    a. F4
    b. F3
    c. F2
    d. F1

20. To prepare a test specimen properly, it's important to _____.
    a. quench the specimen to speed up cooling
    b. grind it crosswise to a smooth finish
    c. remove face or root reinforcement
    d. file the edges to 90 degrees

Answers to odd-numbered Module Review Questions are found in *Appendix A*.

Answers to Section Review Questions

Answer	Section Reference	Objective
Section 1.0.0		
1. b	1.1.0	1a
2. a	1.2.2	1b
Section 2.0.0		
1. a	2.1.2	2a
2. c	2.2.1	2b
Section 3.0.0		
1. a	3.1.2	3a
2. b	3.2.2	3b
3. b	3.3.1	3c
Section 4.0.0		
1. c	4.1.1	4a
2. c	4.2.2	4b
3. b	4.3.5	4c

SMAW – Equipment and Setup

Source: © Miller Electric Mfg. LLC

Objectives

Successful completion of this module prepares you to do the following:

1. Describe SMAW and its associated electrical and safety principles.
   a. Summarize SMAW and the electrical principles behind it.
   b. List SMAW safety procedures.
2. Summarize SMAW equipment.
   a. Identify and describe SMAW machines.
   b. Identify and describe welding cables and connectors.
   c. Identify and describe tools that clean welds.
3. Outline setting up, safely operating, and maintaining SMAW equipment.
   a. Explain how to set up SMAW equipment.
   b. Explain how to start and stop SMAW equipment.
   c. Explain how to maintain SMAW equipment.

Performance Task

Under supervision, you should be able to do the following:

1. Set up a welding machine for SMAW work.

Overview

When trainee welders start learning the different welding processes, they usually begin with shielded metal arc welding (SMAW). This simple and versatile method is the most popular process. It's suitable for many applications and requires relatively basic equipment. This module introduces that equipment, as well as setting it up. It also explores SMAW safety and lays the groundwork for later modules that teach SMAW welding techniques.

NOTE

This module uses US standard and metric units in up to three different ways. This note explains how to interpret them.

Exact Conversions

Exact metric equivalents of US standard units appear in parentheses after the US standard unit. For example: "Measure 18" (45.7 cm) from the end and make a mark."

Approximate Conversions

In some cases, exact metric conversions would be inappropriate or even absurd. In these situations, an approximate metric value appears in parentheses with the ~ symbol in front of the number. For example: "Grip the tool about 3" (~8 cm) from the end."

Parallel but not Equal Values

Certain scenarios include US standard and metric values that are parallel but not equal. In these situations, a slash (/) surrounded by spaces separates the US standard and metric values. For example: "Place the point on the steel rule's 1" / 1 cm mark."

Digital Resources for Welding

Scan this code using the camera on your phone or mobile device to view the digital resources related to this craft.

Industry Recognized Credentials

If you are training through an NCCER-accredited sponsor, you may be eligible for credentials from NCCER's Registry. The ID number for this module is 29107. Note that this module may have been used in other NCCER curricula and may apply to other level completions. Contact NCCER's Registry at 1.888.622.3720 or go to **www.nccer.org** for more information.

You can also show off your industry-recognized credentials online with NCCER's digital badges. Transform your knowledge, skills, and achievements into badges that you can share across social media platforms, send to your network, and add to your resume. For more information, visit **www.nccer.org**.

1.0.0 Shielded Metal Arc Welding

Performance Tasks

There are no Performance Tasks in this section.

Objective

Describe SMAW and its associated electrical and safety principles.

a. Summarize SMAW and the electrical principles behind it.
b. List SMAW safety procedures.

Welders commonly refer to shielded metal arc welding (SMAW) as *stick welding*. It melts and joins metals by heating them with an electric arc. A *welding machine* produces this arc with a large electric current. To create good welds, craftworkers must understand the electrical principles behind SMAW. They must also manage SMAW's hazards safely.

1.1.0 SMAW Principles

SMAW machines create the welding arc with an electric current that flows between a consumable electrode and the workpiece. The arc's high temperature—around 6,000°F (~3,300°C)—melts both the base metal and the electrode. The melted electrode (the filler metal) combines with the molten base metal, forming the *weld pool* (or *puddle*). When it cools and solidifies, it makes the weld joint.

Oxygen in the air, as well as other contaminants, can weaken the weld as it cools. To protect the weld at each step, the rod-shaped electrode has a *flux coating*. The arc's heat melts and vaporizes this clay-like material. Some material forms a cloud of *shielding gas* that protects the weld pool as it forms. The rest sits on top of the weld pool, protecting it as it cools. Later, the welder removes this coating, called *slag*. The flux also helps make the arc more stable.

1.1.1 Basic Electricity

Atoms make up all materials (*Figure 1*). A normal atom has a *nucleus* at its center made from *protons* and usually neutrons. Tiny particles called *electrons* surround the nucleus, swirling around it. Light materials have just a few protons and electrons, while heavy materials have many. Each proton or electron has an *electrical charge* that attracts its opposite. Scientists call the proton's charge *positive* (+) and the electron's *negative* (–).

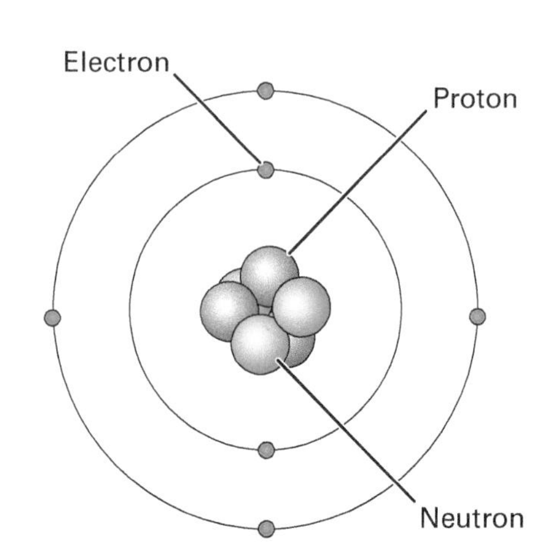

Figure 1 An atom and its particles.

Sometimes electrons leave their atoms and collect in a group. Their individual charges add up, giving a larger negative charge. Atoms with missing electrons can pool their individual charges too, giving a larger positive charge. When two different charge collections are separated, a force develops between them called **voltage** (*Figure 2*). The bigger the charge difference, the greater the voltage. Batteries and generators separate charges to produce a voltage between their two *poles* (terminals).

NOTE

Voltage is measured in units called *volts*. The unit symbol is V.

Voltage: The electrical force between two terminals or points in a circuit. Measured in volts (V).

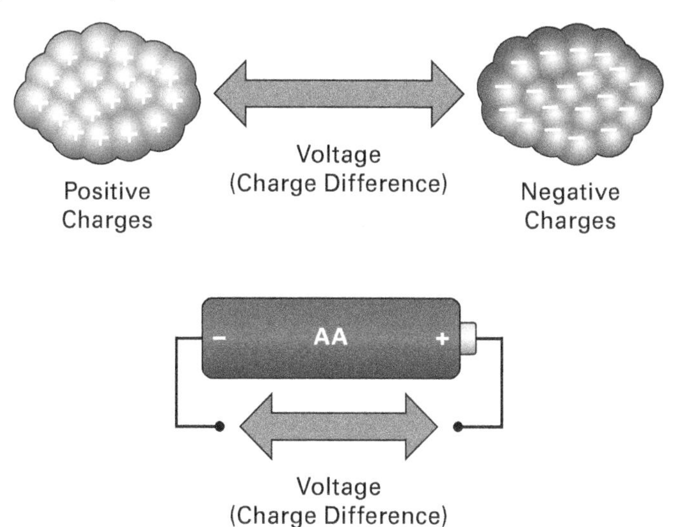

Figure 2 Voltage.

Voltage is the pressure that makes electrons move. If you connect a battery or generator's poles together, voltage makes the electrons move from the negative pole towards the positive one. This steady flow of electrons is called an **electric current** or *amperage*. It can do useful work, including heating the metal during welding. Think of it like water or pressurized gas flowing through a hose.

An *electrical circuit* is a closed loop containing a power supply and something that uses the energy (*Figure 3*). A welding machine is a power supply. Its energy heats the metals to create the weld. The larger the current (amperage), the hotter the metals become and the faster they melt. Welders adjust the amperage to control the welding action.

NOTE

Current is measured in units called *amperes* or *amps*. The unit symbol is A.

Electric current: The electron flow through a circuit. Measured in amperes or amps (A). Sometimes called *amperage*.

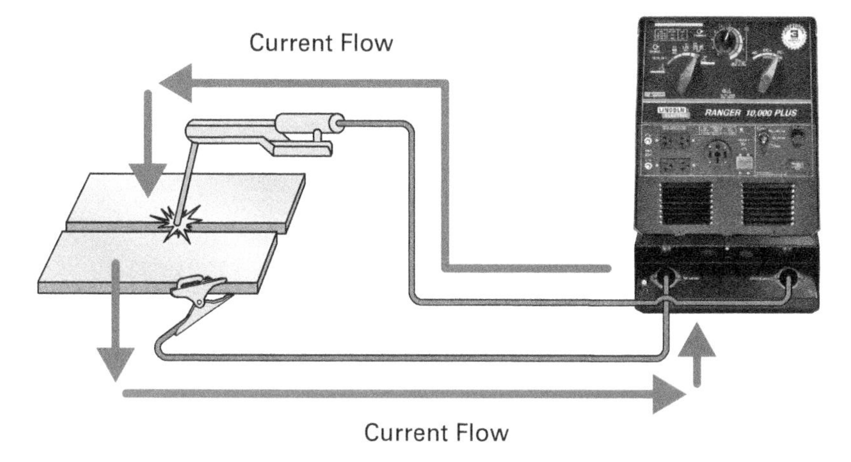

Figure 3 A welding electrical circuit.
Source: The Lincoln Electric Company, Cleveland, OH, USA

1.1.2 The Welding Arc

When setting up a welding machine, the welder attaches one of the machine's terminals to the workpiece. The other terminal connects to the electrode. The welder completes the circuit by very briefly touching the electrode to the workpiece and then slightly pulling it away. This quick current flow changes the air between the electrode and the workpiece. It becomes *ionized*, meaning that it can carry an electric current just like a wire. Electricity continues flowing through the gap between the electrode and the workpiece. This is the arc, which glows intensely from the heat (*Figure 4*).

NOTE

Touching the electrode to the workpiece for too long causes them to stick together. No arc forms, so welding can't happen. Learning to "strike an arc" reliably is an important SMAW skill that trainees must develop.

Figure 4 Striking an arc.
Source: © Miller Electric Mfg. LLC

The welder builds the weld by moving the electrode across the workpiece while maintaining the gap. If the gap becomes too large, the electric current won't flow, and the arc stops. Touching the electrode to the workpiece stops the arc too. If you do this, the electrode usually sticks to the workpiece. Producing a good weld requires maintaining the proper gap between the electrode and workpiece.

WARNING!

If the electrode sticks to the workpiece, never remove your helmet. Freeing the electrode will cause a flash that can injure your eyes. Release the electrode from its holder. Use pliers to break it free from the workpiece.

Striking an Arc

Welders strike an arc in one of two ways. Some briefly tap the electrode on the workpiece. Others lightly scratch it across the workpiece. Beginners often prefer the scratch method. Some welding machines won't strike an arc unless you scratch the electrode. Be aware, however, that most code specifications don't accept arc strikes *outside* the weld area.

1.1.3 Welding Current Types

Welding machines deliver electricity in one of two forms. **Direct current (DC)** electricity flows in one direction only. Batteries produce it, as do some generators and power supplies. These have two terminals marked with + (positive) and – (negative) to identify their **polarity**. Current flows out of the negative terminal and returns by the positive terminal.

Alternating current (AC) electricity changes direction many times per second. Some generators and power supplies produce it. These also have two terminals, but they don't have polarity markings since AC regularly changes direction. Current flows back and forth between the terminals.

Direct current (DC): An electric current that flows in one direction.

Polarity: A terminal's electrical charge, either positive or negative, which determines whether current flows into or away from it.

Alternating current (AC): An electric current that regularly changes direction and size.

AC doesn't just change direction. Its voltage and current regularly rise and fall too. *Figure 5* shows a generator's output. As the generator rotates, voltage and current rise until they reach a maximum value. They then drop to zero. They change direction and rise in the opposite direction to a maximum value. Finally, they drop back to zero. The process then repeats.

NOTE

The number of repetitions per second is *frequency*. It's measured in *hertz*. The unit symbol is Hz.

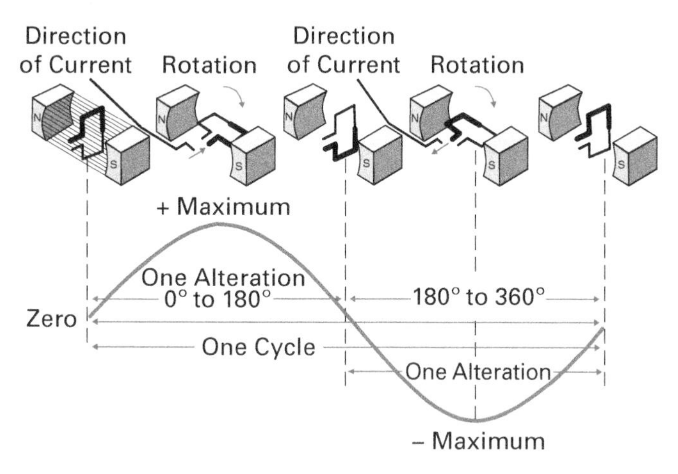

Figure 5 Alternating current.

Commercially supplied AC in the United States goes through 60 repetitions (cycles) every second. It therefore has a frequency of 60 Hz. Many countries, particularly those in Europe, use a slightly lower frequency of 50 Hz.

Some welding machines operate on AC, while others require DC. It depends on the brand and model. The welding processes they support determine whether they output AC, DC, or both. Always use a welding machine designed for the welding process you're using.

1.1.4 SMAW Welding Machines

The SMAW welding process works with either AC or DC. Some machines offer both choices. Many AC machines run on the local electrical supply. Mobile units contain an engine-driven AC generator. A **step-down transformer** in the welding machine reduces the input voltage to a lower and safer value (*Figure 6*). Dropping the voltage *increases* the amperage, which is very desirable for welding.

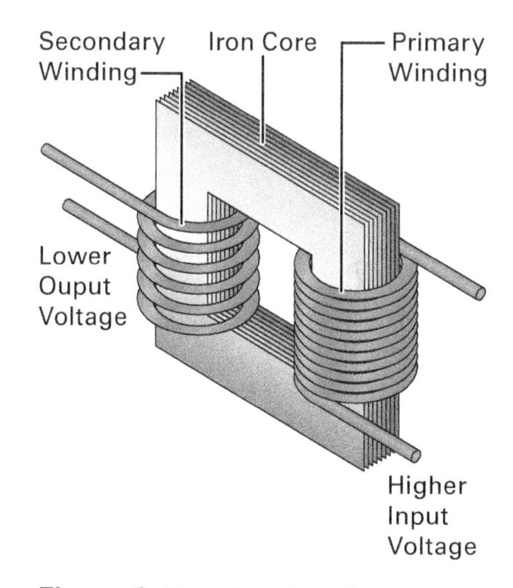

Figure 6 Step-down transformer.

Step-down transformer: An electromagnetic device that changes an alternating current to one with a lower voltage.

WARNING!

Even though welding machines reduce the input voltage to a relatively low value, they still can shock and injure you. Wearing the proper PPE and taking appropriate precautions reduces these risks.

DC welding machines may also run on the local electrical supply. They too contain a step-down transformer that reduces the voltage. But they contain a **rectifier** that converts the AC from the transformer's output to DC. Mobile DC welding machines run on an engine-driven generator.

Rectifier: An electronic device that converts alternating current to direct current.

Inverter welding machines can run on either AC or DC. They contain an electronic controller that pulses the input current on and off very rapidly. The pulsating current then goes into a step-down transformer. By changing the pulse patterns, the controller produces whatever output the welder requires for the job.

Regardless of whether it delivers AC or DC, an SMAW machine must supply *constant current (CC)* power. In other words, the amperage flowing through the electrode and workpiece must remain steady. The voltage can vary, but the current can't. Other welding processes may have different requirements.

NOTE

For safety, some welding machines automatically reduce the open-circuit voltage to a low value after welding stops. When the welder strikes a new arc, the machine returns the voltage to the normal open-circuit value. Check a machine's manual to find out if it has this feature.

1.1.5 Voltage and Current Ratings

When using a constant current welding machine, the welder selects the desired amperage. If the machine isn't welding, its *open-circuit voltage* will be relatively high (50V to 80V). When it's working, the *operating voltage* drops to a lower value (18V to 45V). The actual value depends on the selected current and the distance between the electrode and the workpiece. A voltage-current graph shows the relationship between voltage and current (*Figure 7*).

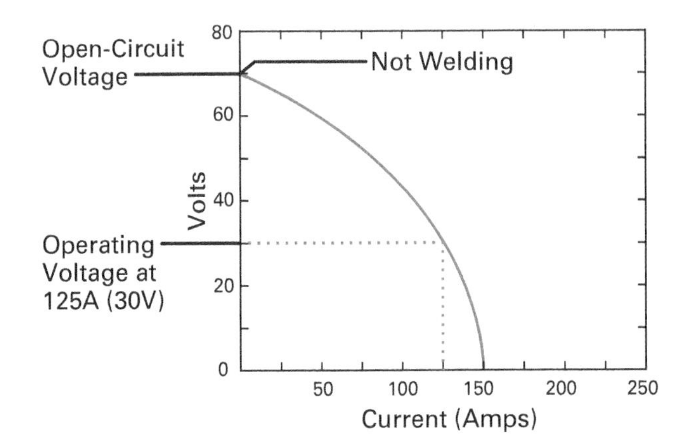

Figure 7 Voltage-current graph.

The welder chooses the amperage based on the weld type, position, and electrode size. Welding currents range from a few dozen amperes to hundreds. Manufacturers rate welding machines by their current capacity. Always choose a machine large enough to handle the job.

1.1.6 Polarity

A DC welding machine's current flows in one direction only. The direction that it flows between the electrode and the workpiece influences the finished weld's properties. For this reason, polarity matters. Welders chose the current direction by connecting the welding cables to the correct terminals on the machine.

Connecting the electrode holder to the positive (+) terminal and the workpiece cable to the negative (–) one gives *direct current electrode positive (DCEP)*. Connecting the electrode holder to the negative (–) terminal and the workpiece cable to the positive (+) one gives *direct current electrode negative (DCEN)*. *Figure 8* shows each arrangement.

Some welding machines have a polarity selector switch. Instead of moving the two cables, you simply change the selector position. AC welding machines don't have polarity since their output current changes direction 60 (or 50) times per second. Performing SMAW work with an AC machine produces welds that are a cross between those produced by DCEP and DCEN.

NOTE

Many welders use the terms *straight polarity* for DCEN and *reverse polarity* for DCEP.

CAUTION

Never flip the polarity selector while welding. This can damage the machine.

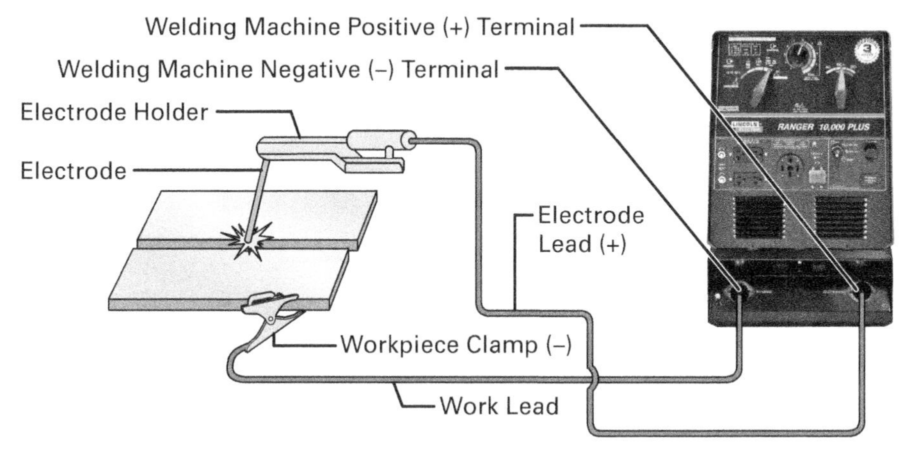

Direct Current Electrode Positive (DCEP) Hookup

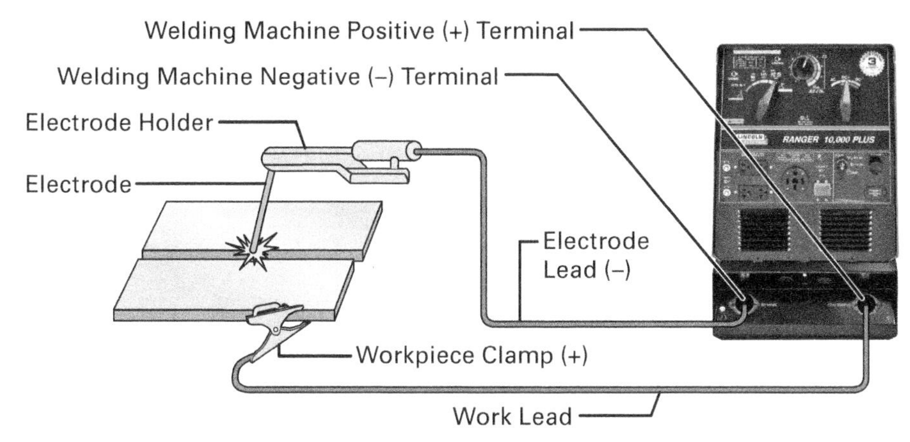

Direct Current Electrode Negative (DCEN) Hookup

Figure 8 DCEP and DCEN.
Source: The Lincoln Electric Company, Cleveland, OH, USA

1.2.0 Safety Practices

Shielded metal arc welding generates significant heat with an electric current. Using a welding machine safely requires following specific procedures. This section summarizes good practices but isn't a complete SMAW safety training. Always complete all training that your workplace requires. Follow all workplace safety guidelines and wear appropriate PPE.

While SMAW machines output relatively low voltages (usually less than 80V), they still can shock you. They can deliver very large electric currents, so they can do a lot of damage and heat metal to very high temperatures. The welding arc emits UV radiation. Treat SMAW equipment with respect.

NOTE

Be sure that you've completed NCCER Module 29101, *Welding Safety*, before continuing.

WARNING!

Energized SMAW equipment is dangerous. Always disconnect the power supply and all other energy sources before servicing or otherwise working with the equipment.

1.2.1 Protective Clothing and Equipment

Welding and cutting tasks are dangerous. Unless you wear all required PPE, you're at risk of serious injury (*Figure 9*). The following guidelines summarize PPE for SMAW work:

- Always wear safety glasses, plus a helmet or hood. The glasses or helmet/hood lens must have the proper shade value for SMAW work (*Figure 10*). Never view the cutting arc directly or indirectly without using the proper lens.
- Wear protective leather and/or flame-retardant clothing along with welding gloves that will protect you from flying sparks and molten metal.
- Wear high-top safety shoes or boots. Make sure the pant leg covers the tongue and lace area. If necessary, wear leather spats or chaps for protection. Boots without laces or holes are better for welding and cutting.
- Wear a 100 percent cotton cap with no mesh material in its construction. The bill should point to the rear or the side with the most exposed ear.

WARNING!

Never wear a cap with a button on top. The conductive metal under the fabric is a safety hazard.

- Wear earplugs to protect your ear canals from sparks. Since welding can be noisy, always wear hearing protection.

Figure 9 PPE suitable for SMAW work.
Source: The Lincoln Electric Company, Cleveland, OH, USA

Arc Welding Processes

Process	Electrode	Amperage	Lens Shade Numbers
			2 3 4 5 6 7 8 9 10 11 12 13 14
Shielded Metal Arc Welding (SMAW)	< 3/32" (2.4 mm)	< 60A	7
	3/32" – 5/32" (2.4 mm – 4.0 mm)	60A – 160A	8 9 10
	5/32" – 1/4" (4.0 mm – 6.4 mm)	160A – 250A	10 11 12
	> 1/4" (6.4 mm)	250A – 550A	11 12 13 14

Lens shade values are based on OSHA minimum values and ANSI/AWS suggested values. Choose a lens shade that's too dark to see the weld zone. Without going below the required minimum value, reduce the shade value until it's visible.

Figure 10 Shielded Metal Arc Welding – Guide to Lens Shade Numbers.

1.2.2 Fire/Explosion Prevention

SMAW produces high temperatures to weld metal. Welding or cutting a vessel or container that once contained combustible, flammable, or explosive materials is hazardous. Residues can catch fire or explode. Before welding or cutting vessels, check whether they contained explosive, hazardous, or flammable materials. These include petroleum products, citrus products, or chemicals that release toxic fumes when heated.

American Welding Society (AWS) F4.1, Safe Practices for the Preparation of Containers and Piping for Welding and Cutting, and *ANSI/AWS Z49.1* describe safe practices for these situations. Begin by cleaning the container to remove any residue. Steam cleaning, washing with detergent, or flushing with water are possible methods. Sometimes you must combine these to get good results.

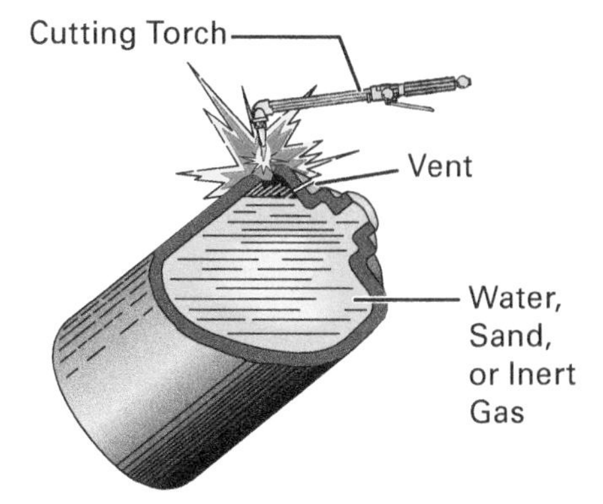

Note: *ANSI Z49.1* and AWS standards should be followed.

Figure 11 "Inerting" a container.

WARNING!

Clean containers only in well-ventilated areas. Vapors can accumulate in a confined space during cleaning, causing explosions or toxic substance exposure.

After cleaning the container, you must formally confirm that it's safe for welding or cutting. *American Welding Society (AWS) F4.1, Safe Practices for the Preparation of Containers and Piping for Welding and Cutting,* outlines the proper procedure. The following three paragraphs summarize the process.

Immediately before work begins, a qualified person must check the container with an appropriate test instrument and document that it's safe for welding or cutting. Tests should check for relevant hazards (flammability, toxicity, etc.). During work, repeated tests must confirm that the container and its surroundings remain safe.

Alternatively, fill the container with an inert material ("inerting") to drive out any hazardous vapors. Water, sand, or an insert gas like argon meets this requirement. Water must fill the container to within 3" (~7.5 cm) or less of the work location. The container must also have a vent above the water so air can escape as it expands (*Figure 11*). Sand must completely fill the container.

When using an inert gas, a qualified person must supervise, confirming that the correct amount of gas keeps the container safe throughout the work. Using an inert gas also requires additional safety procedures to avoid accidental suffocation.

NOTE

A *lower explosive limit (LEL)* gas detector can check for flammable or explosive gases in a container or its surroundings. Gas monitoring equipment checks for specific substances, so be sure to use a suitable instrument. Test equipment must be checked and calibrated regularly. Always follow the manufacturer's guidelines.

WARNING!

Never weld or cut drums, barrels, tanks, vessels, or other containers until they have been emptied and cleaned thoroughly. Residues, such as detergents, solvents, greases, tars, or corrosive/caustic materials, can produce flammable, toxic, or explosive vapors when heated. Never assume that a container is clean and safe until a qualified person has checked it with a suitable test instrument. Never weld or cut in places with explosive vapors, dust, or combustible products in the air.

The following general fire- and explosion-prevention guidelines apply to SMAW work:

- Never carry matches or gas-filled lighters in your pockets. Sparks can ignite the matches or cause the lighter to explode.
- Always comply with all site and employer requirements for hot work permits and fire watches.
- Never use oxygen in place of compressed air to blow off the workpiece or your clothing. Never release large amounts of oxygen. Oxygen makes fires start readily and burn intensely. Oxygen trapped in fabric makes it ignite easily if a spark falls on it. Keep oxygen away from oil, grease, and other petroleum products.
- Remove all flammable materials from the work area or cover them with a fire-resistant blanket.
- Before welding, heating, or cutting, confirm that an appropriate fire extinguisher is available. It must have a valid inspection tag and be in good condition. All workers in the area must know how to use it.
- Prevent fires by maintaining a neat and clean work area. Confirm that metal scrap and slag are cold before disposing of them.

1.2.3 Work Area Ventilation

Vapors and fumes fill the air around their sources. Welding and cutting fumes can be hazardous. Welders often work above the area from which fumes come. Good ventilation and source extraction equipment help remove the vapors and protect the welder. The following lists good ventilation practices to use when welding and cutting:

- Always weld or cut in a well-ventilated area. Operations involving hazardous materials, such as metal coatings and toxic metals, produce dangerous fumes. You must wear a suitable respirator when working with them. Follow your workplace's safety protocols and confirm that your respirator is in good condition.

WARNING!

Cadmium, mercury, lead, zinc, chromium, and beryllium produce toxic fumes when heated. When welding or cutting these materials, wear appropriate respiratory PPE. Supplied-air respirators (SARs) are best for long-term work.

- Never weld or cut in a confined space without a confined space permit. Prepare for working in the confined space by following all safety protocols.
- Set up a confined space ventilation system before beginning work.
- Never ventilate with oxygen or breathe pure oxygen while working. Both actions are very dangerous.

WARNING!

When working around other craftworkers, always be aware of what they're doing. Take precautions to keep your work from endangering them too.

1.2.4 Working with Portable Welding Equipment

Welding machines in the shop are either permanently set up on a platform or roll around on a cart. Welding in the field usually requires portable equipment. These machines may be small enough to lift or come with their own trailer (*Figure 12*).

Figure 12 A large mobile SMAW machine.
Source: The Lincoln Electric Company, Cleveland, OH, USA

Always move, lift, or hoist welding machines carefully, even small ones. Smaller machines can injure feet or hands if they fall or tip over. A trailer-mounted machine could start to roll if it breaks loose or isn't properly chocked. Always follow the proper precautions for moving equipment. Stay alert and pay attention to your surroundings. Never cut corners because you're in a hurry.

Sometimes, moving a machine involves rigging and hoisting. Use the proper equipment and get help when handling large machines. Use tag lines when hoisting since heavy objects can spin or swing. Only those properly qualified should set up rigging.

1.2.5 Working at Heights

Sometimes, welders must work while suspended from a structure. Always wear approved fall arrest and fall protection devices. Inspect your safety equipment each time you use it. Confirm that it's up to date and in excellent condition. Never do a job unless you're properly trained for it. If you're uncertain, speak to your supervisor.

1.2.6 Electrical Hazards

Welding machines generally operate on a much higher voltage than they output. Small units may run on regular 120VAC. Larger machines may require 240VAC, 480VAC, or 575VAC, which are much more hazardous. Regardless of voltage, treat all electrical equipment with the same precautions.

Always know the main circuit breaker's location. When possible, turn it off before connecting or disconnecting the welding machine's power cord. Confirm that the welding machine itself is switched off before connecting or disconnecting the power cord.

Never forget that the welding electrode can arc to any **conductor** that's connected to the welding machine. When the machine is on, treat the electrode with care. Never set it down carelessly or touch it to metal objects accidentally.

Conductor: A material that can carry an electric current.

Wet conditions make working with electricity far more hazardous. It's very easy to be shocked if you're wet or standing on a wet surface. If you must work in wet conditions, wear PPE designed to protect you from electrical shock. Confirm that your equipment is in good condition, without broken insulation or exposed connectors.

SMAW passes an electric current through the workpiece. If it arcs or flows through certain components, it could cause damage. Carefully positioning the workpiece clamp can prevent this problem. Sometimes, you should remove sensitive components before beginning work.

Never allow the welding current to pass through bearings, seals, valves, or lubricated parts. When welding assembled equipment, position the workpiece clamp to avoid these components. Check for possible gaps where the current could arc. Reposition the clamp if necessary.

The welding current can easily damage electrical and electronic components. If necessary, remove them or have an electrician isolate them. If the equipment contains a battery, disconnect the lead connected to the equipment frame (the *ground lead*). Welding current can cause a battery to explode or catch fire. Never weld near a battery. Whenever possible, remove it entirely.

WARNING!

Welding current entering a battery can have serious consequences. Lead acid batteries can explode and shower the work area with corrosive sulfuric acid. Lithium batteries can catch fire and burn vigorously or even explode.

1.2.7 Other Equipment Hazards

Be careful around engine-powered welding machines. Gasoline or diesel engines use flammable fuels. They also generate exhaust that can be toxic in a confined space. Run engines in the open air. Be sure that air currents can't blow the exhaust back into the work area. Be cautious around the engine's moving parts.

1.0.0 Section Review

1. An alternating current changes direction regularly. What else changes?
   a. Electrical resistance
   b. Voltage and current
   c. Voltage and frequency
   d. Frequency

2. What is the correct eye protection shade range for SMAW work?
   a. 2 to 4
   b. 4 to 6
   c. 6 to 8
   d. 7 to 14

2.0.0 Shielded Metal Arc Welding Equipment

Performance Tasks

There are no Performance Tasks in this section.

Objective

Summarize SMAW equipment.

a. Identify and describe SMAW machines.
b. Identify and describe welding cables and connectors.
c. Identify and describe tools that clean welds.

SMAW work requires not only a suitable welding machine but also supporting equipment. This section examines SMAW machines in greater depth. It also introduces the accessories and tools that go with them.

2.1.0 SMAW Machines

Manufacturers classify welding machines by the processes they support. Some welding processes require the current to remain steady during welding—constant current (CC). Others require the voltage to remain steady—constant voltage (CV). *Figure 13* shows voltage-current graphs for each. Manufacturers also classify machines by the current type they output (AC or DC) and the technology that produces it.

Many welding machines use the local electrical supply for power (*Figure 14*). Small machines can plug into a regular 120VAC outlet. These handle only light welding tasks, however, because normal outlets can't deliver much current. Larger machines plug into a 208VAC to 240VAC single- or three-phase outlet. Very large industrial machines require 480VAC or 575 VAC three-phase power. In field situations or places without a suitable electrical supply, welders use engine-driven welding machines. An engine driving a generator produces the power.

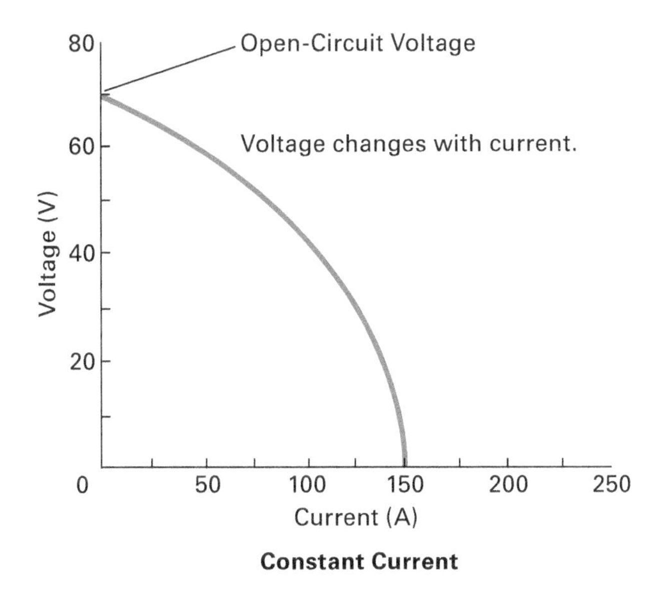

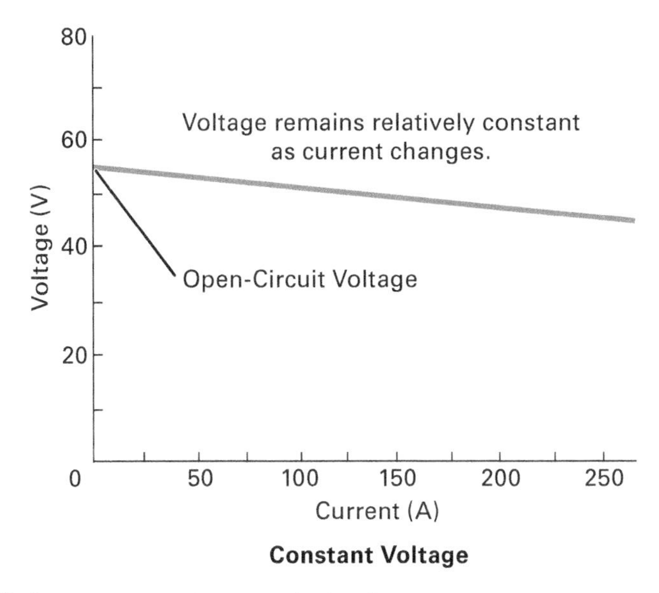

Figure 13 Constant current vs. constant voltage.

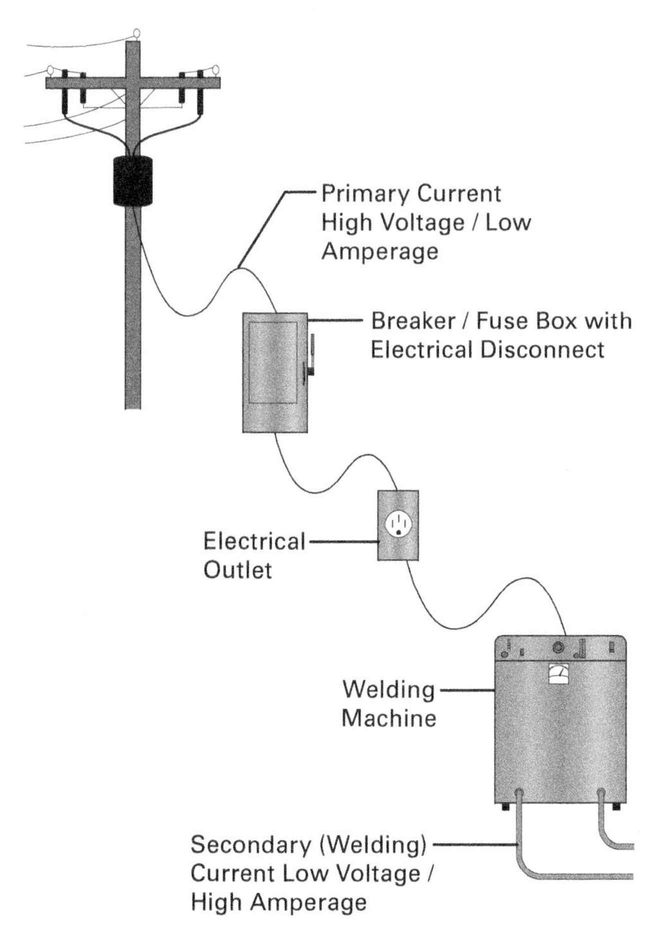

Figure 14 Primary and secondary welding circuits.

WARNING!

A welding machine's primary circuit can kill you. Treat it with care. For safety, all welding machines require a proper electrical ground connection.

As you know, SMAW requires a machine that delivers constant current AC or DC. Several technologies support the SMAW process. The following sections introduce these. Be aware, however, that some welding machines support multiple welding processes. Manufacturers call these *multiprocess machines* (*Figure 15*). To use one for SMAW work, it must have an SMAW mode. If you're uncertain about a welding machine's capabilities, consult its manual.

2.1.1 Transformer Machines

These machines use a step-down transformer to produce a suitable AC output. The transformer's primary coil connects to the electrical supply through a circuit breaker. The transformer reduces the voltage to a suitable level. It also increases the output amperage. The transformer's secondary coil connects to the welding machine's output terminals.

Transformer machines come in many sizes (*Figure 16*). They're not as common as other welding machines. Smaller ones operate on regular 120VAC. Larger units require higher input voltages and three-phase power. All have simple controls: a power switch and an amperage control.

Figure 15 Multiprocess welding machine.
Source: © Miller Electric Mfg. LLC

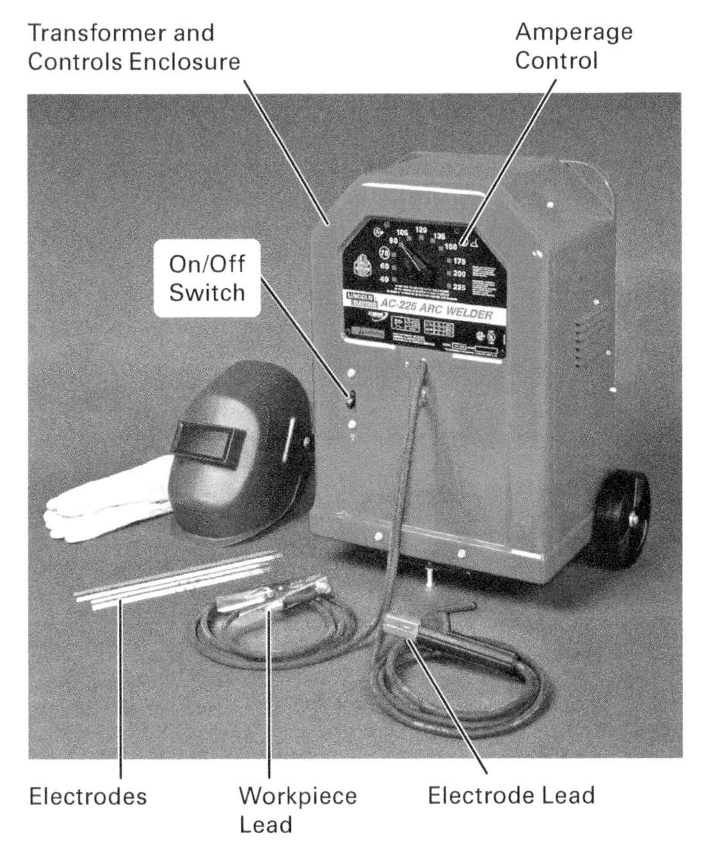

Figure 16 Transformer welding machine.
Source: The Lincoln Electric Company, Cleveland, OH, USA

2.1.2 Transformer-Rectifier Machines

These machines use a step-down transformer to produce a suitable output voltage and current. The transformer's primary coil connects to the electrical supply through a circuit breaker. The transformer reduces the voltage to a suitable level. It also increases the output amperage. The transformer's secondary coil connects to a rectifier that changes the AC to DC. The rectifier connects to the welding machine's output terminals.

Transformer-rectifier machines come in many sizes (*Figure 17*). They're very popular and versatile. Smaller ones operate on regular 120VAC. Larger units require higher input voltages and three-phase power. Some transformer-rectifier units have a rectifier bypass switch. Flipping it turns them into ordinary transformer (AC) machines.

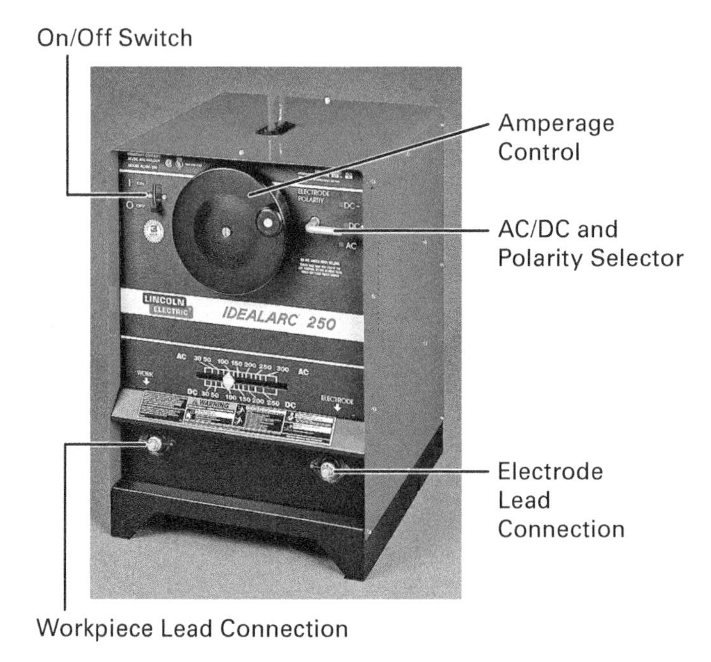

Figure 17 Transformer-rectifier welding machine.
Source: The Lincoln Electric Company, Cleveland, OH, USA

Many have simple controls: a power switch and an amperage control. More sophisticated units have a switch that changes between constant current and constant voltage modes. This feature lets them support multiple welding processes. They may also have a polarity selector switch that eliminates having to swap cables to change between DCEP and DCEN.

2.1.3 Inverter Welding Machines

These welding machines contain sophisticated technology that makes them extremely versatile. They're smaller and lighter than transformer-rectifier machines of the same capacity (*Figure 18*). They also use significantly less energy and support multiple welding processes.

An inverter first converts its input electrical supply from AC to DC. It delivers the DC to high-speed electronic switches that can turn on and off thousands of times per second. An electronic controller pulses the switches in a specific pattern. Their output goes to a small step-down transformer, which reduces the voltage and increases the amperage.

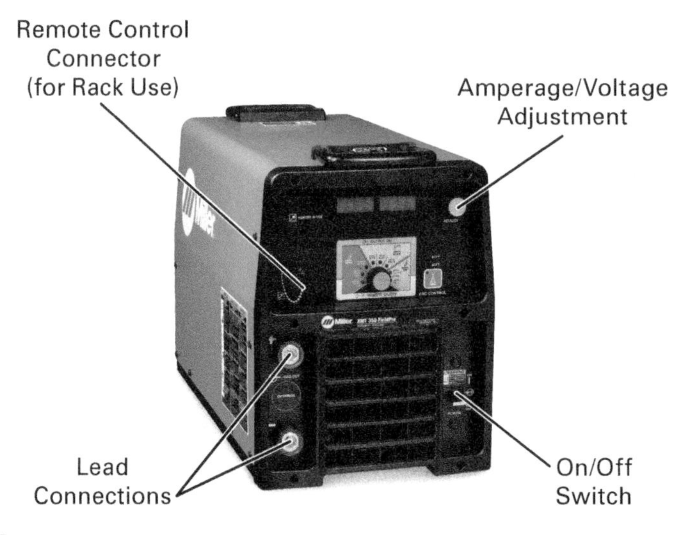

Figure 18 Inverter welding machine.
Source: © Miller Electric Mfg. LLC

Inverter welding machines produce very smooth DC that gives a stable welding arc. They can operate in either constant current or constant voltage mode. Many include advanced controls and handy features that make welding easier. For example, in SMAW mode, they can automatically shut down if the electrode becomes stuck to the workpiece.

2.1.4 Pack Machines

Construction sites sometimes require multiple welders to work close together. Moving individual welding machines to the site wastes time. Connecting their many power cables creates a challenge and perhaps a tangle too. A *pack machine* solves these problems (*Figure 19*).

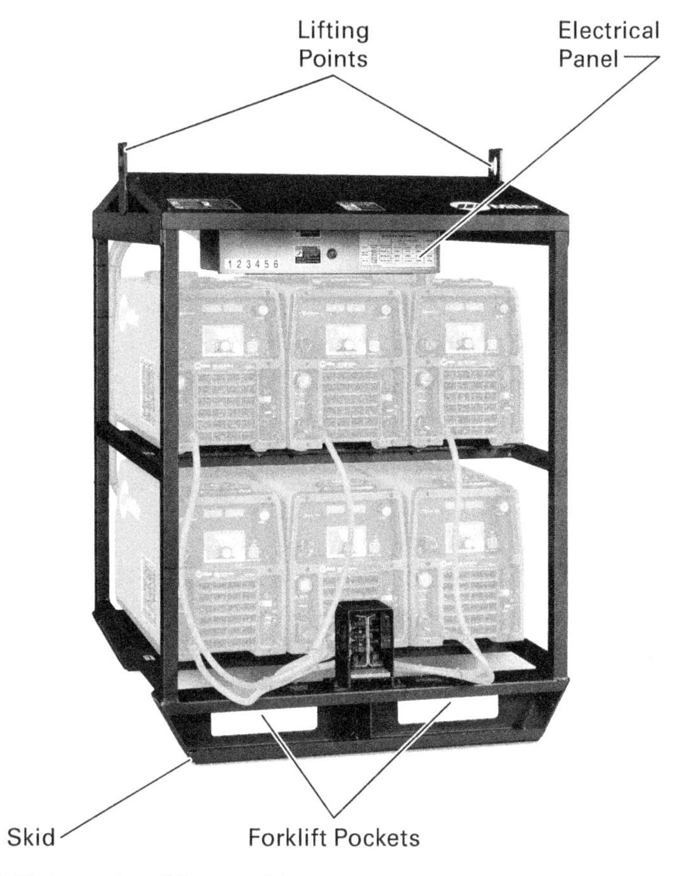

Figure 19 Eight-pack welding machine.
Source: © Miller Electric Mfg. LLC

Pack machines hold four, six, or eight welding machines in a sturdy rack. The rack has skids, forklift pockets, and lifting points, so it's easy to hoist and move around the site. It includes an electrical panel with a master cut-off switch. Some racks include 120VAC outlets for lights or other accessories. A single power cable (usually 480VAC or 575VAC three-phase) supplies all the welding machines. Each has its own controls and welding cables. Some units include remote controls so the welders can adjust their own machines easily.

2.1.5 Engine-Driven Machines

These portable machines include a gasoline or diesel engine turning a generator (*Figure 20*). They bring welding to locations without electrical service. Larger units include a trailer. They offer the same features as regular welding machines but are more complex because of the engine and its supporting equipment.

Figure 20 Engine-driven welding machine.
Source: The Lincoln Electric Company, Cleveland, OH, USA

Governor: A device that limits an engine's speed.

To produce the correct welding current, the generator must spin at the right speed. A **governor** regulates the engine's speed to the proper value. Some engine-powered machines manage the engine speed intelligently. When no one is welding, the governor drops the engine speed to idle ("auto idle"). When welding begins, it switches the engine to a higher speed. This strategy saves fuel and reduces wear and tear.

> **Going Green**
>
> **Welding Machine Maintenance**
>
> When maintaining engine-driven welding machines, avoid spilling oil and fuel. Even small spills can harm the environment. Clean up all spills and properly discard the waste.

Engine-driven welding machines may have an auxiliary generator that produces 120VAC for lighting, tools, or other accessories. When operating in auto idle mode, the auxiliary generator can't deliver as much power. If accessory power demands go up, however, the engine will exit auto idle mode to meet them.

Engine-driven machines have standard welding controls—amperage, polarity, and welding process selectors. But they also include engine controls and gauges.

2.1.6 SMAW Machine Ratings

Duty cycle: The percentage of time a welding or cutting machine can operate without overheating within a 10-minute period.

Manufacturers rate welding machines by amperage and **duty cycle**. A machine's duty cycle is the percentage of time that it can operate during a 10-minute period. For example, if a welding machine has a 40 percent duty cycle, it can weld 4 minutes out of 10. It must cool off for the remaining 6 minutes or it will overheat. The welding and cooling cycles don't have to occur all at once. You can alternate welding and cooling intervals within the 10-minute period.

A welding machine with a 10 to 40 percent duty cycle is light- or medium-duty. An industrial, heavy-duty machine has a duty cycle of at least 60 percent. Some automated welding machines can operate with no cooling cycles at all (100 percent duty cycle).

> **Engines**
>
> An engine-driven welding machine's engine size depends on the machine's capacity. Small machines have single-cylinder engines, while larger machines have four- or six-cylinder engines.

Machines usually can deliver more current than their rated value but at a lower duty cycle. A machine rated at 300A with a 60 percent duty cycle might be able to deliver 375A with a 30 percent duty cycle. Reducing the amperage below the rated value usually increases the duty cycle. Setting a 300A machine with a duty cycle of 60 percent to 200A might raise its duty cycle to 100 percent. Always read the manufacturer's specifications carefully to know what a machine can do. *Figure 21* shows the relationship between amperage and duty cycle.

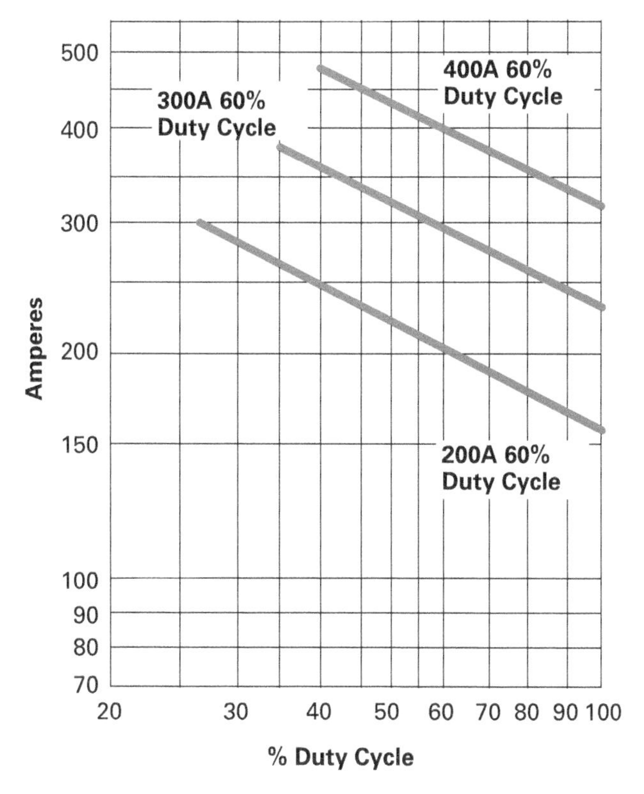

Figure 21 Amperage and duty cycle.

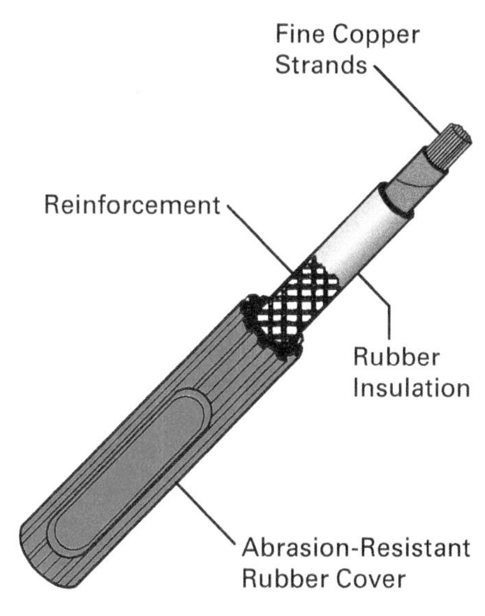

Figure 22 Welding cable.

2.2.0 Welding Cables and Connectors

Cables and connectors link the welding machine to the workpiece and the torch. They come in various styles and capacities. Choosing the correct one depends on the application and the operation's requirements.

2.2.1 Welding Cable

Welding cables are usually thick and heavy so they can carry the large currents that welding demands (*Figure 22*). They're strong and flexible too, built to work in demanding environments. A welding cable has many fine strands of copper wire surrounded by thick rubber insulation. Nylon or polyester strands embedded in the rubber add extra strength.

The more wire a cable contains, the larger its diameter. The larger its diameter, the more amperage it can carry. A number identifies a welding cable's diameter. The smaller the number, the larger the diameter (and amperage capacity). The smallest welding cable is #8, while the largest is #4/0 (pronounced "four-aught").

The correct cable size depends on two factors. First, you must know the amperage that the job requires. Second, you must know the distance between the welding machine and the workpiece. Long distances reduce the electrode voltage. The cable also heats up more.

Most cable manufacturers publish tables to help you select the right cable (*Table 1*). Measure the distance between the welding machine and the workpiece. Double this value since there are two cables (the electrode cable and the workpiece/ground cable). Use the machine's amperage rating for the machine size. For example, a 250A machine that's 75' from the workpiece needs a #2 cable.

Welding Machine Thermal Protection

Most welding machines have a thermal circuit breaker that will shut off the machine if it gets too hot. When the machine has cooled down, the circuit breaker will reset.

CAUTION

Don't forget to double the distance between the welding machine and the workpiece. If you don't do this, you could pick a cable that can't carry the welding current. The cable will run hot, and the weld quality could be poor.

TABLE 1 Welding Cable Sizes

Recommended Cable Sizes for Manual Welding						
		Copper Cable Sizes for Combined Length of Electrode Plus Ground Cable				
Machine Size in Amperes	**Duty Cycle (%)**	**Up to 50'**	**50' to 100'**	**150'**	**150' to 200'**	**200' to 250'**
100	20	#8	#4	#3	#2	#1
180	20	#5	#4	#3	#2	#1
180	30	#4	#4	#3	#2	#1
200	50	#3	#3	#2	#1	#1/0
200	60	#2	#2	#2	#1	#1/0
225	20	#4	#3	#2	#1	#1/0
250	30	#3	#3	#2	#1	#1/0
300	60	#1/0	#1/0	#1/0	#2/0	#3/0
400	60	#2/0	#2/0	#2/0	#3/0	#4/0
500	60	#2/0	#2/0	#3/0	#3/0	#4/0
600	60	#3/0	#3/0	#3/0	#4/0	***
650	60	#3/0	#3/0	#4/0	**	***

** Use Double Strand of #2/0

*** Use Double Strand of #3/0

CAUTION

Always tightly secure connectors to their cables. Loose connectors heat up and don't transfer the welding current reliably. The result could be a damaged cable and poor weld quality.

CAUTION

Place the workpiece clamp over clean, bright metal. Corrosion, paint, or surface coatings will block or reduce the welding current, resulting in a poor weld. If necessary, clean the metal before attaching the clamp.

2.2.2 Lugs and Quick Disconnects

Equipping a welding cable with connectors makes attaching it to the machine, the workpiece, and the electrode holder simpler. Connectors come in several styles.

Lugs connect the cable to the welding machine's terminals. They come in several sizes and crimp onto the cable. Their shape keeps them from pulling off the terminals. *Quick disconnects* make connecting and disconnecting cables fast and simple (*Figure 23*). The DINSE® style clamps onto the cable with a setscrew. They're insulated and require a half-twist to lock or unlock them. Keep quick disconnects clean so they'll join firmly. Tighten terminals properly to ensure maximum current flow.

2.2.3 Workpiece Clamps

Workpiece clamps attach the ground cable to the workpiece (*Figure 24*). They connect to the cable with a setscrew or other mechanical means. Most have spring-loaded jaws that grip the workpiece, forming a mechanical and electrical connection. Some attach with powerful magnets. Manufacturers rate workpiece clamps by amperage. Always choose one rated equal to or higher than the welding machine's capacity.

2.2.4 Electrode Holders

The electrode holder (*SMAW torch* or *stinger*) grips the electrode. A connector on the holder attaches it to the electrode cable. Electrode holders come in several sizes, each with a rated amperage. The larger the amperage rating, the larger the holder will be.

Choose an electrode holder based on the welding current and the electrodes that you're planning to use. *Figure 25* shows two holder styles. The jaw-type holder grips the electrode between spring-loaded jaws. Collet-type holders waste less of the electrode since its end fits directly into the torch.

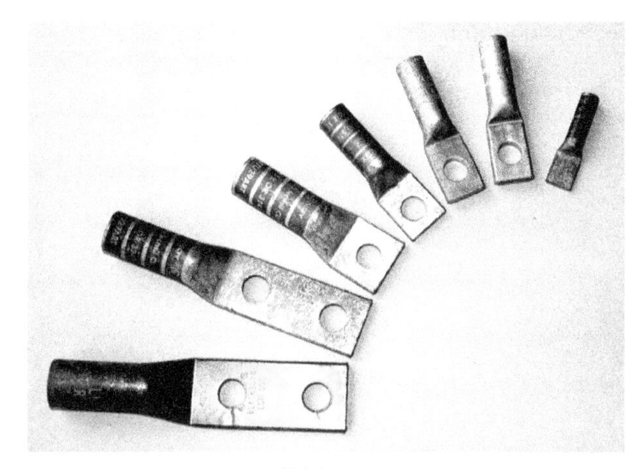

(A) Lugs

(B) Quick Disconnects

Figure 23 Lugs and quick disconnects.

2.3.0 Cleaning Tools

Welders prepare metal for welding by cleaning it. Welding dirty or corroded metal is either impossible or produces poor-quality welds that won't pass inspection. Welders must remove slag from finished welds before they can lay additional beads. They use both hand and powered tools for these tasks.

WARNING!

Always wear eye and hand PPE when preparing or cleaning metal. Finished welds stay hot for a long time. Work carefully to avoid burns. Chipping slag and coatings from finished welds produces flying debris. Wear suitable eye and face protection. Confirm that other workers within range are wearing similar PPE.

2.3.1 Hand Tools

Many welders prefer simple hand tools for cleaning surfaces and welds. Wire brushes remove corrosion, dirt, and paint from metal. Wire brushing metal before welding ensures better electrical contact and a cleaner weld. Files remove thicker coatings and work well on surfaces that aren't flat. Chipping hammers break off slag from finished welds. *Figure 26* shows common welding hand tools.

2.3.2 Powered Tools

Weld slag chippers and needle scalers are pneumatic (air-powered) tools (*Figure 27*). They clean surfaces and remove slag from welds. They're also handy for removing paint and hardened dirt. They don't work very well on corrosion. A weld slag chipper has a single chisel, while a needle scaler has 18 to 20 blunt steel needles. Most weld slag chippers can convert to needle scalers with an attachment.

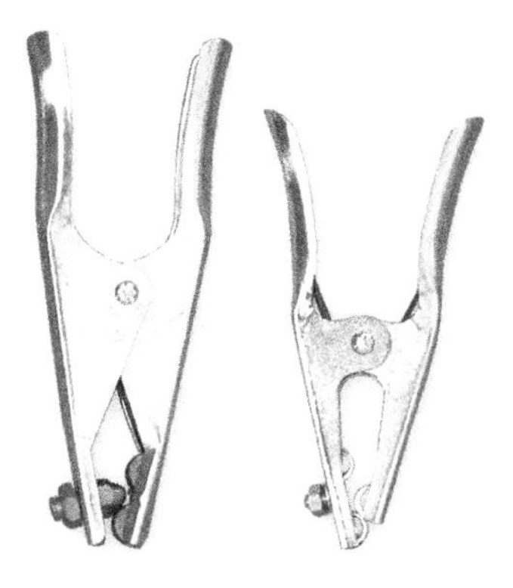

(A) Low-Current Clamps

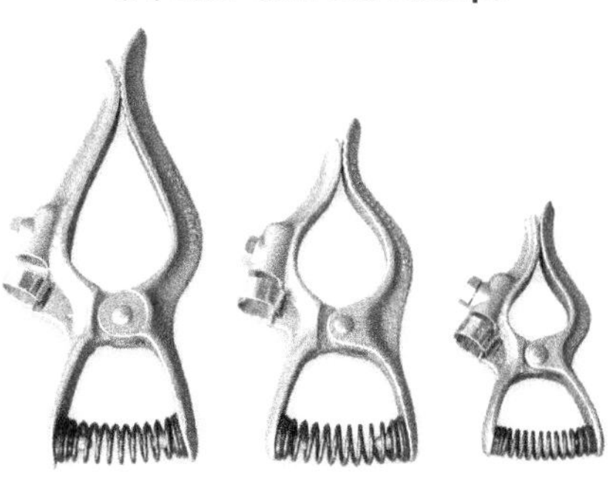

(B) High-Current Clamps

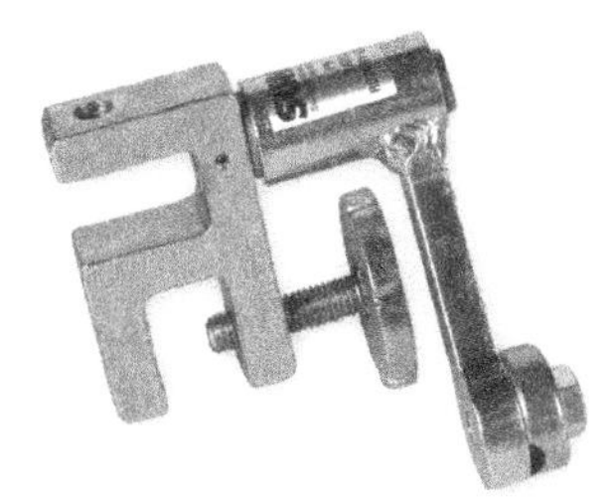

(C) Rotary Ground Clamp

(D) Magnetic Swivel Clamp

Figure 24 Workpiece clamps.
Source: Courtesy of Sumner Manufacturing Co. (24C); PowerBase™ Grounding Magnet, Strong Hand Tools, **www.StrongHandTools.com** (24D)

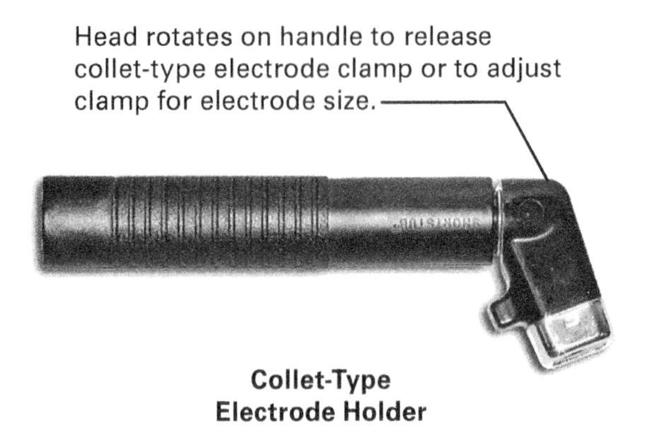

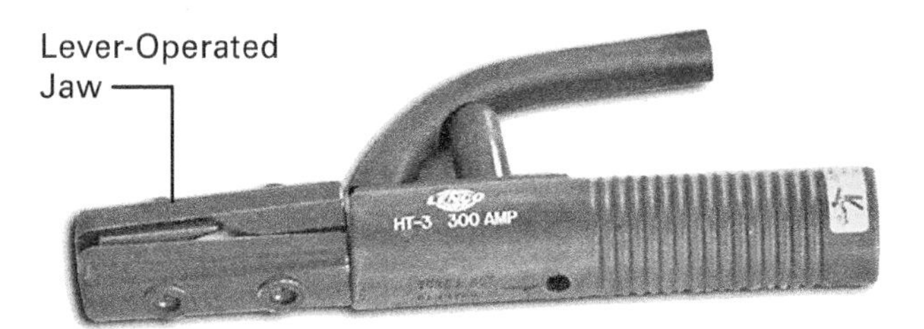

Figure 25 Electrode holders.

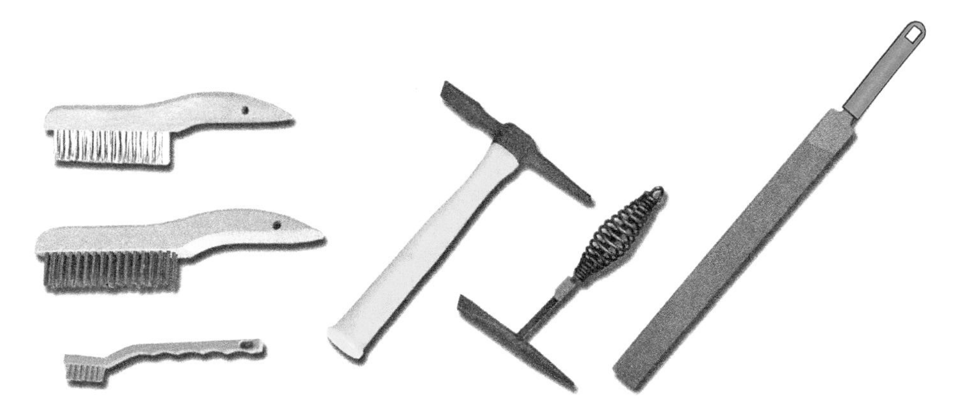

Figure 26 Common welding hand tools.

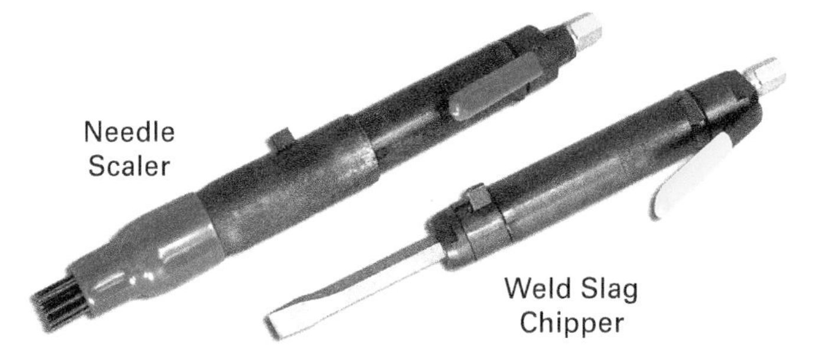

Figure 27 Pneumatic weld slag chipper and needle scaler.

2.0.0 Section Review

1. A large industrial welding machine *most* likely requires a _____.
   a. 120VAC single-phase power supply
   b. 208VAC single-phase power supply
   c. 240VAC single-phase power supply
   d. 480VAC three-phase power supply

2. If the distance between the welding machine and the workpiece is long, you'll need a _____.
   a. double-insulated welding cable
   b. small-diameter welding cable
   c. polyester-core welding cable
   d. large-diameter welding cable

3. Before laying another bead on top of an existing weld, a welder *must* remove _____.
   a. corrosion
   b. slag
   c. paint
   d. dirt

3.0.0 Setting Up Shielded Metal Arc Welding Equipment

Objective

Outline setting up, safely operating, and maintaining SMAW equipment.

a. Explain how to set up SMAW equipment.
b. Explain how to start and stop SMAW equipment.
c. Explain how to maintain SMAW equipment.

Performance Task

1. Set up a welding machine for SMAW work.

SMAW equipment is relatively simple to set up. But SMAW itself creates numerous hazards that affect both the equipment and the welder. Following specific procedures reduces these hazards. The following sections introduce setting up SMAW equipment, starting it up, shutting it down, and maintaining it.

3.1.0 Setting Up

Properly and safely setting up SMAW equipment requires a series of steps. While these are simple, they still require you to work safely and thoughtfully.

3.1.1 Selecting the Equipment

Begin by selecting a welding machine that meets the application's requirements. To support SMAW, a machine must have a constant current mode (AC or DC). If you plan to use a multiprocess welding machine, be sure that it meets this requirement.

Next, verify that the machine can deliver enough amperage for the welds that you plan to produce. An underpowered machine won't work well. The required current depends on the material, its thickness, and the electrodes you'll use.

Finally, confirm that the local electrical supply works for the machine. It must have the correct voltage, phase, and current capacity. If local power isn't available or isn't suitable, you'll need to use an engine-driven welding machine.

3.1.2 Positioning the Welding Machine

Whenever possible, position the welding machine near the workpiece. This keeps the leads short and is more convenient to the welder. Place the machine so it won't be showered by sparks or molten metal. Be sure that it has good air circulation so it can cool itself properly. Try to keep it out of dirty and dusty environments as these can clog the air filter and cooling fan. Don't place the machine in standing water or on a wet surface.

WARNING!

Never operate a welding machine around flammable or explosive gases or fumes. Be sure the air doesn't contain combustible dust or other fire hazards.

If the machine plugs into an outlet, be sure the available ones are the right type for the machine. Never create an adapter or attempt to wire the machine directly into the electrical service. Only qualified electricians may do these tasks. Identify the nearest electrical disconnect in case you must shut down the welding machine quickly.

If you're using an engine-powered welding machine, position it in a safe place. Be sure it has good airflow around it. Place it outside if you're working inside. Verify that exhaust fumes can't find their way back inside or enter a building air intake. Store its fuel in approved containers away from flames or sparks.

3.1.3 Moving Welding Machines

Moving a large welding machine requires care. All are heavy and some are top heavy. These can tip over easily. Machines with their own trailers are relatively easy to move with a truck or tractor. Some machines have wheels and are light enough to roll by hand.

WARNING!

If a welding machine starts to tip over, do not attempt to hold it. Get out of its way and let it fall.

Some welding machines have skid bases and lifting points so a crane or hoist can move them. Look up the machine's weight. Confirm that the crane or hoist and its tackle are rated at more than this value. Never lift the machine by anything other than a designated lifting point. Don't place the lifting hook directly in a lifting eye. Instead, use a suitable sling and shackle. Verify that welding cables and attached equipment are secure. *Figure 28* shows a proper lifting setup.

WARNING!

Always inspect the rigging before using it. Replace anything worn or damaged. Only qualified riggers should set up and execute a lift.

3.1.4 Stringing Welding Cables

Welding cables should comfortably reach the work area, but they shouldn't be so long that you must keep coiling them up. Always string cables to prevent tripping hazards and damage from traffic. Keep cables out of walkways and aisles. If possible, string them overhead. If this isn't feasible, protect cables and foot traffic with boards and ramps (*Figure 29*).

Before stringing a welding cable, check it for damage. Repair cuts or breaks in the insulation. Bare cable can arc to nearby metal surfaces and equipment. This will cause damage and further harm the cable. If you can't properly repair the cable, replace it.

Cranes and Wheeled Welding Machines

On crowded construction sites, cranes often position and place welding machines, even those with wheels. This strategy is more efficient and reduces the site's vehicle traffic.

When stringing cable, be careful of nearby equipment. Keep cable away from moving equipment like overhead cranes or trolleys. You could foul the equipment, or it could damage the cable. Welding cables are heavy. Never use a cable tray to carry a welding cable unless it has the weight capacity for it. Don't support welding cable on instrumentation cable trays or conduits as these can't carry heavy weights. Be especially cautious around temporary cable supports since they can collapse.

Improperly supported cable could tug on connectors or pull the welder off balance. When stringing welding cable overhead or between floors, tie off the cables with rope. Dry rope is relatively nonconductive and won't abrade the welding cables.

CAUTION

When a welding machine is operating, its cables radiate electrical noise. Never place welding cables near instrumentation cables or sensitive electronic equipment. You could cause an equipment malfunction or even significant damage.

3.1.5 Locating the Workpiece Clamp

The welding current can damage sensitive equipment or arc to nearby surfaces as it flows through the workpiece. Welding current can also damage electrical or electronic equipment and batteries. Properly locating the workpiece clamp routes the current through safe places.

Identify the weld zone and position the workpiece clamp so the welding current won't pass through anything sensitive between those two points. Disconnect or remove batteries entirely. Have a qualified electrician isolate any nearby electrical or electronic devices. If you're unsure, ask your supervisor for assistance.

WARNING!

Welding current entering a battery can have serious consequences. Lead acid batteries can explode and shower the work area with corrosive sulfuric acid. Lithium batteries can catch fire and burn vigorously or even explode.

CAUTION

If welding on machinery or other assembled equipment, position the workpiece clamp so the welding current will not pass through seals, bearings, or components that arcing or heat could damage. Isolate any nearby electrical or electronic components from the SMAW circuit.

Lifting Hook

Sling

Shackle

Figure 28 Lifting a welding machine.
Source: The Lincoln Electric Company, Cleveland, OH, USA

Going Green

Recycling

Welding cables are mostly copper, a valuable metal. Never throw old or damaged cables in the trash. Instead, send them to a metal recycling facility that can recover the copper.

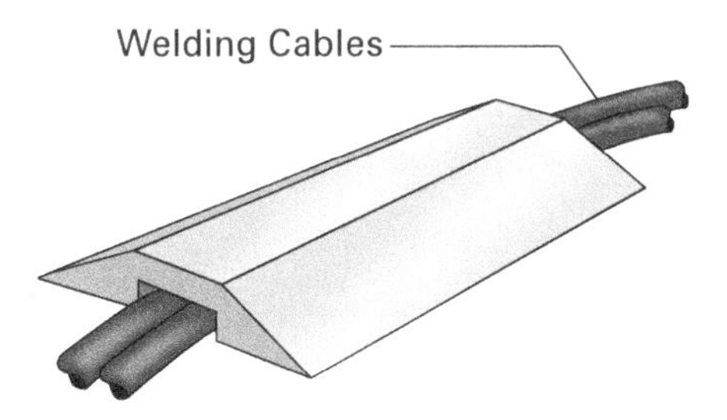

Figure 29 Protecting welding cables.

WARNING!

Never connect the workpiece clamp to a pipe carrying flammable, explosive, or corrosive substances. The welding current could heat the pipe or generate sparks. These could start a fire or cause an explosion.

To lay a quality weld, the workpiece clamp must make good electrical contact with the workpiece. Dirt, paint, coatings, or corrosion could reduce current flow or cause arcing. Surface contaminants could also affect the weld quality and cause defects. Clean surfaces with a wire brush or file until the metal is clean and bright. If the workpiece clamp is scorched, scarred, or pitted, replace it. Some clamps have replaceable contacts.

Case History

An Expensive Mistake

A welder had the job of reattaching a refrigerated trailer's mud flap. The loaded trailer was parked right next to the welding machine, so it looked like an easy job. The welder connected the workpiece clamp to the front of the trailer. He then pulled the electrode cable around to the back and welded the mud flap. The entire job took less than 15 minutes.

When the customer collected the trailer, he noticed that the refrigeration unit was off and its status display blank. A service technician analyzed the problem and discovered that the refrigerator's electronic controller had failed.

Further investigation revealed that the welding current had passed through several sensors connected to the controller. The current entered the controller through the sensor wiring, destroying it. The welding shop had to pay more than $1,000 in parts and labor to replace the controller.

The Bottom Line: If the welder had placed the workpiece clamp close to the weld zone, this problem wouldn't have happened. Always check for nearby electrical and electronic devices. They're very sensitive to welding current.

3.2.0 Starting Up and Shutting Down Equipment

CAUTION

Before starting *any* welding machine, check that the electrode holder is resting on a safe surface. If it's touching a metal surface, it could arc when you start the machine.

Welding machines that connect to an outlet are easy to start. On the other hand, starting an engine-driven machine requires a few more steps. The following sections outline starting up and shutting down both machine types.

3.2.1 Connecting and Starting a Regular Machine

Plug a regular welding machine into an appropriate outlet. Confirm that the outlet supplies the correct voltage and has the correct number of phases. Also confirm that it can supply enough current for the machine. Check the machine's electrical nameplate if you're uncertain.

Light-duty welding machines plug into a 120VAC or 240VAC single-phase outlet. These have three or four prongs. Medium- and heavy-duty machines plug into a 240VAC, 480VAC, or 575VAC three-phase outlet. These have four prongs in one of several arrangements (*Figure 30*). The arrangement matches the outlet's current capacity and prevents you from plugging an inappropriate machine into the outlet. Workstation welding machines may be wired directly to a dedicated circuit.

Figure 30 Single- and three-phase plugs.
Source: The Lincoln Electric Company, Cleveland, OH, USA

WARNING!

Whether you're using a permanently wired machine or one that connects with a plug, always identify the nearest electrical disconnect before starting work. In an emergency, you can cut power to the machine from this point.

If a welding machine doesn't have a plug attached to its cable, have a qualified electrician install one. Alternatively, an electrician can connect it directly to a circuit breaker. Never wire a welding machine yourself.

A machine's startup procedure varies by brand and model. *Figure 31* shows a newer welding machine that uses buttons for most of its controls. Other machines, particularly older models, may use knobs instead.

The following startup sequence applies to the machine in *Figure 31*. Other machines have similar controls and steps. If you're uncertain about a particular machine's controls, consult its manual.

Polarity Selection

When operating in DC mode, you'll need to select the machine's polarity. Some machines have a polarity selector switch. This machine always operates in DC mode and doesn't have a polarity selector. Select its polarity (DCEP or DCEN) by connecting the welding leads to the appropriate terminals.

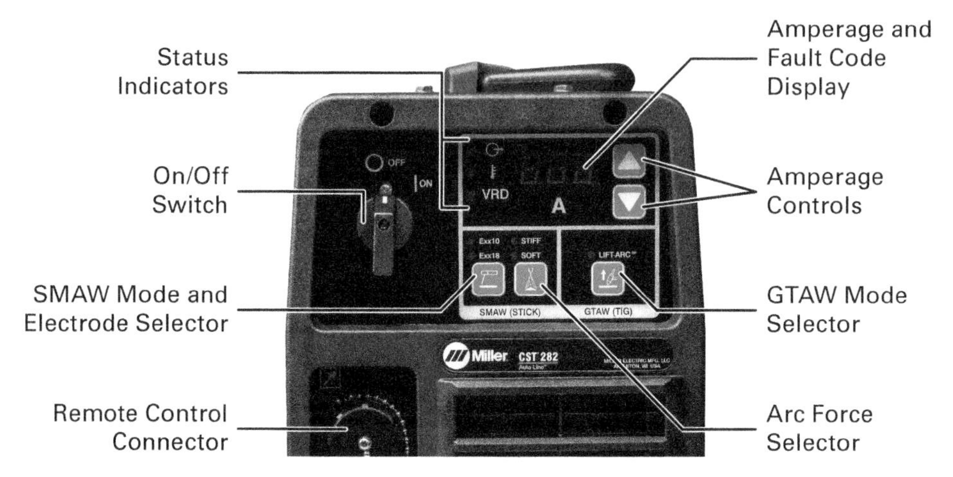

Figure 31 Regular welding machine controls.
Source: © Miller Electric Mfg. LLC

Remote Control

Some machines can use a remote control that lets the welder position the machine away from the workpiece. Remote controls adjust the amperage and control other operations. This machine has a connector for an optional remote control. If you're going to use it, connect it before turning on the machine. Some machines have a switch that enables or disables remote control.

NOTE

If you're not going to use the machine for a while after you finish welding, turn it off. Then unplug it or de-energize the circuit with the disconnect switch.

Power Control

The power switch turns the machine on and off. Turn the machine on so you can complete the setup process. When you're finished welding, switch the machine off.

Welding Process Selection

Many welding machines support several welding processes. This machine has SMAW (stick) and GTAW (TIG) modes. Press the SMAW mode button.

NOTE

Some machines support both AC and DC SMAW modes. If your machine offers this choice, use the current type selector switch to choose between them.

Electrode Selection

This machine lets you specify the electrode's *classification*. Doing this adjusts the machine for the best performance with the chosen electrode. Not all machines have this feature. Press the SMAW mode button until the indicator next to the right electrode classification lights up.

NOTE

NCCER Module 29108, *SMAW Electrodes*, introduces electrode types and explains AWS electrode classifications.

Arc Force Selection

This machine lets you select the *arc force*. Not all machines have this feature. A "soft" arc is less aggressive and gives a smoother weld bead. A "stiff" arc is more aggressive and produces a bigger weld pool. Press the arc force selector button until the indicator next to the desired choice lights up.

Amperage Adjustment

Finally, select the correct welding amperage for the electrode and material thickness. On this machine, press the UP and DOWN arrows until the correct amperage appears on the display. Many machines use a knob to select the amperage. This machine also displays fault codes on the display if something goes wrong.

3.2.2 Starting an Engine-Driven Machine

Starting an engine-driven welding machine requires a few preliminary steps. Some facilities have a pre-start checklist that you must complete and sign. Even if your facility doesn't require this, performing a basic pre-start check is wise. Complete the following checklist:

- *Engine oil* — If the engine has run recently, wait at least 10 minutes. This gives the oil time to drain back into the crankcase. Then check the level with the dipstick or oil sight glass. If the oil is low, add the correct oil grade. Don't overfill.

- *Engine coolant* — If the engine is liquid-cooled, check the coolant level. Ideally, you should check when the engine is cold. If the engine has run recently, let it cool first. Never remove the radiator cap on a hot engine. If the coolant level is low, add the correct coolant. Make sure to seat the radiator cap properly.
- *Engine fuel* — Check the fuel level. Most engines have a fuel sight glass or gauge. If necessary, top up with the correct fuel. Usually, the fuel tank or its cap identifies the proper type. If not, ask your supervisor. Never guess!

WARNING!

Never add fuel to a hot engine as it could ignite. Wait for the engine to cool.

- *Engine battery* — If the battery isn't a sealed type, check the water level in each cell. Add distilled water to any that are low.

WARNING!

Always be very careful around batteries. Lead acid batteries give off explosive hydrogen gas. Be sure that they have enough ventilation so the hydrogen can dissipate quickly. Never weld near them or do any work that generates sparks.

- *Hour meter* — If the machine has an hour meter, record its reading on the checklist. An hour meter shows the machine's total operating hours. Maintenance personnel use these numbers to decide when to perform service tasks.
- *Engine appearance* — Always look over the engine before starting it. If you spot any damage or loose parts, have the problem fixed before continuing.
- *Fuel shutoff valve* — If the machine has one, open the fuel shutoff valve. You'll find it between the fuel tank and the carburetor or fuel-injection intake.

CAUTION

Don't add plain water to a radiator unless the manufacturer specifies it. Water mixed with antifreeze prevents the engine from freezing in cold weather. It also protects the cooling system from corrosion. Adding plain water dilutes the antifreeze, reducing its effectiveness. The engine block, radiator, or water pump could freeze on a cold day.

CAUTION

Adding the wrong fuel to an engine can damage it or even start a fire. Never substitute one fuel for another unless the manufacturer specifically authorizes it.

The engine's startup procedure varies by brand and size. Larger engines have an ignition switch and possibly a starter button. Some start with a key. Small engines may have a pull rope like a lawn mower.

After starting the engine, let it warm up for 5–10 minutes before welding. Complete the machine's setup while you wait. *Figure 32* shows an older engine-driven welding machine's controls.

The following startup sequence applies to the machine in *Figure 32*. Other machines have similar controls and setup steps. If you're uncertain about a particular machine's controls, consult its manual.

Choke

This engine has a choke knob that you should pull out if the engine is cold. If the engine is warm, leave it pushed in. Not all engines have a choke control. Choke operation varies with the engine's brand and model.

NOTE

Some welding machine models can come with one of several engine choices. Always identify your machine's engine and check the manual for the proper choke operation.

Engine Control Switch

Turn the engine control switch to the START position and hold it there until the engine starts. Once the engine is running, turn the switch to the AUTO IDLE or HIGH IDLE position. Push the choke knob fully in. When you're finished welding, turn the engine control switch to the OFF position and wait for the engine to stop.

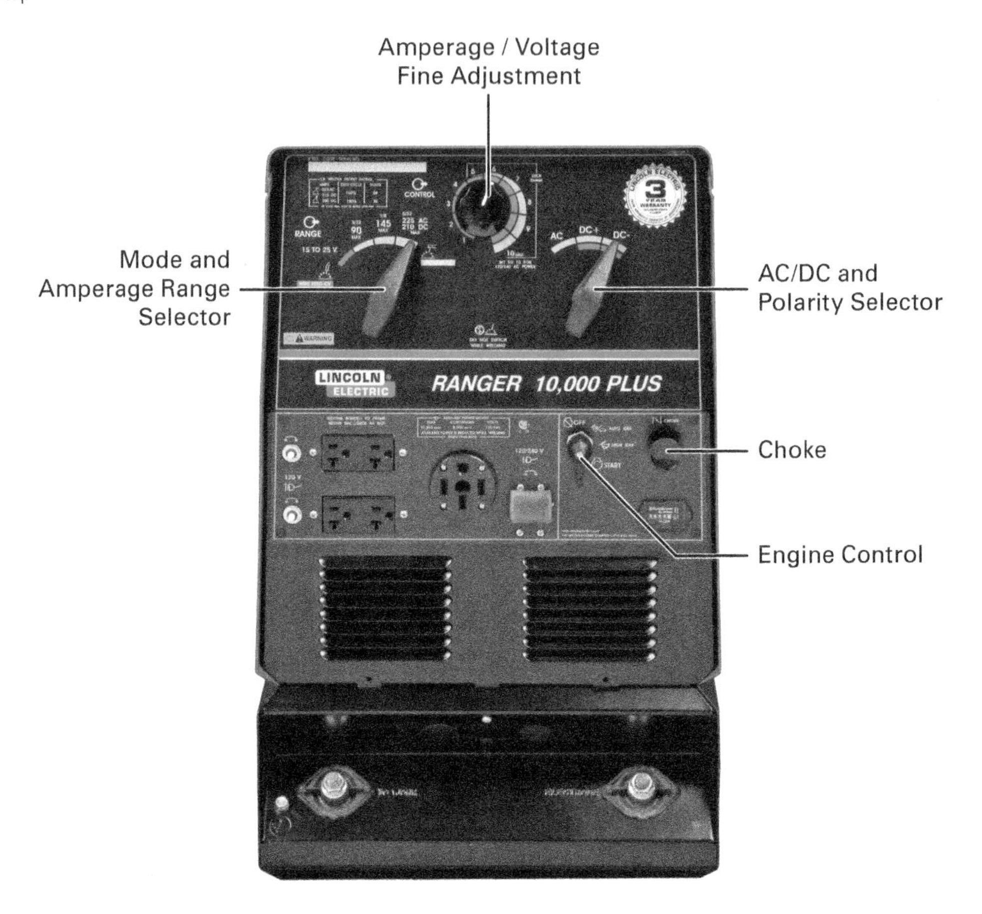

Figure 32 Engine-driven welding machine controls.
Source: The Lincoln Electric Company, Cleveland, OH, USA

Current Type and Polarity Selection

Select the welding current type by turning the selector knob to AC or one of the two DC positions. When operating in DC mode, you'll need to select the machine's polarity. Turn the selector knob to DC+ for DCEP and DC– for DCEN.

Welding Process Selection

Many welding machines support several welding processes. This machine has SMAW (stick), GMAW (MIG), and GTAW (TIG) modes. Turn the mode / amperage range selector to any of the positions marked in blue (constant current). The position marked in red is for constant voltage welding processes.

Amperage Adjustment

Selecting the correct welding amperage is a two-step process. First, select the amperage range by turning the mode / amperage range selector to one of the three range choices marked in blue. Tune the amperage by turning the amperage / voltage fine adjustment knob to any of the positions marked in blue.

NOTE

On this machine, positions marked in red are for constant voltage welding processes like GMAW. In this mode, the amperage / voltage fine adjustment knob sets the welding voltage rather than amperage. Most multiprocess welding machines have similar controls.

3.2.3 Stopping the Engine

When you're finished welding or won't be working for 30 minutes or more, shut down the machine. Turn off the ignition switch and wait for the engine to stop. If there is one, close the fuel valve.

WARNING!

The engine will be hot for some time after you stop it. It can burn you or damage the welding cables if they touch it. A hot engine can also start a fire if it comes near flammable materials. Allow it to cool before storing the machine. Keep the welding cables away from hot surfaces.

3.3.0 Caring for SMAW Equipment

Standard welding machines require relatively little maintenance. You should, however, look over the machine after using it. If you notice damage or wear, have it repaired immediately. Regularly check the machine's air filter and replace it if it's clogged or excessively dirty. If the fan is covered with dust, clean it. Check welding cables, electrode holders, and workpiece clamps for damage. Repair or replace anything that isn't in good condition.

Engine-driven welding machines require regular preventive maintenance to keep them running properly. Most facilities follow a preventative maintenance schedule based on the machine's operating hours. If the working environment is severe, maintenance may be more frequent.

In some companies, welders perform their own maintenance. Others have dedicated maintenance personnel. Never perform tasks for which you're not qualified. Generally, a qualified service technician should handle engine repairs. Typical routine maintenance tasks include the following:

- Checking/changing the filters (air, fuel, and oil)
- Checking/changing the oil
- Checking/changing the coolant
- Maintaining/replacing the spark plugs
- Maintaining the battery
- Replacing worn drive belts
- Cleaning the engine surfaces
- Greasing the undercarriage
- Repacking the wheel bearings

NOTE

Preventative maintenance should follow the manufacturer's guidelines. Consult the welding machine's manual for this information. Always follow your company's specific procedures as well.

Going Green

Engine-Driven Machine Maintenance

Properly maintained engine-driven machines run smoothly and pollute less. If a machine leaks, smokes excessively, or runs badly, have it serviced.

3.0.0 Section Review

1. When lifting a welding machine, connect the lifting eye to the lifting hook with a _____.
   a. cable lug and crimp
   b. sling and shackle
   c. quick disconnect
   d. swivel clamp

2. A welding machine's power cable doesn't have a plug. What should you do?
   a. Splice the cable to another cable that has a suitable plug.
   b. Obtain the correct plug type and install it yourself.
   c. Wire the machine directly to an unused circuit breaker.
   d. Have a qualified electrician install a plug.

3. What usually determines when maintenance personnel perform specific tasks on an engine-driven welding machine?
   a. The machine's general appearance
   b. The machine's current operating hours
   c. The machine's fuel consumption rate
   d. The machine's warm-up time

Module 29107 Review Questions

1. Welders often refer to SMAW as _____.
   a. HF welding
   b. stick welding
   c. gap welding
   d. resistive welding

2. An electric current is a flow of electrons that's measured in _____.
   a. volts (V)
   b. watts (W)
   c. charges (Q)
   d. amperes (A)

3. The number of cycles that an alternating current completes in one second is its _____.
   a. polarity
   b. travel
   c. frequency
   d. current

4. SMAW machines *must* supply _____.
   a. bipolar DC
   b. constant current
   c. regulated AC
   d. constant voltage

5. An electronic device that converts alternating current to direct current is a(n) _____.
   a. conductor
   b. rectifier
   c. alternator
   d. generator

6. What do welding machines contain that reduces their input voltage to the lower value that they deliver to the electrode?
   a. A governor
   b. A transcoder
   c. A bipolar varactor
   d. A step-down transformer

7. The voltage present at the leads when a welding machine is on but not welding is the _____.
   a. open-circuit voltage
   b. operating voltage
   c. zero voltage
   d. floating voltage

8. Which of the following terms applies *only* to DC?
   a. Current
   b. Polarity
   c. Amperage
   d. Voltage

9. How frequently should you inspect fall arrest and fall prevention equipment?
   a. Every week
   b. Every time that you use it
   c. Every month
   d. Every other day that you use it

10. Many transformer-rectifier SMAW machines can deliver _____.
    a. AC welding current only
    b. DC welding current only
    c. AC and DC welding current
    d. high-frequency pulse current

11. What kind of welding machine is much lighter than a transformer-rectifier machine of the same capacity?
    a. A transformer welding machine
    b. An inverter welding machine
    c. An eight-pack welding machine
    d. An AC welding machine

12. A welding machine's duty cycle is based on a _____.
    a. 10-minute period
    b. 15-minute period
    c. 20-minute period
    d. 30-minute period

13. Which of the following welding cables has the *largest* diameter and the *greatest* amperage capacity?
    a. #8
    b. #5
    c. #3
    d. #4/0

14. What can you do to prevent welding current from passing through sensitive parts?
    a. Move the workpiece clamp.
    b. Switch polarity from DCEP to DCEN.
    c. Use an AC welding machine.
    d. Use an inverter welding machine.

15. Machines requiring three-phase power will have a _____.
    a. two-prong plug
    b. three-prong plug
    c. four-prong plug
    d. five-prong plug

Answers to odd-numbered Module Review Questions are found in *Appendix A*.

Answers to Section Review Questions

Answer	Section Reference	Objective
Section 1.0.0		
1. b	1.1.3	1a
2. d	1.2.1; *Figure 10*	1b
Section 2.0.0		
1. d	2.1.0	2a
2. d	2.2.1	2b
3. b	2.3.0	2c
Section 3.0.0		
1. b	3.1.3	3a
2. d	3.2.1	3b
3. b	3.3.0	3c

User Update

Did you find an error? Submit a correction by visiting **https://www.nccer.org/olf** or by scanning the QR code using your mobile device.

Source: The Lincoln Electric Company, Cleveland, OH, USA

Objectives

Successful completion of this module prepares you to do the following:

1. Summarize SMAW electrodes and their classification system.
   a. Describe the AWS filler metal specification system and electrode characteristics.
   b. Describe the four main electrode groups.
2. Summarize how to select, handle, and care for electrodes.
   a. Describe selecting the correct electrode.
   b. Describe properly handling and storing electrodes.

Performance Tasks

This is a knowledge-based module. There are no Performance Tasks.

Overview

Shielded metal arc welding (SMAW) uses an electric current to melt the base metal. At the same time, the electrode delivering the current melts and mixes with the molten base metal. This combination of base and filler metals forms the weld. Since the electrode becomes part of the weld, its properties are extremely important. This module explores SMAW electrodes and the system that classifies them.

Industry Recognized Credentials

If you are training through an NCCER-accredited sponsor, you may be eligible for credentials from NCCER's Registry. The ID number for this module is 29108. Note that this module may have been used in other NCCER curricula and may apply to other level completions. Contact NCCER's Registry at 1.888.622.3720 or go to **www.nccer.org** for more information.

You can also show off your industry-recognized credentials online with NCCER's digital badges. Transform your knowledge, skills, and achievements into badges that you can share across social media platforms, send to your network, and add to your resume. For more information, visit **www.nccer.org**.

NOTE

This module uses US standard and metric units in up to three different ways. This note explains how to interpret them.

Exact Conversions

Exact metric equivalents of US standard units appear in parentheses after the US standard unit. For example: "Measure 18" (45.7 cm) from the end and make a mark."

Approximate Conversions

In some cases, exact metric conversions would be inappropriate or even absurd. In these situations, an approximate metric value appears in parentheses with the ~ symbol in front of the number. For example: "Grip the tool about 3" (~8 cm) from the end."

Parallel but not Equal Values

Certain scenarios include US standard and metric values that are parallel but not equal. In these situations, a slash (/) surrounded by spaces separates the US standard and metric values. For example: "Place the point on the steel rule's 1" / 1 cm mark."

Digital Resources for Welding

Scan this code using the camera on your phone or mobile device to view the digital resources related to this craft.

1.0.0 Shielded Metal Arc Welding Electrodes

Performance Tasks

There are no Performance Tasks in this section.

Objective

Summarize SMAW electrodes and their classification system.

a. Describe the AWS filler metal specification system and electrode characteristics.
b. Describe the four main electrode groups.

NOTE

Other welding processes use electrodes too. But they come in different forms, such as wire rolls. Some processes, such as GTAW, don't consume the electrode.

Shielded metal arc welding (SMAW), commonly called *stick welding*, relies on rod-shaped metal electrodes. These not only carry the welding current to the workpiece, but they also provide the weld's filler metal. Welders call them *sticks* or *rods*. The industry itself calls them *consumables* since the welding process uses them up.

Manufacturers produce SMAW electrodes for many different metals and applications. To produce a good weld, the welder must select the correct electrode. Trainee welders learn to interpret an electrode's classification code, so they can do this task correctly. They also learn how to handle and store electrodes, so the electrodes will perform properly.

1.1.0 Electrodes and the AWS Specification System

NOTE

SMAW produces temperatures around 6,500°F (~3,600°C), which can melt most common metals.

Flux: A material that dissolves or inhibits oxides and protects a weld joint from the atmosphere.

Alloy: A metal containing additional elements that significantly change its properties.

All SMAW electrodes have a metal core with an outer **flux** coating. The core carries the welding current from the electrode holder (*SMAW torch* or *stinger*) to the workpiece. The arc at the electrode's tip melts the base metal, as well as the electrode itself. The electrode's filler metal mixes with the molten base metal, forming the weld.

Meanwhile, the flux coating melts, and some of it vaporizes. The vaporized flux contains gases like carbon dioxide that "shield" the weld. The shielding gases displace the atmosphere (push away the air). It contains oxygen and moisture. Both can harm the weld by causing defects.

Melted flux also flows onto the weld itself, dissolving oxides and other contaminants. These too can harm the weld. As the flux cools, it forms a *slag* above the weld. The slag protects the cooling weld from the atmosphere. It also slows the cooling process.

Some fluxes contain powdered metals. These melt and mix with the filler and base metals. The resulting **alloy** gives a strong weld with the right mechanical properties and appearance. Fluxes also make the welding arc more stable. A stable arc produces a smoother weld with less spatter. *Figure 1* shows the SMAW process.

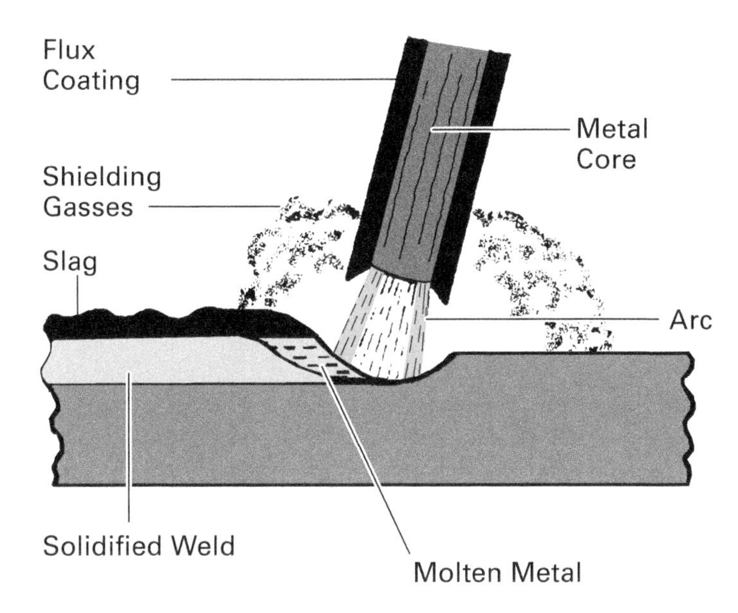

Figure 1 The SMAW process.

1.1.1 The AWS Filler Metal Specification System

The American Welding Society (AWS) creates specifications for many welding consumables, including SMAW electrodes. Specifications begin with the code *AWS A5.xx*, followed by a number in place of the *xx*. Major welding codes cite AWS specifications, and most industries accept them.

Specifications set standards that all welding consumable manufacturers must follow. This guarantees a consistent product, regardless of brand. Welders can select consumables like electrodes and be confident that they'll perform as expected.

A specification defines the consumable's chemical composition. It also defines the deposited weld metal's mechanical properties. Finally, the specification includes a classification code that welders interpret to identify the consumable.

The following are sample AWS specifications associated with SMAW electrodes:

- *A5.1* — Specification for Carbon Steel Electrodes for Shielded Metal Arc Welding
- *A5.3* — Specification for Aluminum and Aluminum Alloy Electrodes for Shielded Metal Arc Welding
- *A5.4* — Specification for Stainless Steel Electrodes for Shielded Metal Arc Welding
- *A5.5* — Specification for Low-Alloy Steel Electrodes for Shielded Metal Arc Welding
- *A5.6* — Specification for Covered Copper and Copper-Alloy Arc Welding Electrodes
- *A5.11* — Specification for Nickel and Nickel-Alloy Welding Electrodes for Shielded Metal Arc Welding
- *A5.15* — Specification for Welding Electrodes and Rods for Cast Iron
- *A5.16* — Specification for Titanium and Titanium-Alloy Welding Electrodes and Rods
- *A5.21* — Specification for Bare Electrodes and Rods for Surfacing
- *A5.24* — Specification for Zirconium and Zirconium-Alloy Welding Electrodes and Rods

NOTE

Some industries require other or additional specifications. These include the American Bureau of Shipping (ABS), the US Coast Guard, the Federal Highway Administration, Military Specifications (MIL), and Lloyds of London. The American Society of Mechanical Engineers (ASME) Boiler Code and the European Norm (EN) electrode specifications closely follow the AWS specifications.

NOTE

The specification's revision year follows its number (*AWS A5.5:2014*, for example). Always refer to a specification's most recent edition.

NOTE

All electrodes used for code-governed work must have the AWS classification printed on them, as well as on their container. Some codes require additional classifications from other standards. For example, ASME code work requires electrodes marked with both the AWS classification and an ASME classification.

Over time, welders become familiar with these and other specifications. For now, the most important are *AWS A5.1* and *AWS A5.5*. These define SMAW electrodes used with carbon steels and low-alloy steels. This module introduces these specifications.

1.1.2 *AWS A5.1* Classification System

The *AWS A5.1* specification, which covers carbon steel electrodes, includes a classification system. It identifies electrodes by a series of letters and numbers. Manufacturers stamp the classification just above the place where the electrode holder grips the electrode (*Figure 2*).

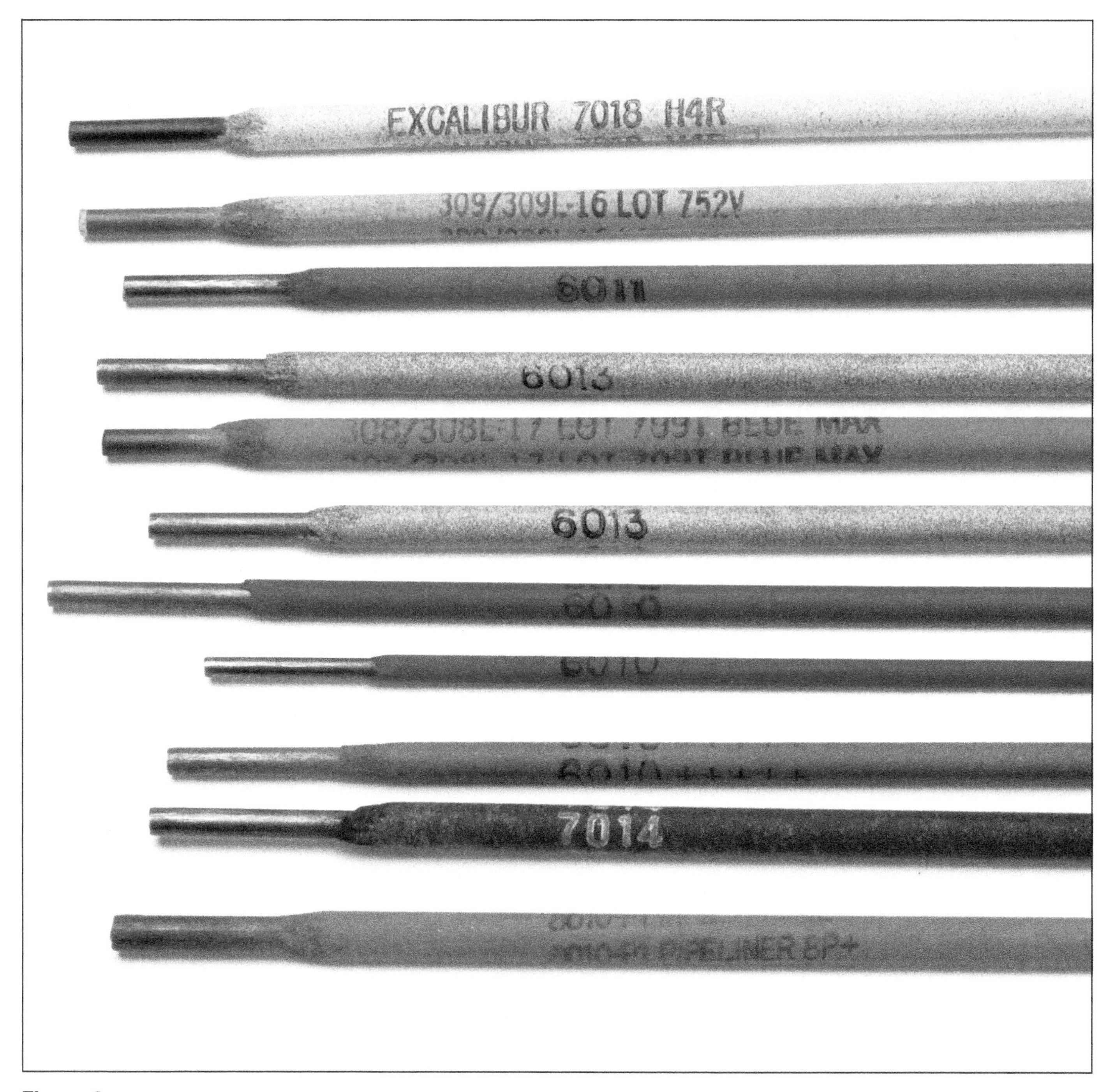

Figure 2 Electrode classifications.
Source: The Lincoln Electric Company, Cleveland, OH, USA

Each classification begins with an E ("electrode"), followed by four or five digits. As *Figure 3* shows, these digits divide into three groups identifying the electrode's important properties.

With four-digit identifiers, the first two digits specify the deposited weld metal's **ultimate tensile strength**. For five-digit identifiers, the first three digits specify the tensile strength. Multiply the number by 1,000 to get the tensile strength in pounds per square inch (psi). For example, if the identifier is E6010, the tensile strength is 60,000 psi (60 × 1,000).

Ultimate tensile strength: The amount of pulling force a material can tolerate before breaking. Often shortened to *tensile strength*.

The third (or fourth) digit identifies the electrode's permissible weld positions:

1 — All positions permissible

2 — Flat and horizontal fillet weld positions only

4 — Flat, horizontal, overhead, and vertical-down positions only

The fourth (or fifth) digit identifies the flux type and its welding current requirements. SMAW welding currents may be AC, direct current electrode positive (DCEP), or direct current electrode negative (DCEN).

Table 1 shows example electrodes covered by the *AWS A5.1* specification. In the *AWS Class* column, the two Xs after the E represent the two- or three-digit tensile strength identifier. The remaining two digits identify the recommended welding position and the flux type. Examine this table carefully until you understand the classification system.

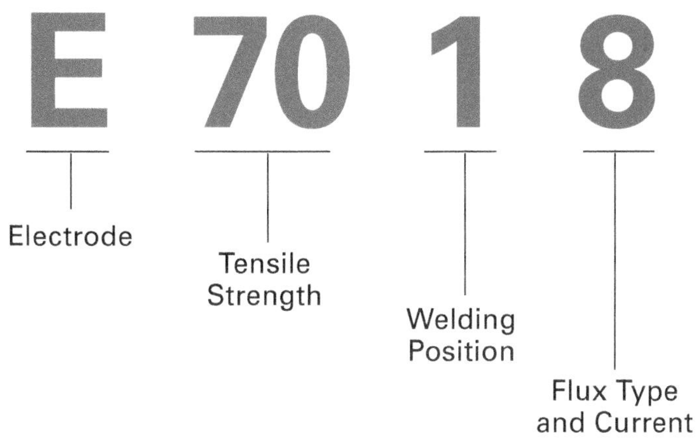

E 70 1 8

Electrode

Tensile Strength

Welding Position

Flux Type and Current

Figure 3 Interpreting an *AWS A5.1* electrode classification.

TABLE 1 Carbon Steel Electrodes

AWS Class	Position	Flux Coating	Current Requirements	Characteristics
EXX10 EXX20	All Flat, horizontal fillet	Cellulose sodium	DCEP	Deep penetration, flat beads
EXX11	All	Cellulose potassium	AC, DCEP	Deep penetration, flat beads
EXX12	All	Titania sodium	AC, DCEN	Medium penetration
EXX13	All	Titania potassium	AC, DCEP, DCEN	Shallow penetration
EXX14 EXX24	All Flat, horizontal fillet	Titania iron powder	AC, DCEP, DCEN	Medium penetration, fast deposit
EXX15	All	Low-hydrogen sodium	DCEP	Moderate penetration
EXX16	All	Low-hydrogen potassium	AC, DCEP	Moderate penetration
EXX27	Flat, horizontal fillet	Iron powder, iron oxide	AC, DCEP, DCEN	Medium penetration
EXX18 EXX28	All Flat, horizontal fillet	Low-hydrogen iron powder	AC, DCEP	Shallow to medium penetration

1.1.3 *AWS A5.5* Classification System

The *AWS A5.5* specification, which covers low-alloy steel electrodes, also includes a classification system. It's almost identical to the *AWS A5.1* system but has a *suffix* of several additional characters after the main classification (*Figure 4*).

The suffix's first character is a letter that identifies the electrode's alloy type. Alloys contain a mix of metals and other elements:

A — carbon-molybdenum alloy steel
B — chromium-molybdenum alloy steel
C — nickel steel alloy
D — manganese-molybdenum alloy steel
G — other alloys with minimal elements
M — military specification

The suffix's second character is a number. It identifies the alloy's chemical composition based on the percentage of the following elements:

- Carbon (C)
- Chromium (Cr)
- Manganese (Mn)
- Molybdenum (Mo)
- Nickel (Ni)
- Phosphorus (P)
- Silicon (Si)
- Sulfur (S)
- Vanadium (V)

Figure 5 shows *AWS A5.5* electrodes listed by alloy content. These may change when the AWS updates the specification. Always consult the most recent edition for the most accurate information.

NOTE

Not all *AWS A5.5* classifications have two characters in the suffix. Some have just one. Others have three characters. The third character, if present, further identifies the alloy's characteristics. Consult the *AWS A5.5* specification for more information.

CAUTION

Select electrodes very carefully. A 7018-B2 electrode is *not* the same as a 7018 electrode. It contains a different alloy that will produce a weld with different chemical and mechanical properties. Never assume that electrodes with similar numbers are interchangeable.

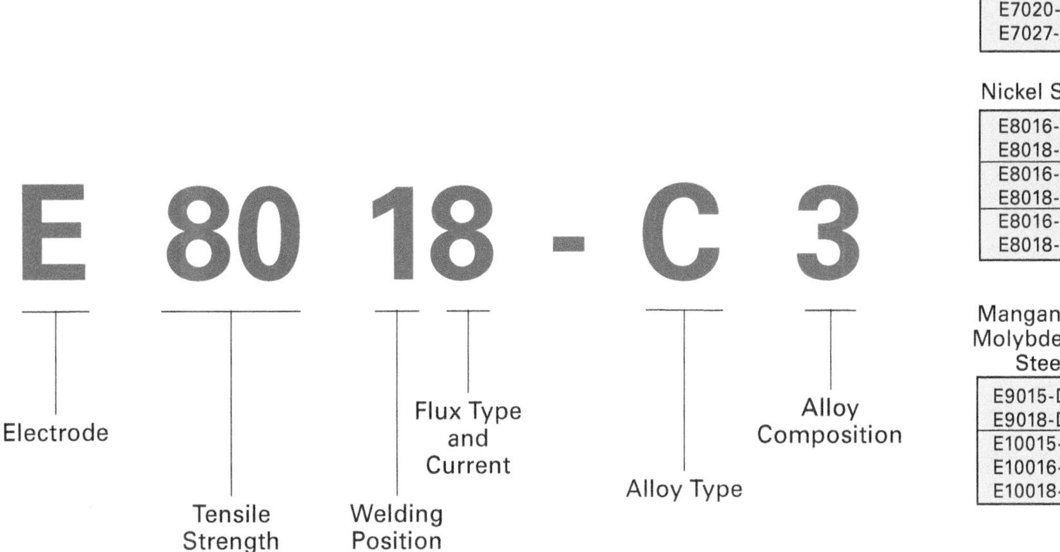

Figure 4 Interpreting an *AWS A5.5* electrode classification.

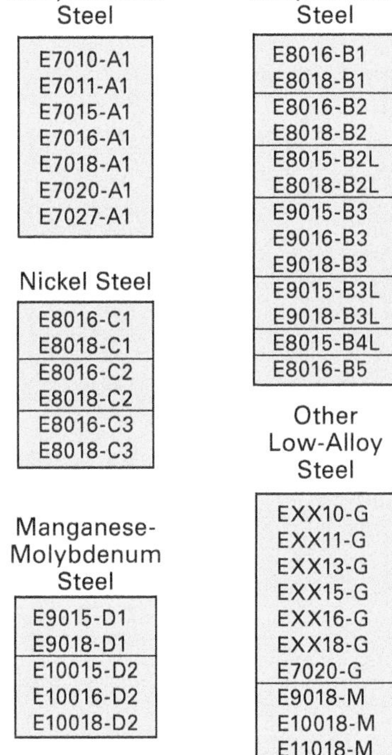

Figure 5 Electrodes listed by alloy.

1.1.4 Manufacturers' Classification Systems

Besides the AWS classification, electrode manufacturers generally print their own unique identifier on the electrode and its container. Consult the manufacturer's catalog or website to interpret these.

Be aware that manufacturers may sell several different electrode brands with the same AWS classification. For example, Hobart offers E6010 electrodes in both its Hobart® 610 and Pipemaster® Pro-60 brands. Lincoln Electric sells E6010 electrodes under four different brands: Fleetweld® 5P®, Fleetweld® 5P+®, Fleetweld® 5P®-RSP, and Pipeliner® 6P+.

Manufacturers offer multiple brands because they provide specific advantages for certain applications. Lincoln Electric's Fleetweld® 5P® works well for general fabrication and maintenance welding, as well as for pipe. But Fleetweld® 5P+® offers additional advantages for pipe. It produces welds with better resistance to **heat-affected zone** cracking.

When a manufacturer offers the same AWS electrode in several brands, consult its catalog or website to understand their differences. When in doubt, check with your supervisor. *Table 2* shows some examples. Remember that manufacturers often change trade names and product specifications.

Heat-affected zone: Unmelted base metal whose structure has been changed by heating.

TABLE 2 Brand Names and AWS Classifications

Manufacturer	AWS Classification			
	E6010	E6011	E7014	E7018
Hobart	Hobart® 610, Pipemaster® Pro-60	Hobart® 335A	Hobart® 14A	—
ESAB	Pipeweld® 6010 Plus, Sureweld® 10P, Sureweld® 10P PLUS, Sureweld® 6010	Sureweld® 6011	Sureweld® 7014	Sureweld® 7018
Lincoln	Fleetweld® 5P®, Fleetweld® 5P+®, Fleetweld® 5P®-RSP, Pipeliner® 6P+	Fleetweld® 35, Fleetweld® 180, Fleetweld® 180-RSP, Murex® 6011C	Fleetweld® 47, Fleetweld® 47-RSP, Murex® 7014	Excalibur® 7018 MR®

1.1.5 SMAW Electrode Sizes

Carbon steel electrodes for SMAW work come in standard sizes ranging from 3/32" to 1/4" (2.4 mm to 6.4 mm) in diameter. The diameter selected depends on the weld and its amperage requirements. Electrodes come in 14" (~36 cm) and 18" (~46 cm) lengths. Manufacturers usually package them in 10 lb (~5 kg) and 50 lb (~25 kg) cans. Some offer them in smaller containers too (*Figure 6*).

The number of electrodes in a can depends on their diameter and length. *Table 3* shows the number per pound for several sizes. While each brand is slightly different, these values are typical.

NOTE

While manufacturers may include metric dimensions in their electrode specifications, they are only approximate values. AWS and ASME specifications are based on US Standard units. Included metric units are a courtesy to metric users, not exact dimensions.

TABLE 3 Number of Electrodes per Pound

	Diameter / Length					
Electrode	3/32" (2.4 mm) / 14" (36 cm)	1/8" (3.2 mm) / 14" (36 cm)	5/32" (4.0 mm) / 14" (36 cm)	3/16" (4.8 mm) / 14" (36 cm)	7/32" (5.6 mm) / 18" (46 cm)	1/4" (6.4 mm) / 18" (46 cm)
6010	30	17	12	8	—	—
6011	25	15	11	7	—	—
6013	25	15	10	7	—	—
7014	24	13	9	6	—	—
7024	—	10	7	—	4	2
7018	32	15	10	7	—	3

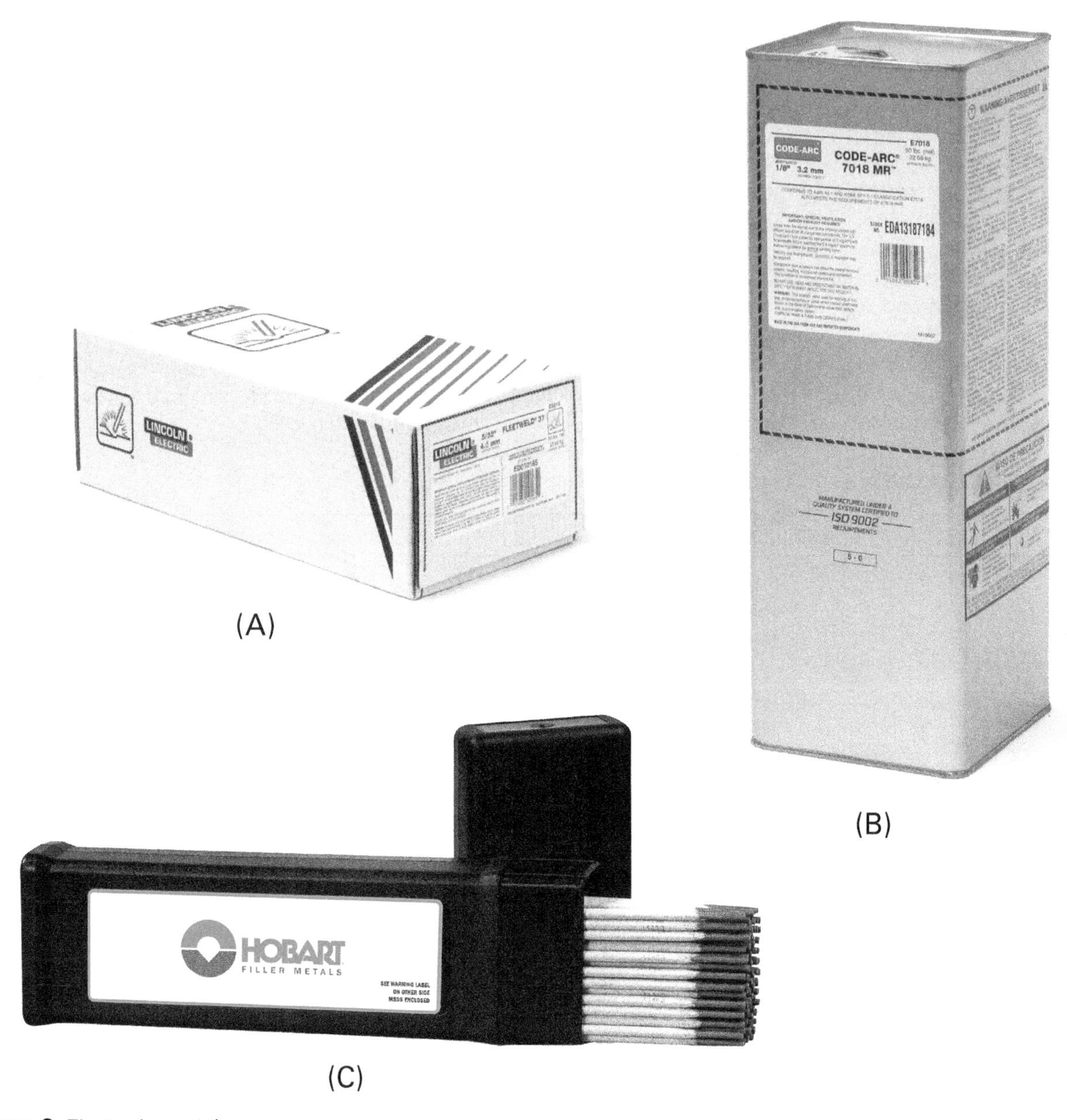

Figure 6 Electrode containers.
Sources: The Lincoln Electric Company, Cleveland, OH, USA (6A–6B); Courtesy of Hobart Brothers LLC (6C)

CAUTION

Meticulously follow the traceability requirements laid out in the jobsite quality control manuals. Not following them can have severe consequences. Be sure that you understand traceability requirements before beginning the job. If you're uncertain, speak to your supervisor, the project engineer, or a welding inspector.

Traceability: Verifying through documentation that the correct materials have been used to produce a weld.

1.1.6 Filler Metal Traceability Requirements

A job's Welding Procedure Specification (WPS) and applicable welding codes normally specify the electrode type for each weld. Being able to prove that the weld meets code specifications and the WPS is crucial. Proper documentation provides **traceability**, proving that the welder used the correct electrode for each weld.

Traceability requirements vary with client requirements and job specifications. Many projects have a quality control documentation package. Some follow the requirements in the *ASME Boiler and Pressure Vessel Code, Section II, Materials*. It lists acceptable base metal and filler metal combinations.

1.2.0 Electrode Groups

SMAW electrodes come in many types, each suitable for specific applications. Welders must understand their differences and select the right ones for the job. Electrodes divide into four main groups that define their key qualities:

F1 — Fast-fill
F2 — Fill-freeze
F3 — Fast-freeze
F4 — **Low-hydrogen**

Low-hydrogen: An electrode type manufactured to contain little or no moisture.

Notice that each group has an "F-number" associated with it. Under *AWS D1.1,* a welder who passes a qualification test with an electrode of a particular F-number automatically qualifies for all lower-numbered electrodes.

For example, a welder might take an AWS qualification test with an F4 electrode. The test requires making a V-groove weld with metal backing in the 3G and 4G positions. Passing this test qualifies the welder to make groove and fillet welds in all positions with an F1, F2, F3, or F4 electrode.

This principle may or may not be true with other codes. It also doesn't necessarily apply across different base metals. For example, under AWS, qualifying with an electrode for stainless steel doesn't qualify the welder for lower-numbered electrodes for carbon steel.

Table 4 shows each electrode group along with example AWS electrode classifications that fit into it. Refer to this table as you learn about each electrode group.

NOTE

Each welding code has its own rules for qualifying with electrodes. Always consult the specific code for qualification rules. Don't assume that one code matches another.

TABLE 4 F-Number and AWS Electrode Classification

Group	AWS Electrode Classification				
F1 (Fast-Fill)	EXX20	EXX24	EXX27	EXX28	
F2 (Fill-Freeze)	EXX12	EXX13	EXX14	EXX22	
F3 (Fast-Freeze)	EXX10	EXX11			
F4 (Low-Hydrogen)	EXX15	EXX16	EXX18	EXX28	EXX48

1.2.1 Fast-Fill (F1) Electrodes

These electrodes have powdered iron in their flux, giving them a high deposition rate. They have shallow penetration, excellent weld appearance, and almost no spatter. Their heavy slag coating removes easily. But they're only suitable for flat welds and horizontal fillet welds.

The first row in *Table 4* shows example electrode classifications. Notice that each has a number *2* in the location that identifies the permissible weld position. Fast-fill electrodes are suitable for production welds on plate thicker than ¼" (~6 mm), as well as flat and horizontal fillets and lap welds.

Electrode Safety Data Sheets

Every electrode container must include a Safety Data Sheet (SDS). *Appendix 29108A* shows an example electrode SDS. Consult the SDS for all safety-related questions.

1.2.2 Fill-Freeze (F2) Electrodes

These electrodes also go by the name *fast-follow electrodes.* They have a medium deposition rate and medium penetration. The weld bead ranges from smooth to evenly rippled. Their medium slag coating removes easily. Many fill-freeze electrodes work in all positions, although some don't.

The second row in *Table 4* shows example electrode classifications. Notice that most but not all have a number *1* in the location that identifies the permissible weld position. They work well with sheet metal and for the following applications:

- General-purpose welding
- Vertical fillet or lap welds
- Sheet metal fillet and lap welds
- Joints having poor fit-up in the flat position

1.2.3 Fast-Freeze (F3) Electrodes

These electrodes have deep penetration. Unfortunately, they also produce a lot of fine spatter. The weld bead is flat with distinct ripples. Their light slag coating resists removal. A fast-freeze electrode works in all positions.

The third row in *Table 4* shows some example electrode classifications. Notice that each has a number *1* in the location that identifies the permissible weld position. A fast-freeze electrode's arc is easy to control, so they're good for overhead and **vertical welding**. Welders often find it difficult to maintain a precise arc gap in these positions. They work well for the following applications:

Vertical welding: Welding with an upward or downward progression.

- Vertical and overhead plate welding
- Pipe welding
- Joints requiring deep penetration, such as square butts
- Sheet metal welds
- Welds on galvanized, plated, or painted surfaces

WARNING!

Many surface coatings and platings, including the zinc in galvanized steel, produce toxic fumes when welded. Some electrodes also produce hazardous fumes. Read the SDS that accompanies the electrodes to determine if they contain hazardous materials. Whenever you work with toxic metals or coatings, use appropriate respiratory PPE. Be sure those around you are also protected.

1.2.4 Low-Hydrogen (F4) Electrodes

Welding medium- and high-carbon steel, as well as steel containing sulfur or phosphorus, creates a special challenge. Hydrogen entering the weld will create defects like porosity and cracks under the weld bead. Hydrogen comes from the water in the air, which breaks down into hydrogen and oxygen during welding.

Low-hydrogen electrodes are very dry and contain nearly no moisture. The manufacturer ships them in **hermetically sealed** (airtight) containers. Once exposed to the air, they begin to absorb moisture. Unless they're used within a specific time limit, they won't produce good welds. To prevent this problem, welders store them in an oven. The hot environment keeps the electrodes dry until they're needed.

Hermetically sealed: An airtight seal.

The proper storage temperature depends on the welding code driving the project. Some codes, for example, specify a minimum temperature of 250°F (~120°C). Others have different rules. You should also consult the electrode manufacturer's guidelines.

Some codes permit "re-baking" electrodes that have been outside the oven for too long. This process drives out moisture and returns the electrodes to their proper condition. The re-bake temperature is much higher than the storage temperature—500°F to 1,000°F (~260°C to 540°C), depending on the code. Manufacturers specify the correct re-bake temperature for each electrode type. Follow these recommendations. Electrodes may be re-baked only once.

CAUTION

Welding codes establish strict guidelines for storing, handling, and re-baking low-hydrogen electrodes. Jobsite quality standards and the WPS reference these codes. Violating them will result in severe penalties. Improperly storing low-hydrogen electrodes leads to poor-quality welds and failed weld inspections. Avoid these problems by reading and understanding your jobsite standards before working with low-hydrogen electrodes.

Despite their special handling requirements, low-hydrogen electrodes offer a lot of advantages. Used for demanding and critical applications, they produce dense welds with excellent **notch toughness** and **ductility**. They're available with fast-fill and fill-freeze characteristics too.

Notch toughness: A material's ability to resist breaking at a point of concentrated stress.

Ductility: A material's ability to be bent, shaped, or stretched without breaking.

The fourth row in *Table 4* shows example electrode classifications. Notice that they can have any number in the permissible weld position location. Be sure to pay attention to this number so you'll know which welding positions are acceptable. Low-hydrogen electrodes work well for the following applications:

- X-ray-quality welds
- Welds requiring excellent mechanical properties/high tensile strength
- Crack-resistant welds in medium- and high-carbon steels
- Welds that resist hot short cracking in phosphorus steels
- Welds that minimize porosity in sulfur steels
- Welds that minimize cracking in thick or highly restrained mild or alloy steels
- Multi-pass welds in all positions

Low-hydrogen electrodes can have special suffixes after the main classification. An M indicates compliance with military codes. An H plus a number identifies the amount of hydrogen in the electrode. The number indicates how many milliliters (mL) of hydrogen will be present in 100 g of weld metal. Finally, an R indicates that the electrode is moisture resistant. An electrode with an R suffix will stay dry longer than a non-R electrode under the same conditions.

Figure 7 shows an E7018-H4R electrode. It has a tensile strength of 70,000 psi (70 × 1,000) and will work in any welding position (1). It has a flux type of 8. It contains 4 mL of hydrogen per 100 g of weld metal. It's moisture resistant.

Figure 7 Low-hydrogen electrode with suffix.
Source: The Lincoln Electric Company, Cleveland, OH, USA

Portable Electrode Storage

On the jobsite, welders sometimes keep moisture-sensitive electrodes in a portable rod oven. The example shown maintains a fixed temperature of 300°F (~150°C). It automatically lifts the electrodes when the welder opens the lid. This feature is handy since welders usually wear gloves. Some manufacturers also offer non-powered rod storage containers that help keep electrodes dry. These contain a chemical that absorbs water from the air.

Source: The Lincoln Electric Company, Cleveland, OH, USA

Electrode Guides

Most electrode manufacturers publish guides to help welders use their electrodes more effectively. These include recommended applications, amperages, and useful tips. Guides are available in printed and digital form. Visit the manufacturer's website to find them.

1.0.0 Section Review

1. What produces the slag coating on a new weld?
   a. Base metal impurities
   b. Electrode metal impurities
   c. The electrode's flux coating
   d. Hydrogen and oxygen from the air

2. An H8 suffix after an electrode's classification means that _____.
   a. there is 8 g of hydrogen in the electrode
   b. there is 8 mL of hydrogen per 100 g of weld metal
   c. the electrode is 8 times drier than a standard electrode
   d. the electrode can be exposed to the air for up to 8 hours

2.0.0 Electrode Selection and Care

Performance Tasks

There are no Performance Tasks in this section.

Objective

Summarize how to select, handle, and care for electrodes.

a. Describe selecting the correct electrode.
b. Describe properly handling and storing electrodes.

Selecting SMAW electrodes requires some thought, since they come in so many types and sizes. While project specifications and codes often specify the electrode, sometimes welders must make their own decisions. Properly caring for electrodes helps them perform properly, producing good welds without defects.

2.1.0 Selecting an Electrode

Many factors influence electrode selection. The welding machine's current type and amperage range set limits. The base metal, the welding position, and the weld type are important too. Of course, experience plays a part. *Table 5* shows some common carbon steel electrodes, along with their permissible welding positions, current types, and applications.

TABLE 5 Carbon Steel Electrode Guidelines

AWS Classification	Position	Current	Applications
E6011	All	AC, DCEP	Good for dirty or rusty steel
E6013	All	AC, DCEP, DCEN	Good for thin steel and poor fit-up; good for vertical-down welding
E6022	Flat, horizontal fillet	AC, DCEP, DCEN	Good for burn-through spot welds, roof decking, and sheet metal
E7014	All	AC, DCEP, DCEN	Good for thin steel
E7018	All	AC, DCEP	Good for high deposition; good for medium- or high-carbon steels

2.1.1 Codes and Welding Procedure Specifications

For many projects, codes and/or WPSs specify the filler metals. Specifications will include the electrode's AWS classification. Some will even specify a manufacturer or specific brand name. Always follow code requirements and WPSs.

2.1.2 Base Metal Type

To produce good welds that perform correctly, the base and filler metals must match chemically and mechanically. If they don't match, the welds could be defective, crack, or fail. To identify the base metal, consult its mill specifications.

If the project is code-driven, consult the appropriate code to determine the correct filler metal. Codes include tables to make this process easier. The *AWS D1.1/D1.1M, Table 5.3, Approved Base Metals for Prequalified WPSs* and *Table 5.4, Filler Metals for Matching Strength for Table 5.3, Groups I, II, III, and IV Materials – SMAW and SAW* are examples. The *ASME Boiler and Pressure Vessel Code, Section II, Parts A, B, and C, Materials* is another.

Electrode and metal manufacturers also offer tables and datasheets matching base metals to appropriate filler metals. Since it's in their best interest to have their products perform well, most do all they can to help welders make the right choices.

Common sense helps welders avoid poor choices too. Welding mild steel with a high tensile strength electrode like an E12018 (120,000 psi) is a bad idea. The weld is stronger than the base metal, so it won't give. Instead, it transfers stress to the heat-affected zone along the weld, which causes cracking. An E7018 (70,000 psi) is a better choice because the weld and base metal have similar strength.

Using a mild steel electrode like an E6010 (60,000 psi) to weld high tensile strength steel is also a mistake. The base metal is much stronger than the weld, so the weld will carry most of the stress and probably fail prematurely.

2.1.3 Base Metal Thickness

Normally, the electrode's diameter should be less than the base metal's thickness. For sheet metal, a 3⁄32" (2.4 mm) E6013 electrode works well since it has shallow penetration. Thick materials require a low-hydrogen electrode since it produces welds with good ductility. To minimize joint preparation, a deep penetration E6010 or E7010 electrode can be used for the root pass.

2.1.4 Base Metal Surface Condition

Always clean the base metal before welding. Remove surface corrosion, platings, coatings, paint, dirt, and oils. Use a deep penetrating electrode, such as an E6010, when this isn't possible.

WARNING!

Many surface coatings and platings, including the zinc in galvanized steel, produce toxic fumes when welded. Read the metal's SDS to determine if it contains hazardous materials. Whenever you work with toxic metals or coatings, use appropriate respiratory PPE. Be sure those around you are also protected. If you aren't sure about the metal, don't guess. Ask your supervisor for assistance.

2.1.5 Welding Position

Welding position affects electrode selection since some can't be used in certain positions. For example, the E7024 and E7028 only work in the flat and horizontal fillet weld positions. Never weld in vertical or overhead positions with electrodes larger than 3⁄16" (4.8 mm). For low-hydrogen and EXX14 electrodes, the limit is 5⁄32" (4 mm) in the vertical or overhead position. Whenever possible, position the joint for flat welding, so you can use the largest possible electrode.

2.1.6 Joint Design

Joint type heavily influences electrode selection. The E7024 or E7028 works well for lap and T- joints in the flat position. E6010 electrodes provide the deep penetration butt joints require. Shallow penetration electrodes like the E6012 or E6013 help with poor fit-up joints.

To use a low-hydrogen electrode with an open root joint, you must follow a special procedure. First, create the root pass and all additional passes. Next, back-gouge the root side to reach sound metal. Finally, produce the back weld. Alternatively, use metal backing or an E6010 electrode for the root weld. Then fill the weld with the low-hydrogen electrode.

2.1.7 Welding Current

All electrodes work with DC. Many will work with AC, although they may work better with DC. Some, such as the E6010 and E7010, won't work with AC at all. For the greatest versatility, use an SMAW machine that delivers a DC welding current.

Large-diameter electrodes require a high amperage. If you plan to use them, be sure that the welding machine can deliver enough current. Remember that many machines have a reduced duty cycle when operating near their maximum amperage. You may have to pause frequently to let the machine cool down.

2.2.0 Filler Metal Storage and Control

SMAW electrodes must stay dry. The moment you open the container, the electrodes' flux coating starts absorbing moisture from the air. Local conditions play a significant role. An electrode will stay dry longer on a summer's day in Arizona than it will on a rainy day in Louisiana. In as little as half an hour, some electrodes will start producing weld defects. Some, like under-bead cracking, will be invisible.

Low-hydrogen electrodes absorb moisture faster than any other type. The base metals welded with low-hydrogen electrodes are more susceptible to defects too. For these reasons, handle and store low-hydrogen electrodes carefully.

2.2.1 Code Requirements

All welding codes have special rules for low-hydrogen electrodes. The manufacturer must ship them in airtight containers. After opening the container, you must place them in an oven set to a code-specified minimum temperature—usually at least 250°F (~120°C).

When you remove them from the oven, they have a limited working time that usually ranges between 30 minutes and 10 hours. For example, *AWS D1.1/D1.1M, Table 7.1, Allowable Atmospheric Exposure of Low-Hydrogen Electrodes*, lists permissible exposure times. After this time, you must discard the electrode. Other codes may have different rules. Some codes permit re-baking the electrode to drive out the accumulated moisture. The oven must be hotter than 500°F (260°C) to accomplish this.

The jobsite quality standards document usually specifies the exact procedures for handling low-hydrogen filler metal. Follow these procedures exactly. If they're unclear, check with your supervisor since misusing low-hydrogen electrodes will produce weld defects.

CAUTION

The exposure time tables included in the various welding codes are guidelines only. The local climate and weather conditions may reduce these times. Excessive moisture will produce weld defects. Always follow your jobsite quality standards. Not following your jobsite's electrode handling and storage standards can have severe consequences. Never expose low-hydrogen electrodes to air for longer than necessary.

2.2.2 Receiving Electrodes

Inspect all electrodes when they arrive. Low-hydrogen, stainless steel, and nickel alloy electrodes must come in airtight containers. Reject containers with broken seals. Mark them "Not to Be Used" and notify your supervisor.

All other electrodes must come in moisture-resistant containers. Inspect the container for damage. If the container is damaged or even dented, examine the electrodes. If their coatings are intact and haven't contacted water or oil, the electrodes are usable. Mark unacceptable electrodes "Not to Be Used" and notify your supervisor.

Every container must list its weight, the electrode classification, and the electrode size. If a container doesn't have this information, mark it "Not to Be Used" and notify your supervisor.

NOTE

When in doubt, discard questionable electrodes.

2.2.3 Storing Filler Metal

Store filler metal in a dry environment. Consult the applicable code for the correct temperature, which varies with the electrode type. Protect filler metals from **condensation** by placing them on a pallet or shelf rather than on the floor. Store them so they won't get damaged. Unless they're still in their sealed containers, low-hydrogen, stainless steel, and nickel-alloy electrodes must be kept in an oven.

Condensation: Water vapor in the air turning back to its liquid form when it touches a cool surface.

WARNING!

Storage ovens and their contents are extremely hot. Even portable ovens can be very hot inside. To avoid burns, wear appropriate PPE, including gloves rated for temperatures 500°F (260°C) or higher. Re-baking requires much higher temperatures than does regular storage. Wear gloves rated for temperatures up to 1,000°F (~540°C) when working with re-baked electrodes. Be careful when adding or removing materials.

2.2.4 Storage Ovens

Electrode storage ovens are electrically powered and thermostatically controlled. Set the temperature to the electrode manufacturer's recommendation. Be careful since some ovens can be set hotter than is safe for many electrodes. Typical storage temperatures range from 250°F to 300°F (~120°C to 150°C).

Portable ovens hold smaller quantities of electrodes, perhaps just a few pounds. They keep the electrodes dry but close to the actual work. Some jobsite quality standards or WPSs require welders to use portable ovens. Always transfer electrodes from the portable oven back to the main storage oven when you're finished. *Figure 8* shows several oven styles.

CAUTION

Don't put cellulose electrodes, such as E6010 and E6011, in ovens. The heat will damage their flux coating.

Avoiding Mix-Ups

To avoid mix-ups, keep only one electrode type in the oven. If you must store multiple types in one oven, keep them on separate shelves. Label the shelves to prevent mistakes.

CAUTION

Never turn off or unplug a storage oven unless it's empty. Don't connect an oven to a switched outlet that could be turned off accidentally.

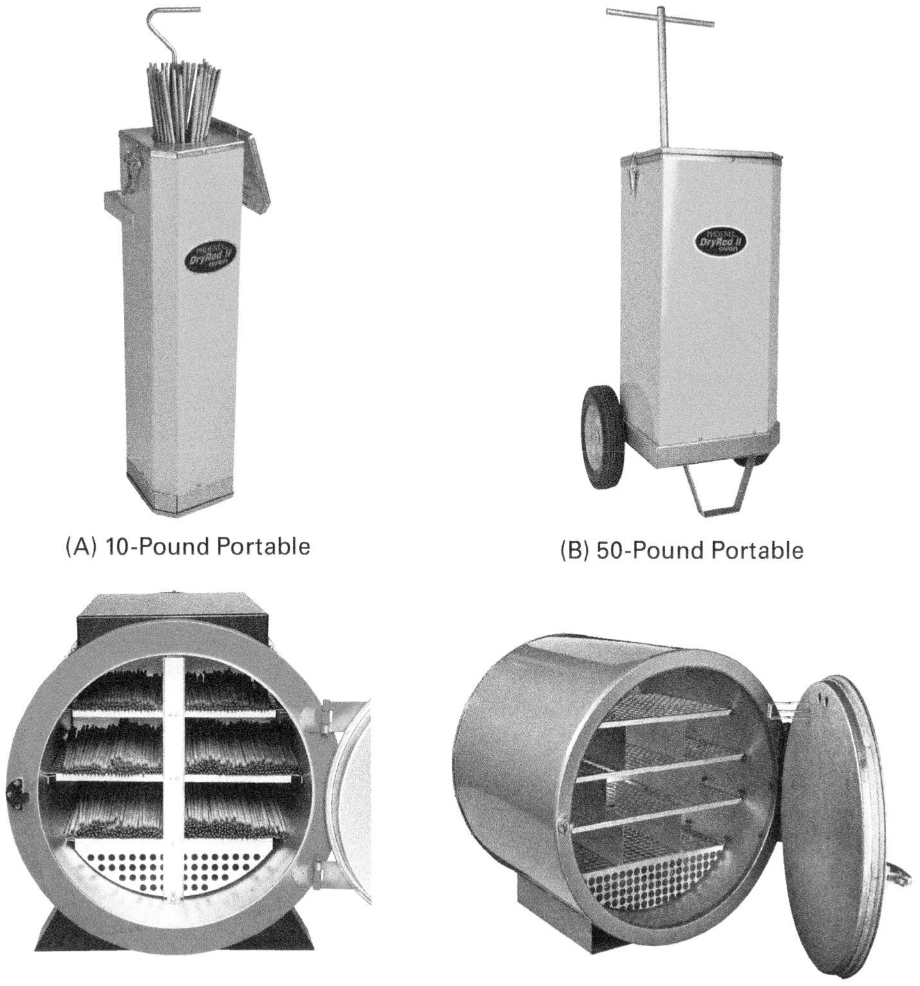

(A) 10-Pound Portable

(B) 50-Pound Portable

(C) Bench Model

(D) Large Bench Model

Figure 8 Electrode storage ovens.
Sources: Phoenix International Inc. (8A–8B, 8D); The Lincoln Electric Company, Cleveland, OH, USA (8C)

NOTE

Any time that electrodes aren't in an oven counts as exposure time.

CAUTION

Don't re-bake electrodes unless your jobsite quality standards permit it. Improperly re-baked electrodes can cause weld defects. Never re-bake electrodes more than once. Discard them instead. Always discard electrodes exposed to water. Check with your supervisor if you're uncertain about electrode handling.

2.2.5 Exposure Time and Drying

Once you remove low-hydrogen, stainless steel, or nickel alloy electrodes from the oven, they have a limited working time. Look up the electrode's exposure time so you can remove only as many electrodes as you can reasonably expect to use. Remember, wet or humid weather reduces the exposure time limit. Pushing the limit is unwise. If possible, keep electrodes in a portable oven until you're ready to weld.

Some quality control groups regularly collect electrode samples from the jobsite and test them for moisture. Based on this information, they may adjust the permissible exposure time. If samples show too much moisture, however, welds may have to be replaced, a frustrating and costly situation.

Jobsite standards vary in what they permit for electrodes that exceed the maximum exposure time. Some standards allow them to be re-baked in an oven set to a temperature much higher than the regular storage temperature. Other standards specify discarding them.

2.0.0 Section Review

1. Which of the following electrodes would be the *best* choice for medium- or high-carbon steels?
   a. E6011
   b. E6013
   c. E7014
   d. E7018

2. If jobsite standards permit re-baking electrodes, how many times may you do it?
   a. One time
   b. Two times
   c. Three times
   d. Four times

Module 29108 Review Questions

1. What carries the welding current from the SMAW torch to the workpiece?
   a. The base metal
   b. The flux coating
   c. The shielding gas
   d. The electrode core

2. In SMAW work, what pushes away oxygen and hydrogen from the molten weld?
   a. The shielding gases
   b. The welding arc blast
   c. The weld spatter
   d. The plasma envelope

3. The American Welding Society (AWS) filler metal specification system sets manufacturing standards for _____.
   a. welding consumables
   b. SMAW machines
   c. welding chemicals
   d. base metals

4. The number placed at the end of a complete AWS code specification identifier indicates the _____.
   a. electrode type discussed in the specification
   b. specification's revision year
   c. number of tables in the specification
   d. specification's original publication date

5. What is the tensile strength of the filler metal from an E7018 electrode?
   a. 1,800 psi
   b. 7,000 psi
   c. 70,000 psi
   d. 80,000 psi

6. All-position electrodes that provide deep penetration belong to which electrode group?
   a. F1 (fast-fill)
   b. F2 (fill-freeze)
   c. F3 (fast-freeze)
   d. F4 (low-hydrogen)

7. Electrodes that must be kept very dry but that produce welds for critical or demanding applications belong to which electrode group?
   a. F1 (fast-fill)
   b. F2 (fill-freeze)
   c. F3 (fast-freeze)
   d. F4 (low-hydrogen)

8. To produce a good weld without defects, which of the following *must* be true?
   a. The base metal must have a higher tensile strength than the fill metal.
   b. The fill metal must have a higher tensile strength than the base metal.
   c. The base and fill metals must match chemically and mechanically.
   d. The electrode must be heated to 250°F (~120°C) right before welding the joint.

9. The exposure time for low-hydrogen electrodes typically ranges from _____.
   a. ½ to 4 hours
   b. ½ to 10 hours
   c. 2 to 5 hours
   d. 3 to 6 hours

10. Low-hydrogen electrodes are usually re-baked at a temperature _____.
    a. between 150°F and 250°F (~66°C and 120°C)
    b. between 250°F and 300°F (~120°C and 150°C)
    c. greater than 500°F (260°C)
    d. greater than 1,400°F (760°C)

Answers to odd-numbered Module Review Questions are found in *Appendix A*.

Answers to Section Review Questions

Answer	Section Reference	Objective
Section 1.0.0		
1. c	1.1.0	1a
2. b	1.2.4	1b
Section 2.0.0		
1. d	2.1.0; *Table 5*	2a
2. a	2.2.5	2b

MODULE 29109

SMAW – Beads and Fillet Welds

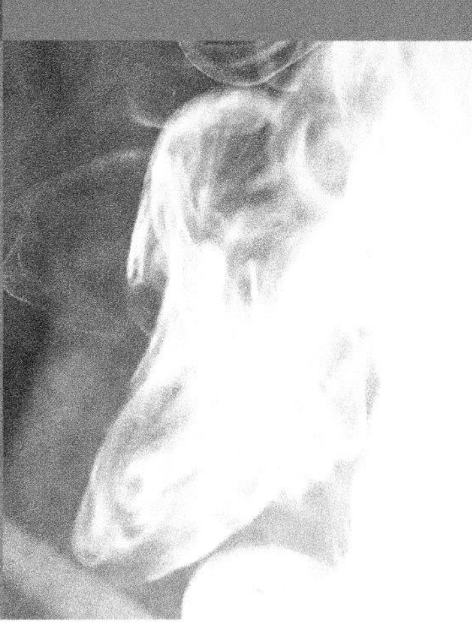

Source: iStock@Funtay

Objectives

Successful completion of this module prepares you to do the following:

1. Summarize fillet welds and preparing to weld them.
   a. Identify fillet welds and their properties.
   b. Outline preparing for fillet welding.
2. Summarize basic SMAW techniques, as well as how to create beads and fillets.
   a. Describe how to strike and control the arc to produce a weld pass.
   b. Describe producing stringer and weave beads, as well as surfacing welds.
   c. Describe fillet welding in all positions.

Performance Tasks

Under supervision, you should be able to do the following:

1. Safely set up SMAW equipment for fillet welding.
2. Strike an arc.
3. Use E6010/E6011 and E7018 electrodes to make stringer, weave, and surfacing beads.
4. Use E6010/E6011 and E7018 electrodes to make flat fillet (1F) welds.
5. Use E6010/E6011 and E7018 electrodes to make horizontal fillet (2F) welds.
6. Use E6010/E6011 and E7018 electrodes to make vertical fillet (3F) welds.
7. Use E6010/E6011 and E7018 electrodes to make overhead fillet (4F) welds.

Overview

The time has come to start welding. You'll start with the shielded metal arc welding (SMAW) process and the simplest weld type—the fillet. To produce this weld, you must learn to strike an arc and control it. You'll then use it to make beads, the basic building blocks of welds. Finally, you'll create fillet welds in all positions.

NOTE

This module uses US standard and metric units in up to three different ways. This note explains how to interpret them.

Exact Conversions

Exact metric equivalents of US standard units appear in parentheses after the US standard unit. For example: "Measure 18" (45.7 cm) from the end and make a mark."

Approximate Conversions

In some cases, exact metric conversions would be inappropriate or even absurd. In these situations, an approximate metric value appears in parentheses with the ~ symbol in front of the number. For example: "Grip the tool about 3" (~8 cm) from the end."

Parallel but not Equal Values

Certain scenarios include US standard and metric values that are parallel but not equal. In these situations, a slash (/) surrounded by spaces separates the US standard and metric values. For example: "Place the point on the steel rule's 1" / 1 cm mark."

Digital Resources for Welding

Scan this code using the camera on your phone or mobile device to view the digital resources related to this craft.

Industry Recognized Credentials

If you are training through an NCCER-accredited sponsor, you may be eligible for credentials from NCCER's Registry. The ID number for this module is 29109. Note that this module may have been used in other NCCER curricula and may apply to other level completions. Contact NCCER's Registry at 1.888.622.3720 or go to **www.nccer.org** for more information.

You can also show off your industry-recognized credentials online with NCCER's digital badges. Transform your knowledge, skills, and achievements into badges that you can share across social media platforms, send to your network, and add to your resume. For more information, visit **www.nccer.org**.

1.0.0 Shielded Metal Arc Welding Preparations

Performance Task

1. Safely set up SMAW equipment for fillet welding.

Objective

Summarize fillet welds and preparing to weld them.

a. Identify fillet welds and their properties.
b. Outline preparing for fillet welding.

Fillet welds are one of the most basic weld types. They're relatively easy to produce and require minimal base metal preparation. Producing good fillet welds requires properly setting up the welding equipment and understanding its safety issues.

1.1.0 Fillet Welds

A fillet weld has a right-triangle profile. It forms T, lap, and corner joints. Welding symbols specify a fillet weld's size and location. A *convex* fillet weld bows outwards like a ball's surface. A *concave* fillet weld bows inwards like the inside of a bowl. Welders describe fillet welds with the following terminology (*Figure 1*):

Weld face — The weld's exposed surface on the side where the welder produced it.

Weld toes — The points at the weld face where the weld and base metals meet.

Weld root — The point where the back of the weld extends farthest into the joint.

Joint root — The place in the joint where the weld members are closest together.

Leg — The distance from the joint root to the weld toe.

Size — The leg lengths of the largest right triangle that can fit within the weld profile.

Actual throat — The shortest distance from the weld root to its face.

Effective throat — The actual throat value minus the weld face's convexity.

Theoretical throat — The perpendicular distance from the hypotenuse of the profile's imaginary right triangle to the joint root.

Welding codes accept fillet welds with equal or unequal legs (*Figure 2*). Codes permit slightly convex or slightly concave faces. Faces must be relatively uniform, although slight non-uniformity is acceptable too. The following rules determine the maximum permissible convexity:

- If the weld size or individual surface bead is ≤ $^5/_{16}$" (8 mm), the maximum convexity is $^1/_{16}$" (1.6 mm).
- If the weld size or individual surface bead is > $^5/_{16}$" (8 mm) and < 1" (25 mm), the maximum convexity is $^1/_8$" (3.2 mm).

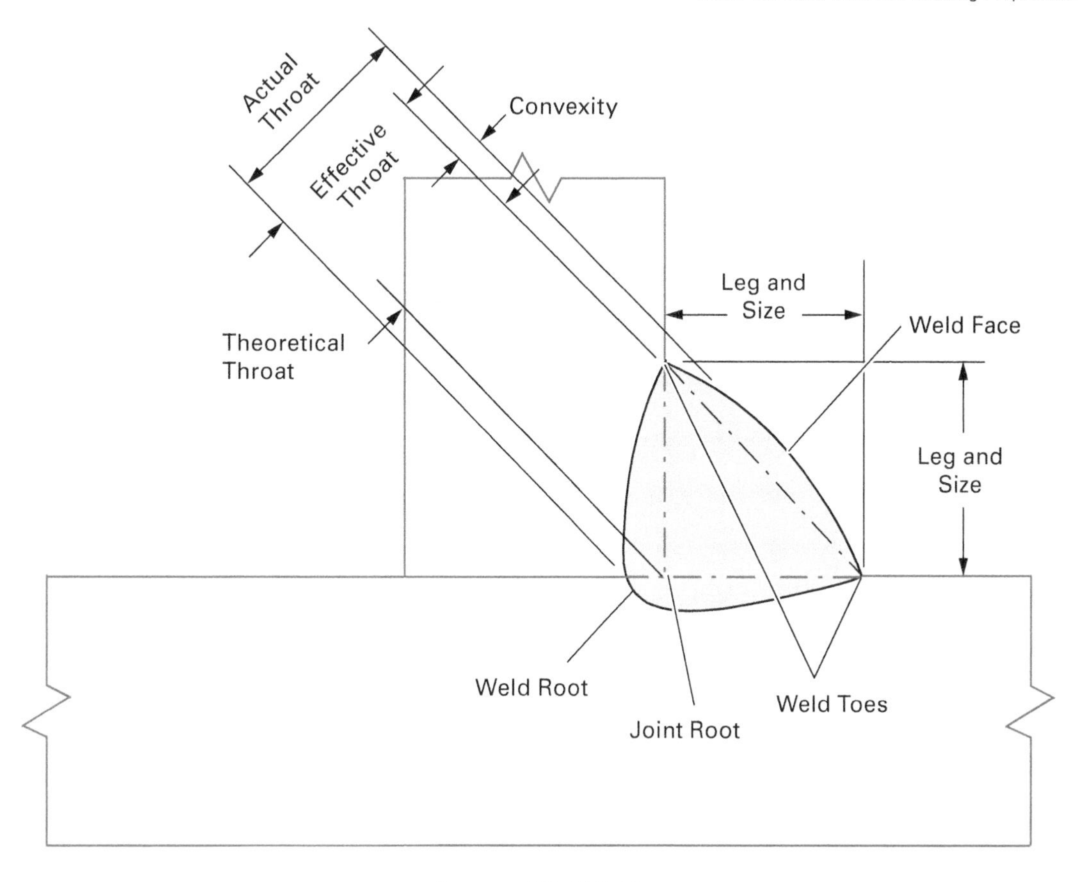

Convex Fillet Weld

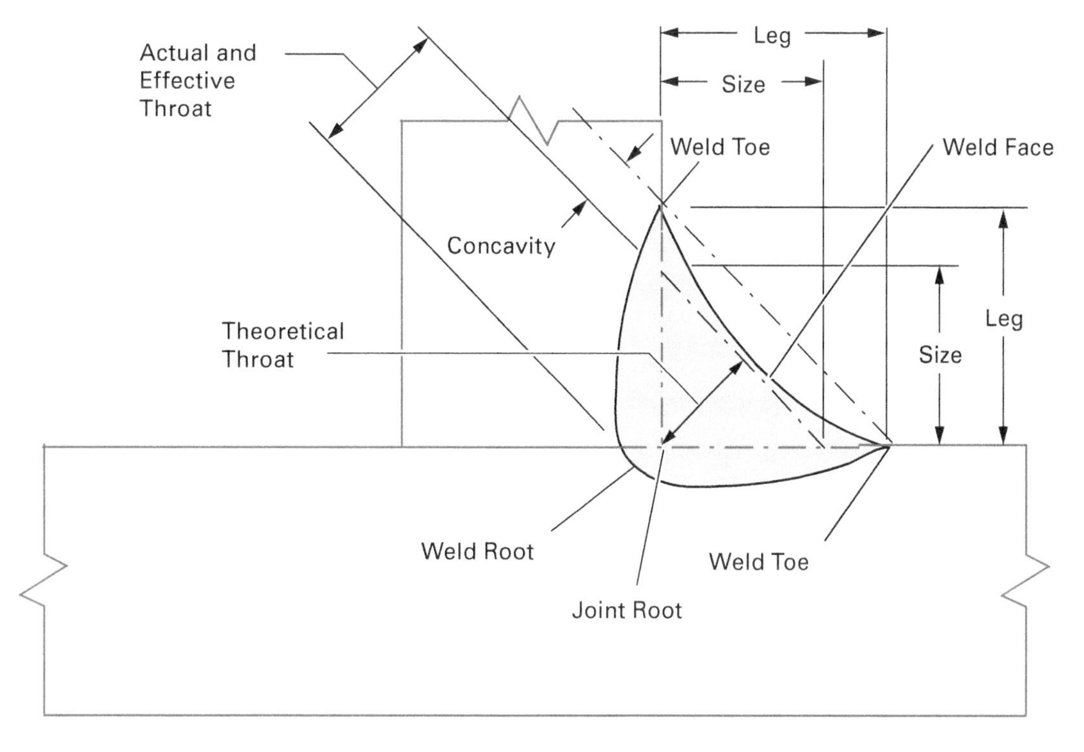

Concave Fillet Weld

Figure 1 Convex and concave fillet welds.

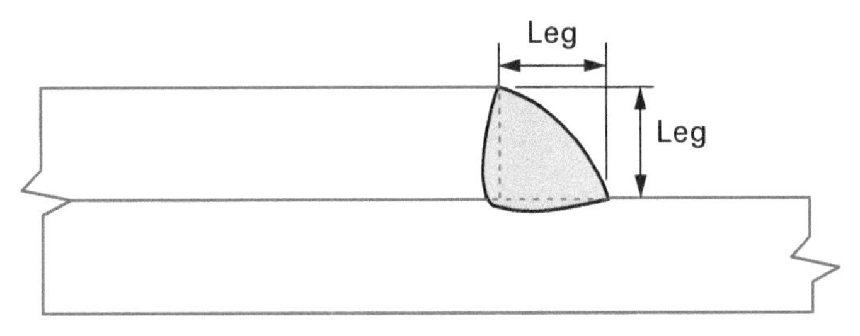

Equal Leg Fillet Weld

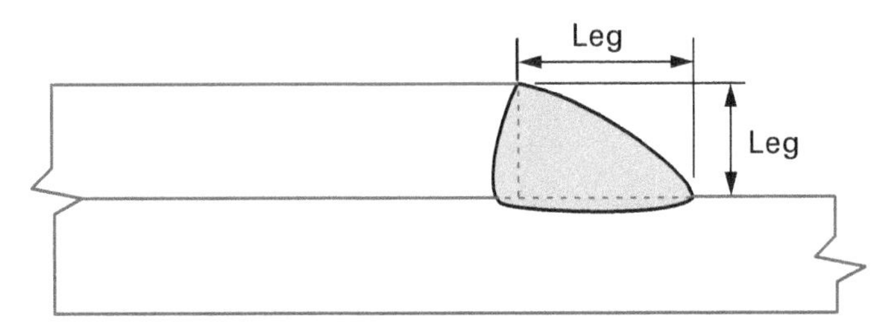

Unequal Leg Fillet Weld

Figure 2 Equal and unequal leg fillet welds.

- If the weld size or individual surface bead is ≥ 1" (25 mm), the maximum convexity is $^{3}/_{16}$" (4.8 mm).

Any of the following profile defects make a fillet weld unacceptable (*Figure 3*):

- Insufficient throat
- Excessive convexity
- Excessive **undercut**
- Overlap
- Insufficient leg
- Incomplete fusion

Undercut: A discontinuity caused by the welder's melting a groove into the base metal at the weld toe.

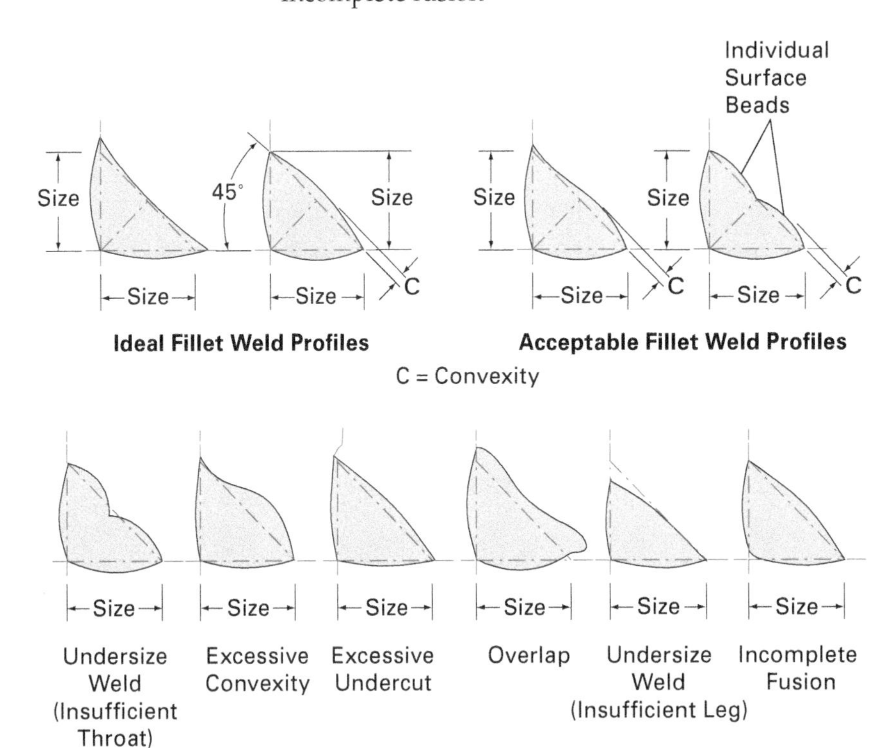

Figure 3 Acceptable and unacceptable fillet weld profiles.

Fillet welds require minimal base metal preparation. But the weld area must be clean and free from dross. Dross from cutting operations will cause discontinuities in the weld. Codes require removing it before welding.

1.2.0 Welding Preparations

This module explains preparing and producing fillet welds in all positions. You'll use E6010/E6011 and E7018 (low-hydrogen) electrodes and practice on **weld coupons**. What you learn here will prepare you for acquiring welder qualifications, as well as producing other weld types.

You'll need to prepare before you can weld a fillet. You must set up the welding area, equipment, and metal. The following sections explain how to do these tasks safely.

NOTE

The dimensions and specifications in this module represent welding codes in general, not a specific code. Always follow the proper codes for your jobsite. Also follow the applicable Welding Procedure Specification (WPS).

1.2.1 Safety Preparations

Cutting and welding equipment produces high temperatures, sparks, and many other hazards. Using equipment safely requires following specific procedures. This section summarizes good practices but isn't a complete safety training. Always complete all training that your workplace requires. Follow all workplace safety guidelines and wear appropriate PPE (*Figure 4*).

NOTE

Be sure that you've completed NCCER Module 29101, *Welding Safety*, before continuing.

Weld coupons: Metal pieces that welders use for practice or to produce test welds for qualifications.

Figure 4 Welding PPE suitable for SMAW work.
Source: © Miller Electric Mfg. LLC

1.2.2 Protective Clothing and Equipment

Welding and cutting tasks are dangerous. Unless you wear all required PPE, you're at risk of serious injury. The following guidelines summarize PPE for SMAW work:

- For cutting, always wear close-fitting goggles or safety glasses, plus a full face shield, helmet, or hood. The goggles, glasses, face shield, or helmet/hood lens must have the proper shade value for the type of cutting work (*Figure 5*). Never cut without using the proper lens.
- For welding, always wear safety glasses, plus a helmet or hood. The glasses or helmet/hood lens must have the proper shade value for the welding process (*Figure 5*). Never view the cutting arc directly or indirectly without using the proper lens.

WARNING!

Don't open your welding helmet/hood lens and expose your eyes when grinding or cleaning a weld. Either use a helmet or hood with a clear lens under the shaded lens or an auto-darkening lens. Alternatively, remove the helmet or hood and use a full face shield.

Arc Welding Processes

Process	Electrode	Amperage	Lens Shade Numbers (2 3 4 5 6 7 8 9 10 11 12 13 14)
Shielded Metal Arc Welding (SMAW)	< 3/32" (2.4 mm)	< 60A	7
	3/32" – 5/32" (2.4 mm – 4.0 mm)	60A – 160A	8 9 10
	5/32" – 1/4" (4.0 mm – 6.4 mm)	160A – 250A	10 11 12
	> 1/4" (6.4 mm)	250A – 550A	11 12 13 14
Gas Metal Arc Welding (GMAW) and Flux Cored Arc Welding (FCAW)		< 60A	7
		60A – 160A	10 11
		160A – 250A	10 11 12
		250A – 500A	10 11 12 13 14
Gas Tungsten Arc Welding (GTAW)		< 50A	8 9 10
		50A – 150A	8 9 10 11 12
		150A – 500A	10 11 12 13 14
Plasma Arc Welding (PAW)		< 20A	6 7 8
		20A – 100A	8 9 10
		100A – 400A	10 11 12
		400A – 800A	11 12 13 14
Carbon Arc Welding (CAW)			14

Arc Cutting Processes

Process	Electrode	Amperage	Lens Shade Numbers (2 3 4 5 6 7 8 9 10 11 12 13 14)
Plasma Arc Cutting (PAC)		< 20A	4
		20A – 40A	5
		40A – 60A	6
		60A – 80A	8
		80A – 300A	8 9
		300A – 400A	9 10 11 12
		400A – 800A	10 11 12 13 14
Air-Carbon Arc Cutting (A-CAC)	Light	< 500A	10 11 12
	Heavy	500A – 1,000A	11 12 13 14

Gas Processes

Process	Type	Plate Thickness	Lens Shade Numbers (2 3 4 5 6 7 8 9 10 11 12 13 14)
Oxyfuel Welding (OFW)	Light	< 1/8" (3 mm)	4 5
	Medium	1/8" – 1/2" (3 mm – 13 mm)	5 6
	Heavy	> 1/2" (13 mm)	6 7 8
Oxyfuel Cutting (OC)	Light	< 1" (25 mm)	3 4
	Medium	1" – 6" (25 mm – 150 mm)	4 5
	Heavy	> 6" (150 mm)	5 6
Torch Brazing (TB)			3 4
Torch Soldering (TS)			2

Lens shade values are based on OSHA minimum values and ANSI/AWS suggested values. Choose a lens shade that's too dark to see the weld zone. Without going below the required minimum value, reduce the shade value until it's visible. If possible, use a lens that absorbs yellow light when working with fuel gas processes.

Figure 5 Guide to lens shade numbers.

- Wear protective leather and/or flame-retardant clothing along with welding gloves that will protect you from flying sparks and molten metal.
- Wear high-top safety shoes or boots. Make sure the pant leg covers the tongue and lace area. If necessary, wear leather spats or chaps for protection. Boots without laces or holes are better for welding and cutting.
- Wear a 100 percent cotton cap with no mesh material in its construction. The bill should point to the rear or the side with the most exposed ear. If you must wear a hard hat, use one with a face shield.

WARNING!

Never wear a cap with a button on top. The conductive metal under the fabric is a safety hazard.

- Wear earplugs to protect your ear canals from sparks. Since welding and cutting can be noisy, always wear hearing protection.

1.2.3 Fire/Explosion Prevention

All processes in this module can produce high temperatures. Welding or cutting a vessel or container that once contained combustible, flammable, or explosive materials is hazardous. Residues can catch fire or explode. Before welding or cutting vessels, check whether they contained explosive, hazardous, or flammable materials. These include petroleum products, citrus products, or chemicals that release toxic fumes when heated.

American Welding Society (AWS) F4.1, Safe Practices for the Preparation of Containers and Piping for Welding and Cutting, and *ANSI/AWS Z49.1* describe safe practices for these situations. Begin by cleaning the container to remove any residue. Steam cleaning, washing with detergent, or flushing with water are possible methods. Sometimes you must combine these to get good results.

WARNING!

Clean containers only in well-ventilated areas. Vapors can accumulate in a confined space during cleaning, causing explosions or toxic substance exposure.

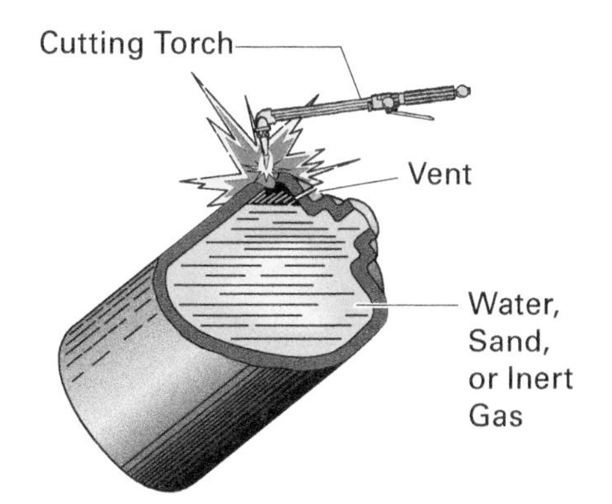

Note: *ANSI Z49.1* and AWS standards should be followed.

Figure 6 "Inerting" a container.

After cleaning the container, you must formally confirm that it's safe for welding or cutting. *American Welding Society (AWS) F4.1, Safe Practices for the Preparation of Containers and Piping for Welding and Cutting*, outlines the proper procedure. The following three paragraphs summarize the process.

Immediately before work begins, a qualified person must check the container with an appropriate test instrument and document that it's safe for welding or cutting. Tests should check for relevant hazards (flammability, toxicity, etc.). During work, repeated tests must confirm that the container and its surroundings remain safe.

Alternatively, fill the container with an inert material ("inerting") to drive out any hazardous vapors. Water, sand, or an inert gas like argon meets this requirement. Water must fill the container to within 3" (~7.5 cm) or less of the work location. The container must also have a vent above the water so air can escape as it expands (*Figure 6*). Sand must completely fill the container.

When using an inert gas, a qualified person must supervise, confirming that the correct amount of gas keeps the container safe throughout the work. Using an inert gas also requires additional safety procedures to avoid accidental suffocation.

NOTE

A *lower explosive limit (LEL)* gas detector can check for flammable or explosive gases in a container or its surroundings. Gas monitoring equipment checks for specific substances, so be sure to use a suitable instrument. Test equipment must be checked and calibrated regularly. Always follow the manufacturer's guidelines.

WARNING!

Never weld or cut drums, barrels, tanks, vessels, or other containers until they have been emptied and cleaned thoroughly. Residues, such as detergents, solvents, greases, tars, or corrosive/caustic materials, can produce flammable, toxic, or explosive vapors when heated. Never assume that a container is clean and safe until a qualified person has checked it with a suitable test instrument. Never weld or cut in places with explosive vapors, dust, or combustible products in the air.

The following general fire- and explosion-prevention guidelines apply to cutting and SMAW work:

- Never carry matches or gas-filled lighters in your pockets. Sparks can ignite the matches or cause the lighter to explode.
- Always comply with all site and employer requirements for hot work permits and fire watches.
- Never use oxygen in place of compressed air to blow off the workpiece or your clothing. Never release large amounts of oxygen. Oxygen makes fires start readily and burn intensely. Oxygen trapped in fabric makes it ignite easily if a spark falls on it. Keep oxygen away from oil, grease, and other petroleum products.
- Remove all flammable materials from the work area or cover them with a fire-resistant blanket.
- Before welding, heating, or cutting, confirm that an appropriate fire extinguisher is available. It must have a valid inspection tag and be in good condition. All workers in the area must know how to use it.
- Never release a large amount of fuel gas into the work environment. Some gases (propane and propylene) accumulate on the floor, while others (acetylene and natural gas) accumulate near the ceiling. Either can ignite far from the release point. Acetylene will explode at lower concentrations than other fuel gases.
- Prevent fires by maintaining a neat and clean work area. Confirm that metal scrap and slag are cold before disposing of them.

1.2.4 Work Area Ventilation

Vapors and fumes fill the air around their sources. Welding and cutting fumes can be hazardous. Welders often work above the area from which fumes come. Good ventilation and source extraction equipment help remove the vapors and protect the welder. The following lists good ventilation practices to use when welding and cutting:

- Always weld or cut in a well-ventilated area. Operations involving hazardous materials, such as metal coatings and toxic metals, produce dangerous fumes. You must wear a suitable respirator when working with them. Follow your workplace's safety protocols and confirm that your respirator is in good condition.

WARNING!

Cadmium, mercury, lead, zinc, chromium, and beryllium produce toxic fumes when heated. When welding or cutting these materials, wear appropriate respiratory PPE. Supplied-air respirators (SARs) work best for long-term work.

- Never weld or cut in a confined space without a confined space permit. Prepare for working in the confined space by following all safety protocols.
- Set up a confined space ventilation system before beginning work.
- Never ventilate with oxygen or breathe pure oxygen while working. Both actions are very dangerous.

WARNING!

When working around other craftworkers, always be aware of what they're doing. Take precautions to keep your work from endangering them.

Going Green

Vapor Extraction Systems

Protect the environment by using a vapor extraction system that filters out the welding fumes. This strategy is superior to simply releasing the fumes into the environment.

1.2.5 Preparing the Welding Area

To practice fillet welds, you'll need a welding table, bench, or stand (*Figure 7*). It should have a way to place welding coupons out of position.

The following steps outline setting up the welding area:

Step 1 Check the work area ventilation. If necessary, use fans, windows, or source extraction equipment.

Step 2 Check the area for fire hazards. Remove any flammable materials. Cover with a welding blanket anything flammable that you can't move.

Step 3 Locate the nearest fire extinguisher. Confirm that it's charged and that you know how to use it.

Step 4 Position the welding machine near the welding table.

Step 5 Set up welding screens around the welding area.

Figure 7 Welding station with source extraction.
Source: The Lincoln Electric Company, Cleveland, OH, USA

1.2.6 Electrodes

These welding exercises require E6010/E6011 and E7018 electrodes. Use a diameter of $^{3}/_{32}$" (2.4 mm), $^{1}/_{8}$" (3.2 mm), or $^{5}/_{32}$" (4.0 mm). Remember that E7018 electrodes must stay dry. Leave them in their storage oven until you're ready to weld. Remove only as many as you'll use during the session.

Keep electrodes in a pouch or holder so they don't get damaged. Never leave them loose on the table. They can fall on the floor and be damaged or create a slip hazard. Have a suitable metal container nearby to hold discarded electrode stubs.

NOTE

The electrode sizes are recommendations only. Check with your instructor for the correct size to use.

WARNING!

Don't throw electrode stubs on the floor. They can roll under your feet, causing accidents.

Electrode Housekeeping

Electrode stubs on the floor can roll under your feet, creating a serious hazard. Many employers discipline employees who throw stubs on the floor. Some even make this practice a dismissible offense. Always place electrode stubs in a suitable container. Good welders keep their workspaces clean and tidy.

NOTE

If a DC welding machine isn't available, an AC machine will work. You'll need to use E6011 electrodes instead of E6010. E7018 electrodes perform properly in constant current AC mode. In cases where arc blow is a problem, switching from DC to AC may solve the problem.

NOTE

These values are general guidelines only. Amperage recommendations vary by manufacturer, welding position, current type, and electrode brand. If possible, refer to the WPS for guidance. If a WPS isn't available, follow the electrode manufacturer's guidelines.

1.2.7 Preparing the Welding Machine

Welders will use many different welding machine brands and models. Not all will have manuals with them. Learn to recognize welding machine types and their controls, so you can use them safely and correctly (*Figure 8*).

The following steps outline setting up a regular welding machine:

Step 1 Locate the nearest power disconnect so you can shut down the machine quickly.

Step 2 Plug in the machine.

Step 3 Select SMAW mode (constant current DC operation).

Step 4 If the machine has a polarity selector, set it to DCEP. If not, connect the leads for DCEP.

Step 5 Connect the workpiece lead's clamp to the workpiece.

Step 6 Set the amperage for the electrode type and size. *Table 1* shows typical settings.

TABLE 1 Typical Electrode Sizes and Amperages

Electrode	Size	Amperage
E6010	3/32"	40A to 80A (DCEP)
E6010	1/8"	70A to 130A (DCEP)
E6010	5/32"	90A to 165A (DCEP)
E6011	3/32"	40A to 90A (AC)
E6011	1/8"	65A to 120A (AC)
E6011	5/32"	115A to 150A (AC)
E7018	3/32"	70A to 110A (DCEP) 80A to 120A (AC)
E7018	1/8"	90A to 160A (DCEP) 100A to 160A (AC)
E7018	5/32"	130A to 210A (DCEP) 140A to 210A (AC)

Step 7 Confirm that the electrode holder *isn't* resting on a grounded surface.

Step 8 Turn on the welding machine when you're ready to weld.

If you're using an engine-driven welding machine, you must place it in a safe, properly ventilated space. Startup procedures vary, depending on the machine's engine model. Consult NCCER Module 29107, *SMAW – Equipment and Setup*, for more detail when working with engine-driven welding machines.

Rod Holders

Welders refer to electrodes as *rods*. They keep them in various rod holders. Leather pouches are popular. Sealable holders keep electrodes from getting wet. Unlike heated rod holders (portable ovens), they don't keep low-hydrogen electrodes moisture-free. They do, however, slow moisture absorption, which is beneficial.

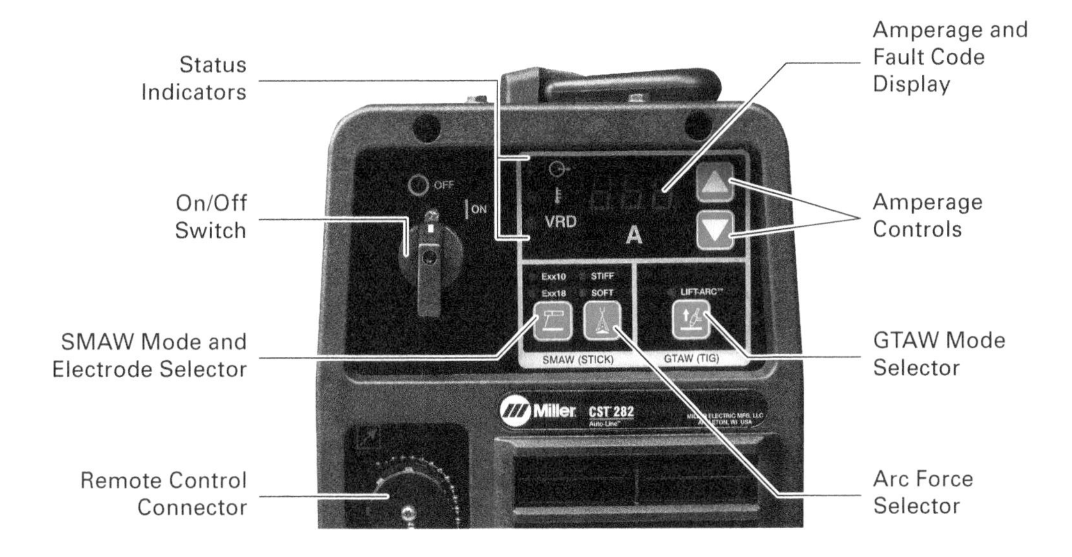

(A) Regular Welding Machine Controls

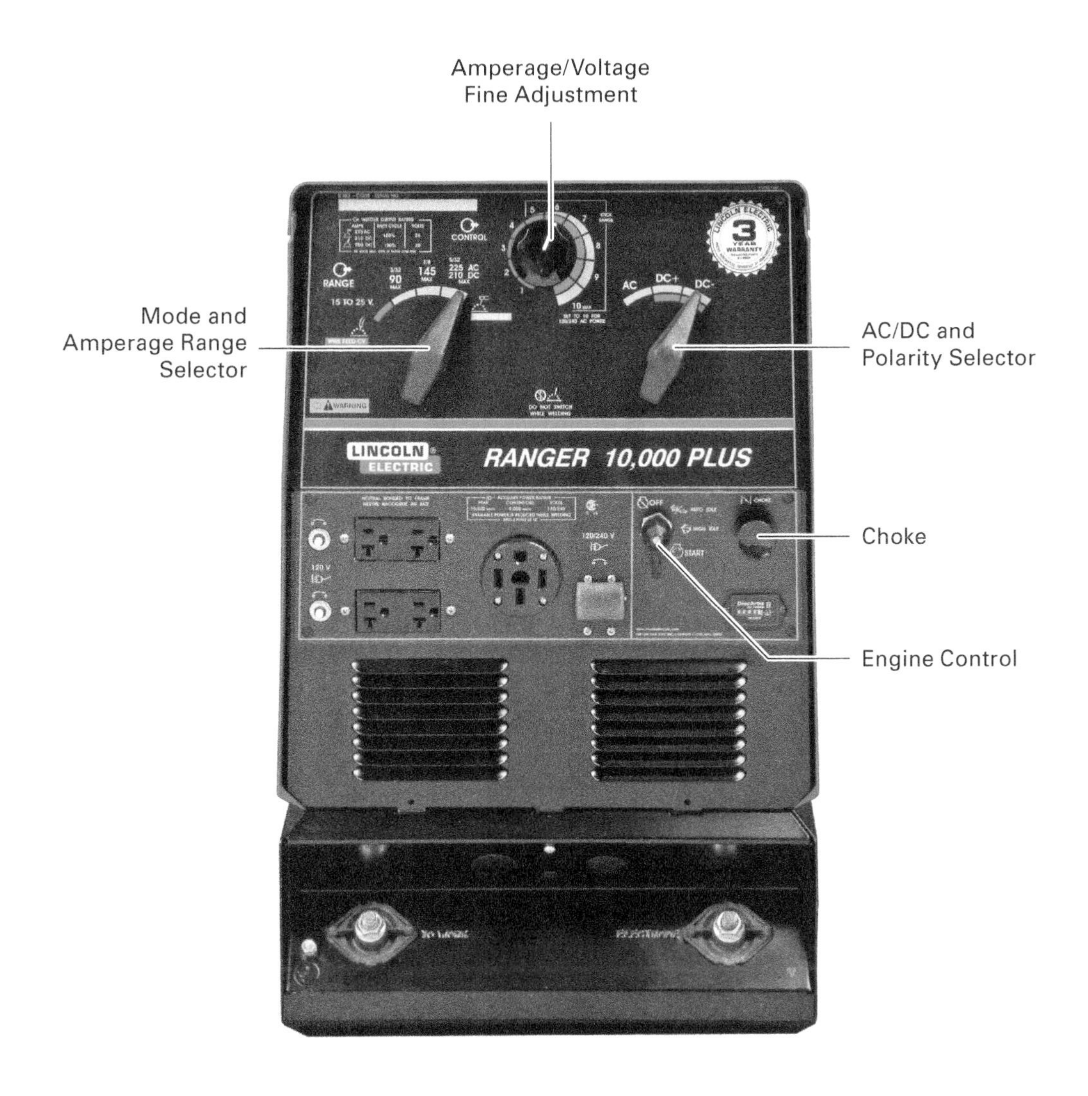

(B) Engine-Driven Welding Machine Controls

Figure 8 Regular and engine-driven welding machines.

Sources: © Miller Electric Mfg. LLC (10A); The Lincoln Electric Company, Cleveland, OH, USA (10B)

WARNING!

Remember that the workpiece lead (ground lead) isn't always grounded. It could have the welding machine's full open-circuit voltage between it and any grounded object. Always be cautious with both welding leads.

WARNING!

When you're finished welding, always remove the electrode from its holder. This prevents arcing, which can cause burns and arc flash.

1.2.8 Preparing the Weld Coupons

If possible, use ¼" to ¾" (~6 mm to 19 mm) thick carbon steel for weld coupons. You'll need coupons for practicing striking an arc, making beads, and building pads. You'll also need two coupons for practicing fillet welds. Clean the steel plates with a wire brush or grinder to remove scale and corrosion.

WARNING!

Be extremely careful with powered wire brushes. Verify that the brush's speed rating matches or exceeds the tool's maximum speed. Loose wires can fly off during cleaning. Regardless of the tool type, always wear appropriate PPE when cleaning metal.

NOTE

To save material, your instructor may select a smaller size for fillet weld practice.

For the fillet weld coupons, cut one plate to 3" × 6" (7.6 cm × 15.2 cm). The other should be 4" × 6" (10.2 cm × 15.2 cm). *Figure 9* shows the coupons assembled to produce a fillet weld. The other practice coupons may be any convenient size.

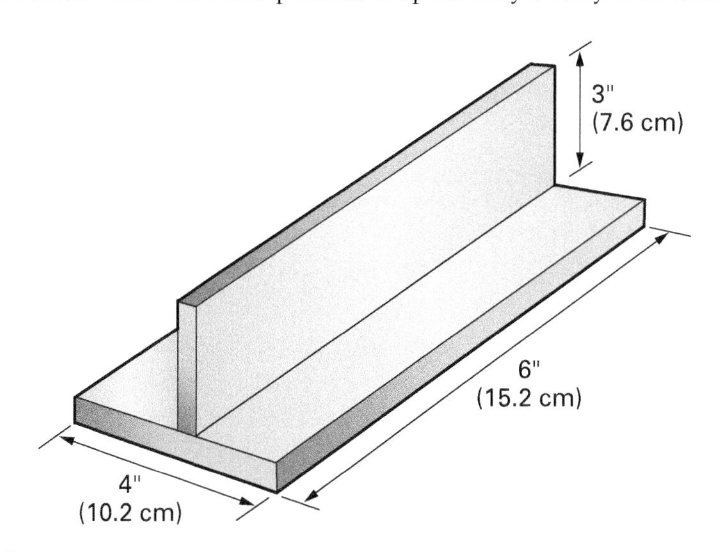

Figure 9 Fillet weld coupons.

Going Green

Welding Coupons

Reuse or recycle welding coupons rather than discarding them. Doing this reduces unnecessary waste and expense. Cut up the larger used coupons into smaller ones. Practice running beads on smaller used coupons.

1.0.0 Section Review

1. A fillet weld's face may be flat, concave, or _____.
   a. profiled
   b. convex
   c. ellipsoid
   d. convoluted

2. Electrode stubs should be _____.
   a. discarded on the floor for later collection
   b. welded together to form practice electrodes
   c. discarded in a metal container
   d. placed in water to reduce the risk of fire

2.0.0 Basic Shielded Metal Arc Welding Techniques and Fillet Welds

Objective

Summarize basic SMAW techniques, as well as how to create beads and fillets.

a. Describe how to strike and control the arc to produce a weld pass.
b. Describe producing stringer and weave beads, as well as surfacing welds.
c. Describe fillet welding in all positions.

Performance Tasks

2. Strike an arc.
3. Use E6010/E6011 and E7018 electrodes to make stringer, weave, and surfacing beads.
4. Use E6010/E6011 and E7018 electrodes to make flat fillet (1F) welds.
5. Use E6010/E6011 and E7018 electrodes to make horizontal fillet (2F) welds.
6. Use E6010/E6011 and E7018 electrodes to make vertical fillet (3F) welds.
7. Use E6010/E6011 and E7018 electrodes to make overhead fillet (4F) welds.

With the equipment set up, the electrodes selected, and the coupons prepared, you're ready to begin welding. First, you'll learn to strike and control the arc. Once you've mastered this technique, you'll practice creating weld beads. From these, you'll start producing fillet welds in all positions.

2.1.0 Striking and Controlling an Arc

Welders refer to the process that starts the weld as *striking an arc*. They do this by briefly touching the electrode to the workpiece. This action starts the electric current flowing, which rapidly heats the electrode tip and the air around it. The ionized (electrically conductive) air can carry an electric current just like a wire.

As the tip moves away from the workpiece, the electric current continues flowing through the ionized air, making it even hotter. This is the welding arc, which creates the weld by melting the base metal and the electrode. The arc also melts the electrode's flux coating, producing the cloud of shielding gases around the weld.

WARNING!

When you're finished welding, always remove the electrode from its holder. This prevents arcing, which can cause burns and arc flash.

2.1.1 Techniques for Striking an Arc

Although striking an arc looks easy, it requires practice to do it reliably. Touching the electrode to the workpiece for too long causes it to stick. Pulling it too far away from the workpiece extinguishes the arc. Trainees learn several ways to strike an arc reliably.

General Procedure and Arc Length

Touch the electrode tip to the base metal. Immediately, lift it to produce the correct arc length. The *arc length* is the distance between the metal electrode core's tip and the base metal. As a rule, the arc length should equal the core's diameter. In other words, when using a $\frac{1}{8}$" (3.2 mm) electrode, the arc length should be about $\frac{1}{8}$".

Some electrodes, like the E6013 and E7018, have thick flux coatings. The electrode core is slightly recessed. With these, reduce the visible arc length since it's longer than it appears (*Figure 10*).

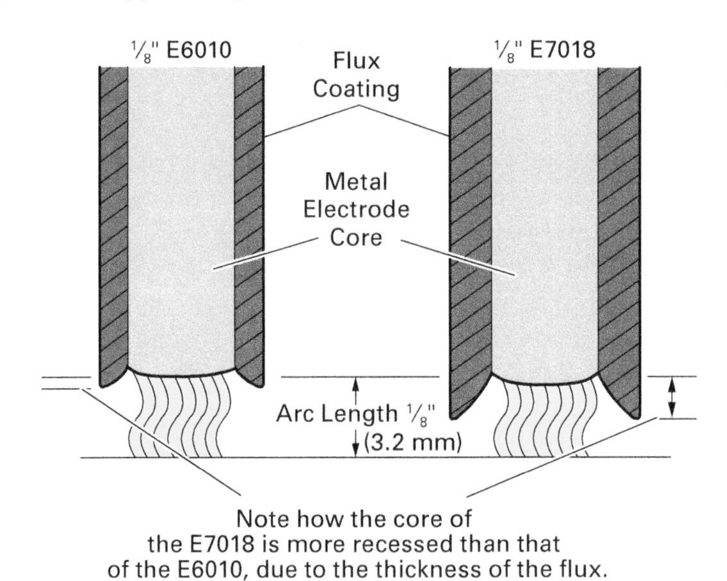

Figure 10 Arc length.

Scratching Method

The scratching method is the easiest way to strike an arc. Trainee welders usually learn it first. It's like striking a match. Simply draw the electrode across the base metal. When the arc starts, lift the electrode to produce the correct arc length (*Figure 11*).

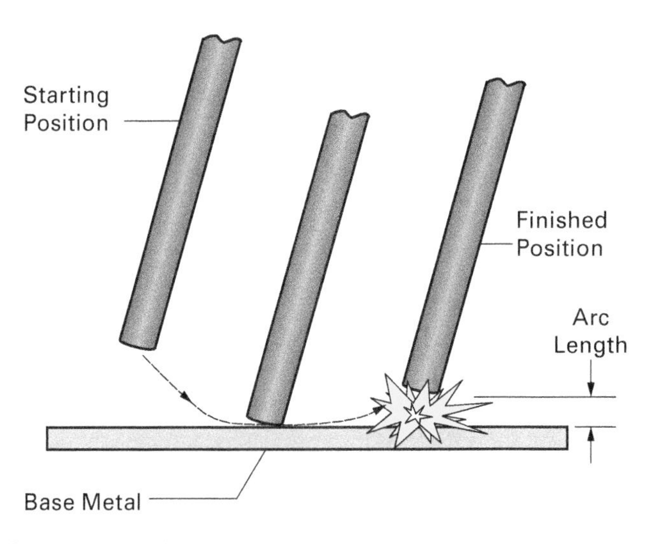

Figure 11 Scratching method.

Tapping Method

The tapping method takes more practice, but it produces better results. Scratching the electrode across the base metal can leave an **arc strike** trail. Most welding codes treat these as defects unless the weld bead has re-melted and consumed them. The tapping method is also better when using a transformer-rectifier welding machine.

Arc strike: A discontinuity caused by arc initiation and visible as a melted spot outside the weld area.

To strike an arc with the tapping method, position the electrode ¼" to ½" (~6 mm to 13 mm) past the point where the weld will begin. Lightly tap the electrode tip to the base metal and lift it to produce the correct arc length. As the arc stabilizes, move the electrode back to the weld starting point (*Figure 12*). Be sure to re-melt the filler metal deposited when you struck the arc. This metal wasn't protected from oxidation and contamination since the flux hadn't formed the gaseous shield yet.

NOTE

Some inverter welding machines won't strike an arc if you use the tapping method. When not welding, these machines reduce their output voltage to a very low value. They increase it when the welder touches the electrode to the workpiece. The quick tap is too brief to trigger this change. You must use the scratching method with these machines.

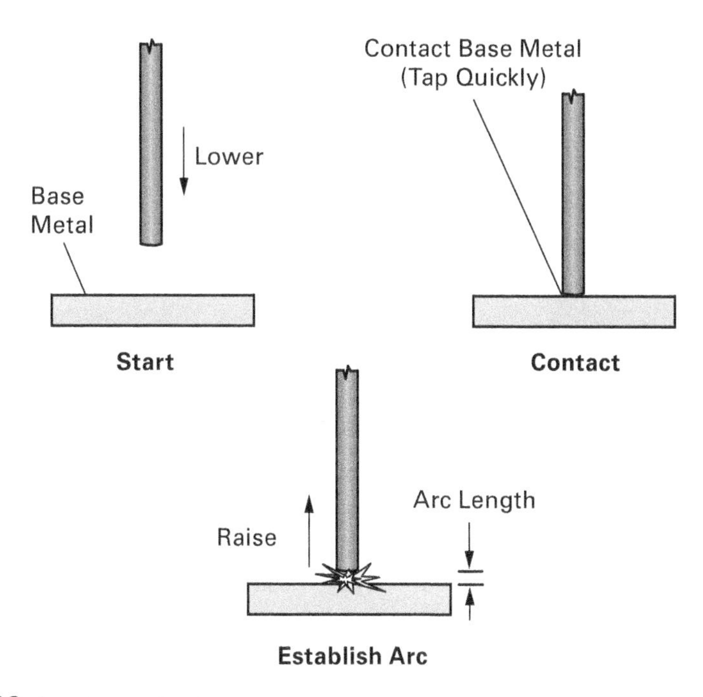

Figure 12 Tapping method.

Tap-Striking Accurately

To tap-strike the electrode at the right spot, rest the electrode like a pool cue on your gloved finger. Angle the rod correctly. Then move the electrode quickly down and up to strike an arc. This method works best if you have a helmet with an auto-darkening lens or can lower a conventional helmet by nodding your head.

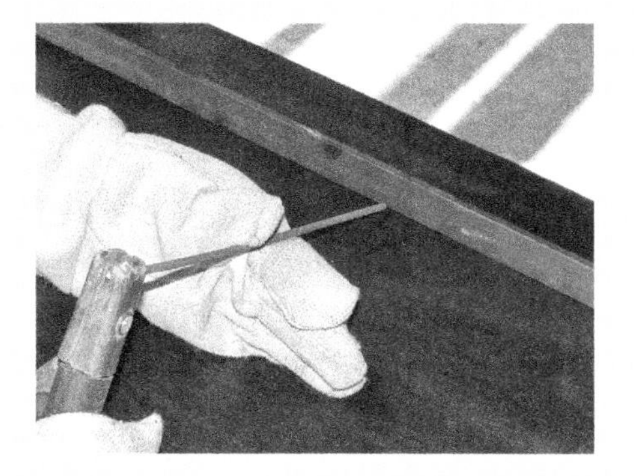

2.1.2 Practicing Striking and Extinguishing the Arc

Striking the arc correctly is a fundamental skill. Place a practice coupon flat on the welding table. Practice both striking methods with $^{3}/_{32}$" (2.4 mm), $^{1}/_{8}$" (3.2 mm), or $^{5}/_{32}$" (4.0 mm) E6010/E6011 electrodes. Keep practicing until you can strike and maintain the arc consistently. To extinguish the arc, quickly lift the electrode away from the workpiece.

Trainees usually experience two challenges when practicing. The first is the electrode tip's sticking to the base metal. Scratching or tapping too slowly can cause this problem. Maintaining too short an arc length can also cause the electrode to stick. New welders can prevent sticking by setting the amperage to the upper end of the recommended range.

Regardless of the cause, if the electrode sticks, try to free it by quickly moving the electrode holder from side to side. If this fails, release the electrode from the holder. Then break the electrode free with pliers.

NOTE

Releasing the electrode from the holder usually causes arcing. Over time, arcing erodes the jaws. Regularly inspect them for pitting or gouging damage and replace them as required.

The second problem is that the arc will extinguish abruptly. Raising the electrode too high above the base metal is one cause. Not steadily feeding the electrode into the arc as the rod melts is another. Over time, you'll gain experience in controlling the arc length, so this problem won't happen as often.

Once you feel confident with E6010/E6011 electrodes, start practicing with E7018 electrodes. Striking an arc with them is harder because they stick to the base metal more easily. Keep practicing with E7018 electrodes until you can strike and maintain the arc consistently.

Reusing a Stuck Electrode

Before reusing an electrode that you've had to break free from the base metal, inspect the tip. If the flux coating is cracked, burned, or missing, try to re-form the tip. Strike an arc on a piece of scrap. Hold a long arc for a few moments. If this technique corrects the problem, you've avoided wasting an electrode. If you can't successfully re-form the tip, discard the electrode.

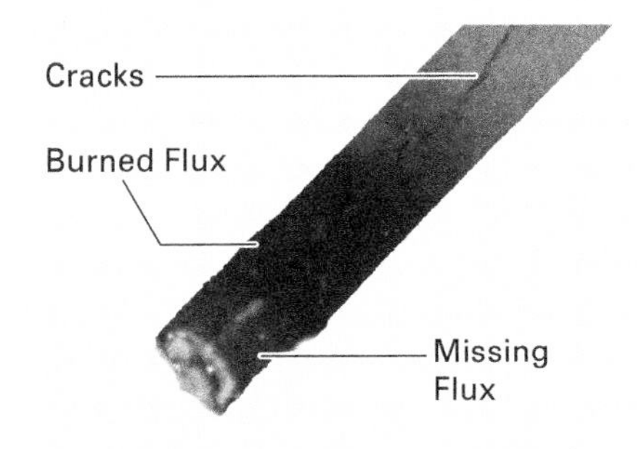

Striking with a Cold Electrode

Sometimes, a cold, partially consumed electrode forms a flux "cup" around the tip. The cup prevents the core from touching the base metal, so the arc won't strike. Scratching or tapping harder than usual may break the cup and solve the problem. Unfortunately, these strategies can also break off chunks of flux.

An electrode with missing flux won't strike easily. It will also produce an unstable arc that makes poor welds. Some codes don't permit welders to use electrodes with damaged flux coatings. A better fix is to draw the electrode tip across a medium-coarse file to remove the flux cup.

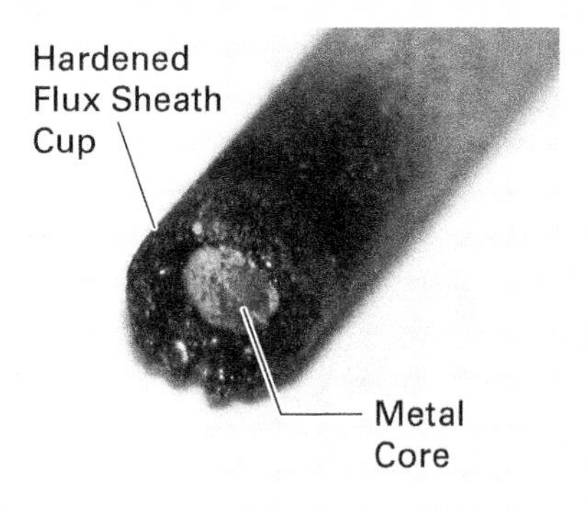

2.1.3 Electrode Angles

As you learn different welding techniques, you'll notice that the instructions usually tell you how to angle your electrode. Electrode angles are crucial for producing good welds. In many cases, a welding technique requires you to change angles during the procedure. There are two electrode angles (*Figure 13*).

The **travel angle** refers to the electrode's angle as seen from the side. In other words, it's in the plane of the electrode's motion. The travel angle is the angle between the **electrode axis** and a line perpendicular to the **weld axis**. If the electrode axis itself is perpendicular to the weld axis, the travel angle is 0 degrees. Travel angles between 10 and 15 degrees are common.

Travel angle: The electrode's forward or backward angle with respect to the groove or weld face.

Electrode axis: A line drawn along the electrode's length and through its center.

Weld axis: A line drawn along the weld's length and through its center.

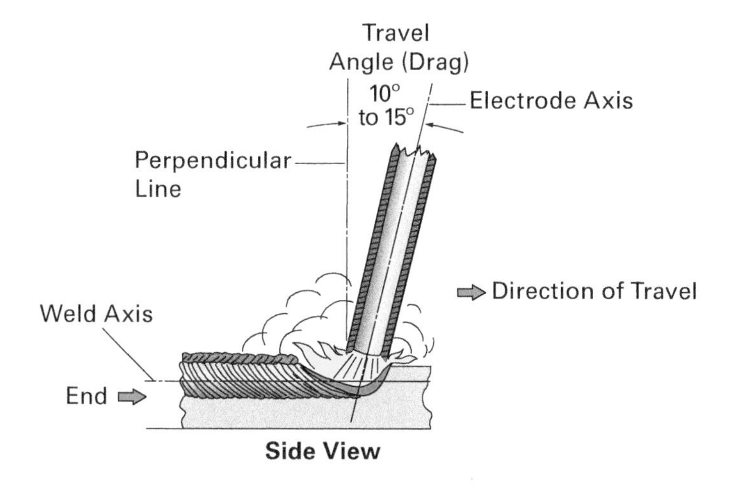

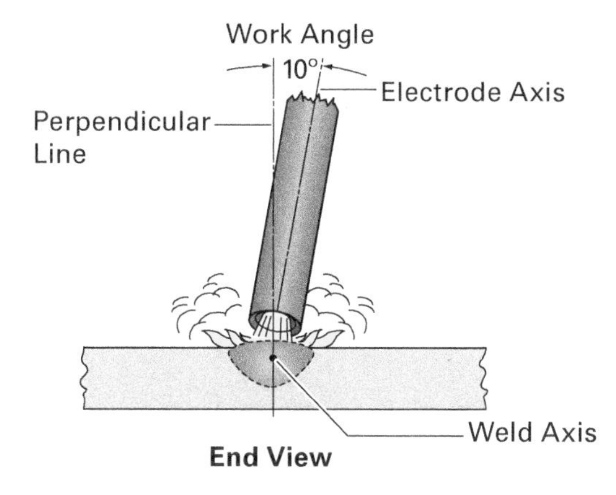

Figure 13 Electrode angles.

When producing a weld, the welder may lean the electrode backward and push it along the weld axis. In this case, the travel angle is called the **push angle**. In other cases, the welder may pull the electrode along the weld axis. In this case, the travel angle is called the **drag angle**. In SMAW work, most welds require dragging rather than pushing. Vertical welds are the exception and have push angles. *Figure 14* illustrates both.

The **work angle** refers to the electrode's angle as seen from the end. In other words, it's in the plane perpendicular to the electrode's motion. The work angle is the angle between the electrode axis and a line perpendicular to the weld face. If the electrode axis is perpendicular to the weld face, the work angle is 0 degrees. Many welds require a 0-degree work angle. In some cases, however, the welder will tilt the electrode to one side. Work angles range between 0 and 25 degrees.

Push angle: The travel angle when the electrode points in the same direction as the bead's progression.

Drag angle: The travel angle when the electrode points in the direction opposite the bead's progression.

Work angle: The electrode's side-to-side angle with respect to the groove or weld face.

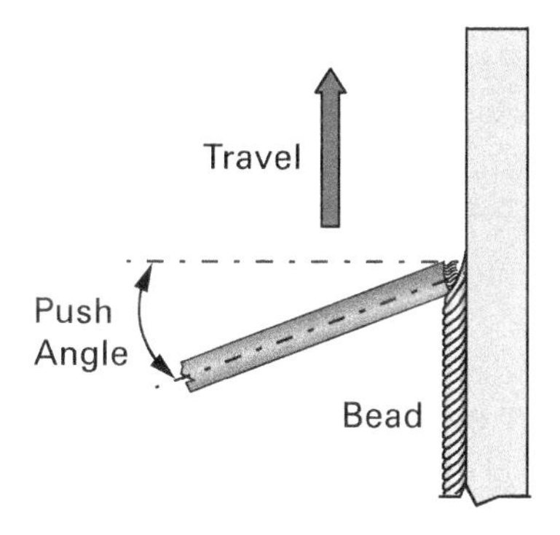

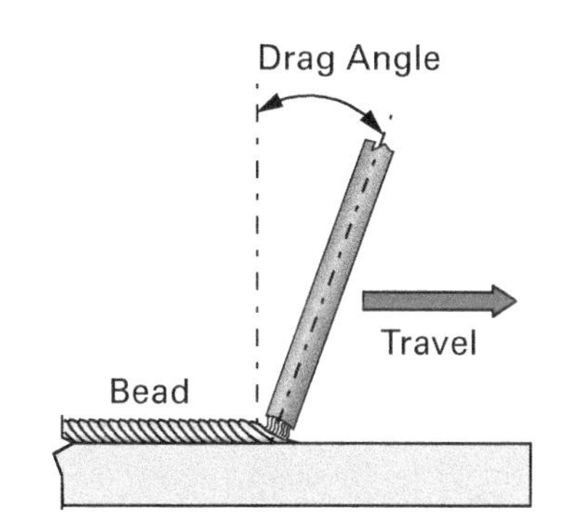

Figure 14 Push and drag angles.

Work Angles Clarified

Viewed perpendicularly to the electrode's motion, the work angle is the electrode's side-to-side angle. It's formed between the electrode axis and a line perpendicular to the weld face. For groove welds, that line is perpendicular to the base metal surface. So, when the work angle is 0 degrees, the electrode axis is perpendicular (90 degrees) to the base metal surface.

Fillet welds are a bit more confusing. For a T-joint, the weld face isn't in the same plane as either base metal component. A line perpendicular to the weld face usually forms a 45-degree angle with each component. So, when the work angle is 0 degrees, the electrode axis is at a 45-degree angle to each base metal surface.

Some welding diagrams show the correct work angle for laying each weld bead. These are relative to the line perpendicular to the weld face. In the figure, beads 1 and 5 have 0-degree work angles. Beads 2 and 4 have work angles tilted 20 degrees from the perpendicular line. Beads 3 and 6 also require tilting 20 degrees, but in the opposite direction.

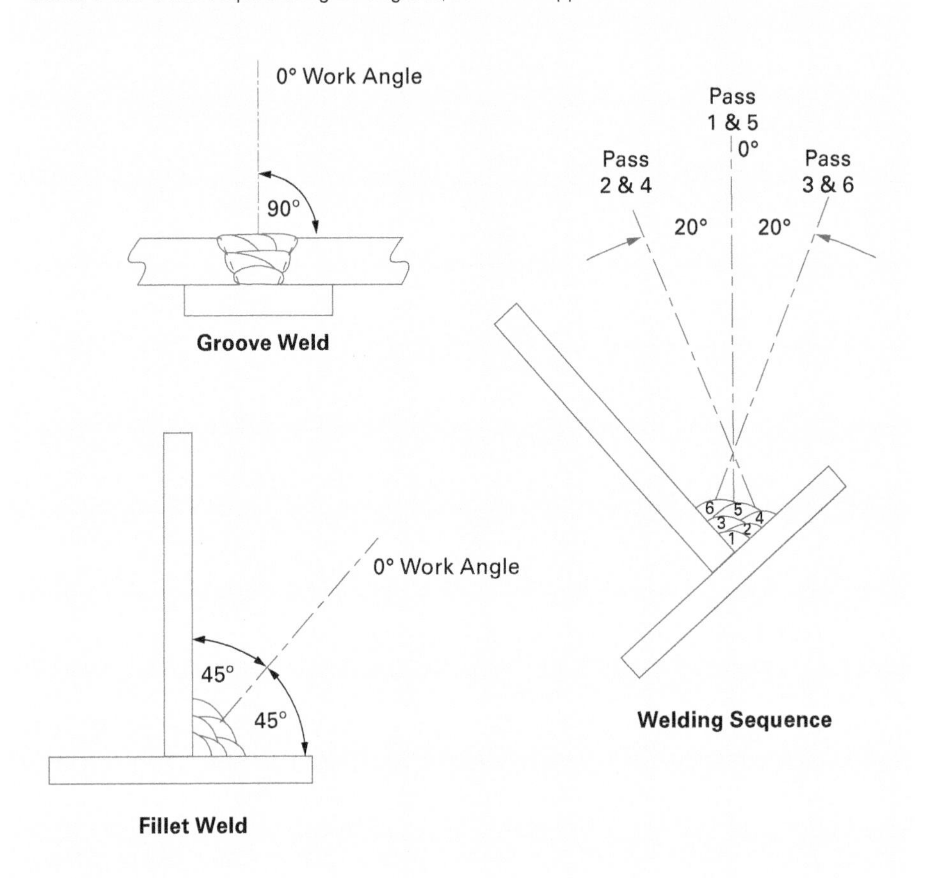

2.1.4 Arc Blow

When a DC welding current flows through the workpiece, it creates a constant magnetic field. Sometimes, this field concentrates in one spot and becomes strong enough to affect the welding arc. When the welder moves into this area, the arc will suddenly swerve. This behavior, **arc blow**, causes defects like spatter and porosity. Ferrous metal parts or welding jigs that become permanently magnetized can cause arc blow too.

The following strategies help combat arc blow:

- If permitted by the WPS and jobsite quality specifications, change to AC operation. You may need to use a different electrode and adjust the amperage. Confirm that these are acceptable before switching.
- If possible, join the workpieces with small welds at the ends and in the middle.
- Use a shorter arc length. Magnetic fields affect short arcs less.
- Change the travel angle to compensate for the arc blow (*Figure 15*).

NOTE

AC welding machines rarely cause arc blow. The rapidly changing current produces a changing magnetic field that doesn't affect the arc.

Arc blow: Magnetic forces that deflect the welding arc from its intended path.

- Reposition the workpiece lead clamp. This changes the current path through the workpiece, which changes the magnetic field.
- If possible, weld towards the workpiece lead clamp.
- If possible, reduce the amperage. This reduces the magnetic field's strength.
- Wrap the workpiece lead around the workpiece several times to counteract the magnetic field. Be sure that it's far enough away from the hot metal.

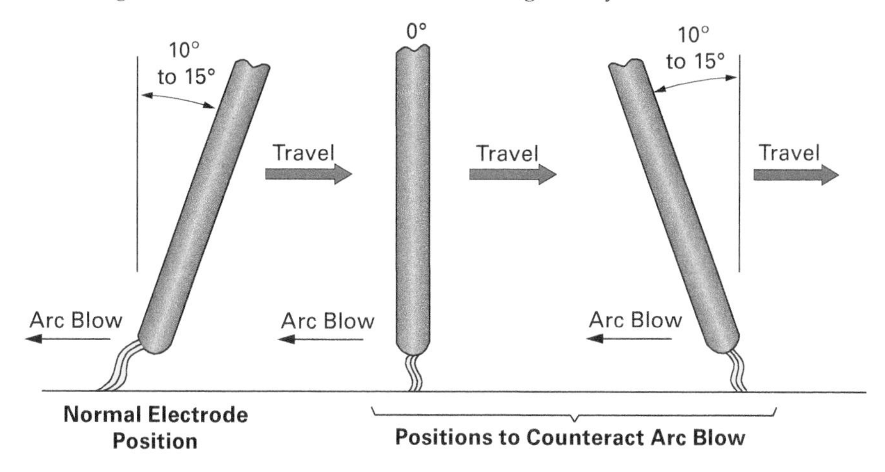

Figure 15 Controlling arc blow.

2.1.5 Terminating a Weld

The way the welder terminates (ends) a weld bead affects the weld's quality. Terminations leave a crater (depression) at the end of the weld. Welding codes require welders to fill the crater completely. This can be difficult if the termination is at a plate edge. Heat builds up here, making it hard to fill the crater.

Producing long welds requires several electrodes. Properly terminating a bead and then restarting (resuming) it with a fresh electrode can be tricky. Again, the way the welder terminates the first bead is important.

The following steps outline producing a good termination with the SMAW process:

Step 1 As you approach the end of the weld, begin to angle the electrode perpendicularly to the base metal (0 degrees). Slow the forward travel.

Step 2 Stop forward movement about $\frac{1}{8}$" (~3 mm) from the end of the weld. Slowly tilt the electrode about 10 degrees toward the start of the weld.

Step 3 Move about $\frac{1}{8}$" (~3 mm) toward the weld. Break the arc when you've filled the crater.

Step 4 Inspect the termination. The crater should be filled to the full cross section of the weld.

Step 5 When you're finished welding, *always* remove the electrode from its holder. This prevents arcing. Discard the stub in the proper container.

Figure 16 shows the complete termination process.

NOTE

If the termination *isn't* the end of the weld, don't completely fill the crater. A shallow crater provides a good place to resume the weld. The two welds will blend more smoothly.

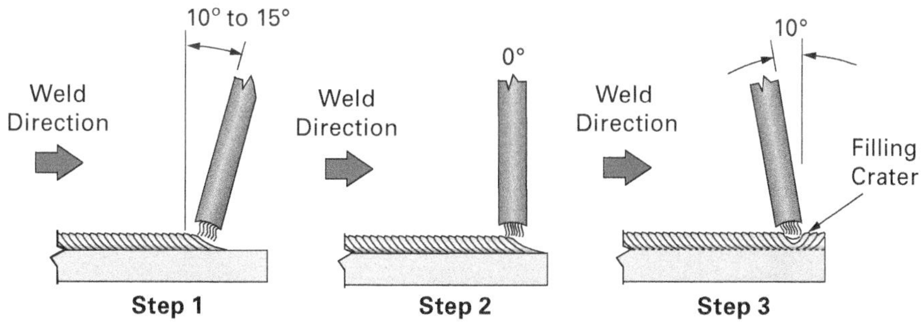

Figure 16 Weld termination.

2.1.6 Restarting a Weld

Restart: The point where a new weld bead smoothly continues from the previous one. Sometimes called a *tie-in*.

A **restart**, or *tie-in*, is the point where one weld bead ends and another begins. Learning to make good restarts is crucial because most SMAW welds will have at least one. The restart must blend smoothly into the existing weld and not stand out. Poor restarts create weld defects.

The following steps outline producing a good restart for all weld bead types:

Step 1 When the electrode is almost used up, quickly increase the travel speed to taper the weld for ¼" to ⅜" (~6 mm to 10 mm). Break the arc.

Step 2 Remove the electrode stub from the holder. This prevents arcing. Discard the stub in the proper container.

Step 3 Chip any slag from the tapered section and the crater.

Step 4 Install a new electrode in the holder. Restrike the arc $\frac{1}{16}$" to ⅛" (~2 mm to 3 mm) in front of the crater and in line with the weld.

NOTE

Welding codes don't allow arc strikes outside of the weld area. Striking just in front of the old bead incorporates the strike into the new bead.

Step 5 Move the electrode back to the tapered weld section. Develop the weld pool with a slight circular motion. Maintain the correct arc length and electrode angles.

Step 6 As soon as the bead achieves the proper width, start moving forward.

Step 7 When you're finished welding, *always* remove the electrode from its holder. This prevents arcing. Discard the stub in the proper container.

Figure 17 shows the restart process.

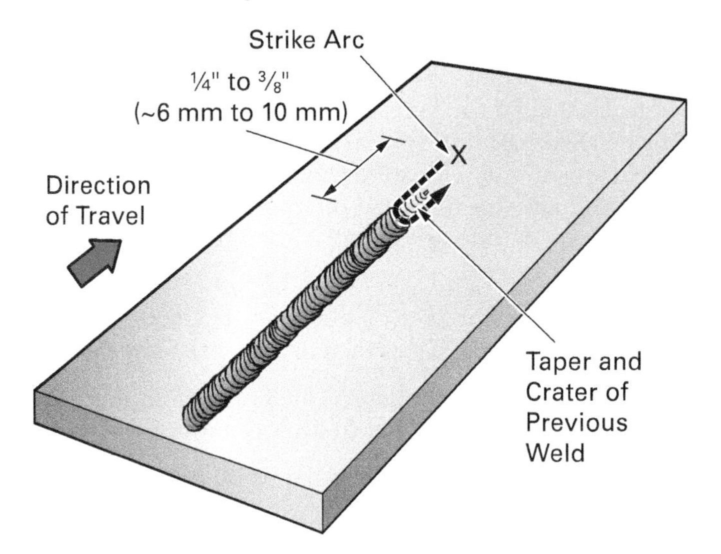

Figure 17 Making a restart.

Arc Problems

When welding two workpieces, welders sometimes struggle with the arc's wandering back and forth between the parts. Similarly, the arc may deposit weld metal on just one piece. An arc that is too long can create this problem. It will tend to jump to the closest workpiece. This problem happens more often with small-diameter electrodes.

If the arc won't deposit metal on one workpiece, confirm that the workpiece is in the welding circuit. Something may be blocking the current flow. The workpiece clamp may be positioned incorrectly.

To fix the problem, either clamp a piece of metal across both parts or weld on a metal table. Alternatively, weave the electrode back and forth across both workpieces to tie them together electrically.

Constant-Current Machines

Welding machines operating in SMAW mode deliver constant current AC or DC. Some machines hold the welding current very close to the selected amperage regardless of arc length. These generally produce better welds since the welding current remains very steady.

Droop-current machines ("droopers") don't maintain as reliable a current flow. Striking the arc and changing the arc length cause the current to rise and fall. These machines don't always produce as consistent a weld since the welding current isn't steady. On the other hand, the welder can exploit this effect to control the weld pool's temperature. Simply raising or lowering the electrode produces a significant change.

Weld Tabs

Welding *run-on* (start) and *run-off* (stop) tabs to the workpiece is an alternative way to eliminate start and stop points. The run-on tab gives the arc time to stabilize and achieve the right penetration. The run-off tab provides a place for the termination crater. Run-on/run-off tabs are particularly useful on multi-pass groove welds. They're also good for welds produced with low-hydrogen or fast-fill electrodes. After completing the weld, the welder removes the tabs, leaving a good weld.

When practicing terminations and restarts, inspect the transition between beads. A properly made restart won't have any sign of the termination crater. If the restart shows undercut, you didn't spend enough time on the taper or crater. Prevent undercut on just one side by using more side-to-side motion as you move into the taper. If the restart has a hump, you spent too much time before resuming forward motion. With practice, you'll soon be making perfect terminations and restarts.

2.2.0 Basic Beads

Welders create welds from *beads*. A weld may be just one bead or many. In *multi-pass welds*, the welder lays down a sequence of beads (passes) in a specific arrangement. Some passes have specific names. For example, in a multi-pass weld, the *root pass* is the first bead. It unites the base metal at the joint root. The *cover pass / cap pass* completes the weld. *Fill passes* fit between the root and cover passes. WPSs show the pass arrangement and order for multi-pass welds.

Welders produce welds from various bead types. To make a **weave bead**, the welder moves the electrode in a side-to-side pattern while moving forward. Pattern choice depends on the application and welding position. By contrast, when making a **stringer bead**, the welder minimizes side-to-side motion while moving forward.

Weave bead: A weld bead made by oscillating the electrode in one of several patterns.

Stringer bead: A weld bead made without any significant oscillating motion.

Weave and stringer beads create a variety of welds and are also useful for resurfacing. *Figure 18* shows examples of each. Drawing symbols, welding codes, WPSs, and jobsite quality specifications outline bead requirements, including width. Always follow these. For example, sometimes the WPS won't permit weave beads for certain passes.

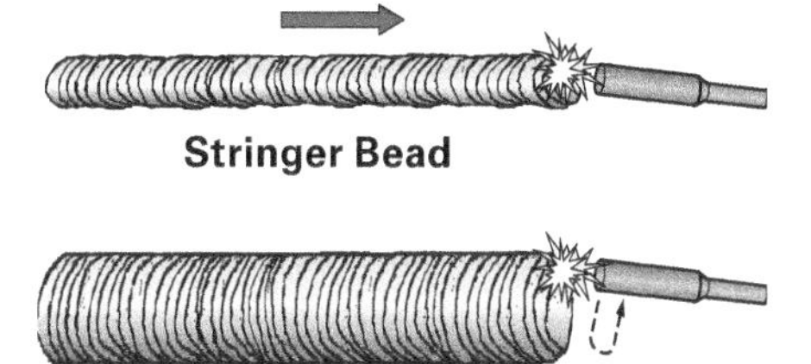

Figure 18 Weave and stringer beads.

2.2.1 Beginning to Weld

Whenever possible, work in a relaxed, comfortable position. Trainee welders sometimes struggle with balance when they first start viewing the world through a dark lens. If you have this problem, lean against a support or widen your stance. These positions reduce fatigue and improve safety. Eventually, you'll become comfortable with limited vision and have no trouble keeping your balance.

To produce excellent welds, you must watch the *entire* work area, not just the arc. Monitor the molten weld pool, the weld buildup at the pool's trailing edge, and the cooling slag. Pay attention to the area around the weld too. Learn to monitor the arc by sound rather than sight. Each electrode type produces a different sound. Arc length and welding position affect the sound too. Experienced welders rely on these cues so they can focus on the big picture rather than just the arc.

Did You Know?

It's Like Driving

Learning to weld is a bit like learning to drive. When you started driving, you had many things to worry about. You had to watch the road, listen to the engine, and keep an eye on your speed. You also had to work the pedals, select gears, operate the turn signal, and find the wiper switch when it rained. Over time, however, you stopped thinking about these things and just got in the car and drove to your destination. You still did these tasks, but your brain handled them automatically.

When you start welding, you'll probably worry about the arc. Successfully striking it and keeping it going will require significant attention. You probably won't have much leisure to see the rest of the weld. For this reason, your first welds might not be very good. Over time, however, your brain will start doing some things automatically, like guiding the weld pool down the right path. Your welds will improve, and you'll start enjoying your work.

Before you start running beads, review basic safety procedures. Make sure you're wearing the correct PPE. Check your eye protection. Confirm that your helmet has a lens with the right shade value for SMAW work (7 to 14). Locate the nearest fire extinguisher.

Some practice instructions may tell you to cool weld coupons by **quenching** them in water. As you learned in NCCER Module 29105, *Base Metal Preparation*, rapidly cooling hot metal hardens it. It also makes it brittle and creates internal stresses that can produce cracks. Never quench production welds or qualification test coupons! Only quench practice coupons.

CAUTION

Always quench in a bucket or quench tank rather than a sink or water fountain. Slag can clog drains.

Quenching: Rapidly cooling a metal by plunging it into a liquid.

2.2.2 Practicing Stringer Beads with E6010/E6011 Electrodes

Practice running stringer beads in the flat position with $^{3}/_{32}$" (2.4 mm), $^{1}/_{8}$" (3.2 mm), or $^{5}/_{32}$" (4.0 mm) E6010/E6011 electrodes. After striking the arc, tilt the electrode to a 10- to 15-degree drag angle. Use a 0-degree work angle.

A whipping (stepping) motion helps control the weld pool when you're depositing a stringer bead. Move the electrode up and forward by about $^{1}/_{4}$" (~6 mm). Then move down and backward by about $^{3}/_{16}$" (~5 mm). Pause after the backward motion to deposit the weld metal. The travel distance can vary according to the application.

NOTE

Never use a whipping motion with E7018 (low-hydrogen) electrodes.

Lengthening and advancing the arc lets the weld pool cool. Lowering the arc temperature reduces metal transfer from the electrode. Advancing the arc preheats the base metal ahead of the weld. This burns off contaminants. *Figure 19* shows the whipping technique.

Listen as you create the weld. If you're using the right arc length, travel speed, and electrode angles, you'll hear a distinct "frying" sound. Make small changes to these factors and notice how the sound changes.

The following steps outline producing a stringer bead:

Step 1 Strike the arc. Move the electrode back slightly to the correct starting point. Position the electrode at a 10- to 15-degree drag angle. Use a 0-degree work angle.

Step 2 Hold the arc steady until the weld pool has a width between two and three times the electrode core's diameter.

Step 3 Slowly move the arc forward with a whipping motion. Keep the arc length consistent. Remember, the electrode is melting and getting shorter as it adds filler metal to the weld. You'll have to compensate to maintain the right arc length.

Step 4 Create a bead 2" to 3" (~5 cm to 8 cm) long. Break the arc by quickly lifting the electrode straight up. This will leave a crater at the termination.

Step 5 Chip, clean, and brush the weld.

Step 6 Make a restart and continue welding to the end of the plate. Properly terminate the finished weld.

Step 7 Chip, clean, and brush the weld.

Step 8 When you're finished welding, *always* remove the electrode from its holder. This prevents arcing. Discard the stub in the proper container.

NOTE

Expect to use about twice as many inches of electrode for each inch of bead that you produce.

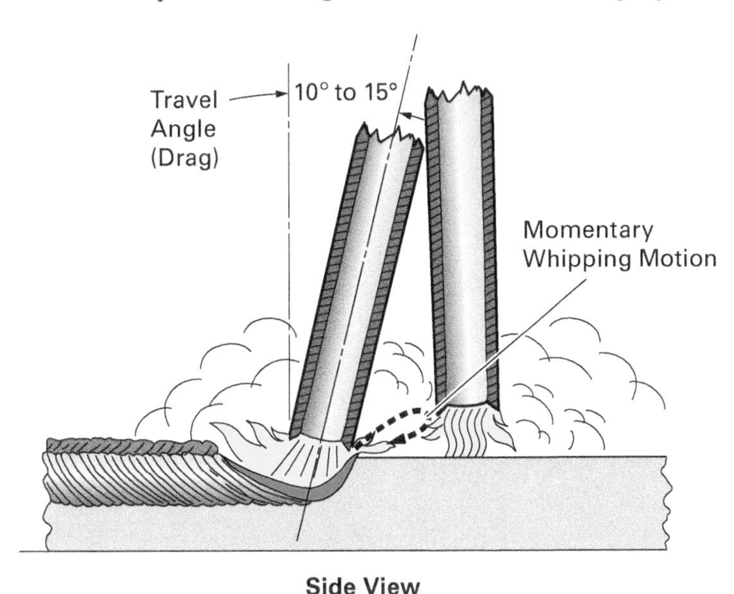

Figure 19 Whipping motion.

Good stringer beads have the following qualities:

- Straight to within ⅛" (~3 mm)
- Uniform bead face
- Crater and restarts filled to the full weld cross section
- Smooth, flat transition with complete fusion at the weld toes
- No porosity
- No excessive undercut at the toes
- No inclusions
- No cracks
- No overlap

Figure 20 shows acceptable and unacceptable stringer beads. It also explains the causes behind the unacceptable beads. Have your instructor inspect your work and offer feedback. Keep practicing until you can consistently make good stringer beads.

NOTE

Always follow the applicable code, WPS, or jobsite quality specifications when producing and evaluating weld beads.

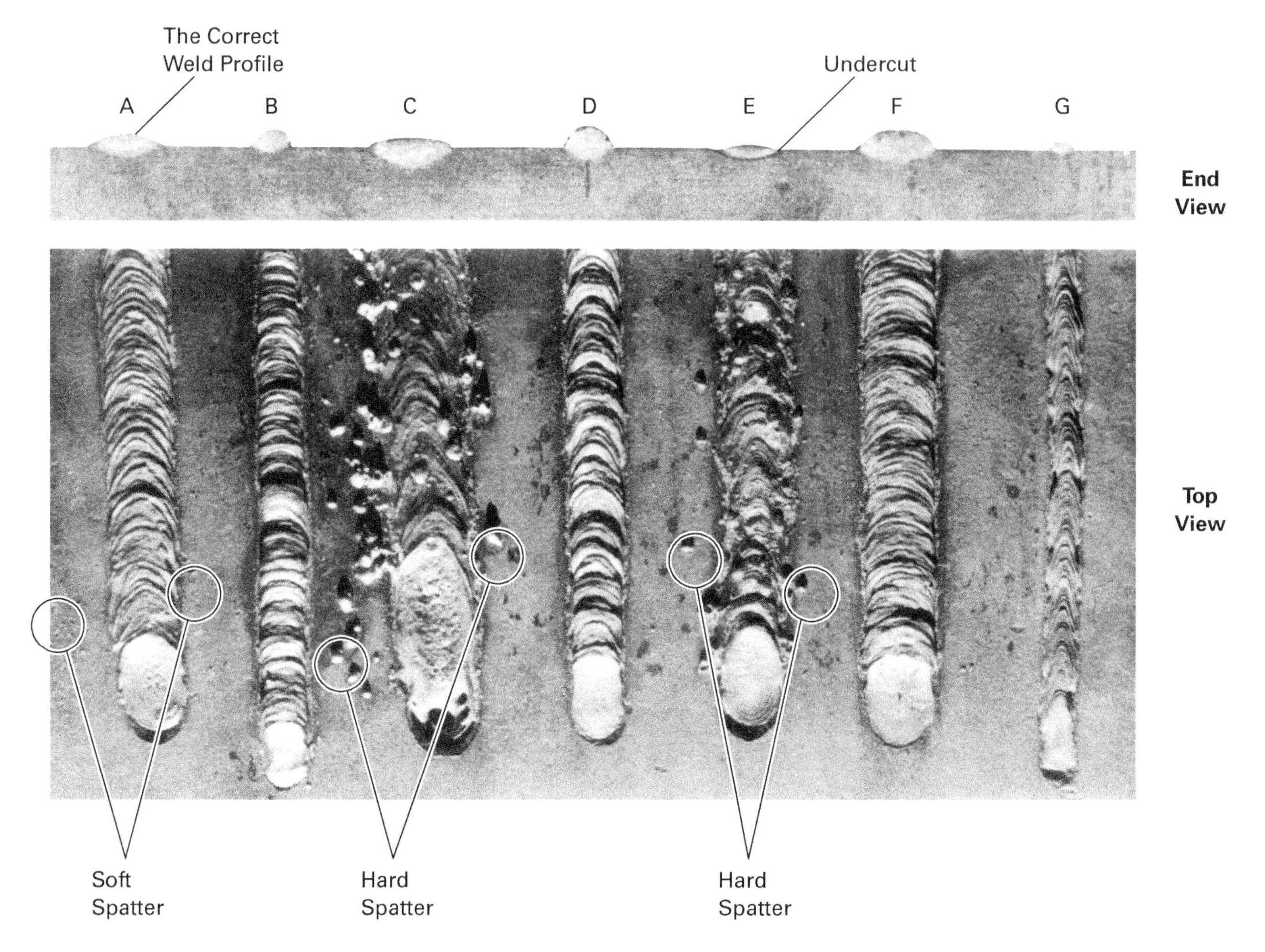

A = Correct current, arc length, and travel speed. Note the easily removed spatter (soft spatter).

B = Current set too low.

C = Current set too high. Note the hard spatter (spatter firmly bonded to base material that must be ground or chiseled off). Note the pointed ends of the bead indicating the weld pool was too hot and cooled too slowly. Impurities are usually trapped in the weld due to the slow cooling.

D = Arc length too short (narrow, high bead caused by arc pressure).

E = Arc length too long. Note the hard spatter and bead undercut of the edges.

F = Travel speed too slow (wide high bead).

G = Travel speed too fast. Note the pointed ends of bead.

Figure 20 Acceptable and unacceptable stringer beads.
Source: Permission granted by the American Welding Society

2.2.3 Practicing Stringer Beads with E7018 Electrodes

Practice running stringer beads in the flat position with $\frac{3}{32}$" (2.4 mm), $\frac{1}{8}$" (3.2 mm), or $\frac{5}{32}$" (4.0 mm) E7018 electrodes. After striking the arc, tilt the electrode to a 10- to 15-degree drag angle. Use a 0-degree work angle.

Use a technique similar to that outlined for E6010/E6011 electrodes, but don't whip the electrode. With low-hydrogen electrodes, the arc should never leave the weld pool. The visible arc should be shorter than with E6010/E6011 electrodes. Remember, the thicker flux coating hides the arc's true length. If the arc leaves the weld pool or is too long, weld defects like porosity or hydrogen embrittlement will appear. Move the arc within the weld pool to control the bead shape.

2.2.4 Practicing Weave Beads with E6010/E6011 Electrodes

Practice running weave beads in the flat position with 3⁄32" (2.4 mm), 1⁄8" (3.2 mm), or 5⁄32" (4.0 mm) E6010/E6011 electrodes. After striking the arc, tilt the electrode to a 10- to 15-degree drag angle. Use a 0-degree work angle.

As you move the electrode forward, also move it side to side in a weave pattern. Possible patterns include zigzags, Js, crescents, boxes, circles, and figure 8s (*Figure 21*). Slow down or pause slightly at the edges to ensure tie-in between the weld toes and the base metal. The pause flattens the weld, giving it the proper profile. Zigzags require longer pauses than other patterns.

Listen as you create the weld. If you're using the right arc length, travel speed, and electrode angles, you'll hear a distinct "frying" sound. Make small changes to these factors and notice how the sound changes.

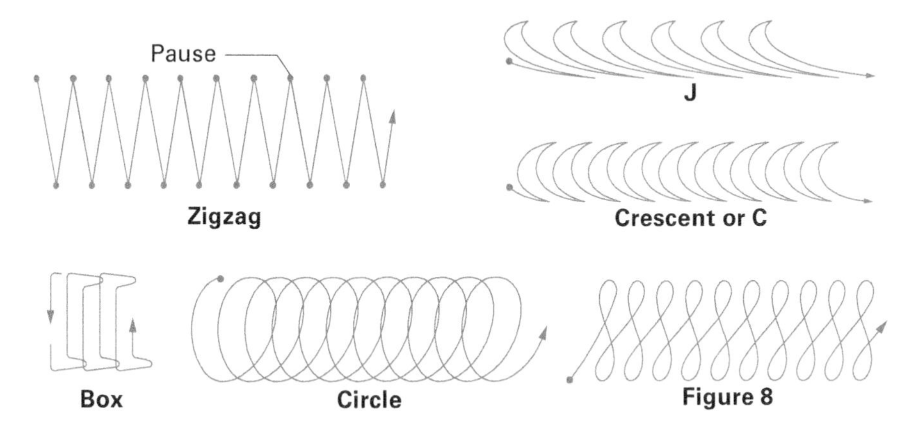

Figure 21 Weave patterns.

The following steps outline producing a weave bead:

Step 1 Strike the arc. Move the electrode back slightly to the correct starting point. Position the electrode at a 10- to 15-degree drag angle. Use a 0-degree work angle.

Step 2 Hold the arc steady until the weld pool has a width between two and three times the electrode core's diameter.

Step 3 Slowly move the arc forward while making the weaving pattern. Keep the arc length consistent. Remember, the electrode is melting and getting shorter as it adds filler metal to the weld. You'll have to compensate to maintain the right arc length.

NOTE

Weave beads can be four to six times the electrode core's diameter. Expect to use several inches of electrode for each inch of bead that you produce.

Step 4 Create a bead 2" to 3" (~5 cm to 8 cm) long. Break the arc by quickly lifting the electrode straight up. This will leave a crater at the termination.

Step 5 Chip, clean, and brush the weld.

Step 6 Make a restart and continue welding to the end of the plate. Properly terminate the finished weld.

Step 7 Chip, clean, and brush the weld.

Step 8 When you're finished welding, *always* remove the electrode from its holder. This prevents arcing. Discard the stub in the proper container.

Good weave beads have the following qualities:

- No wider than four to six times the electrode core's diameter
- Straight to within 1⁄8" (~3 mm)
- Uniform bead face
- Crater and restarts filled to the full weld cross section
- Smooth, flat transition with complete fusion at the weld toes

- Minimal or no porosity (code dependent)
- No excessive undercut at the toes
- No inclusions
- No cracks
- No overlap

NOTE

Always follow the applicable code, WPS, or jobsite quality specifications when producing and evaluating weld beads.

Figure 22 shows an acceptable weave bead. Have your instructor inspect your work and offer feedback. Keep practicing until you can consistently make good weave beads.

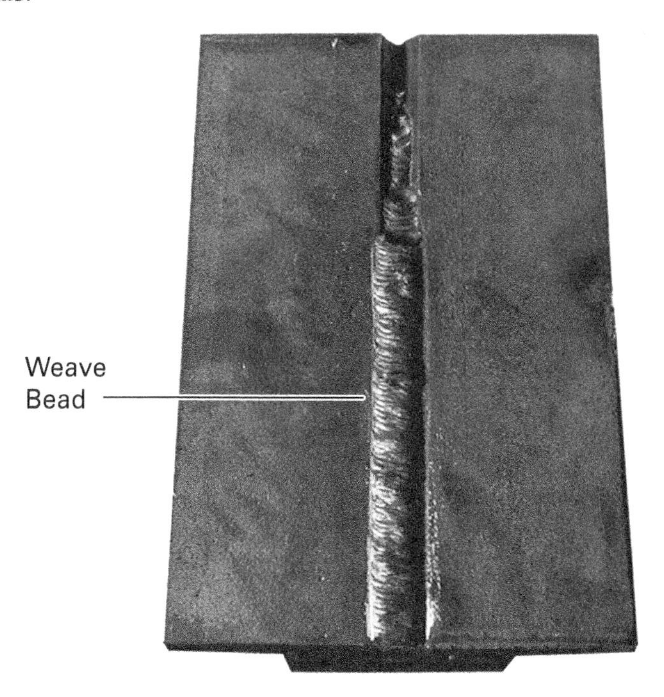

Figure 22 Weave bead.

2.2.5 Practicing Weave Beads with E7018 Electrodes

Practice running weave beads in the flat position with $^3/_{32}$" (2.4 mm), $^1/_8$" (3.2 mm), or $^5/_{32}$" (4.0 mm) E7018 electrodes. After striking the arc, tilt the electrode to a 10- to 15-degree drag angle. Use a 0-degree work angle. When producing weave beads with E7018 electrodes, don't use the circle, box, or figure 8 patterns.

Use a technique similar to that outlined for E6010/E6011 electrodes. With low-hydrogen electrodes, the arc should never leave the weld pool. The visible arc should be shorter than with E6010/E6011 electrodes. Remember, the thicker flux coating hides the arc's true length. If the arc leaves the weld pool or is too long, weld defects like porosity or hydrogen embrittlement will appear. Move the arc within the weld pool to control the bead shape.

Weave Bead Width

Stringer beads are usually two to three times wider than the electrode core's diameter. Weave beads, however, can be up to four to six times the electrode core's diameter. They should never exceed this value.

2.2.6 Practicing Surfacing Beads with E6010/E6011 Electrodes

Surfacing is a technique in which the welder deposits parallel beads on a surface. The beads overlap, producing a raised, relatively flat surface. Surfacing welds can build up a surface that's too thin. The result is a *pad* (*Figure 23*).

Always clean surfacing beads very carefully to prevent slag inclusions. Keep the overlap as consistent as possible so the finished surface is relatively flat. Irregular surfacing beads will require machining before the new surface is usable.

Practice running surfacing beads in the flat position with $^3/_{32}$" (2.4 mm), $^1/_8$" (3.2 mm), or $^5/_{32}$" (4.0 mm) E6010/E6011 electrodes. After striking the arc, tilt the electrode to a 10- to 15-degree drag angle. Use a 0-degree work angle for the first bead. For later beads, use a 10- to 15-degree work angle (*Figure 24*). If the coupon gets too hot between passes, quench it in water.

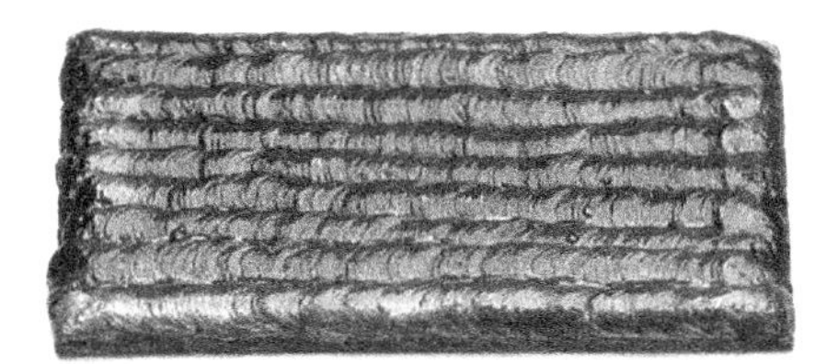

Figure 23 A pad created with surfacing beads.
Source: The Lincoln Electric Company, Cleveland, OH, USA

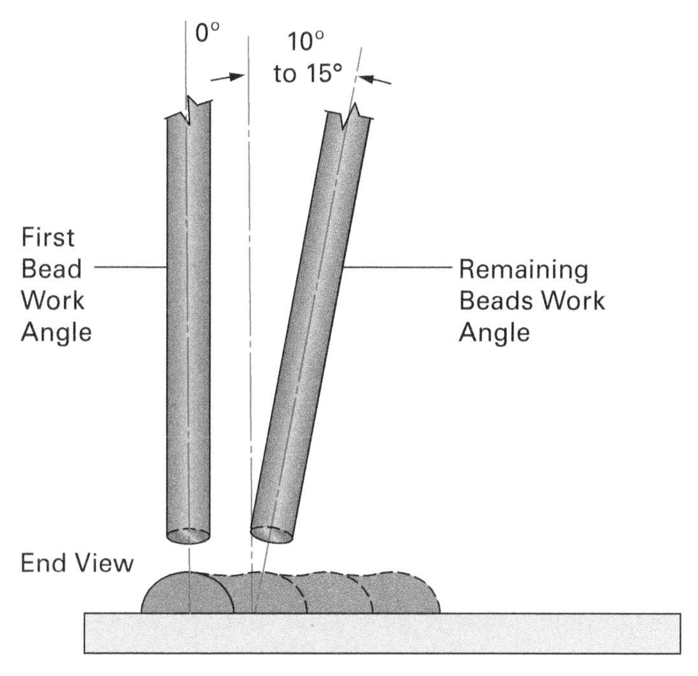

Figure 24 Work angles for surfacing beads.

WARNING!

Wear gloves and handle hot coupons with pliers. Be careful of the steam rising from coupons when you quench them. It can burn unprotected skin.

The following steps outline producing surfacing beads:

Step 1 Using soapstone, mark out a 4" (~10 cm) square on the practice coupon. You'll build the pad starting along one side.

Step 2 Strike the arc. Move the electrode back slightly to the correct starting point. Position the electrode at a 10- to 15-degree drag angle. Use a 0-degree work angle.

Step 3 Hold the arc steady until the weld pool has a width between two and three times the electrode core's diameter.

Step 4 Deposit a stringer bead along one side of the square.

Step 5 Chip, clean, and brush the weld.

Step 6 Deposit another stringer bead parallel to the first and slightly overlapping it. Change to a 10- to 15-degree work angle.

Step 7 Chip, clean, and brush the weld.

Step 8 Continue adding beads until you've completed the pad.

Step 9 When you're finished welding, *always* remove the electrode from its holder. This prevents arcing. Discard the stub in the proper container.

Figure 25 shows acceptable and unacceptable surfacing beads. Have your instructor inspect your work and offer feedback. Keep practicing until you can consistently make a pad. Once you've made a good pad with stringer beads, create another with weave beads.

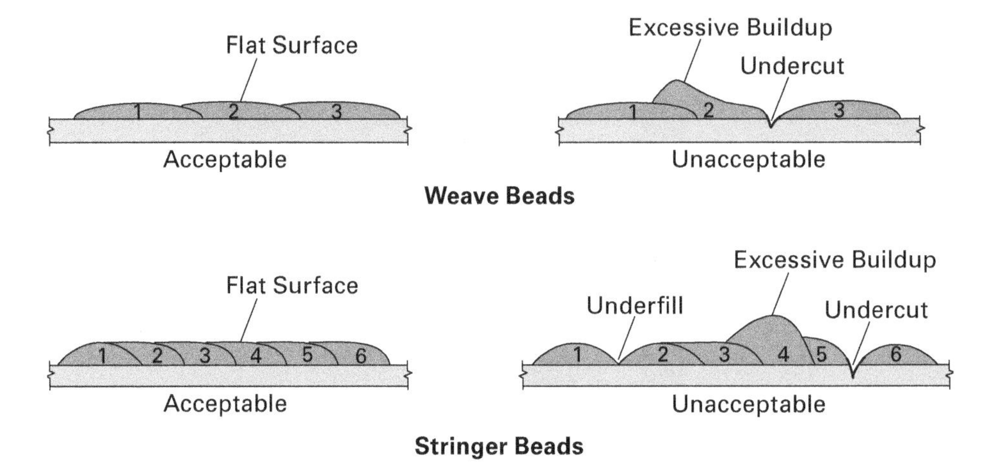

Figure 25 Acceptable and unacceptable surfacing beads.

2.2.7 Practicing Surfacing Beads with E7018 Electrodes

Practice running surfacing beads in the flat position with 3⁄32" (2.4 mm), 1⁄8" (3.2 mm), or 5⁄32" (4.0 mm) E7018 electrodes. After striking the arc, tilt the electrode to a 10- to 15-degree drag angle. Use a 0-degree work angle for the first bead. For later beads, use a 10- to 15-degree work angle. If the coupon gets too hot between passes, quench it in water.

WARNING!

Wear gloves and handle hot coupons with pliers. Be careful of the steam rising from coupons when you quench them. It can burn unprotected skin.

Produce a pad using stringer beads. Then produce another using weave beads. Use a technique similar to that outlined for E6010/E6011 electrodes, but don't whip the electrode. With low-hydrogen electrodes, the arc should never leave the weld pool. The visible arc should be shorter than with E6010/E6011 electrodes. Remember, the thicker flux coating hides the arc's true length. If the arc leaves the weld pool or is too long, weld defects like porosity or hydrogen embrittlement will appear. Move the arc within the weld pool to control the bead shape.

FLASH

Learning the memory aid **FLASH** helps you produce good welds:

F (fit-up) — Correctly align and tack weld the base material together according to your instructor's specifications.

L (length of arc) — Maintain the right distance between the electrode and the base metal. This distance usually equals the electrode core's diameter.

A (angle) — Two angles are critical: the *work angle* and the *travel angle*.

S (speed) — Travel speed, in inches per minute (IPM), affects the weld's width.

H (heat) — The amperage setting controls the heat. The proper value depends on the electrode diameter, base metal type, base metal thickness, and welding position.

2.3.0 Fillet Welds

Now that you've gotten some experience producing beads, it's time to create a weld. You'll begin with the fillet weld, a relatively easy type. They're common in lap and T-joints. Welders can make fillet welds in all positions, provided they use the right electrodes. The weld position identifier for plate depends on the weld type and workpiece orientation (*Figure 26*). All fillet weld positions end in F ("fillet"). Fillet weld positions for plate are flat (1F), horizontal (2F), vertical (3F), and overhead (4F).

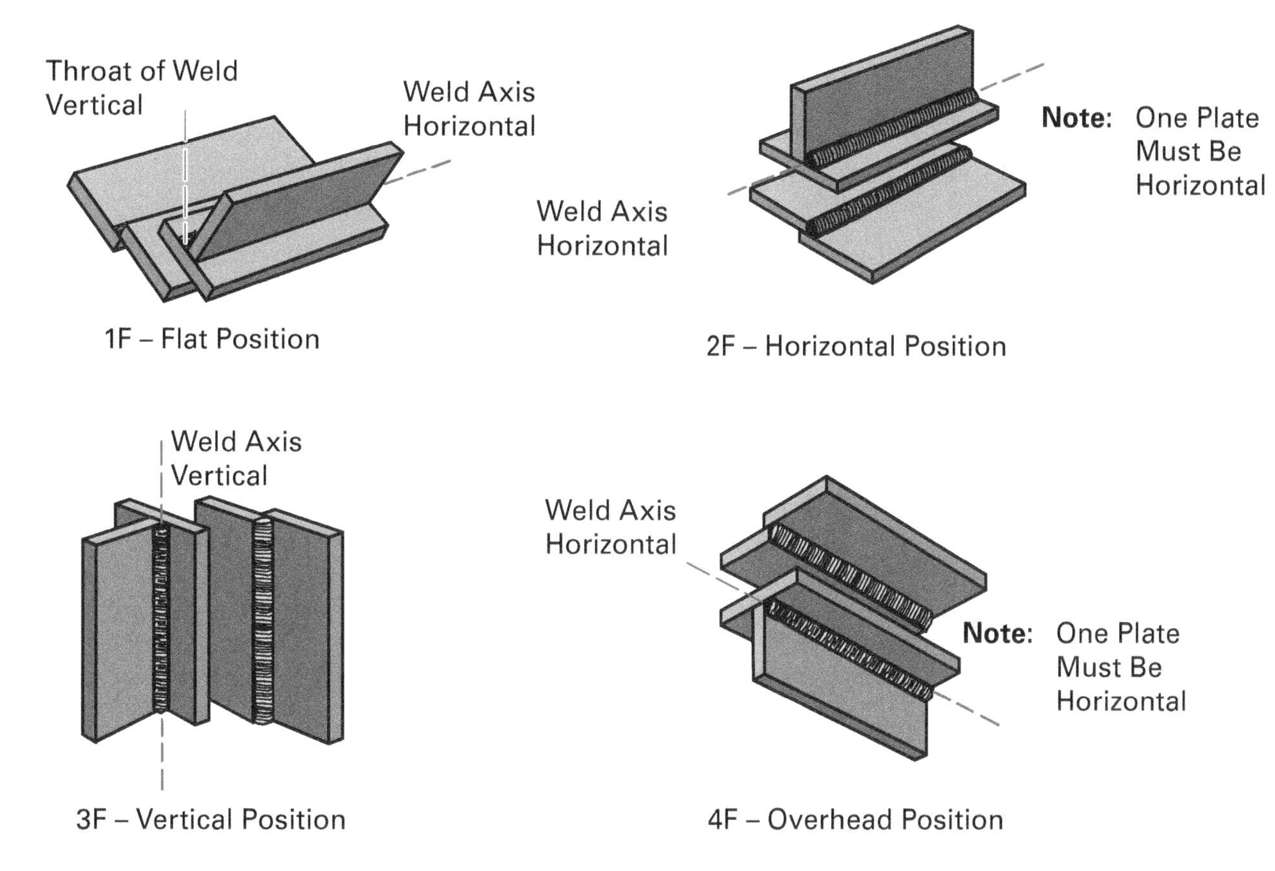

Figure 26 Fillet weld positions for plate.

The welder can incline the weld axis up to 15 degrees in the 1F and 2F positions. The AWS standard specifies a permissible inclination in the 3F and 4F positions that depends on the rotational position.

Good fillet welds in all positions have the following qualities:

- Uniform size to within $\pm^1/_{16}$" (1.6 mm)
- Uniform face
- Profile acceptable to the applicable code
- Crater and restarts filled to the full weld cross section
- Smooth, flat transition with complete fusion at the weld toes
- Limited or no porosity (code dependent)
- No excessive undercut at the toes
- No inclusions
- No cracks
- No overlap

NOTE

Welders use the position identifiers (1F, 2F, 3F, and 4F) when referring to qualification tests. When making production welds, they don't use them. Instead, they use the terms "flat," "horizontal," "vertical," and "overhead" to identify position.

NOTE

Always follow the applicable code, WPS, or jobsite quality specifications when producing and evaluating fillet welds.

2.3.1 Tack Welding

When welders assemble a weldment, they need a way to hold its pieces together. Sometimes, they use clamps, jigs, or fixtures. Often, however, they use small welds, called **tack welds**, to keep everything correctly aligned (*Figure 27*).

Think of tack welds as temporary. They don't add strength to the final weld. They just keep things from moving. Sometimes, the welder incorporates them into the actual weld. Alternatively, the welder may remove them. Tack welds can temporarily secure an assembly in a specific welding position. They can also attach temporary braces and fixtures.

Tack welds: Small welds made to hold a weldment's parts in the proper alignment before welding.

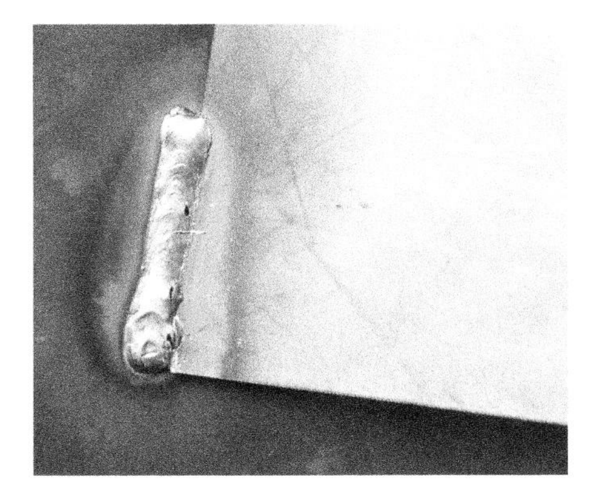

Figure 27 Tack weld.
Source: iStock@Temir Shintemirov

E6010/E6011 electrodes are common choices for making tack welds. They penetrate deeply and leave easily removable slag. Sometimes, however, the WPS or jobsite quality specifications require specific electrode types for tack welds. Welders often save electrode stubs for making tack welds.

The following steps outline producing a tack weld:

Step 1 Position the parts in the proper relationship. If necessary, clamp them together.

Step 2 Strike the arc on one part.

Step 3 Hold the arc steady until the weld pool widens to about twice the electrode core's diameter.

Step 4 Slowly move the arc onto the second part. The weld pool should bridge both parts.

Step 5 Weld along the joint far enough to produce the required holding strength. Tack welds $\frac{1}{4}$" to $\frac{1}{2}$" (~6 mm to 13 mm) long are typical.

Step 6 Break the arc by quickly lifting the electrode straight up.

Step 7 Chip, clean, and brush the weld.

Step 8 When you're finished welding, *always* remove the electrode from its holder. This prevents arcing. Discard the stub in the proper container.

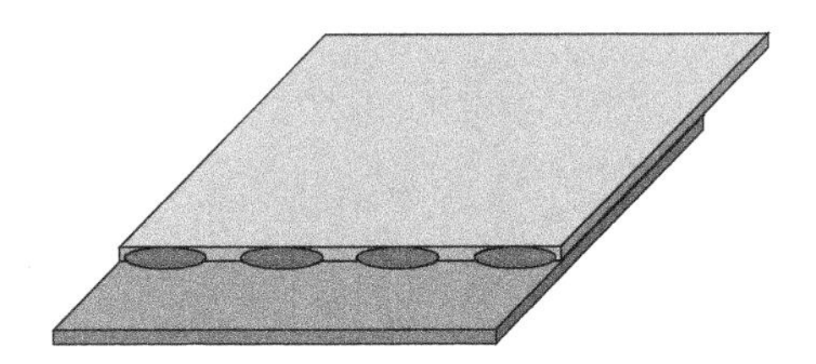

Figure 28 Practicing tack welds.

Finished tack welds are circular or look like short beads. They shouldn't be too large. A good way to practice them is to arrange two scrap coupons in a lap joint arrangement. Make a series of tack welds along the edge (*Figure 28*).

To make a T-joint fillet weld, you'll need to assemble the coupons with tack welds. Arrange them as *Figure 29* shows. Then produce a tack weld at each end where the coupons come together.

Tacking and Aligning Workpieces

To tack weld two workpieces, create a weld about $\frac{1}{2}$" (~13 mm) long on one side. Use a hammer or similar tool to adjust the parts' alignment. Create a second tack weld on the opposite side. Finally, create two more tack welds at the opposite end. Intermediate tack welds every 5" to 6" (~13 cm to 15 cm) will reduce lengthwise distortion.

2.3.2 Practicing Fillet Welds with E6010/E6011 Electrodes (1F Position)

NOTE

When practicing in the flat position, ask your instructor which bead type to use.

Practice fillet welds in the 1F (flat) position with $\frac{3}{32}$" (2.4 mm), $\frac{1}{8}$" (3.2 mm), or $\frac{5}{32}$" (4.0 mm) E6010/E6011 electrodes. The root pass must always be a stringer bead. The remaining passes may either be stringer beads or weave beads. Assemble two coupons into a T-joint arrangement and create a multi-pass weld. Use a 10- to 15-degree drag angle. Use a 0-degree work angle for the root pass. The work angle for the remaining passes depends on the bead type and pass number. Increase or decrease the travel speed to control the weld metal buildup.

Follow these steps to practice making fillet welds in the 1F position using stringer beads:

Figure 29 Assembled fillet weld coupon.

Step 1 Tack weld two coupons together to form a T-joint (*Figure 29*). Clean the tack welds.

Step 2 Place the assembly on the welding table in the 1F position. If necessary, use clamps to hold it in place.

Step 3 Run the first bead along the joint root. Use a 10- to 15-degree drag angle. Use a 0-degree work angle. Deposit the stringer bead with a whipping motion. Push the arc into the root. The root must fuse together, or a notch will appear on the bead's leading edge.

Step 4 Chip, clean, and brush the root pass.

Step 5 Run the second bead along the bottom toe of the first weld. Deposit the stringer bead with a slight **oscillation** or stepping motion. The bead should overlap about 75 percent of the root pass. Use the electrode work angle shown in *Figure 30*.

Oscillation: A regular, repetitive side-to-side motion.

Step 6 Chip, clean, and brush the bead.

Step 7 Run the remaining stringer beads along the toes of the underlying beads. The beads should overlap by about 50 percent. Chip, clean, and brush each bead. *Figure 30* shows all welding passes with their approximate work angles.

Step 8 When you're finished welding, *always* remove the electrode from its holder. This prevents arcing. Discard the stub in the proper container.

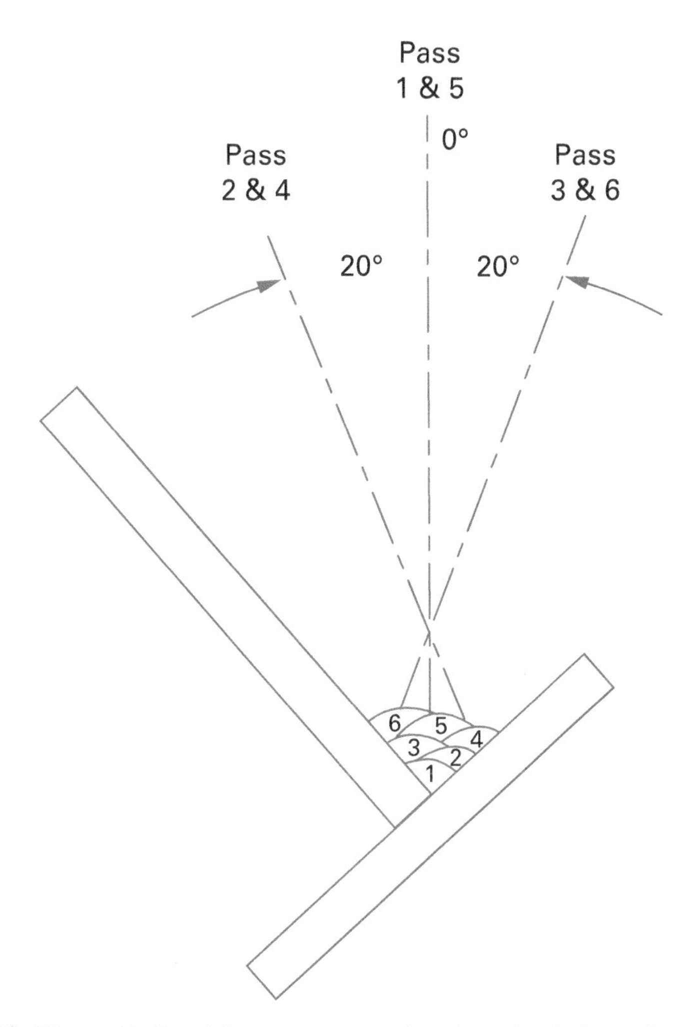

Figure 30 Fillet weld 1F welding sequence and work angles (stringer, E6010/E6011).

Follow these steps to practice making fillet welds in the 1F position using weave beads:

Step 1 Tack weld two coupons together to form a T-joint (*Figure 29*). Clean the tack welds.

Step 2 Place the assembly on the welding table in the 1F position. If necessary, use clamps to hold it in place.

Step 3 Run the first bead along the joint root. Use a 10- to 15-degree drag angle. Use a 0-degree work angle. Deposit the stringer bead with a whipping motion. Push the arc into the root. The root must fuse together, or a notch will appear on the bead's leading edge.

Step 4 Chip, clean, and brush the root pass.

Step 5 Run the remaining weave beads using the pattern that your instructor specifies. Use a 0-degree work angle. *Figure 31* shows all welding passes. Move slowly across the weld face, pausing at each toe to fill the crater. A slight whip controls the weld pool when you reach the toe. Chip, clean, and brush each pass as you complete it.

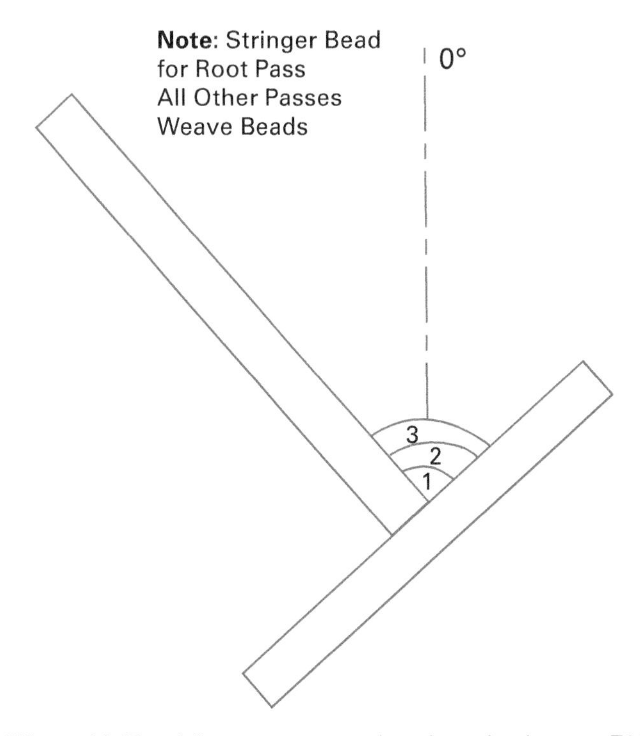

Figure 31 Fillet weld 1F welding sequence and work angles (weave, E6010/E6011).

Step 6 When you're finished welding, *always* remove the electrode from its holder. This prevents arcing. Discard the stub in the proper container.

Examine the finished weld and compare its profile to the examples in *Figure 3*. Have your instructor inspect your work and offer feedback. Keep practicing until you can consistently make good fillet welds.

Preferred Fillet Weld Contours

Flat or slightly convex faces are best for single-pass, weave-bead fillet welds where two workpieces meet at an angle (not lap joints). This arrangement distributes weld stresses more evenly throughout the fillet and workpieces.

2.3.3 Practicing Fillet Welds with E7018 Electrodes (1F Position)

Practice fillet welds in the 1F (flat) position with $^{3}/_{32}$" (2.4 mm), $^{1}/_{8}$" (3.2 mm), or $^{5}/_{32}$" (4.0 mm) E7018 electrodes. Use the same general technique as that outlined for E6010/E6011 electrodes. When producing the root pass' stringer bead, don't whip the electrode. Instead, just drag it along with minimal side-to-side motion. With low-hydrogen electrodes, the arc should never leave the weld pool. The visible arc should be shorter than with E6010/E6011 electrodes. Remember, the thicker flux coating hides the arc's true length.

2.3.4 Practicing Fillet Welds with E6010/E6011 Electrodes (2F Position)

Practice fillet welds in the 2F (horizontal) position with $^{3}/_{32}$" (2.4 mm), $^{1}/_{8}$" (3.2 mm), or $^{5}/_{32}$" (4.0 mm) E6010/E6011 electrodes. All passes should be stringer beads. Assemble two coupons into a T-joint arrangement and create a multi-pass weld. Use a 10- to 15-degree drag angle. Use a 0-degree work angle for the root pass. For later passes, adjust the work angle as required. Increase or decrease the travel speed to control the weld metal buildup.

Follow these steps to practice making fillet welds in the 2F position:

Step 1 Tack weld two coupons together to form a T-joint (*Figure 29*). Clean the tack welds.

Step 2 Place the assembly on the welding table in the 2F position. If necessary, use clamps to hold it in place.

Step 3 Run the first bead along the joint root. Use a 10- to 15-degree drag angle. Use a 0-degree work angle. Deposit the stringer bead with a whipping motion. Push the arc into the root. The root must fuse together, or a notch will appear on the bead's leading edge.

Step 4 Chip, clean, and brush the root pass.

Step 5 Run the second stringer bead along the bottom toe of the first weld. Use a slight oscillation or stepping motion. The bead should overlap about 75 percent of the root pass. Use the electrode work angle shown in *Figure 32*.

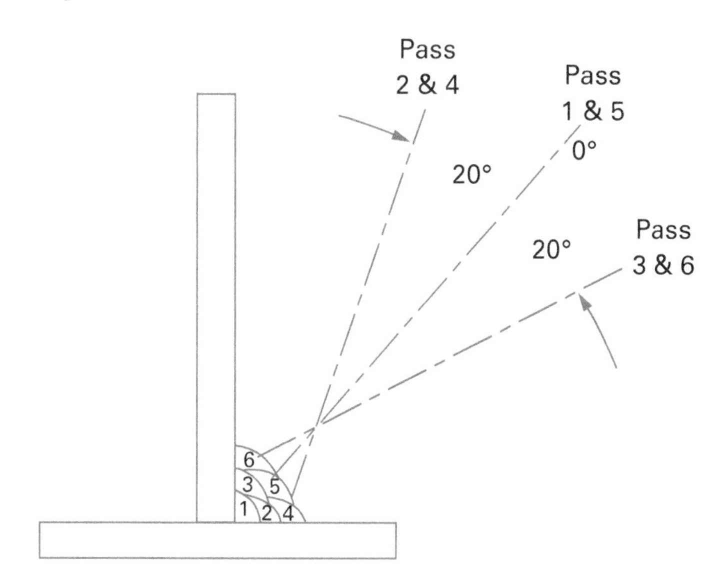

Figure 32 Fillet weld 2F welding sequence and work angles (E6010/E6011).

Step 6 Chip, clean, and brush the bead.

Step 7 Run the remaining stringer beads along the toes of the underlying beads. The beads should overlap by about 50 percent. Chip, clean, and brush each bead. *Figure 32* shows all welding passes with their approximate work angles.

Step 8 When you're finished welding, *always* remove the electrode from its holder. This prevents arcing. Discard the stub in the proper container.

Examine the finished weld and compare its profile to the examples in *Figure 3*. Have your instructor inspect your work and offer feedback. Keep practicing until you can consistently make good fillet welds.

2.3.5 Practicing Fillet Welds with E7018 Electrodes (2F Position)

Practice fillet welds in the 2F (horizontal) position with $^{3}/_{32}$" (2.4 mm), $^{1}/_{8}$" (3.2 mm), or $^{5}/_{32}$" (4.0 mm) E7018 electrodes. Use the same general technique as that outlined for E6010/E6011 electrodes. When producing the stringer beads, don't whip the electrode. Instead, just drag it along with minimal side-to-side motion. With low-hydrogen electrodes, the arc should never leave the weld pool. The visible arc should be shorter than with E6010/E6011 electrodes. Remember, the thicker flux coating hides the arc's true length.

2.3.6 Practicing Fillet Welds with E6010/E6011 Electrodes (3F Position)

Welders call this weld a *vertical-up fillet weld*. It requires a different technique than fillet welds produced in the other positions. You'll use a push angle instead of a drag angle. Either stringer or weave beads are acceptable. Be sure to use the same type for all passes. Weave beads are common when welding with E6010/E6011 electrodes. They're also good for welding carbon steel with E7018 electrodes. Stringer beads work better when welding alloy steels with E7018 electrodes.

NOTE

When practicing in the vertical position, ask your instructor which bead type to use.

Practice fillet welds in the 3F (vertical) position with 3⁄32" (2.4 mm), 1⁄8" (3.2 mm), or 5⁄32" (4.0 mm) E6010/E6011 electrodes. Assemble two coupons into a T-joint arrangement and create a multi-pass weld. Use a 0- to 10-degree push angle. Use a 0-degree work angle for the root pass. The work angle for the remaining passes depends on the bead type and pass number. Increase or decrease the travel speed to control the weld metal buildup.

Follow these steps to make a vertical-up fillet weld using stringer beads:

Step 1 Tack weld two coupons together to form a T-joint (*Figure 29*). Clean the tack welds.

Step 2 Tack weld the assembly in the vertical position.

Step 3 Run the first stringer bead along the joint root, starting from the bottom. Use a 0- to 10-degree push angle. Use a 0-degree work angle. Build the bead with a whipping motion. Quickly raise the electrode about 1⁄4" (~6 mm) and then drop it back into the weld pool. Pause in the weld pool.

Step 4 Chip, clean, and brush the root pass.

Step 5 Run the remaining stringer beads. *Figure 33* shows all welding passes with their approximate work angles. Chip, clean, and brush each pass as you complete it.

Step 6 When you're finished welding, *always* remove the electrode from its holder. This prevents arcing. Discard the stub in the proper container.

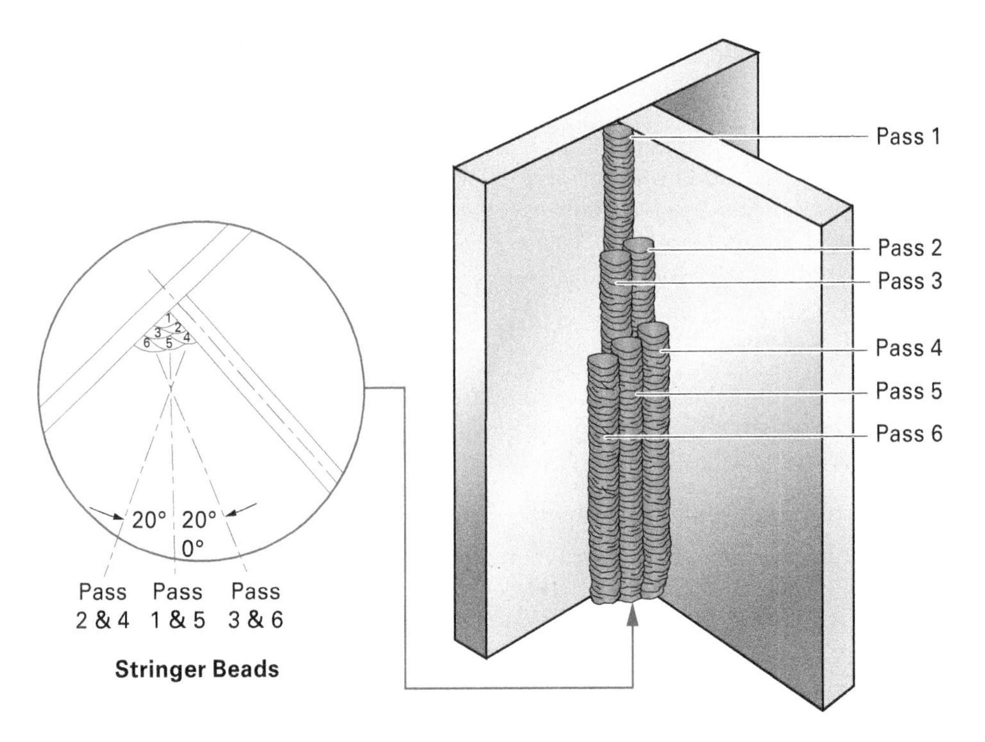

Figure 33 Fillet weld 3F welding sequence and work angles (stringer, E6010/E6011).

Follow these steps to make a vertical-up fillet weld using weave beads:

Step 1 Tack weld two coupons together to form a T-joint (*Figure 29*). Clean the tack welds.

Step 2 Tack weld the assembly in the vertical position.

Step 3 Run the first weave bead along the joint root, starting from the bottom. Use a 0- to 10-degree push angle. Use a 0-degree work angle. Build the bead with a whipping motion. Quickly raise the electrode about $^{1}/_{4}$" (~6 mm) and then drop it back into the weld pool. Pause in the weld pool.

Step 4 Chip, clean, and brush the root pass.

Step 5 Run the remaining weave beads. Use a 0-degree work angle. *Figure 34* shows all welding passes. Notice that each has a different weave pattern. Move slowly across the weld face, pausing at each toe to fill the crater. A slight whip controls the weld pool when you reach the toe. Chip, clean, and brush each pass as you complete it.

Step 6 When you're finished welding, *always* remove the electrode from its holder. This prevents arcing. Discard the stub in the proper container.

Examine the finished weld and compare its profile to the examples in *Figure 3*. Have your instructor inspect your work and offer feedback. Keep practicing until you can consistently make good fillet welds.

2.3.7 Practicing Fillet Welds with E7018 Electrodes (3F Position)

Practice fillet welds in the 3F (vertical) position with $^{3}/_{32}$" (2.4 mm), $^{1}/_{8}$" (3.2 mm), or $^{5}/_{32}$" (4.0 mm) E7018 electrodes. Either stringer or weave beads are acceptable. Use a 0- to 10-degree push angle. Use a 0-degree work angle for the root pass. The work angle for the remaining passes depends on the bead type and pass number.

Use the same general technique as that outlined for E6010/E6011 electrodes. When producing stringer beads, don't whip the electrode. Instead, just drag it along with minimal side-to-side motion. With low-hydrogen electrodes, the arc should never leave the weld pool. The visible arc should be shorter than with E6010/E6011 electrodes. Remember, the thicker flux coating hides the arc's true length.

For the root pass, use a quick side-to-side motion. Move the electrode about $^{1}/_{8}$" (~3 mm) without removing the arc from the weld pool. Pause slightly at each toe to fill the crater and prevent undercut.

When producing stringer beads, use the sequence in *Figure 33*, which includes the approximate work angles. Chip, clean, and brush each pass as you complete it.

When producing weave beads, use the pattern that your instructor specifies. Use a 0-degree work angle. *Figure 35* shows all welding passes. Move slowly across the face, adjusting the travel speed to control the weld metal buildup. Chip, clean, and brush each pass as you complete it.

Vertical Weave Beads

To help control undercut, you may wish to use a triangular weave bead pattern. Pausing the electrode at the edges fills the previous undercut. This motion creates undercut at the existing weld pool, but the next weave will fill it.

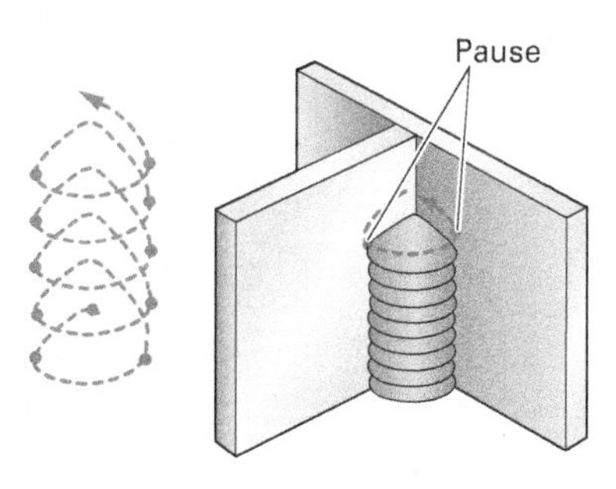

Triangular Weave Pattern

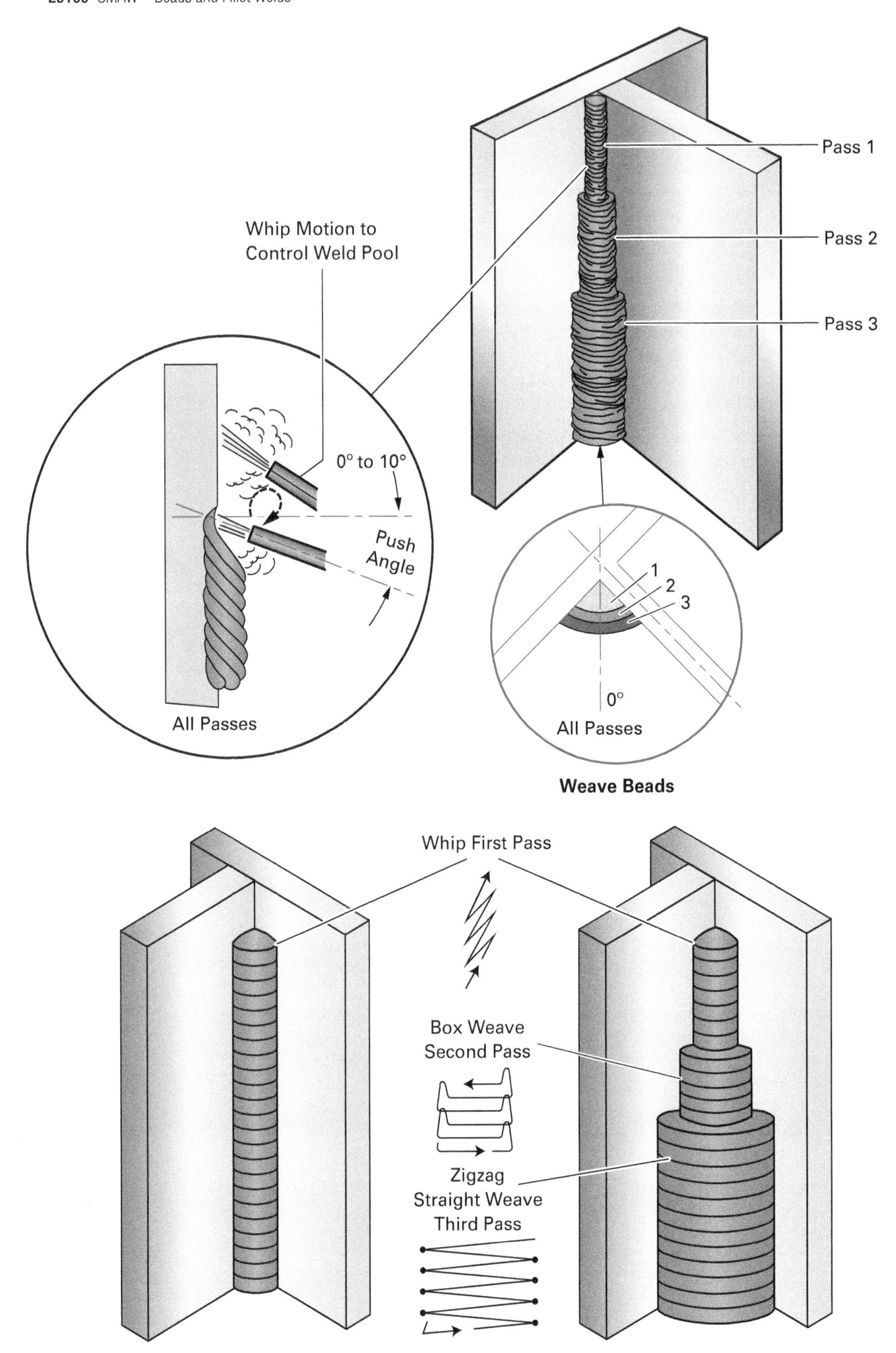

Figure 34 Fillet weld 3F welding sequence and work angles (weave, E6010/E6011).

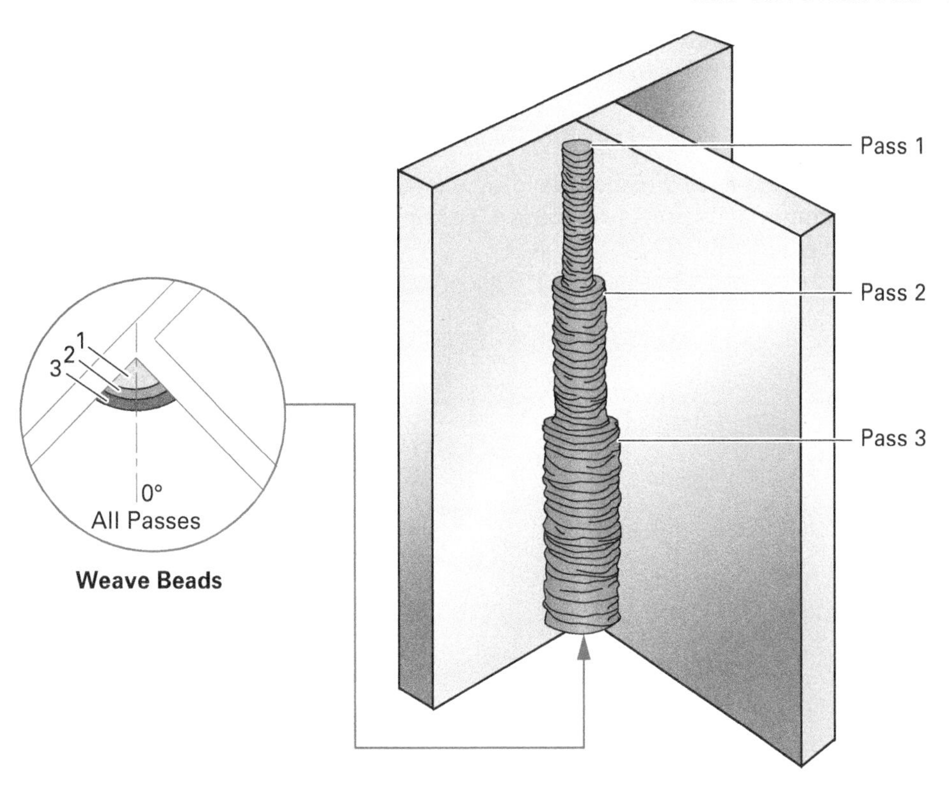

Figure 35 Fillet weld 3F welding sequence and work angles (weave, E7018).

2.3.8 Practicing Fillet Welds with E6010/E6011 Electrodes (4F Position)

Practice fillet welds in the 4F (overhead) position with $^{3}/_{32}$" (2.4 mm), $^{1}/_{8}$" (3.2 mm), or $^{5}/_{32}$" (4.0 mm) E6010/E6011 electrodes. All passes should be stringer beads. Assemble two coupons into a T-joint arrangement and create a multi-pass weld. Fillets should be convex. Use a 10- to 15-degree drag angle. Use a 0-degree work angle for the root pass. For later passes, adjust the work angle as required. Increase or decrease the travel speed to control the weld metal buildup.

WARNING!

Avoid standing directly under the coupons while welding. Hot slag and molten metal can drop on you, causing severe burns. Besides regular PPE, wear the additional protective clothing required for overhead work.

Follow these steps to practice making fillet welds in the 4F position:

Step 1 Tack weld two coupons together to form a T-joint (*Figure 29*). Clean the tack welds.

Step 2 Tack weld the assembly in the overhead position.

Step 3 Run the first bead along the joint root. Use a 10- to 15-degree drag angle. Use a 0-degree work angle. Deposit the stringer bead with a slight oscillation to tie in the weld at the toes.

Step 4 Chip, clean, and brush the root pass.

Step 5 Run the second stringer bead along the bottom toe of the first weld. Use a slight oscillation. The bead should overlap about 75 percent of the root pass. Use the electrode work angle shown in *Figure 36*.

Step 6 Chip, clean, and brush the bead.

Step 7 Run the remaining stringer beads along the toes of the underlying beads. The beads should overlap by about 75 percent. Chip, clean, and brush each bead. *Figure 36* shows all welding passes with their approximate work angles.

Step 8 When you're finished welding, *always* remove the electrode from its holder. This prevents arcing. Discard the stub in the proper container.

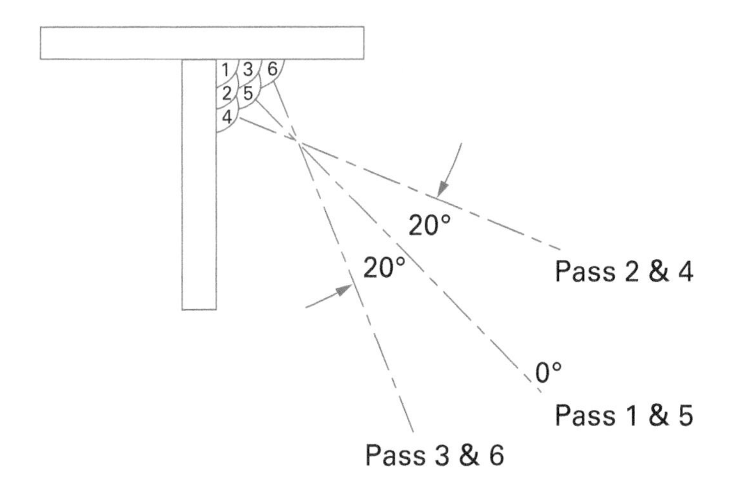

Figure 36 Fillet weld 4F welding sequence and work angles (E6010/E6011).

Examine the finished weld and compare its profile to the examples in *Figure 3*. Have your instructor inspect your work and offer feedback. Keep practicing until you can consistently make good fillet welds.

2.3.9 Practicing Fillet Welds with E7018 Electrodes (4F Position)

Practice fillet welds in the 4F (overhead) position with $^3/_{32}$" (2.4 mm), $^1/_8$" (3.2 mm), or $^5/_{32}$" (4.0 mm) E7018 electrodes. Use the same general technique as that outlined for E6010/E6011 electrodes. When producing the stringer beads, don't whip the electrode. Instead, just drag it along with minimal side-to-side motion. With low-hydrogen electrodes, the arc should never leave the weld pool. The visible arc should be shorter than with E6010/E6011 electrodes. Remember, the thicker flux coating hides the arc's true length.

T-Joint Facts

In T-joints, the welding heat dissipates more rapidly in the thicker parts. The same is usually true for the non-butting part. In either case, concentrating the arc in these places can compensate for heat loss.

The most common defect in T-joints is undercut on the joint's vertical member. Using a J-weave usually eliminates the problem. If the problem persists, however, angle the arc slightly towards the vertical part. This will force more metal into the bead at the top edge.

2.0.0 Section Review

1. To make a restart, restrike the arc _____.
   a. in front of the crater and in line with the weld
   b. behind the crater and in line with the weld
   c. in front of the crater and to one side of the weld
   d. behind the crater and to one side of the weld

2. When moving the electrode with a whipping motion, the lengthened arc _____.
   a. significantly increases the welding current
   b. takes advantage of E7018 electrode characteristics
   c. deposits additional hydrogen into the weld pool
   d. lets the weld pool cool and reduces metal transfer

3. Which fillet weld position requires a push angle rather than a drag angle?
   a. 1F
   b. 2F
   c. 3F
   d. 4F

Module 29109 Review Questions

1. A fillet weld face that curves inward is _____.
   a. convex
   b. improperly reinforced
   c. concave
   d. incompletely fused

2. What should you *always* do when preparing weld coupons?
   a. Clean them thoroughly to remove corrosion.
   b. Preheat them with an oxyfuel torch to relieve stress.
   c. Grind all edges to a 45-degree angle.
   d. Acid-etch them to remove cadmium.

3. Which method of striking an arc can leave undesirable discontinuities outside the weld area?
   a. The tapping method
   b. The hop-skip method
   c. The scratching method
   d. The inverter method

4. Tilting the electrode to the left or right in the plane perpendicular to its plane of motion adjusts the _____.
   a. drag angle
   b. travel angle
   c. push angle
   d. work angle

5. Experienced welders monitor the arc by _____.
   a. sound
   b. color
   c. odor
   d. brightness

6. When creating stringer beads with E7018 (low-hydrogen) electrodes, you should _____.
   a. whip the electrode vigorously to control the weld pool
   b. avoid the zigzag weave pattern
   c. strike the arc by tap-striking the flux coating rather than the core
   d. never whip the electrode

7. To build up a surface with a pad, create _____.
   a. surfacing beads
   b. perpendicular passes
   c. spaced weave beads
   d. open stringer beads

8. What do welders use to hold weldment pieces in the proper alignment when they're welding them together?
   a. Surfacing beads
   b. Tack welds
   c. Scratch welds
   d. Button beads

9. When producing a T-joint fillet weld in the 1F position, use a _____.
   a. 0- to 5-degree push angle
   b. 10- to 15-degree push angle
   c. 10- to 15-degree drag angle
   d. 45-degree drag angle

10. Which welding position often requires extra PPE because hot slag and molten metal can drop as you work?
    a. 1F
    b. 2F
    c. 3F
    d. 4F

Answers to odd-numbered Module Review Questions are found in *Appendix A*.

Answers to Section Review Questions

Answer	Section Reference	Objective
Section 1.0.0		
1. b	1.1.0	1a
2. c	1.2.6	1b
Section 2.0.0		
1. a	2.1.6	2a
2. d	2.2.2	2b
3. c	2.3.6	2c

User Update

Did you find an error? Submit a correction by visiting **https://www.nccer.org/olf** or by scanning the QR code using your mobile device.

Joint Fit-Up and Alignment

Objectives

Successful completion of this module prepares you to do the following:

1. Summarize fit-up and alignment tools.
   a. Identify and describe weldment positioning equipment.
   b. Identify and describe plate alignment tools.
   c. Identify and describe pipe and flange alignment tools.
   d. Identify and describe fit-up gauges and measuring devices.
2. Summarize managing weldment distortion.
   a. Explain weldment distortion and its causes.
   b. Explain the tools and techniques that control weldment distortion.

Performance Tasks

Under supervision, you should be able to do the following:

1. Use specialized tools to fit up plate and pipe joints.
2. Use gauges and other measuring tools to check joints for proper fit-up and alignment.

Overview

Good welders work hard to produce excellent welds. A properly welded assembly depends on several factors. One is the welder's skill and experience. Another is the assembly's *fit-up* and *alignment*. These terms refer to putting the weldment's parts together and positioning them properly. Only then can the welder join them with appropriate welds. This module introduces the tools and techniques that welders use to fit up and align the weldment.

Industry Recognized Credentials

If you are training through an NCCER-accredited sponsor, you may be eligible for credentials from NCCER's Registry. The ID number for this module is 29110. Note that this module may have been used in other NCCER curricula and may apply to other level completions. Contact NCCER's Registry at 1.888.622.3720 or go to www.nccer.org for more information.

You can also show off your industry-recognized credentials online with NCCER's digital badges. Transform your knowledge, skills, and achievements into badges that you can share across social media platforms, send to your network, and add to your resume. For more information, visit www.nccer.org.

NOTE

This module uses US standard and metric units in up to three different ways. This note explains how to interpret them.

Exact Conversions

Exact metric equivalents of US standard units appear in parentheses after the US standard unit. For example: "Measure 18" (45.7 cm) from the end and make a mark."

Approximate Conversions

In some cases, exact metric conversions would be inappropriate or even absurd. In these situations, an approximate metric value appears in parentheses with the ~ symbol in front of the number. For example: "Grip the tool about 3" (~8 cm) from the end."

Parallel but not Equal Values

Certain scenarios include US standard and metric values that are parallel but not equal. In these situations, a slash (/) surrounded by spaces separates the US standard and metric values. For example: "Place the point on the steel rule's 1" / 1 cm mark."

Digital Resources for Welding

Scan this code using the camera on your phone or mobile device to view the digital resources related to this craft.

1.0.0 Joint Fit-Up and Alignment Tools

Performance Tasks

1. Use specialized tools to fit up plate and pipe joints.
2. Use gauges and other measuring tools to check joints for proper fit-up and alignment.

Objective

Summarize fit-up and alignment tools.

a. Identify and describe weldment positioning equipment.
b. Identify and describe plate alignment tools.
c. Identify and describe pipe and flange alignment tools.
d. Identify and describe fit-up gauges and measuring devices.

A weldment's safety depends on its joints. Codes and specifications often govern joint design and setup. To satisfy these requirements, welders use special tools to position and align weldment components. Measuring tools then confirm proper alignment.

1.1.0 Weldment Positioning Equipment

Positioning the weldment's components is the first phase of fit-up and alignment. Welders can select from many tools to assist them. Some are specialized, while others are common workplace tools. Always check tools carefully for damage, problems, or missing parts. Never use a tool outside its intended purpose. Clean and return all tools when you're finished with them.

Be careful about cross-contamination. Carbon steel particles on stainless steel can cause weld defects and other problems. Never use carbon steel tools to position or hold stainless steel components. Check positioning equipment for oily residue before using it. Clean if necessary.

1.1.1 Hydraulic Jacks

Hydraulic jacks support heavy components and are easily adjustable (*Figure 1*). During welding, protect the jack from spatter. If the ram is off the floor, secure it with rope or chain. Check jacks for leaking hydraulic fluid. Clean up leaks before welding since hydraulic fluid is flammable. It can also contaminate welds.

WARNING!

Never weld anything to a hydraulic jack. Don't allow the welding current to pass through a hydraulic jack. The heat generated can cause the jack to fail and drop the load. This can cause damage, injury, or even a fatality.

1.1.2 Chain Hoists, Chain Falls, and Come-Alongs

Chain hoists, chain falls, and come-alongs give welders a mechanical advantage for lifting or moving heavy parts. Many assume that these devices are identical, but they're not. A chain hoist or chain fall lifts a load vertically. It uses pulleys and a looped chain to produce the mechanical advantage. The operator raises or lowers the load by pulling the chain. Critically, the load can't fall accidentally. The only way to lower the load is by pulling the chain. Powered chain hoists/falls work the same way.

Come-alongs, on the other hand, move the load horizontally. They use pulleys and a looped chain or cable to give the mechanical advantage. The operator works a ratcheted lever to move the load. A release mechanism lets the chain or cable run out freely. For this reason, never lift a vertical load with a come-along since the load can fall accidentally.

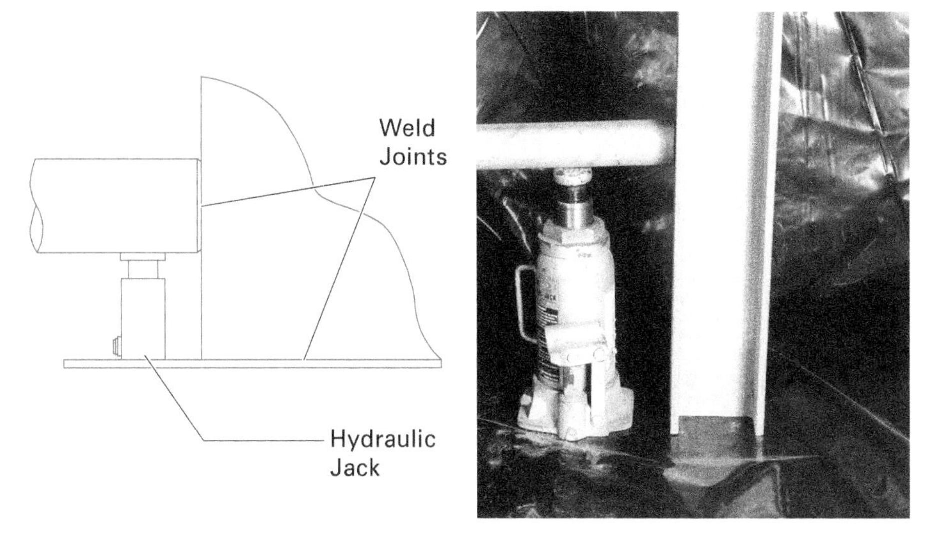

Figure 1 Using a hydraulic jack for joint fit-up.

WARNING!

Manufacturers use many terms to describe lifting and moving equipment. Always understand your equipment's capabilities and limitations before using it. Never vertically lift a load with any mechanism equipped with a free-running release. Read the manufacturer's instructions before using unfamiliar equipment. Misusing these machines can cause damage, injury, or even a fatality.

Welders use chain hoists/falls to lift and position weldment parts (*Figure 2*). Always hoist components using approved slings and rigging. Attach the rigging near the object's center of gravity (balance point).

WARNING!

Rigging and hoisting require proper training. Mistakes can cause damage, injury, or even a fatality. Unless you're properly qualified, never set up rigging yourself.

WARNING!

Before lifting any component, add up the part's weight, the rigging's weight, and the hoist's weight. The total is the *full load*. Confirm that whatever you plan to use to support the hoist can carry the full load. Inspect all rigging before using it. Never use anything outside its intended purpose.

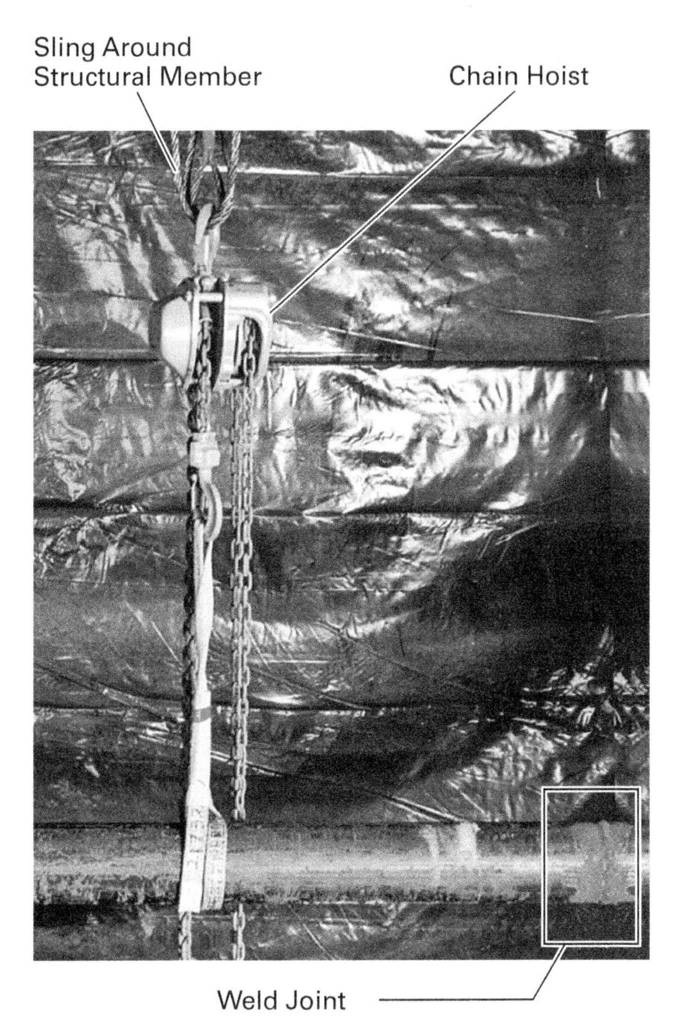

Figure 2 Chain hoist lifting a pipe into position.

CAUTION

Don't weld a lifting eye or other attachment point to a building structural member. Welding can weaken or damage it. Never attach hoisting equipment to a structural member unless you're sure that it can carry the full load safely.

Attach the hoist to a support or structural member that can carry the full load. Building supports like I-beams may be suitable. Never attach a chain hoist/fall to piping, conduits, ducts, or raceways. These can't support loads and could collapse without warning.

Come-alongs are appropriate for horizontally positioning components. Sometimes you may need to use more than one machine for a complex move. In all cases, confirm that the equipment can handle the load before using it.

When welding supported components, always attach the workpiece clamp directly to the part that you're welding. Don't connect the clamp to the chain hoist/fall or come-along. The welding current could cause equipment failure and an accident.

Chain Collectors

Many chain hoists/falls come with a container that collects the chain loops as you lift the load. If you're using a hoist with this feature, make sure the container is properly aligned. If the collector isn't set up correctly, it could dump the chain on you, causing injury. Always wear appropriate PPE, including a hard hat, when using lifting equipment.

1.1.3 Pipe Jack Stands and Rollers

Pipe jack stands support pipes for fit-up and welding (*Figure 3*). They usually have either a V-shaped head or a set of rollers. The jack height and sometimes the head width are adjustable.

Figure 3 Pipe jack stands and rollers.
Source: Courtesy of Sumner Manufacturing Co.

Welders use jack stands and rollers to fabricate piping assemblies in the shop. These support the pipes during welding and make repositioning simple. Welding pipe from above is the most convenient position. Rollers make it possible for the welder to work from this position, rotating the pipe while producing the weld. Jack stands and rollers aren't as helpful for working on existing systems in the field. Craftworkers may use them, however, for fabricating subassemblies.

WARNING!

Never field-fabricate jack stands and roller assemblies. This equipment can fail, causing damage, injury, or even a fatality. It may also lack required safety features. Commercially built equipment has a certified load capacity and complies with relevant safety standards.

When welding supported pipes, always attach the workpiece clamp directly to the pipe that you're welding. Don't connect the clamp to the jack stand or roller assembly. The welding current could arc between the roller and the pipe. Current passing through bearings can damage them.

WARNING!

Confirm that jack stands and rollers can carry the anticipated load. Also be sure that the supports are rated for the pipes they're carrying. For example, OSHA requires using four-legged jack stands for pipes over 10" (DN250).

Case History

Field-Fabricated Failure

On May 12, 1999, a mechanic jacked up a transit bus so he could work on its brakes. He supported the rear of the bus with a pair of jack stands and removed the back wheels. A few minutes later, the bus moved forward, slipping off the jack stands and pinning the mechanic underneath. He died about an hour later from severe abdominal trauma.

California Fatality Assessment and Control Evaluation (CA/FACE) investigated and discovered two important facts. The mechanic had forgotten to chock the bus's front wheels. He was also using a pair of field-fabricated jack stands. The stands didn't have a certified load rating. They also lacked an important basic safety feature. They had flat tops instead of a lip, which would have reduced the likelihood of the load slipping off.

This accident underscores an important point: never use field-fabricated jack stands. The cost savings just aren't worth the risk. Commercial jack stands have a clearly identified load rating and comply with appropriate safety standards.

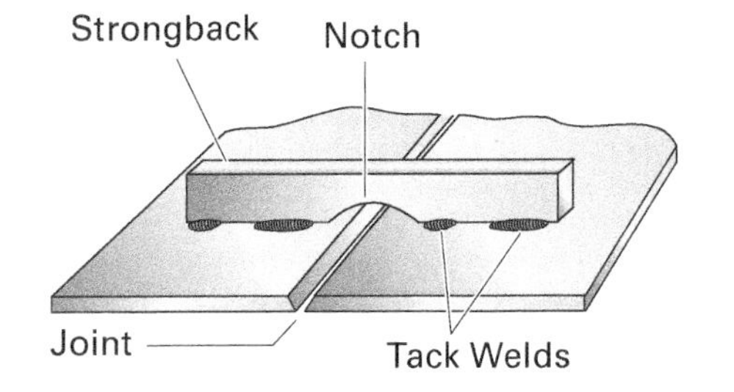

Figure 4 Strongback.

1.2.0 Plate Alignment Tools

After fit-up, welders must ensure that components stay aligned. For smaller plate-type components, tack welds are enough. Large assemblies or thick plates may require something stronger or more sophisticated.

1.2.1 Field-Fabricated Clamping Tools

Strongbacks are custom fabricated from heavy bar stock (*Figure 4*). They attach to the plates on either the root or face side. Notches in the strongback provide access to the weld joint. Tack welds or clamps hold the strongback in place.

When possible, tack weld strongbacks on only one side to make removal easier. After finishing the job, use a grinder to cut the tack welds or reduce them. A hammer blow then breaks them free. Grind down the tack welds flush with the base metal. Don't gouge the base metal or reduce its thickness.

CAUTION

Be careful when using a hammer to break a strongback free. You can cause cracking within the base metal. Sometimes, a chunk of base metal will tear free. Welders call this *lamellar tearing*.

Clips, *yokes*, and *wedges* offer another holding method. Craftworkers fabricate them for specific jobs. Welders tack weld clips to one plate's edge. They place a wedge on the other plate and drive it under the clip. Yokes work similarly. A plate with the yoke attached sits under the assembly. A slotted plate sits on top with the yoke passing through the slot. The wedge pulls everything tight (*Figure 5*).

For greater holding force, welders use *dogs*—wedges and yokes combined with a strongback. The strongback straddles the joint. Yokes welded to the plates straddle the strongback. Wedges driven between the yokes and strongback force everything together.

Alternatively, bolts welded to the plates can provide the clamping force. The welder makes the strongback from angle iron or channel stock. Holes drilled in the strongback match the bolts. After placing the strongback over the joint with the bolts passing through the holes, the welder installs and torques nuts and washers over the bolts. The welder removes the bolts with a grinder or torch when finished. *Figure 6* shows both methods.

1.2.2 Commercial Clamping Tools

Sometimes, commercial clamping tools are best for holding plates in the correct alignment. Welders can choose from many styles. As always, confirm that they're compatible with the base metals and won't contaminate them.

Figure 7 shows one example. The tool's yoke straddles the top surface of both plates. The gap plate fits between the plates, setting the correct root opening. The root bar slides through the slot in the gap plate. When the welder turns the threaded rod, it draws the root bar upwards, pulling the plates into alignment. If the plates are different thicknesses, the root bar's adjustment screw compensates to maintain good alignment.

NOTE

Some codes require that all temporary attachments be made from materials compatible with the base and filler metals. The attachments must also be traceable. Only qualified welders may weld them into place and must follow a WPS. If you're uncertain about a project's code requirements, speak to your supervisor.

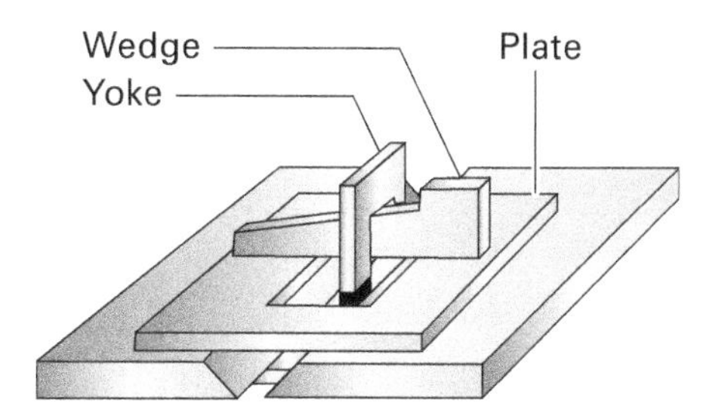

Figure 5 Yoke and wedge.

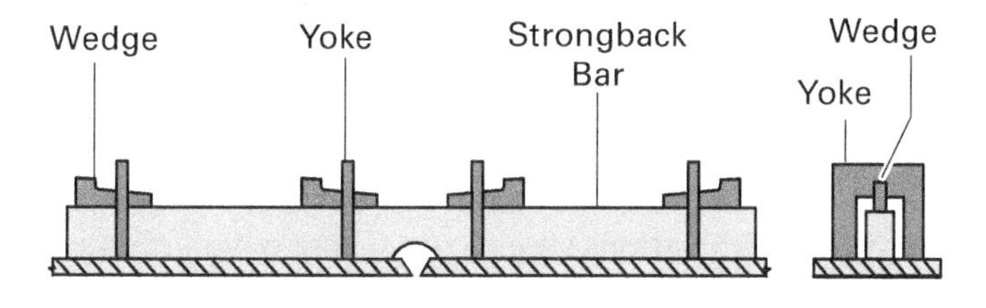

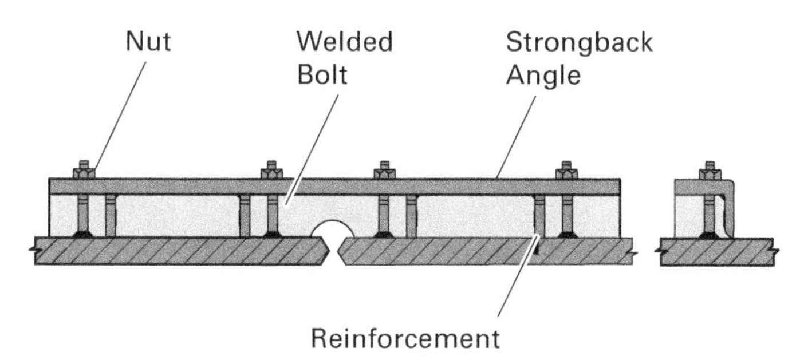

Figure 6 Combining a strongback with other clamping tools.

Field-Fabricated Tool ID

Always clearly mark field-fabricated tools like strongbacks, clips, yokes, and wedges. Unidentified tools may find their way into the recycle bin. To prevent cross-contamination, also identify a field-fabricated tool's metal.

No-Spark Alignment Tools

Industries that work with flammable or explosive chemicals avoid tools made from ferrous metals. These can spark easily, making them a constant hazard. When preparing field-fabricated clamping devices, craftworkers usually choose nonferrous metals like brass instead.

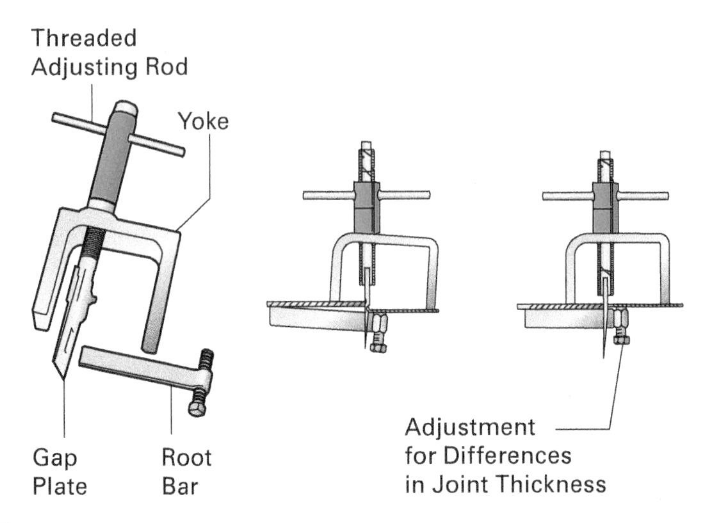

Figure 7 Plate alignment tool.

1.3.0 Pipe and Flange Fit-Up Tools

Keeping pipes properly aligned after fit-up presents some unique challenges. Unlike plates, they may have to rotate during welding. Manufacturers offer many tools to maintain alignment under these conditions.

1.3.1 Chain Clamps

These mechanisms grip the pipe with a wrap-around chain. Some have a single chain, while others use a pair. Chains come in several styles. Some are like bicycle chains, while others are like regular link chains. A jackscrew pulls the chain tightly around the pipe. Some clamps use one jackscrew, while others use two (*Figure 8*).

The single-chain, single-jackscrew clamp in *Figure 8* can even re-form an out-of-round pipe up to Schedule 40 thickness. It will clamp heavier pipes but can't re-form them. The double-jackscrew clamp in *Figure 8* can re-form thicker pipes.

Chain clamps come with various accessories. These include spacers to set and maintain an accurate root gap. Auxiliary chain attachments support elbows, Ts, and flanges (*Figure 8*). An adjustment crank tensions the chain to hold the fitting.

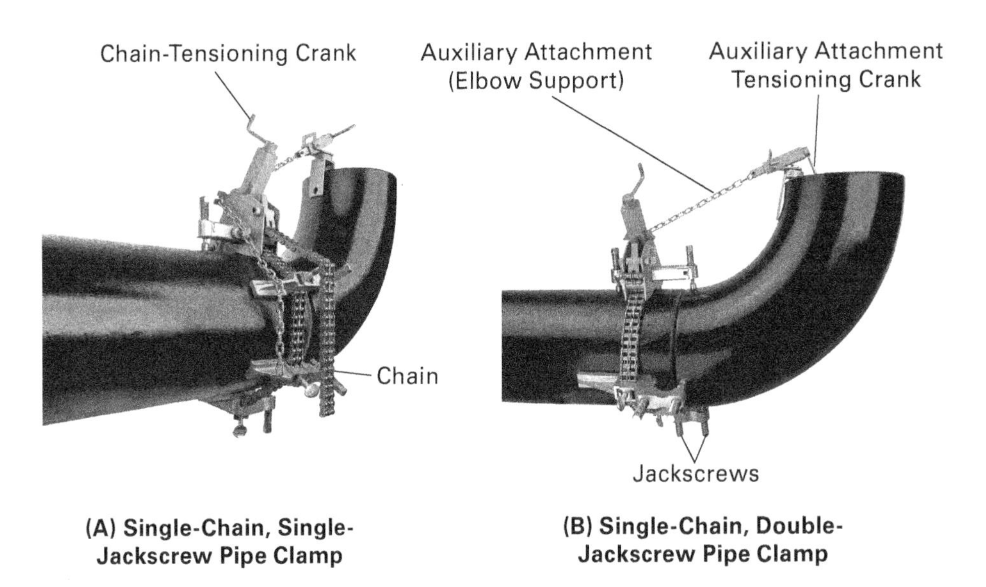

Figure 8 Chain clamps.
Source: Courtesy of Mathey Dearman

1.3.2 Cage and Rim Clamps

Preparing pipes for welding usually requires cutting and grinding their ends so they mate properly. Chain clamps can secure the pipes as they rest upon a roller-equipped jack stand. Alternatively, *cage clamps* may be better for this task.

These clamps wrap around the pipe and grip it tightly when the craftworker operates the locking mechanism. Older clamps must be hammered into the locked position. Modern designs use a ratchet-operated screw mechanism or a locking handle. *Figure 9* shows several cage clamp styles, along with their accessories.

The crossbars make up the cage sides. They're usually replaceable. The straight crossbars flare out at the ends to make sliding the pipe into place easier. The arched crossbars provide better access to the weld joint.

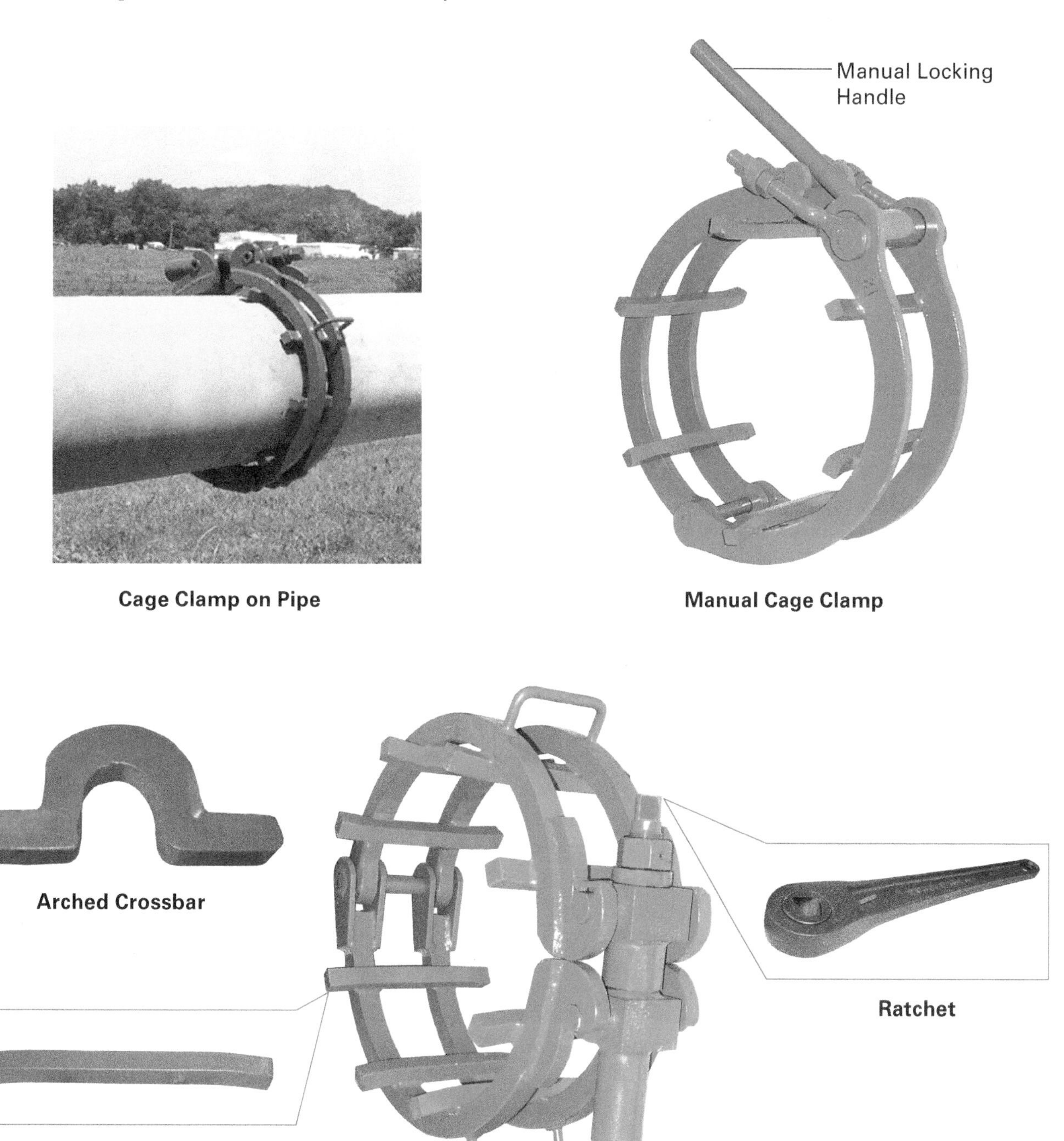

Figure 9 Cage clamps.
Source: Courtesy of Mathey Dearman

Cage clamps are good for securing two straight pipe sections, but they won't hold a straight section and a flange. A *rim clamp* is the right tool for this task (*Figure 10*). Welders use them for jobs in which the clamp must stay in place for the entire preparation and welding process. The clamp fits around the pipe's rim and holds the flange in the proper position. These tools can also re-form thick pipes.

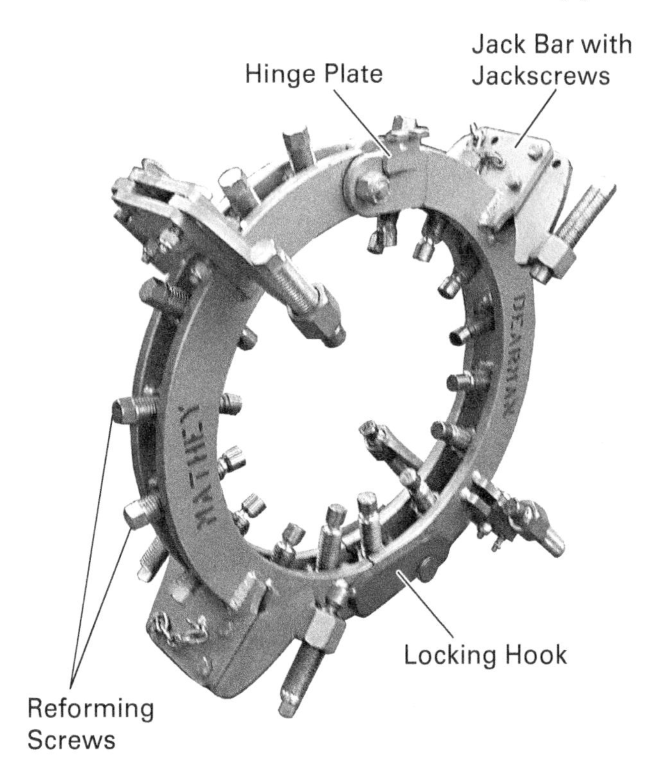

Figure 10 Rim clamp.
Source: Courtesy of Mathey Dearman

1.3.3 Small-Diameter Pipe Clamping Devices

Standard chain clamps can be too bulky for smaller pipes. Some workplaces build clamps from angle iron. The pipe rests in the angle. C-clamps or wires secure the pipe. Small blocks hold the angle in position.

Commercial clamps offer better accuracy and easier adjustability (*Figure 11*). These have three-point jaws that clamp and align the pipe when the pipefitter turns the clamping screws. Stainless steel models won't contaminate stainless steel pipes.

Consumable inserts: Prefabricated filler metal components designed to fuse into the weld root and become part of the weld.

1.3.4 Pipe Pullers

Pipe pullers are like chain clamps (*Figure 12*). They pull sections of large-diameter pipe together for welding. They're common in petrochemical facilities, which often modify their piping systems. Sometimes, **consumable inserts** fit between the pipe joint faces. These establish the correct root opening. During welding, they fuse into the joint.

NOTE

Some project specifications don't permit pipe pullers.

1.3.5 Flange Alignment Tools

Pipe flanges require careful aligning before welding. An improperly aligned flange makes assembly difficult. It can leak or put stress on nearby equipment. To check a flange's alignment before tack welding it into place, insert pins into the bolt holes (*Figure 13*). Measuring instruments use these pins as reference points. Alignment fixtures and jigs also use the bolt holes as reference points (*Figure 14*).

CAUTION

When coupling new pipes to existing equipment, such as pumps and valves, alignment is crucial. Slightly misaligned flanges put great stress on the equipment. Pump bearings could fail, or components could crack. Use positioning and alignment tools to produce excellent alignment before welding.

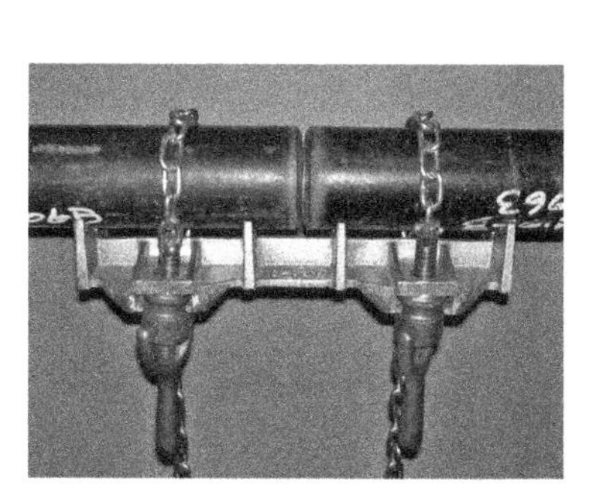

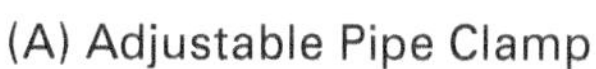

(A) Adjustable Pipe Clamp (B) ULTRA™ Clamp

Figure 11 Pipe clamps for smaller pipe.
Sources: Courtesy of H&M Pipe Beveling Machine Company, Inc. (11A); Courtesy of Sumner Manufacturing Co. (11B)

CAUTION

Be careful around flanges. Never damage the face or sealing surface. Flange repairs can be difficult and costly.

Think About It

Root Openings—Fitting and Aligning

During fit-up and alignment, why is it necessary to establish and maintain a precise root opening between two pipes?

Figure 12 Pipe puller.
Source: Courtesy of Mathey Dearman

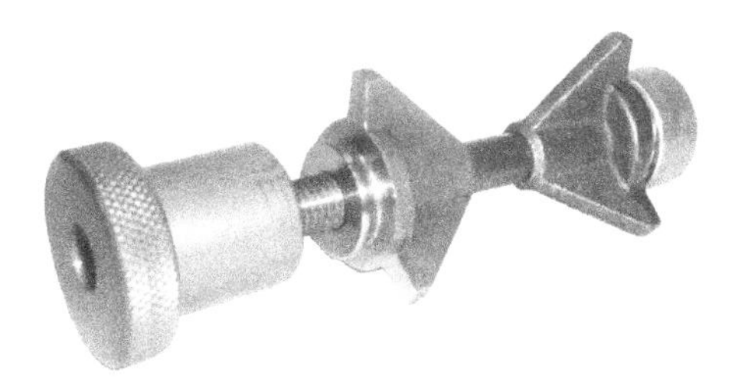

Figure 13 Flange pins.

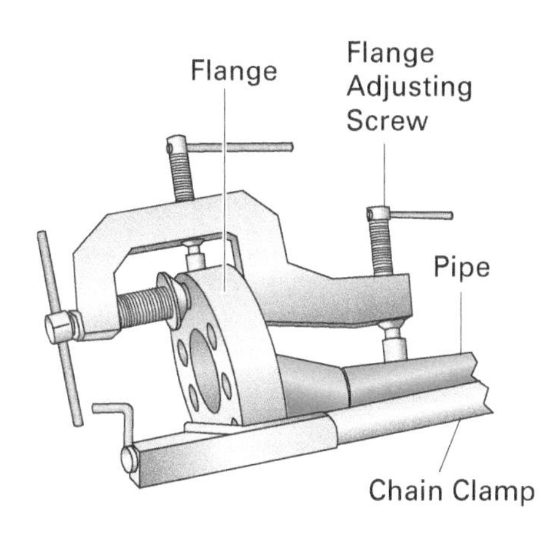

Figure 14 Flange alignment tool.

Flanges

Pipes in large industrial installations connect to equipment and each other through *flanges*. Bolts through the flanges' bolt circles pull them together, producing a good seal. Flanges come in many styles and seal in several ways. Some meet specific application requirements or work in demanding conditions. Regardless of the type or application, all flanges require careful installation to work properly.

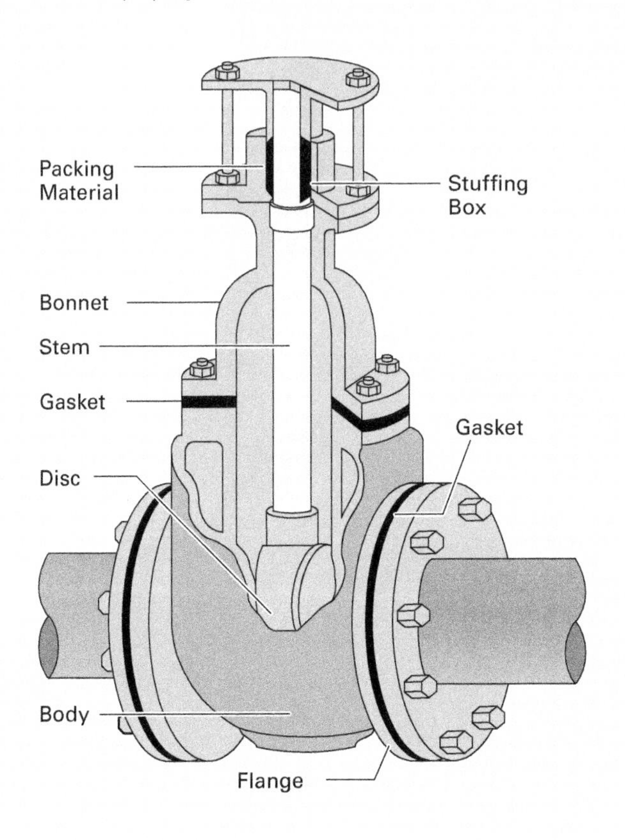

1.4.0 Fit-Up Gauges and Measuring Devices

After fit-up, but before welding, the welder must check the joint to confirm that it meets all requirements. Usually, a WPS or the jobsite quality specifications outline these. Gauges and other measuring devices provide objective evidence that the joint meets the requirements.

As with positioning and alignment tools, be careful not to contaminate the weldment with the measuring tool. Don't use carbon steel tools on stainless steel. Using stainless steel tools with carbon steel parts usually isn't a problem. If a measuring tool is oily, clean it first.

1.4.1 Straightedges

Straightedges are metal tools with precisely machined edges (*Figure 15*). They come in many styles and lengths. Some have measurement marks like a ruler. Welders use straightedges to check joint alignment or to scribe straight lines on components.

Handle straightedges carefully. Never do anything that could damage the machined edge. Be careful not to heat a straightedge since heat can warp it. Placing it on a hot surface or near a welding arc is a bad idea. Regularly check straightedges by sighting along the edge. Discard a warped or damaged straightedge.

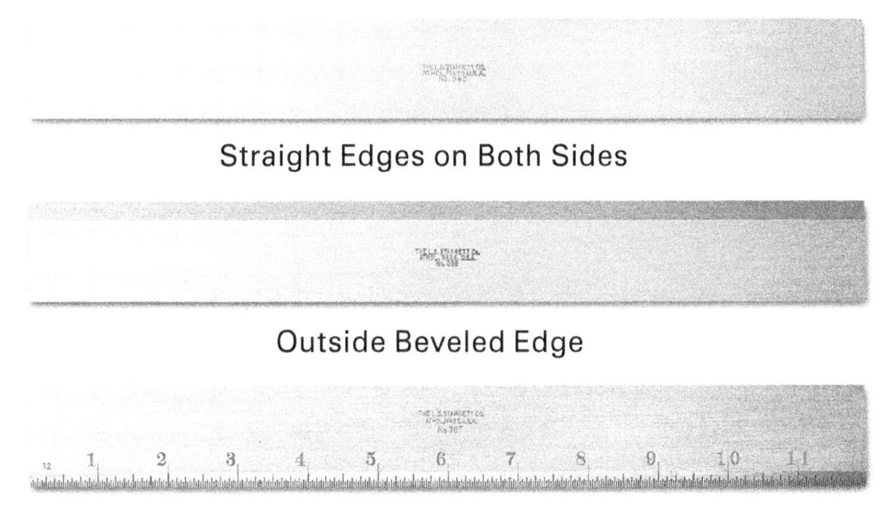

Figure 15 Straightedges.
Source: The L.S. Starrett Company

1.4.2 Squares

Squares are L-shaped tools with a precise 90-degree angle. Welders use squares to measure angles and check alignments. A pipefitter's square is like a carpenter's square (*Figure 16*). These come in several sizes. Many have useful information stamped onto the blade.

Combination squares are smaller but much more versatile. They come with several removable heads that slide along the blade. Some heads have two precisely machined angles—90 and 45 degrees. Others can rotate to any angle. These are useful for measuring and laying out angles. V-shaped heads can find a shaft or other round object's center. Some combination squares include a built-in scriber.

As with straightedges, protect squares from heat. Don't let welding spatter land on them either. It makes using the tool more difficult and can cause inaccurate measurements. Keep carbon steel squares free from rust.

1.4.3 Levels

After fit-up and alignment, welders may check to see if weldments are **level** or **plumb**. A level object is parallel to the earth's surface. A plumb object is perpendicular to it. Tools called *levels* check for both.

Level: A horizontal line parallel to the earth's surface. Also, a tool that checks whether a surface is level or plumb.

Plumb: A vertical line perpendicular to the earth's surface.

Levels range in size from a few inches to a few feet long (*Figure 17*). Welders use models made from metal or plastic. Some have magnetic bases. All contain one or more glass vials partially filled with a colored liquid. A bubble in the liquid moves as the level tilts. The tool is level (or plumb) when the bubble is centered. *Figure 18* shows a level resting on flange pins to check a pipe flange's alignment.

Some levels include a vial set at a 45-degree angle. This can check 45-degree angles. A few levels even have a rotating vial with a protractor scale. These can check any angle. Many combination squares include a built-in level.

Handle levels carefully since the vials break easily. Protect them from spatter too. Metal particles stuck to the precision edges prevent you from placing the level firmly in contact with the measured surface.

NOTE

When using a straightedge, square, or level, be sure that the machined edge fully contacts the surface it's resting on.

A Level in Your Pocket

Don't forget that smartphones come with a built-in level app. You can measure angles and check for levelness just by pulling out your phone. Remember, however, that most smartphone cases don't have precision edges. For best results, remove the phone from its case before placing it on the surface that you're measuring.

Not all jobsites permit smartphones due to safety or security concerns. Before using an app on the jobsite, confirm that smartphones are permitted.

Plumb Bobs

Another way to check whether an object is plumb is to hang a *plumb bob* next to it. This tool is just a heavy weight attached to a string. Gravity makes it hang perpendicularly to the earth's surface. Compare the object to the string to check for plumbness.

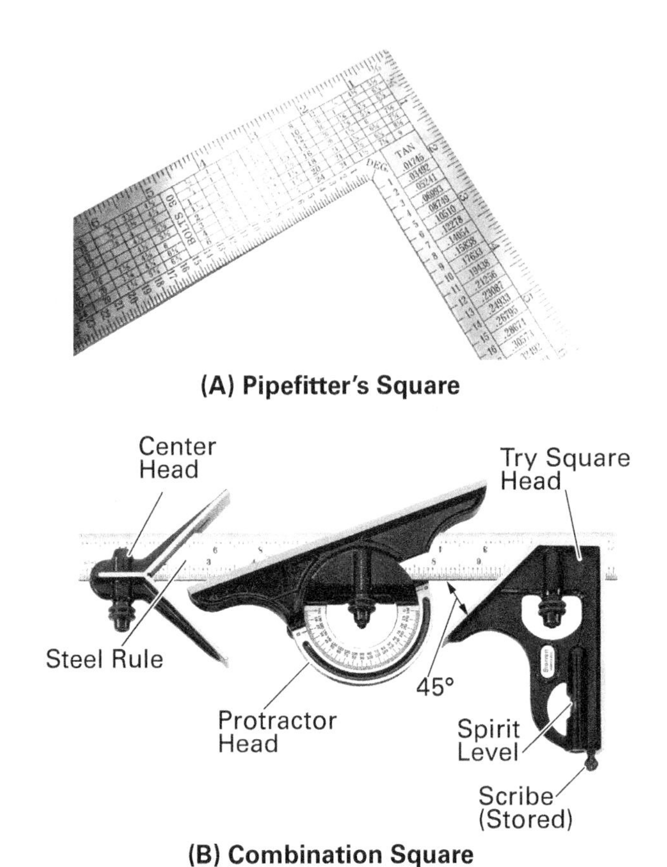

Figure 16 Squares.
Sources: Courtesy of Mathey Dearman (16A); The L.S. Starrett Company (16B)

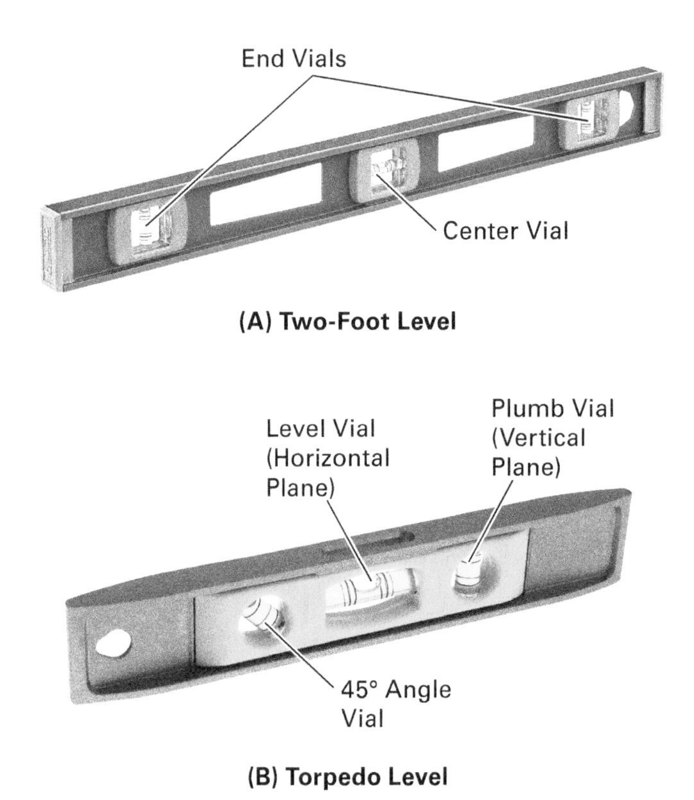

Figure 17 Levels.
Sources: Ridge Tool Company (17A); Image property of Stanley Black & Decker. Used with permission. (17B)

1.4.4 Hi-Lo Gauges

When assembling pipes, welders need to check for proper vertical and/or horizontal alignment. Similarly, some applications require precisely aligned plates. Pipes or plates not lined up correctly can create many problems. **Joint mismatch** (high-low) is the alignment difference between two pipes or plates. A tool called a *Hi-Lo gauge* measures it.

Joint mismatch: Alignment discrepancy in pipe sections or plates. Also called *high-low*.

Hi-Lo gauges have two tapered alignment stops (prongs) that slide back and forth against two scales (*Figure 19*). To measure misalignment, slide the stops until they're firmly against the two surfaces. Read the correct scale to see the difference between the stops.

For example, to check two pipes' internal alignment, insert the gauge's stops into the root opening between the pipes. Adjust the tool until one stop touches the inside of each pipe (*Figure 20*). The reading on the scale indicates the alignment difference, either in inches or millimeters.

NOTE

Check pipe alignment twice—once after positioning the pipes and again after tack welding them together.

Hi-Lo gauges can measure other welding dimensions too. These include the joint root opening, material thickness, and weld reinforcement. To measure the root opening, insert the $^1/_{16}$" (1.6 mm) thick tapered stops into the opening. Visually confirm that the stops just barely fit.

To measure thickness, place the material on a smooth, flat surface. Rest one tool foot on the surface. Rest the other on the material's surface. Be sure that both feet fully contact the surfaces. Read the scale to find the height difference, which equals the material's thickness.

Measuring reinforcement is similar. Place one foot on the base metal surface. Place the other on the reinforcement. Be sure both feet fully contact the surfaces. Read the scale to find the difference in height, which equals the reinforcement (*Figure 21*).

Figure 18 Checking a pipe flange with a level.
Source: Courtesy of Mathey Dearman

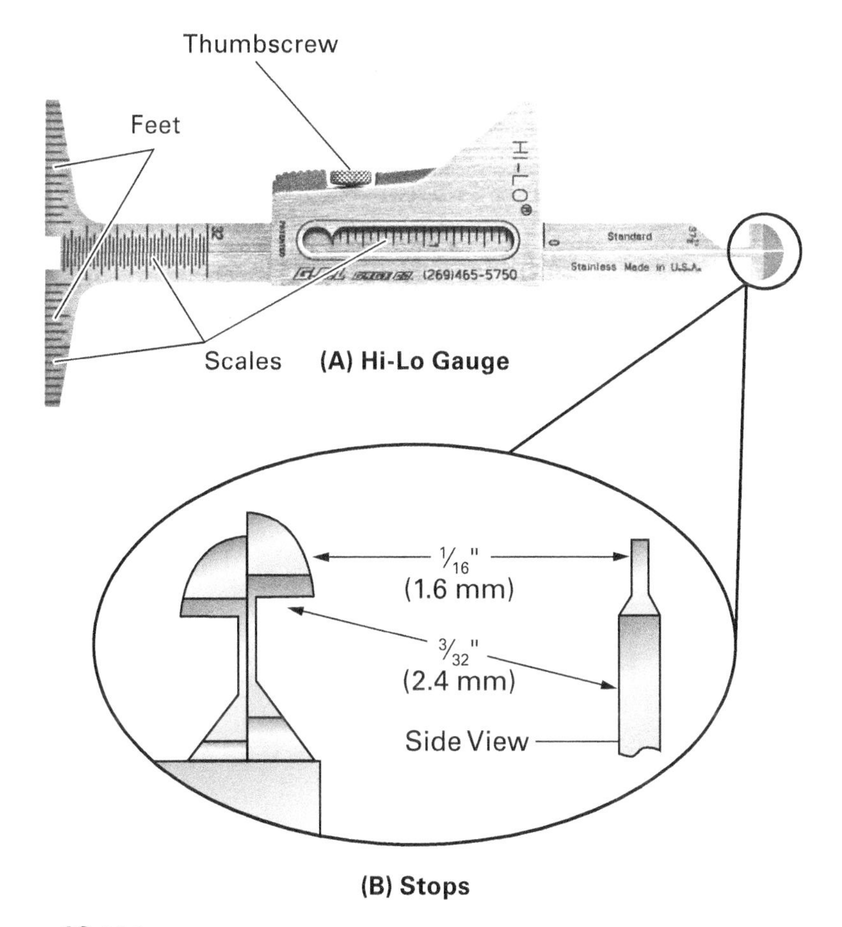

Figure 19 Hi-Lo gauge.
Source: G.A.L. Gage Co

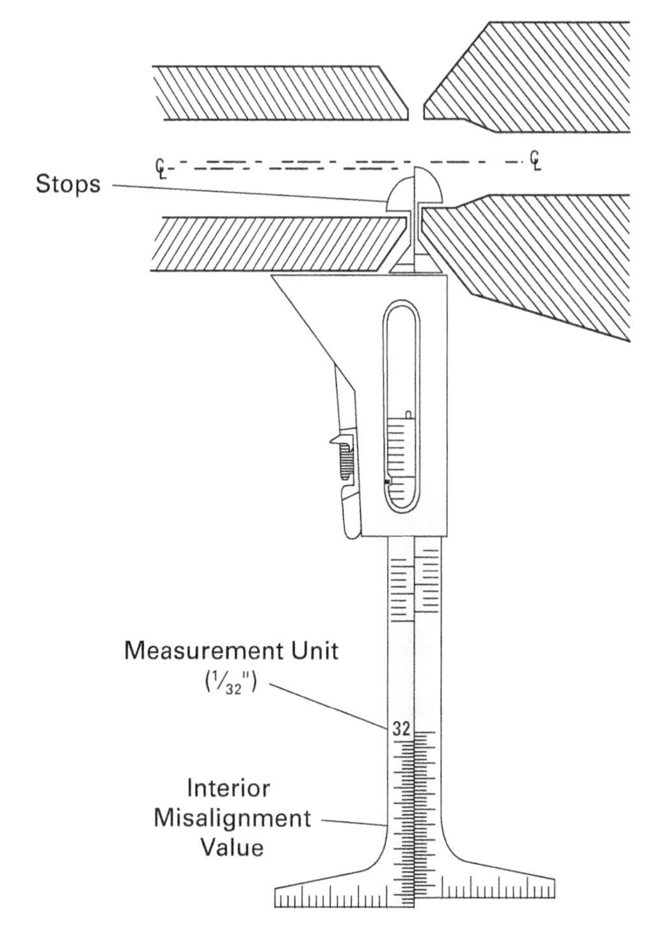

Figure 20 Measuring pipe internal misalignment.

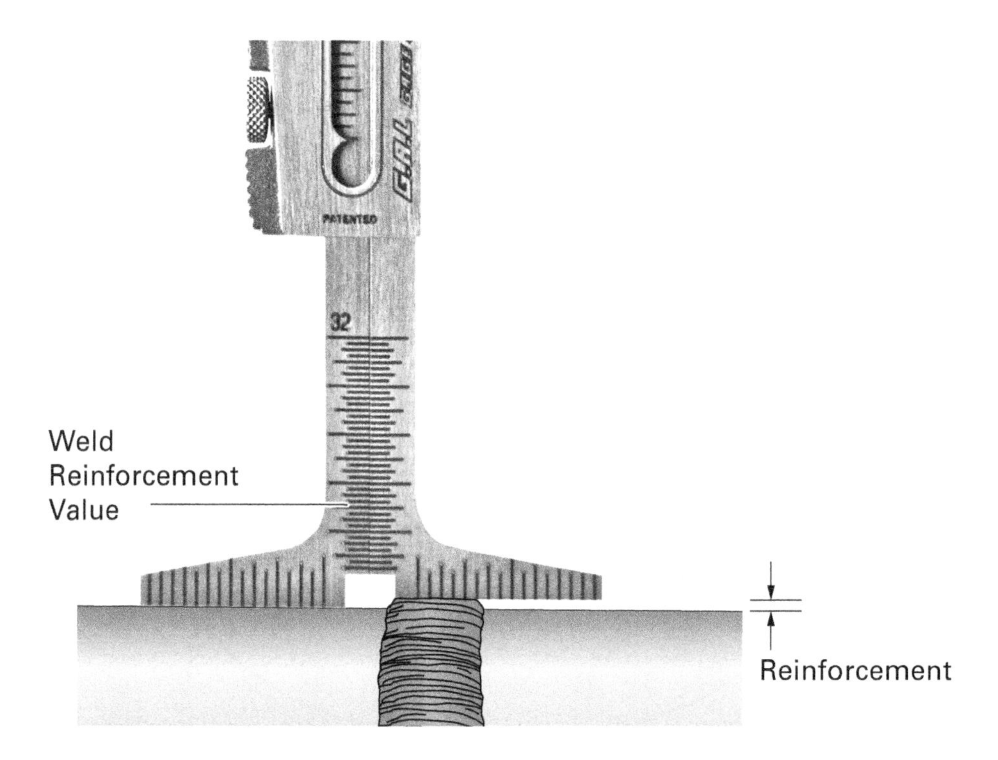

Figure 21 Measuring reinforcement.
Source: G.A.L. Gage Co

1.0.0 Section Review

1. When vertically lifting components, *never* use a _____.
   a. come-along
   b. chain hoist
   c. chain fall
   d. sling

2. Clips, yokes, and wedges are generally _____.
   a. purchased in bulk
   b. field-fabricated
   c. included with the base metal
   d. made from carbon fiber

3. A simple shop-built clamp made from angle iron can help align _____.
   a. heavy plates
   b. large pipes
   c. small pipes
   d. sheet metal

4. Hi-Lo gauges measure _____.
   a. whether an object is level or plumb
   b. the difference between two surfaces
   c. whether two objects are perpendicular
   d. 45-degree angles

2.0.0 Weldment Distortion

Objective

Summarize managing weldment distortion.

a. Explain weldment distortion and its causes.

b. Explain the tools and techniques that control weldment distortion.

Performance Tasks

There are no Performance Tasks in this section.

All metals change as they heat up and cool down. Heating makes them grow in each dimension—called *expansion*. Cooling makes them shrink—called *contraction*. Each metal responds differently, but all expand and contract with temperature changes.

During welding, components expand from the heat. This behavior can cause undesirable changes in the weldment's shape—called *distortion*. Alignment tools like clamps control or redirect expansion to manage distortion.

Dimensional changes during the welding process create another problem. When metals expand and contract, they generate forces called *stresses*. These stresses can alter the weldment's shape or even damage it.

Welders can't eliminate expansion, contraction, distortion, and stress. They can manage them, however, so the welds that they produce meet the project specifications. This section explores distortion and the best ways to manage it.

2.1.0 Distortion

Distortion happens when metals can't expand and contract freely. Heating a metal bar causes it to expand uniformly in every dimension. When it cools, it contracts uniformly. Even though its size changes, its shape stays the same (*Figure* 22).

On the other hand, if something prevents the bar from moving in one dimension during expansion or contraction, its shape will change. The bar in *Figure* 23 is constrained in the lengthwise dimension. When heated, it tries to expand in all directions but can't. Instead, the expansion forces reshape the object. Its width and height increase. When it cools down, it contracts but doesn't return to the original shape. In other words, it's now distorted.

All welding processes cause non-uniform expansion and contraction. During welding, the molten weld and base metals expand. The solid base metal surrounding the weld pool (the *heat-affected zone*) isn't as hot, so it doesn't expand as much. Since it's solid, it also constrains the molten weld. The cooler base metal outside the heat-affected zone expands even less. It, in turn, constrains the heat-affected zone.

As the weld cools, the molten metals solidify and try to contract. Since the surrounding base metal constrains the cooling weld, the weld can't shrink freely. Instead, it develops stresses within itself. It also pulls on the heat-affected zone. If the weldment isn't clamped down, its shape may change (*Figure 24*). This shape change may or may not be significant.

If the weldment is clamped, those stresses will stretch the metal. Small stresses stretch the metal like a spring but don't damage it. When the welder releases the clamps, the weldment may distort from these stresses.

Larger stresses can cause the metal to *yield* (permanently change shape). Ductile metals can tolerate more stretching than brittle metals. All metals have a limit—their *ultimate tensile strength*. If the stresses exceed the ultimate tensile strength, the metal will crack.

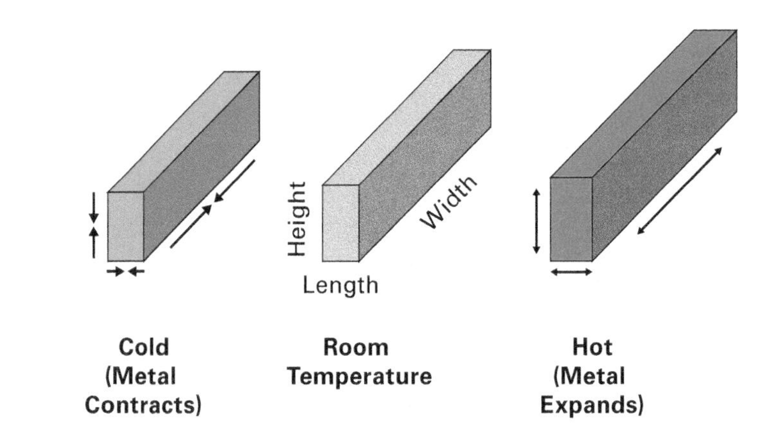

Figure 22 Uniform expansion and contraction.

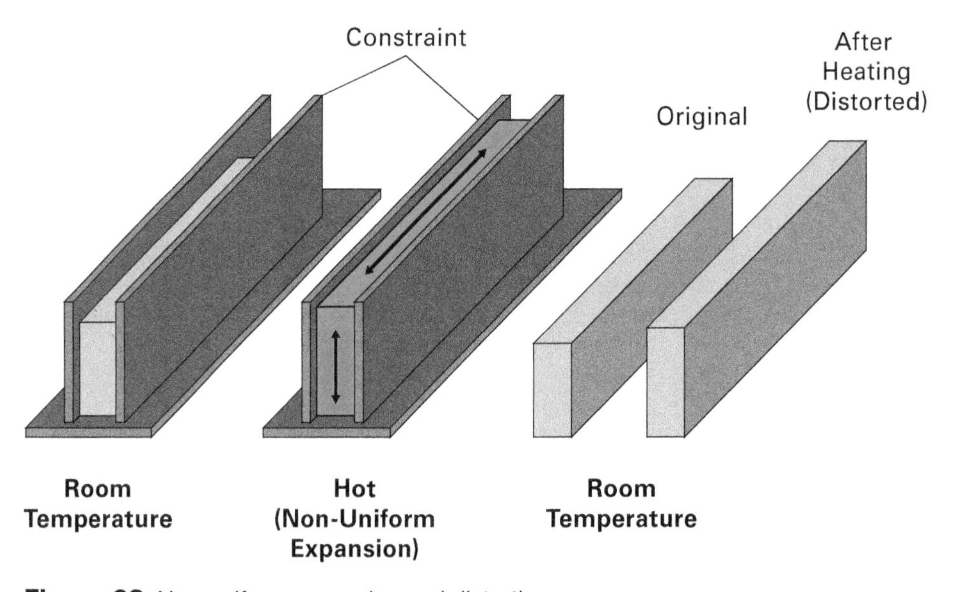

Figure 23 Non-uniform expansion and distortion.

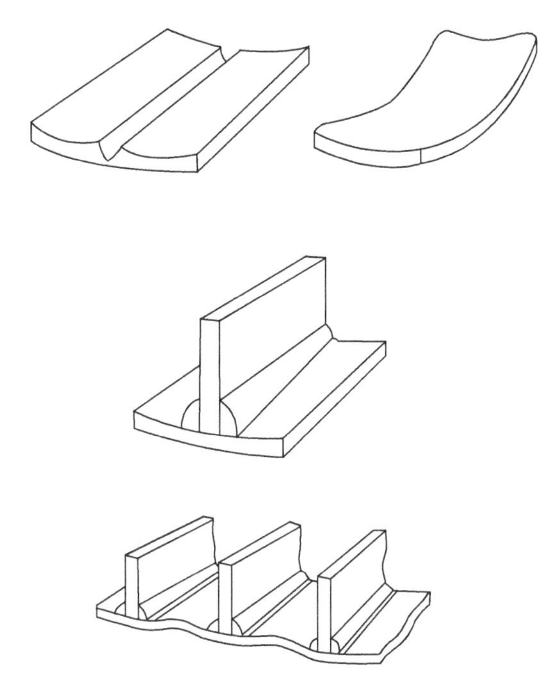

Figure 24 Stress and distortion in welds.

Even when the metals don't visibly distort, problems may be lurking inside. Stresses can become trapped within the weld joint and base metal. This **residual stress** is present even at normal temperatures. It may be harmless, cause distortion, or create problems that emerge over time.

Residual stress: Heat-induced stress remaining in a weldment at normal temperatures.

Many factors contribute to distortion and stress. Fit-up, alignment, and the welding process are major factors, as is the welder's technique. The base and weld metals' mechanical properties significantly influence the outcome, too.

Each metal expands and contracts by a different amount for the same temperature change. Aluminum, for example, expands and contracts more than carbon steel does. Metals also transfer heat differently. Some take a long time to heat up and cool down. Others change very quickly. Aluminum heats up and cools down faster than carbon steel.

To compare metals, you'll need to consult a table listing each metal's properties. A metal's **coefficient of linear thermal expansion** indicates how its length changes in response to a 1°C temperature change. Bigger numbers mean larger changes. A metal's **thermal conductivity value** indicates how readily the metal transfers heat. Bigger numbers mean better heat transfer. *Table 1* shows these values for several familiar metals.

Coefficient of linear thermal expansion: A number indicating by how much a material's length changes in response to a 1°C temperature change.

Thermal conductivity value: A number indicating how readily a material transfers heat.

TABLE 1 Thermal Expansion and Conductivity for Common Metals

Metal	Coefficient of Linear Thermal Expansion (10^{-6}/°C)	Thermal Conductivity (W/mK)
Aluminum	21–24	220–240
Copper	16–16.7	342–413
Iron	11.3–12	28–94
Steel (carbon)	10.8–12.5	36–54
Steel (stainless)	9.9–17.3	14

If a weldment's parts and the weld metal are different, they'll expand and contract at different rates. These differences can cause distortion and create stresses. Understanding how each metal behaves helps welders manage these problems more effectively.

Latent Distortion

Residual stresses don't always cause trouble right away. Sometimes, a weldment looks fine after it cools down. Then a craftworker machines the part and suddenly it distorts. What happened? The problem is that the weldment had some residual stress locked up inside.

Before machining, the metal was strong enough to resist the stress. After the craftworker removed some metal, the residual stress was strong enough to change the weldment's shape. The distortion is *latent* (hidden) because it doesn't show up immediately. Only under the right conditions does it surface and cause problems.

Welders sometimes relieve this locked-in stress by heating and cooling the metal in a specific way. Some mechanical techniques can relieve or at least reduce it. Usually, the WPS will specify the appropriate method.

2.2.0 Managing Distortion

Distortion is complex and has many contributing factors. Welders use numerous techniques and tools to manage it. The following sections outline several of these.

2.2.1 Clamping and Bracing

Ambient temperature: The temperature of an object's surroundings.

Sometimes just clamping the weldment's parts controls distortion adequately. Since heat-induced stresses can be large, select sturdy clamping devices. Wait until the weld has cooled down to **ambient temperature** before removing the clamps. Be prepared for some distortion since the weldment may have residual stresses.

Clamping a metal that isn't very ductile can cause problems. As the metal cools, it shrinks, producing stresses. Since it can't stretch enough to accommodate the shrinkage stress, it cracks instead. When working with these metals, welders may heat the metal after completing the weld. They then allow it to cool slowly, relieving the stresses developed in and around the weld.

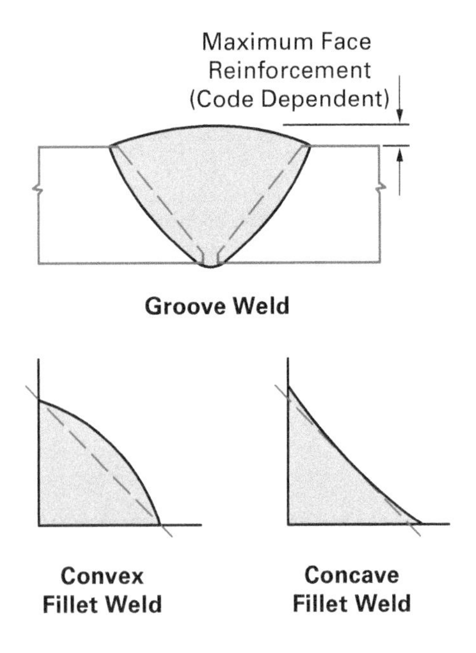

Figure 25 Face reinforcement profiles.

2.2.2 Tack Welding

Sometimes, simple tack welds can keep everything aligned during welding. Welders need to use enough tack welds and position them correctly. A joint that starts closing up probably doesn't have enough tack welds, or they may be too small. Combining tack welds with a clamping device may produce better results.

Sometimes, welders remove the tack welds after they finish the job. They must do this carefully without damaging the weld or the base metal. In other cases, the tack welds become part of the final weld. Only qualified welders may produce permanent tack welds or those for code-driven jobs. Since the tack welds combine with the base and filler metals, they must follow the same WPS or jobsite quality specifications.

2.2.3 Weld Material and Distortion

When a weld cools, the weld metal shrinks, stressing the joint. The more weld metal in the joint, the larger the stresses become. Properly sizing the joint reduces stress and distortion. It also saves metal and time.

Excessive face reinforcement doesn't increase a weld's strength, but it does increase its stresses. For this reason, codes prohibit excessive reinforcement. Usually, codes permit a fillet weld's face to be flat, slightly convex, or slightly concave. With groove welds, a maximum face reinforcement of $^1/_8$" (3.2 mm) is common. *Figure 25* shows acceptable face reinforcement profiles.

NOTE

These guidelines are general. Always follow the applicable code, WPS, or jobsite quality specifications.

Proper joint preparation and fit-up also optimize the weld. Open-root groove welds usually have root openings between $^1/_{16}$" and $^1/_8$" (1.6 mm and 3.2 mm). A root face (land) between $^1/_{16}$" and $^1/_8$" (1.6 mm and 3.2 mm) controls melt-through. Bevels should have an angle between 30 and $37^1/_2$ degrees (a 60- to 75-degree groove angle). Grooves with larger angles require more weld metal to complete the weld. Ideally, the bevel angle should be just large enough to give the welder good root access. The GMAW and GTAW processes require larger openings than SMAW since their nozzles are larger than rod-type electrodes. *Figure 26* shows an open-root V-groove weld preparation.

When possible, use double V-grooves rather than single V-grooves. They use half the weld metal. They also have less distortion since their stresses counteract each other. For thick joints, U-groove and J-groove welds require even less weld metal than V-groove or bevel groove welds. They do, however, require more preparation to produce their curved profile. *Figure 27* shows different groove styles.

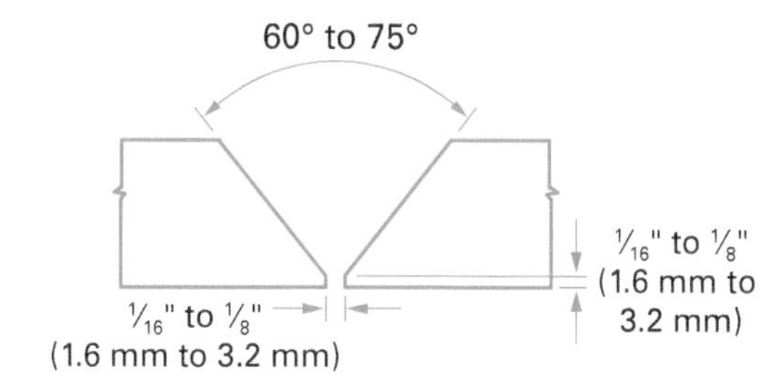

Figure 26 Open-root V-groove weld preparation.

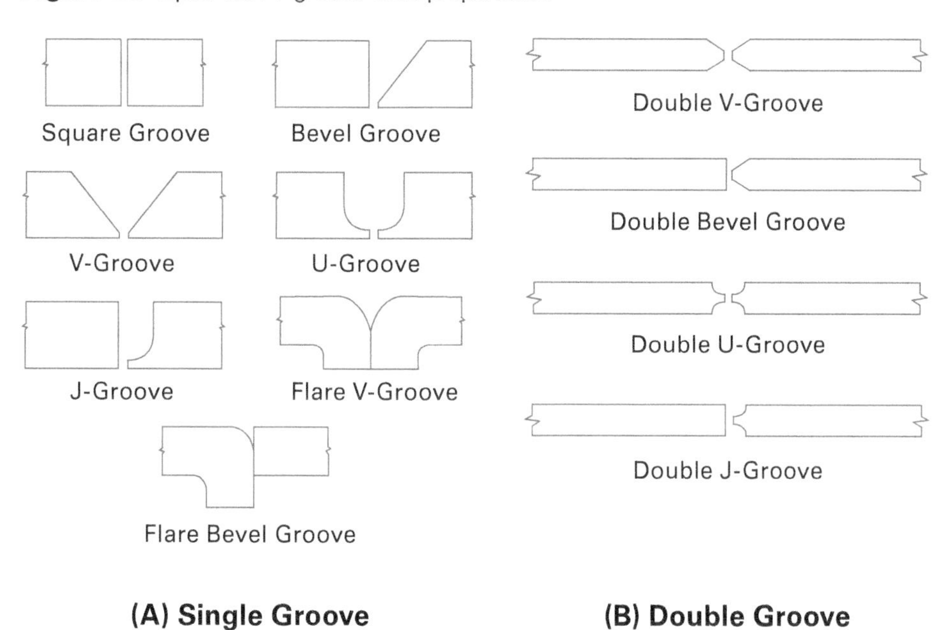

Figure 27 Groove weld styles.

2.2.4 Plate Groove Welds with Backing

Using a backing with a groove weld can help reduce distortion. Backing materials include metal strips, as well as fiberglass, flux-coated, and ceramic tapes. Some backing materials come in convenient rolls that the welder cuts to length and applies.

If permitted by the WPS or jobsite quality specifications, the backing may become part of the finished weld. In these cases, it must be compatible with the base and weld metals. In other cases, the welder removes the backing after completing the weld. For example, copper backing strips are common with stainless steel base metals.

Groove welds with metal backing have different joint preparations than open-root groove welds (*Figure 28*). The root opening is wider—often ¼" (6.4 mm). It doesn't need a root face (land) since the backing prevents melt-through. Instead, it tapers to a knife edge that meets the backing. Bevel angles are shallower—22½ degrees (a 45-degree groove angle).

Groove welds with tape backing are also different (*Figure 28*). The root opening is usually no wider than 3/16" (4.8 mm). The root face (land) is also usually no wider than 3/16" (4.8 mm). Bevels should have an angle of 30 degrees (a 60-degree groove angle).

2.2.5 Pipe Groove Welds with Gas Backing

Welding pipes can be uniquely challenging. Alignment is critical, particularly the inside diameter since it affects flow. In some cases, counter-boring can solve minor internal alignment problems. It's not suitable for bigger problems since it reduces the pipe wall's thickness by too much.

Open-root joints can have a problem with *root reinforcement*—metal extending past the root opening. Excessive root reinforcement could stress the joint, causing distortion. It can also interfere with the flow, creating turbulence or a pressure drop.

Unfortunately, controlling the root reinforcement is difficult since the welder can't easily access the inside of the pipe. Applying a barrier-type backing, for example, is difficult. Instead, the welder may use *gas backing*. This method involves filling the pipe with a gas like argon, nitrogen, or carbon dioxide. The backing gas behaves like a physical backing and helps control melt-through. It also protects the exposed weld metal from corrosion during the root pass.

Managing the gas requires special techniques. The goal is to keep the pipe full of gas, not to pressurize it. For small or short pipes, taping the ends or capping it with metal discs is fine. Larger pipes require a pair of inflatable bladders or soluble plugs around the weld joint (*Figure 29*). After completing the weld, the welder removes the bladders or dissolves the plugs with water.

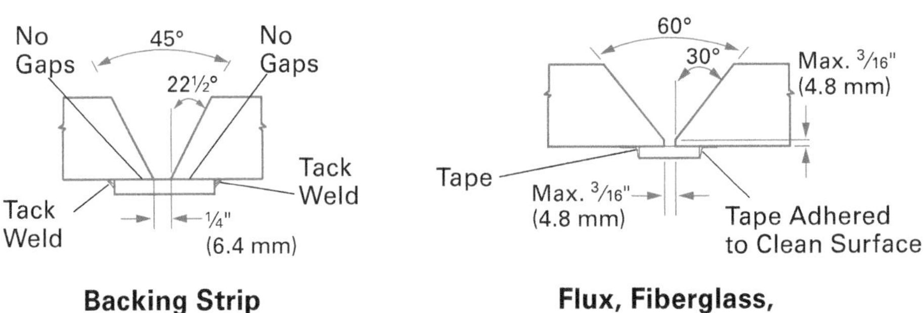

Figure 28 Groove welds with backing.

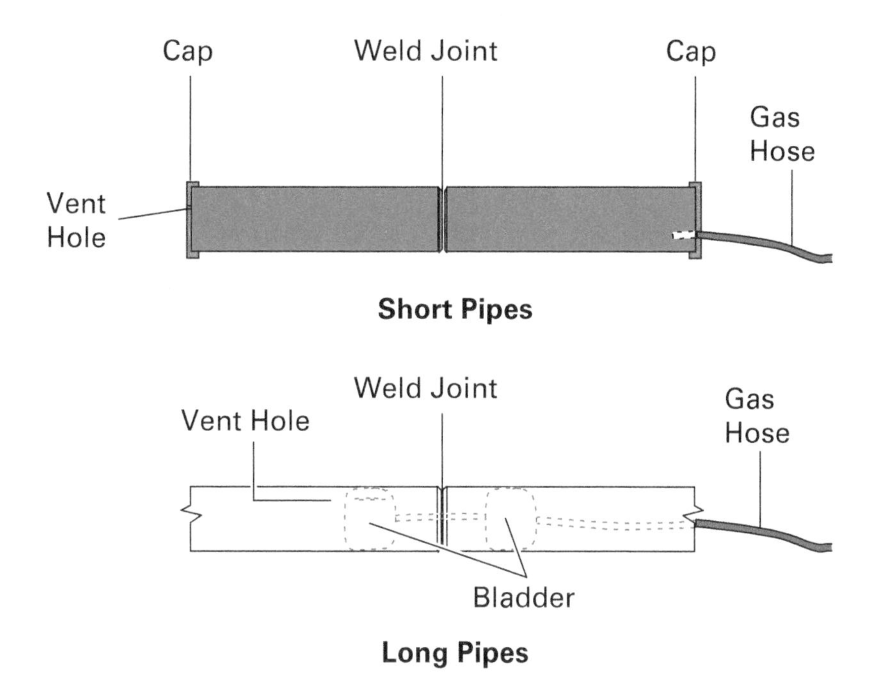

Figure 29 Gas backing techniques.

Backing gas choice depends on the base metal. Some metals require a mixture of gases rather than a single gas. Consult the WPS or jobsite quality specifications for guidance. In all cases, the pipe must be vented to prevent pressure from building up inside.

WARNING!

Although backing gases usually aren't toxic, they don't support life either. They can displace the oxygen-containing air from a confined space. Before using a backing gas, follow all required precautions against asphyxiation. Confirm adequate ventilation, especially if working in a confined space. In many cases, forced ventilation is the only way to ensure safe oxygen levels. Sometimes, a supplied air respirator (SAR) is required.

Gas-backed open-root welds commonly have root openings of $^{1}/_{8}$" (3.2 mm). The root face (land) is also $^{1}/_{8}$" (3.2 mm). Bevels are 30 degrees (a 60-degree groove angle).

2.2.6 Pipe Groove Welds with Backing Rings

While it's difficult to apply a barrier-type backing to a pipe weld, it's not impossible. Backing rings are metal strips rolled up to fit inside the pipe (*Figure 30*). Some are split to make insertion easier and to accommodate pipe that isn't perfectly round.

Many rings have nubs around the outside that establish the correct root opening. After tack welding the pipe, the welder knocks the nubs off with a chipping hammer. Other ring styles have pegs, buttons, or indents. Some are removable, while others fuse into the finished weld.

When using a backing ring, root openings are commonly $^{1}/_{8}$" (3.2 mm). The root face (land) is also $^{1}/_{8}$" (3.2 mm). Bevels are between 30 and $37^{1}/_{2}$ degrees (a 60- to 75-degree groove angle). If the ring has nubs, the root faces should touch them. *Figure 31* shows joint preparations for two backing ring styles.

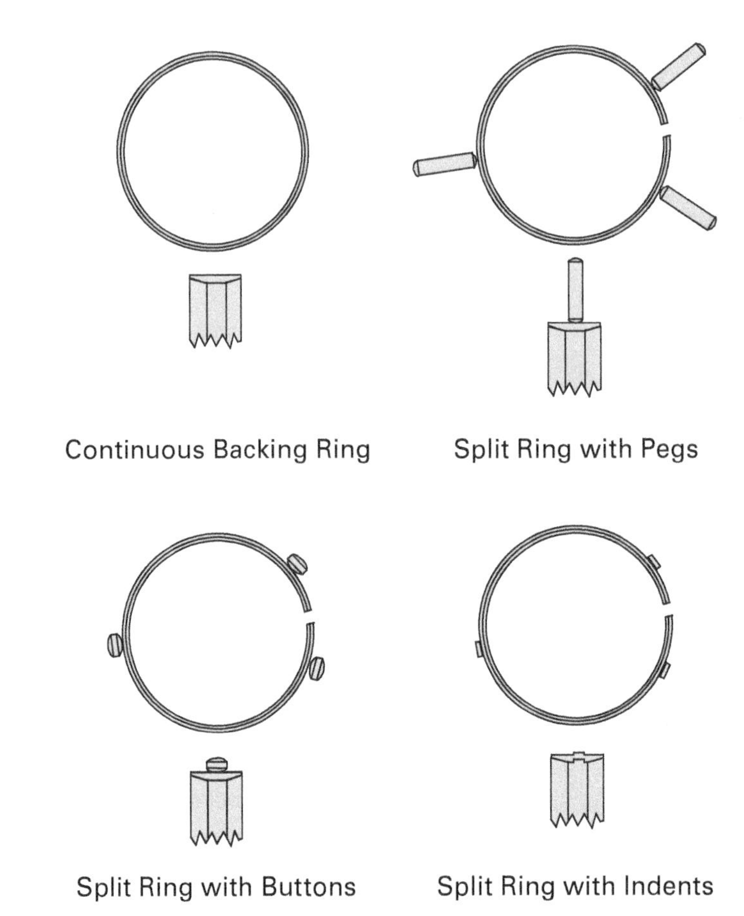

Figure 30 Backing rings.

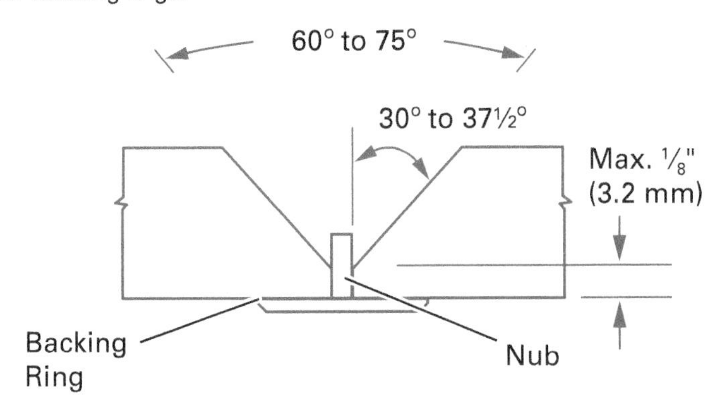

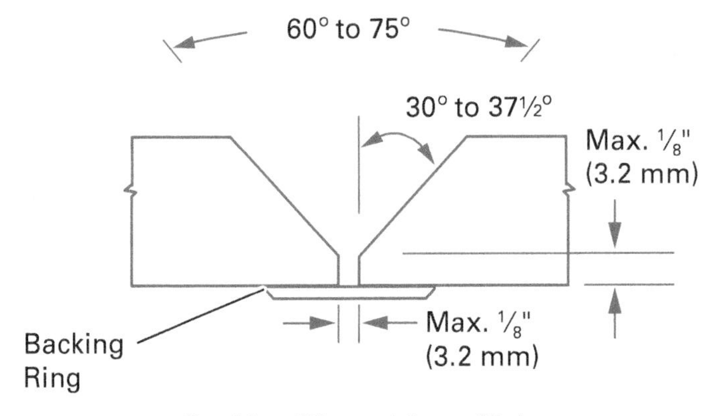

Figure 31 Joint preparation with backing ring.

2.2.7 Pipe Groove Welds with Consumable Inserts

Consumable inserts fit into the root opening and set its spacing. As their name suggests, they fuse into the finished weld. Obviously, they must be compatible with the base and weld metals. Often, welders back them with a gas. They're available in many styles, each identified by a class number (*Figure 32*). Some come as coils, while others come preformed as rings.

Consumable inserts usually require a V-groove or J-groove joint preparation (*Figure 33*). Typical V-grooves have a $^1/_{16}$" (1.6 mm) root face (land). The bevel angle is $37^1/_2$ degrees (a 75-degree groove angle). Typical J-grooves also have a $^1/_{16}$" (1.6 mm) root face (land). The J's bevel angle is about 20 degrees.

NOTE

Welders usually tack weld the consumable insert into place. The WPS or jobsite quality specifications provide the tacking guidelines.

2.2.8 Pipe Fillet Welds with Socket Joints

Socket joints often connect smaller pipes—5" (DN125) or less. Each fitting has a prefabricated socket that the pipe slides into. All common fittings (elbows, flanges, couplings, and reducers), as well as valves, come with sockets. A fillet weld joins the pipe to the socket. This method makes assembly quick and reliable.

To prevent stress, distortion, and cracking, welders and pipefitters must not bottom out the pipe in the socket. Codes usually specify a gap between $^1/_{16}$" and $^1/_8$" (1.6 mm and 3.2 mm). Several methods can achieve this.

One way is to scribe the pipe a specific distance from the end. Measure from the scribed line to the socket face, accounting for the socket depth and the desired gap (*Figure 34*).

As an example, assume that the scribed mark is 3" from the pipe's end. The socket depth is 1" and the gap must be $^1/_8$". The pipe's end, therefore, should penetrate $^7/_8$" into the socket to leave the correct gap (1" – $^1/_8$" = $^7/_8$"). When this is the case, the scribed mark will be $2^1/_8$" from the socket face (3" – $^7/_8$" = $2^1/_8$").

NOTE

Always check the measurement again after tack welding the pipe in place.

A simpler way to measure and scribe the lines is to use a Hi-Lo gauge. Bottom out the pipe and scribe a line along it where it enters the socket. Position the Hi-Lo tool with one foot on the socket and the other on the pipe. Be sure both feet fully contact the surfaces. Note the scribed line's location along the foot's scale. Retract the pipe until the scribed line has moved the required gap distance. Scribe another line on the pipe where it enters the socket (*Figure 35*). Before welding, confirm that the second line is at the socket face.

NOTE

Always confirm that the second scribed line is still at the socket face after tack welding the pipe in place.

Another way to set the gap is to insert a *gap ring* into the socket. It establishes the proper gap and flexes to accommodate expansion during welding (*Figure 36*). The ring itself becomes a permanent part of the joint. Gap rings come in many sizes. They're a simple and convenient way to set the correct gap.

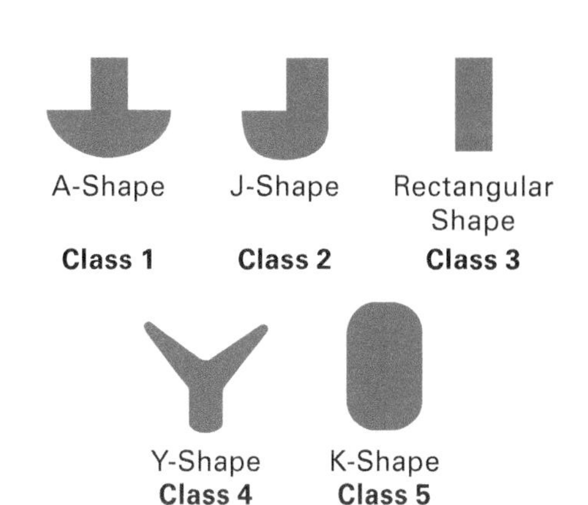

Figure 32 Consumable inserts.

2.2.9 Welding Sequences and Techniques

The *welding sequence*—the order that the welder places the beads—plays a big role in controlling distortion. The right sequence causes shrinkage forces to work against each other, reducing stress and distortion.

The WPS may specify the welding sequence (*Figure 37*). If not, the welder will have to plan it. Sometimes welds don't require a continuous bead. Intermittent beads are often perfectly acceptable for stiffeners, brackets, and braces. This technique also works for some fillet welds. Alternating sides while welding the beads balances the stresses better (*Figure 37*). Welders working in pairs ("buddy welding") can produce these sequences quickly.

In *backstep welding*, the weld progresses from left to right. The welder, however, produces the beads in short increments from right to left (*Figure 38*). This technique reduces distortion by minimizing and interrupting heat input.

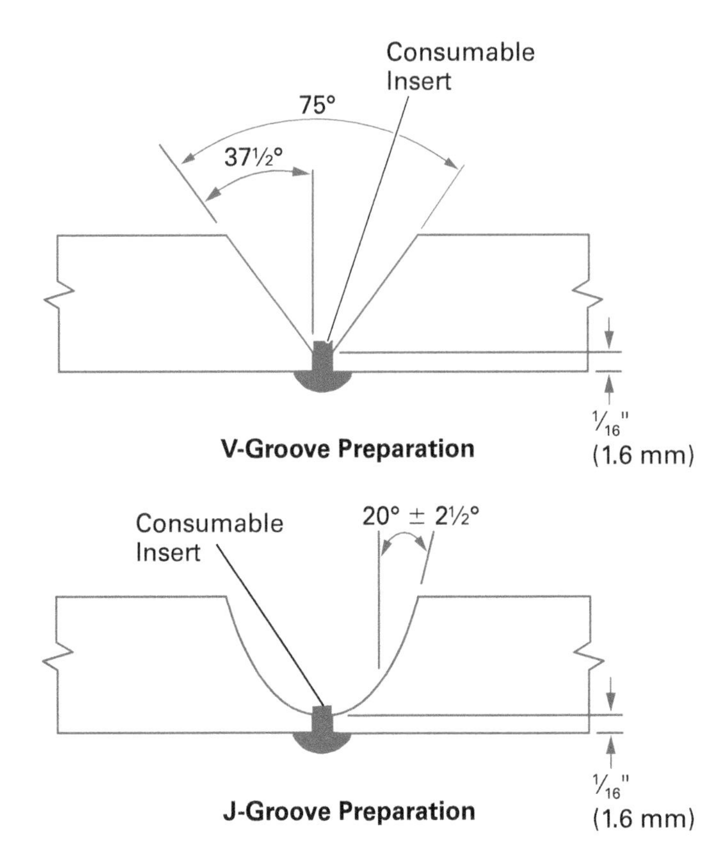

Figure 33 Joint preparation for consumable inserts.

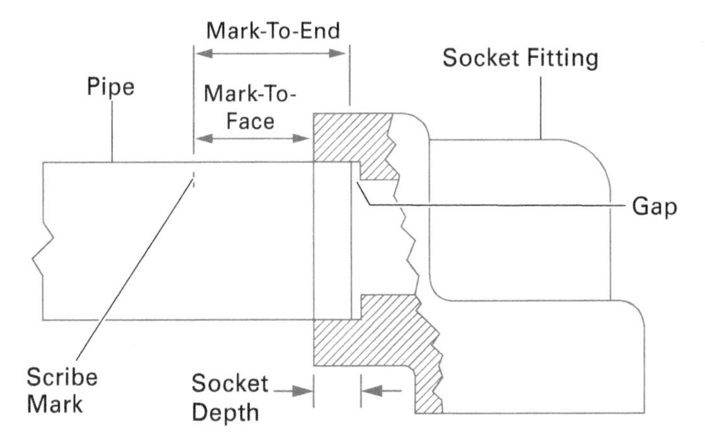

Figure 34 Setting a socket weld gap by measuring.

Alternating Welds

When possible, place alternating welds directly across from the matching weld on the opposite side. For long welds, this may not be appropriate. It could leave long runs unwelded. In these cases, stagger the welds from side to side.

Sometimes, these techniques aren't possible, or you must complete one side first. In these situations, use a joint preparation that requires more weld metal on the second side. Its greater shrinkage will balance the first side's distortion forces.

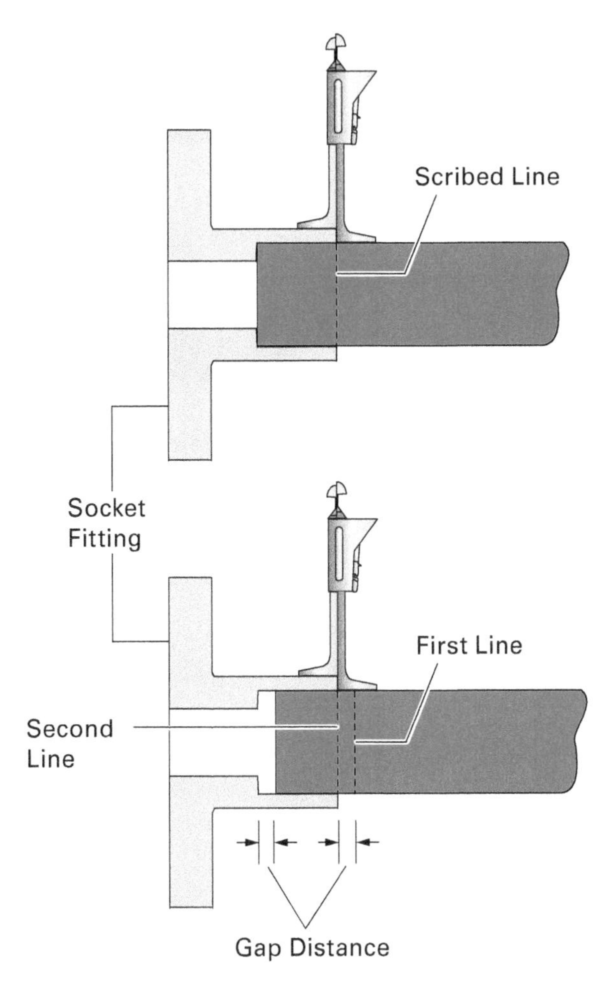

Figure 35 Setting a socket weld gap with a Hi-Lo gauge.

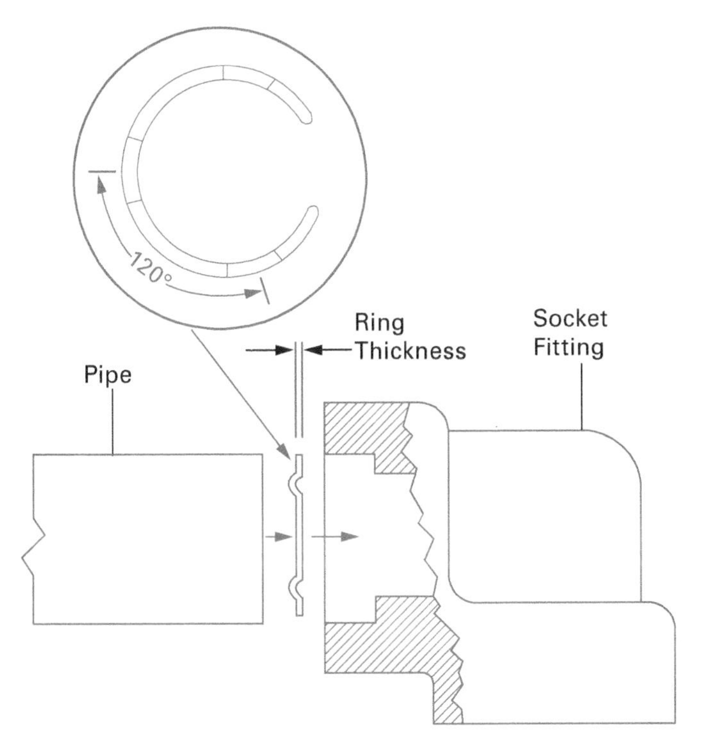

Figure 36 Setting a socket weld gap with a gap ring.

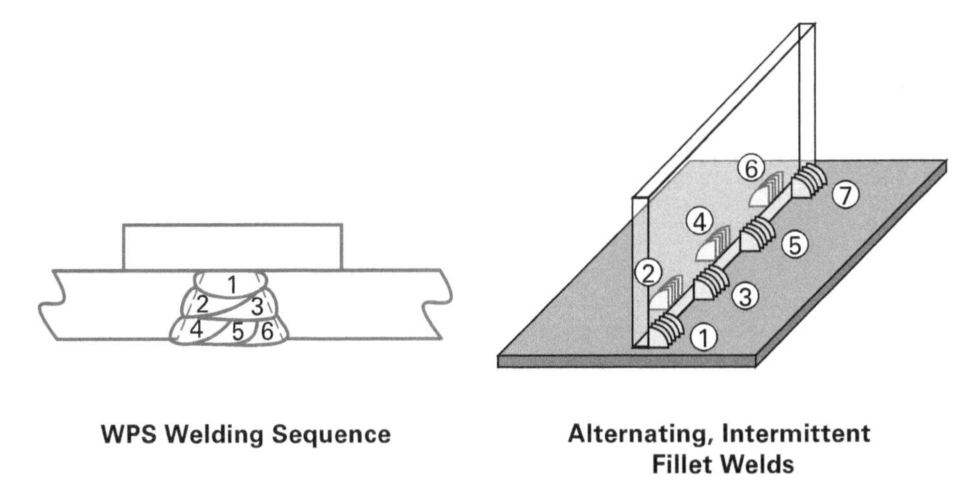

Figure 37 Welding sequences.

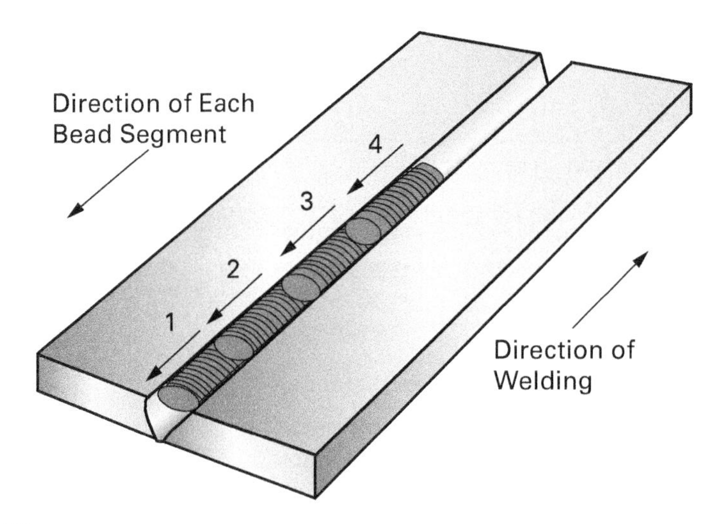

Figure 38 Backstep welding.

2.2.10 Heat Treatments

Using *heat treatment* techniques is another way to manage stress and distortion. These strategies involve selectively applying heat at specific times and then controlling the cooling. Heating before welding is *preheating*, while heating afterwards is *postheating*. The WPS or the jobsite quality specifications will identify the correct time and place for heat treatment. Other modules will cover this topic in detail.

2.2.11 Checking the Joint

Joint preparation, fit-up, and alignment directly affect the completed weld. Many problems, stresses, and distortions come from not checking before welding. Correct problems as you discover them. If you aren't sure how to handle an issue, speak to your supervisor.

The following steps outline checking the joint:

Step 1 Identify the WPS or jobsite quality specifications covering the weld. Go over the information carefully.

Step 2 Confirm that you're certified/qualified to create the weld.

Step 3 Verify that the welding process you're planning to use matches the specifications.

Step 4 Check the base metal type and grade. Confirm that it matches the specifications.

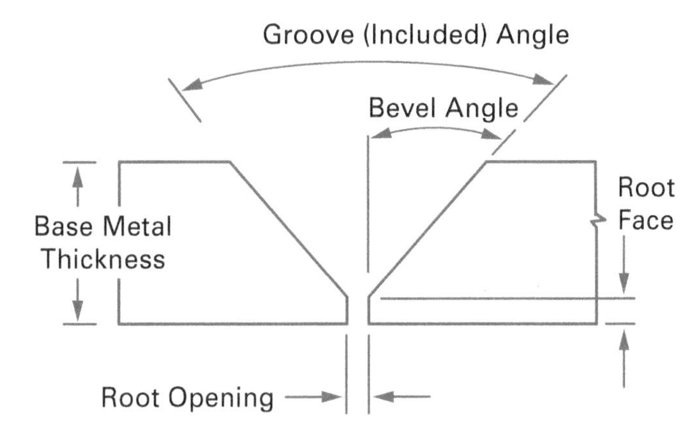

Figure 39 Joint preparation.

Step 5 Check that the consumables (type and size) match the specifications.

Step 6 Verify that you can properly store heated consumables if the specifications require them.

Step 7 Examine the joint surfaces for contamination, rust, grease, and oil. Confirm that they're clean and ready for welding.

Step 8 Check the joint surface for cracks or laminations.

Step 9 Confirm that the joint preparation follows the specifications (*Figure 39*):

- Groove type
- Root opening
- Root face (land)
- Groove angle / Bevel angle / Bevel radius
- Base metal thickness

Step 10 If present, check backing strips, backing rings, or consumable inserts. Everything must fit tightly against the joint. The root opening must be correct.

Step 11 Check that weldment parts are level, plumb, or square.

Step 12 Examine all tack welds. They should be clean and feathered.

Step 13 Confirm that you have a plan for managing preheating, temperature control between passes, and postheating if the specifications require them.

Step 14 Identify the specified welding sequence. If there is no specified sequence, work out an appropriate plan.

Step 15 Confirm that the welding machine's settings match the specifications.

Inspect as You Go

As you work, periodically pause and inspect your work. Schedule these inspection times strategically so everything that you need to check is clearly visible. Check the following items as applicable:

- Compliance with the WPS or jobsite quality specifications
- Proper preheat temperature
- Root pass
- Other weld layers
- Proper postheat temperature

2.0.0 Section Review

1. A clamped weldment that distorts when you release the clamping device likely has a problem with _____.
   a. latent heat
   b. joint preparation
   c. improper tack welds
   d. residual stress

2. What is an appropriate way to protect the root pass from corrosion in an open-root pipe groove weld?
   a. Use gas backing.
   b. Apply a pipe clamp.
   c. Use a knife edge root preparation.
   d. Apply a ring clamp.

Module 29110 Review Questions

1. Always attach a chain hoist to a suitable support such as a _____.
   a. conduit tray
   b. pipe rack
   c. structural member
   d. cable raceway

2. If a joint requires significant holding force, welders may combine wedges and yokes with a _____.
   a. strongback
   b. cage clamp
   c. hydraulic jack
   d. chain fall

3. What is the main advantage of cage clamps with arched crossbars?
   a. They're stronger than regular cage clamps.
   b. They give better access to the weld joint.
   c. They can work with many pipe sizes.
   d. They can position plates around the pipes.

4. To see if a pipe flange is level before welding it, check it with a level supported by _____.
   a. pins
   b. clips
   c. jack stands
   d. straightedges

5. Which tool can confirm that a weldment component is plumb?
   a. Level
   b. Flange pins
   c. Square
   d. Straightedge

6. Hi-Lo gauges are useful for checking _____.
   a. pipe internal alignment
   b. bevel angles
   c. groove angles
   d. weld face overlap

7. What can cause a weldment component's shape to change?
   a. Uniform expansion or contraction
   b. Torsional friction
   c. Non-uniform expansion or contraction
   d. Locked-in heat

8. A material that transfers heat easily will have a large _____.
   a. residual stress
   b. thermal conductivity value
   c. weldment flexure
   d. latent heat of fusion

9. Which of the following causes greater stress when a joint contains more weld metal?
   a. Arc blow
   b. Magnetism
   c. Shrinkage
   d. Thermal conductivity

10. What must you *never* do when creating a pipe socket weld?
    a. Support the pipe on a jack stand.
    b. Scribe a line on the pipe.
    c. Insert a gap ring into the socket.
    d. Bottom out the pipe in the socket.

Answers to odd-numbered Module Review Questions are found in *Appendix A*.

Answers to Section Review Questions

Answer	Section Reference	Objective
Section 1.0.0		
1. a	1.1.2	1a
2. b	1.2.1	1b
3. c	1.3.3	1c
4. b	1.4.4	1d
Section 2.0.0		
1. d	2.1.0; 2.2.1	2a; 2b
2. a	2.2.5	2b

User Update

Did you find an error? Submit a correction by visiting **https://www.nccer.org/olf** or by scanning the QR code using your mobile device.

MODULE 29111

SMAW – Groove Welds with Backing (Plate)

Source: The Lincoln Electric Company, Cleveland, OH, USA

Objectives

Successful completion of this module prepares you to do the following:

1. Summarize groove weld types and preparing to weld them.
   a. Identify groove welds and their properties.
   b. Outline preparing for making groove welds with backing.
2. Summarize techniques for making V-groove welds with backing.
   a. Describe making V-groove welds with backing in the 1G and 2G positions.
   b. Describe making V-groove welds with backing in the 3G and 4G positions.

Performance Tasks

Under supervision, you should be able to do the following:

1. Safely set up SMAW equipment for groove welding.
2. Use E7018 electrodes to make flat (1G) V-groove welds with backing.
3. Use E7018 electrodes to make horizontal (2G) V-groove welds with backing.
4. Use E7018 electrodes to make vertical (3G) V-groove welds with backing.
5. Use E7018 electrodes to make overhead (4G) V-groove welds with backing.

Overview

Groove welds come in several styles and serve many purposes. The standard V-groove weld is popular and versatile. Producing a good V-groove weld, however, requires knowledge and proper technique. This module helps you create good V-groove welds with backing in all positions.

Industry Recognized Credentials

If you are training through an NCCER-accredited sponsor, you may be eligible for credentials from NCCER's Registry. The ID number for this module is 29111. Note that this module may have been used in other NCCER curricula and may apply to other level completions. Contact NCCER's Registry at 1.888.622.3720 or go to **www.nccer.org** for more information.

You can also show off your industry-recognized credentials online with NCCER's digital badges. Transform your knowledge, skills, and achievements into badges that you can share across social media platforms, send to your network, and add to your resume. For more information, visit **www.nccer.org**.

NOTE

This module uses US standard and metric units in up to three different ways. This note explains how to interpret them.

Exact Conversions

Exact metric equivalents of US standard units appear in parentheses after the US standard unit. For example: "Measure 18" (45.7 cm) from the end and make a mark."

Approximate Conversions

In some cases, exact metric conversions would be inappropriate or even absurd. In these situations, an approximate metric value appears in parentheses with the ~ symbol in front of the number. For example: "Grip the tool about 3" (~8 cm) from the end."

Parallel but not Equal Values

Certain scenarios include US standard and metric values that are parallel but not equal. In these situations, a slash (/) surrounded by spaces separates the US standard and metric values. For example: "Place the point on the steel rule's 1" / 1 cm mark."

Digital Resources for Welding

Scan this code using the camera on your phone or mobile device to view the digital resources related to this craft.

1.0.0 Groove Welds

Performance Task

1. Safely set up SMAW equipment for groove welding.

Objective

Summarize groove weld types and preparing to weld them.

a. Identify groove welds and their properties.
b. Outline preparing for making groove welds with backing.

Welders can join two parts with a groove weld between them. They can also make groove welds in part openings. Groove welds apply to all five basic joint types: butt, lap, T, edge, and corner. Simply fitting the parts together creates some groove welds. Other groove welds require preparation steps. *Complete joint penetration (CJP)* welds go completely through the base metal. *Partial joint penetration (PJP)* welds don't go completely through.

This module introduces V-groove welds with backing. A small gap separates the two parts. A strip made from one of several materials bridges the gap. Welders often use these welds to join plate. V-groove welds with backing are relatively easy to produce. Trainee welders learn to make them first before moving on to more challenging welds.

1.1.0 Groove Weld Styles

NOTE

Not every groove style works well with backing. This module focuses on those that do.

A groove weld's name comes from its shape. Welds can have one or two grooves. Single-groove styles include square, bevel, V, U, J, flare V, and flare bevel. Double-groove styles are bevel, V, U, and J. *Figure 1* shows examples of each.

If the parts being joined don't have the right shape (profile), the welder must create it. Most joint configurations start with square (90-degree) edges. A grinder can produce beveled (angled) grooves. Preparing the edges cleans the metal. It also increases the weld size. Welders must produce grooves with good root access. Grooves should never be larger than necessary. The groove preparation's main purpose is to give the welder access to the weld root to achieve the designed weld penetration.

NOTE

This module includes dimensions with both US standard and metric measurements. Welding symbols in drawings, however, usually show US standard units only.

Sometimes, welders combine a fillet weld with a groove weld. This strategy maintains good strength but requires less metal. *Figure 2* shows several combination weld examples.

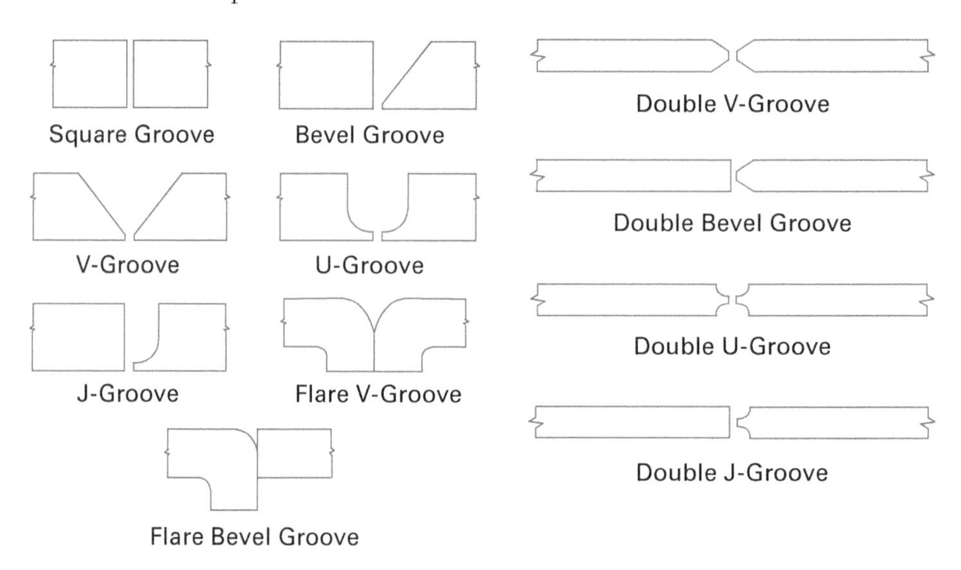

Figure 1 Single- and double-groove welds.

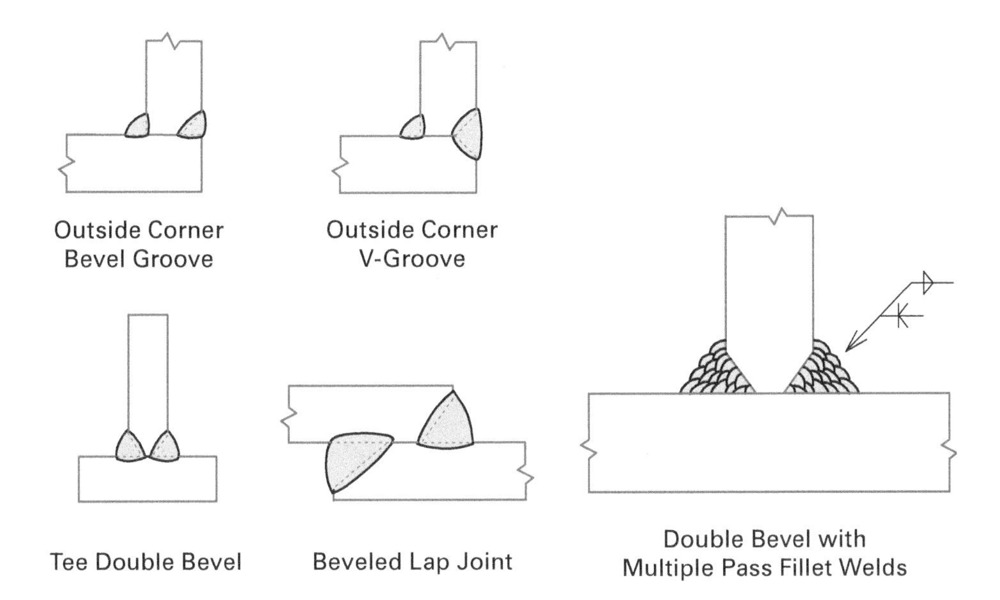

Figure 2 Combination welds.

1.1.1 Groove Weld Terminology

Welders describe weld features and dimensions with a specialized vocabulary. Learning it will help you communicate clearly and accurately with other welders. The following terms describe groove welds with backing features (*Figure 3*):

Weld face — The weld's exposed surface on the side where the welder produced it.

Weld root — The point where the back of the weld extends farthest into the joint.

Weld toes — The points at the weld face where the weld and base metals meet.

Face reinforcement — Weld metal extending beyond the base metal on a groove weld's face side.

Groove weld size — The depth that the weld penetrates the joint.

The following terms apply to the dimensions of groove welds with metal backing (*Figure 4*):

Joint preparation — The cut or machined groove shape that penetrates the base metal.

Groove face — Any groove surface before welding.

Bevel face — The groove's angled surface.

Groove angle — The included angle between the two groove faces.

Bevel angle — The angle between the bevel face and a line perpendicular to the base metal.

Joint root — The place in the joint where the weld members are closest together.

Root opening — The distance between the two members at the root.

1.1.2 Weld Backings

Backings are materials placed behind a weld's root opening (*Figure 5*). Single-groove welds don't require them. They'll be stronger and of higher quality, however, if you use a backing. Backings also make welding the root easier. They can be metal strips that match the base metal.

Alternatively, backings can be other metals, ceramics, tapes, or even a weld. The welder may or may not remove these after finishing. Sometimes, an inert gas functions as a backing even though it isn't a solid barrier.

NOTE

Groove welds with metal backings taper to a knife edge that meets the backing. Other backing types often have a *root face*. This is a flattened area at the end of the groove weld's bevel. Welders often call it the *land*.

1.2.0 Welding Preparations

NOTE

The dimensions and specifications in this module represent welding codes in general, not a specific code. Always follow the proper codes for your jobsite. Also follow the applicable Welding Procedure Specification (WPS) and jobsite quality specifications.

NOTE

Be sure that you've completed NCCER Module 29101, *Welding Safety*, before continuing.

This module explains preparing and producing V-groove welds with metal backing in all positions. You'll use E7018 (low-hydrogen) electrodes. What you learn here will prepare you for acquiring welder qualifications, as well as producing other groove weld types.

You'll need to prepare before you can weld a V-groove joint with backing. You must set up the welding area, equipment, and metal. The following sections explain how to do these tasks safely.

1.2.1 Safety Preparations

Cutting and welding equipment produces high temperatures, sparks, and many other hazards. Using equipment safely requires following specific procedures. This section summarizes good practices but isn't a complete safety training. Always complete all training that your workplace requires. Follow all workplace safety guidelines and wear appropriate PPE (*Figure 6*).

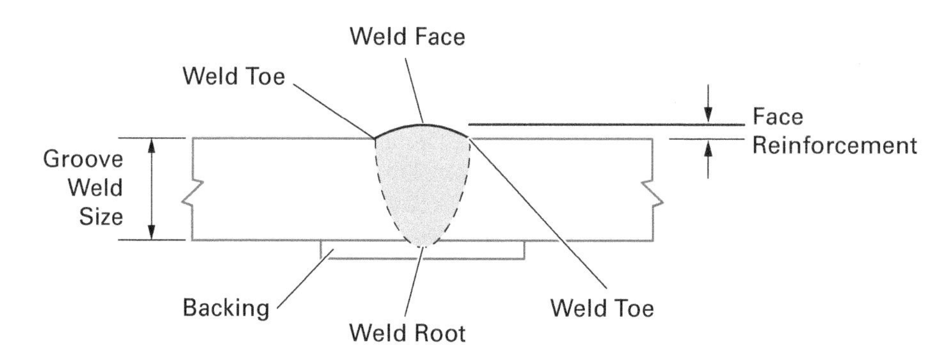

Figure 3 Groove welds with backing features.

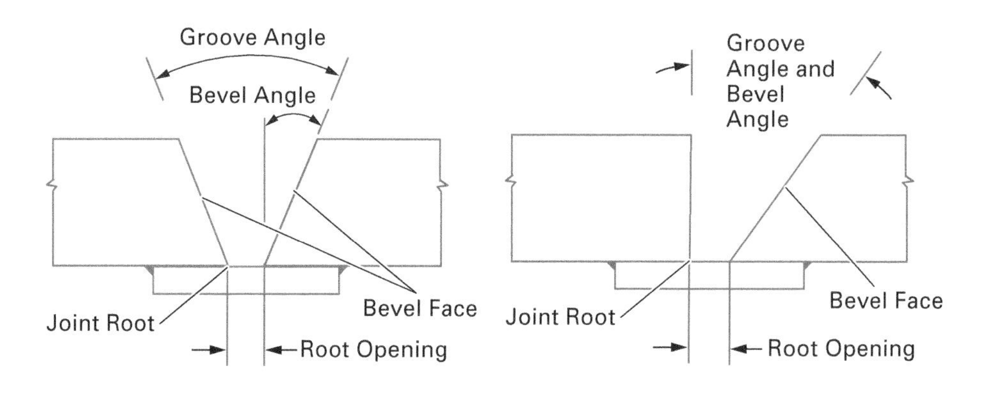

Figure 4 Groove welds with backing dimensions.

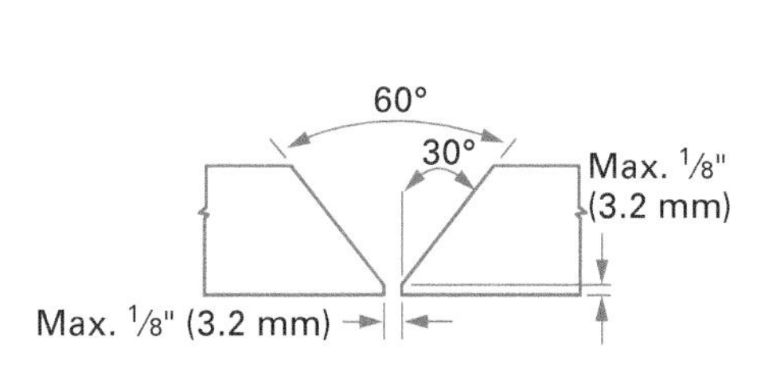

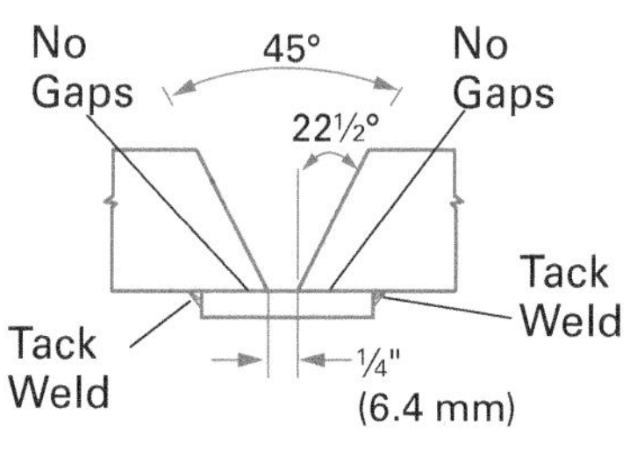

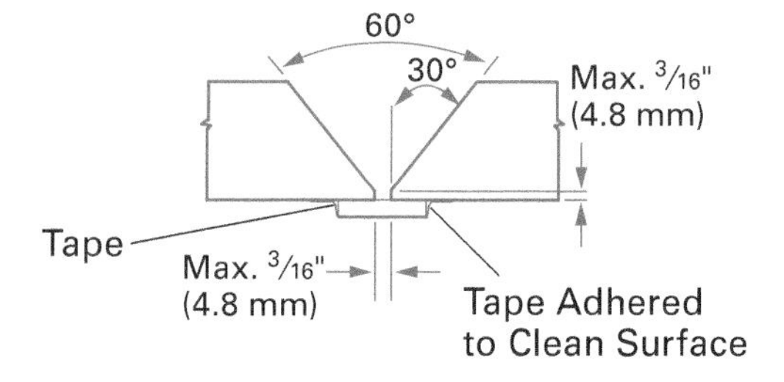

Figure 5 Groove weld backings.

1.2.2 Protective Clothing and Equipment

Welding and cutting tasks are dangerous. Unless you wear all required PPE, you're at risk of serious injury. The following guidelines summarize PPE for SMAW work:

- For cutting, always wear close-fitting goggles or safety glasses, plus a full face shield, helmet, or hood. The goggles, glasses, face shield, or helmet/hood lens must have the proper shade value for the type of cutting work (*Figure 7*). Never cut without using the proper lens.
- For welding, always wear safety glasses, plus a helmet or hood. The glasses or helmet/hood lens must have the proper shade value for the welding process (*Figure 7*). Never view the cutting arc directly or indirectly without using the proper lens.

Figure 6 Welding PPE suitable for SMAW work.
Source: © Miller Electric Mfg. LLC

WARNING!

Don't open your welding helmet/hood lens and expose your eyes when grinding or cleaning a weld. Either use a helmet or hood with an auto-darkening lens or a clear lens under a flip-up shaded lens. Alternatively, remove the helmet or hood and use a full face shield.

- Wear protective leather and/or flame-retardant clothing along with welding gloves that will protect you from flying sparks and molten metal.
- Wear high-top safety shoes or boots. Make sure the pant leg covers the tongue and lace area. If necessary, wear leather spats or chaps for protection. Boots without laces or holes are better for welding and cutting.
- Wear a 100 percent cotton cap with no mesh material in its construction. The bill should point to the rear or the side with the most exposed ear. If you must wear a hard hat, use one with a face shield.

WARNING!

Never wear a cap with a button on top. The conductive metal under the fabric is a safety hazard.

- Wear earplugs to protect your ear canals from sparks. Since welding and cutting can be noisy, always wear hearing protection.

1.2.3 Fire/Explosion Prevention

All processes in this module can produce high temperatures. Welding or cutting a vessel or container that once contained combustible, flammable, or explosive materials is hazardous. Residues can catch fire or explode. Before welding or cutting vessels, check whether they contained explosive, hazardous, or flammable materials. These include petroleum products, citrus products, or chemicals that release toxic fumes when heated.

American Welding Society (AWS) F4.1, Safe Practices for the Preparation of Containers and Piping for Welding and Cutting, and *ANSI/AWS Z49.1* describe safe practices for these situations. Begin by cleaning the container to remove any residue. Steam cleaning, washing with detergent, or flushing with water are possible methods. Sometimes you must combine these to get good results.

WARNING!

Clean containers only in well-ventilated areas. Vapors can accumulate in a confined space during cleaning, causing explosions or toxic substance exposure.

Arc Welding Processes

Process	Electrode	Amperage	Lens Shade Numbers (2 3 4 5 6 7 8 9 10 11 12 13 14)
Shielded Metal Arc Welding (SMAW)	< 3/32" (2.4 mm)	< 60A	7
	3/32" – 5/32" (2.4 mm – 4.0 mm)	60A – 160A	8 9 10
	5/32" – 1/4" (4.0 mm – 6.4 mm)	160A – 250A	10 11 12
	> 1/4" (6.4 mm)	250A – 550A	11 12 13 14
Gas Metal Arc Welding (GMAW) and Flux Cored Arc Welding (FCAW)		< 60A	7
		60A – 160A	10 11
		160A – 250A	10 11 12
		250A – 500A	10 11 12 13 14
Gas Tungsten Arc Welding (GTAW)		< 50A	8 9 10
		50A – 150A	8 9 10 11 12
		150A – 500A	10 11 12 13 14
Plasma Arc Welding (PAW)		< 20A	6 7 8
		20A – 100A	8 9 10
		100A – 400A	10 11 12
		400A – 800A	11 12 13 14
Carbon Arc Welding (CAW)			14

Arc Cutting Processes

Process	Electrode	Amperage	Lens Shade Numbers (2 3 4 5 6 7 8 9 10 11 12 13 14)
Plasma Arc Cutting (PAC)		< 20A	4
		20A – 40A	5
		40A – 60A	6
		60A – 80A	8
		80A – 300A	8 9
		300A – 400A	9 10 11 12
		400A – 800A	10 11 12 13 14
Air-Carbon Arc Cutting (A-CAC)	Light	< 500A	10 11 12
	Heavy	500A – 1,000A	11 12 13 14

Gas Processes

Process	Type	Plate Thickness	Lens Shade Numbers (2 3 4 5 6 7 8 9 10 11 12 13 14)
Oxyfuel Welding (OFW)	Light	< 1/8" (3 mm)	4 5
	Medium	1/8" – 1/2" (3 mm – 13 mm)	5 6
	Heavy	> 1/2" (13 mm)	6 7 8
Oxyfuel Cutting (OC)	Light	< 1" (25 mm)	3 4
	Medium	1" – 6" (25 mm – 150 mm)	4 5
	Heavy	> 6" (150 mm)	5 6
Torch Brazing (TB)			3 4
Torch Soldering (TS)			2

Lens shade values are based on OSHA minimum values and ANSI/AWS suggested values. Choose a lens shade that's too dark to see the weld zone. Without going below the required minimum value, reduce the shade value until it's visible. If possible, use a lens that absorbs yellow light when working with fuel gas processes.

Figure 7 Guide to lens shade numbers.

After cleaning the container, you must formally confirm that it's safe for welding or cutting. *American Welding Society (AWS) F4.1, Safe Practices for the Preparation of Containers and Piping for Welding and Cutting*, outlines the proper procedure. The following three paragraphs summarize the process.

Immediately before work begins, a qualified person must check the container with an appropriate test instrument and document that it's safe for welding or cutting. Tests should check for relevant hazards (flammability, toxicity, etc.). During work, repeated tests must confirm that the container and its surroundings remain safe.

Alternatively, fill the container with an inert material ("inerting") to drive out any hazardous vapors. Water, sand, or an inert gas like argon meets this requirement. Water must fill the container to within 3" (~7.5 cm) or less of the work location. The container must also have a vent above the water so air can escape as it expands (*Figure 8*). Sand must completely fill the container.

When using an inert gas, a qualified person must supervise, confirming that the correct amount of gas keeps the container safe throughout the work. Using an inert gas also requires additional safety procedures to avoid accidental suffocation.

NOTE

A *lower explosive limit (LEL)* gas detector can check for flammable or explosive gases in a container or its surroundings. Gas monitoring equipment checks for specific substances, so be sure to use a suitable instrument. Test equipment must be checked and calibrated regularly. Always follow the manufacturer's guidelines.

WARNING!

Never weld or cut drums, barrels, tanks, vessels, or other containers until they have been emptied and cleaned thoroughly. Residues, such as detergents, solvents, greases, tars, or corrosive/caustic materials, can produce flammable, toxic, or explosive vapors when heated. Never assume that a container is clean and safe until a qualified person has checked it with a suitable test instrument. Never weld or cut in places with explosive vapors, dust, or combustible products in the air.

The following general fire- and explosion-prevention guidelines apply to cutting and SMAW work:

- Never carry matches or gas-filled lighters in your pockets. Sparks can ignite the matches or cause the lighter to explode.
- Always comply with all site and employer requirements for hot work permits and fire watches.
- Never use oxygen in place of compressed air to blow off the workpiece or your clothing. Never release large amounts of oxygen. Oxygen makes fires start readily and burn intensely. Oxygen trapped in fabric makes it ignite easily if a spark falls on it. Keep oxygen away from oil, grease, and other petroleum products.
- Remove all flammable materials from the work area or cover them with a fire-resistant blanket.
- Before welding, heating, or cutting, confirm that an appropriate fire extinguisher is available. It must have a valid inspection tag and be in good condition. All workers in the area must know how to use it.
- Never release a large amount of fuel gas into the work environment. Some gases (propane and propylene) accumulate on the floor, while others (acetylene and natural gas) accumulate near the ceiling. Either can ignite far from the release point. Acetylene will explode at lower concentrations than other fuel gases.
- Prevent fires by maintaining a neat and clean work area. Confirm that metal scrap and slag are cold before disposing of them.

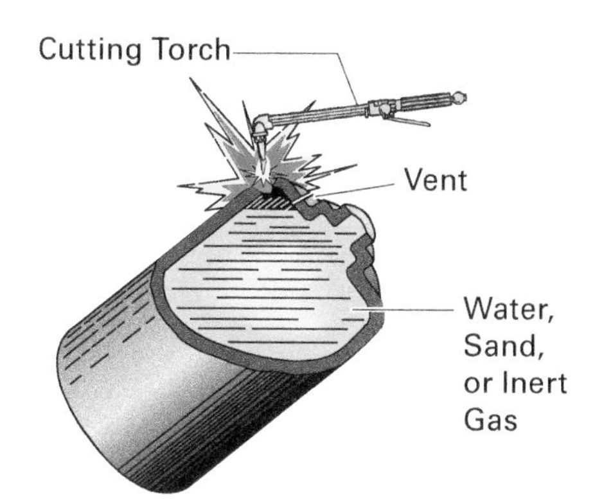

Note: *ANSI Z49.1* and AWS standards should be followed.

Figure 8 "Inerting" a container.

1.2.4 Work Area Ventilation

Vapors and fumes fill the air around their sources. Welding and cutting fumes can be hazardous. Welders often work above the area from which fumes come. Good ventilation and source extraction equipment help remove the vapors and protect the welder. The following lists good ventilation practices to use when welding and cutting:

- Always weld or cut in a well-ventilated area. Operations involving hazardous materials, such as metal coatings and toxic metals, produce dangerous fumes. You must wear a suitable respirator when working with them. Follow your workplace's safety protocols and confirm that your respirator is in good condition.

WARNING!

Cadmium, mercury, lead, zinc, chromium, and beryllium produce toxic fumes when heated. When welding or cutting these materials, wear appropriate respiratory PPE. Supplied-air respirators (SARs) work best for long-term work.

- Never weld or cut in a confined space without a confined space permit. Prepare for working in the confined space by following all safety protocols.
- Set up a confined space ventilation system before beginning work.
- Never ventilate with oxygen or breathe pure oxygen while working. Both actions are very dangerous.

WARNING!

When working around other craftworkers, always be aware of what they're doing. Take precautions to keep your work from endangering them.

Going Green

Vapor Extraction Systems

Protect the environment by using a vapor extraction system that filters out the welding fumes. This strategy is superior to simply releasing the fumes into the environment.

1.2.5 Preparing the Welding Area

To practice making V-groove welds with backing, you'll need a welding table, bench, or stand (*Figure 9*). It should have a way to place welding coupons out of position.

Figure 9 Welding station with source extraction.
Source: The Lincoln Electric Company, Cleveland, OH, USA

The following steps outline setting up the welding area:

Step 1 Check the work area ventilation. If necessary, use fans, windows, or source extraction equipment.

Step 2 Check the area for fire hazards. Remove any flammable materials. Cover with a welding blanket anything flammable that you can't move.

Step 3 Locate the nearest fire extinguisher. Confirm that it's charged and that you know how to use it.

Step 4 Position the welding machine near the welding table.

Step 5 Set up welding screens around the welding area.

1.2.6 Electrodes

These welding exercises require E7018 electrodes. Use a diameter of $^3/_{32}$" (2.4 mm), $^1/_8$" (3.2 mm), or $^5/_{32}$" (4.0 mm). Remember that E7018 electrodes must stay dry. Leave them in their storage oven until you're ready to weld. Remove only as many as you'll use during the session.

Keep electrodes in a pouch or holder so they don't get damaged. Never leave them loose on the table. They can fall on the floor and be damaged or create a slip hazard. Have a suitable metal container nearby to hold discarded electrode stubs.

NOTE

The electrode sizes are recommendations only. Check with your instructor for the correct size to use.

WARNING!

Don't throw electrode stubs on the floor. They can roll under your feet, causing accidents.

Electrode Housekeeping

Electrode stubs on the floor can roll under your feet, creating a serious hazard. Many employers discipline employees who throw stubs on the floor. Some even make this practice a dismissible offense. Always place electrode stubs in a suitable container. Good welders keep their workspaces clean and tidy.

During critical welding jobs, some companies require welders to turn in their electrode stubs at the end of the shift. The number of stubs must match the number of electrodes they received at the start of the shift.

The following general guidelines apply to welding with low-hydrogen electrodes:

- Don't whip the electrode.
- Maintain a short arc. Low-hydrogen electrodes rely on the slag to protect the molten metal. They don't produce a heavy gaseous shield.
- Remove all slag between passes.
- Restart by striking the arc ahead of the crater. Move quickly back into the crater and then proceed as usual. This technique welds over the arc strike. It eliminates porosity and gives a smoother restart.
- When running weave beads, use a slight side-to-side motion. Pause briefly at the weld toe to tie in and flatten the bead.
- Never use electrodes with chipped or missing flux.
- When welding vertically, set the amperage at the lower end of the recommended range. This gives better weld pool control.

Running E7018 Electrodes

Moving the electrode from side to side makes a weave bead. Many different patterns will produce a weave bead, including circles, crescents, and zigzags. Pausing at the edges flattens out the weld, giving it the proper profile.

1.2.7 Preparing the Welding Machine

Welders will use many different welding machine brands and models. Not all will have manuals with them. Learn to recognize welding machine types and their controls, so you can use them safely and correctly (*Figure 10*).

The following steps outline setting up a regular welding machine:

Step 1 Locate the nearest power disconnect so you can shut down the machine quickly.

Step 2 Plug in the machine.

Step 3 Select SMAW mode (constant current DC operation).

Step 4 If the machine has a polarity selector, set it to DCEP. If not, connect the leads for DCEP.

Step 5 Connect the workpiece lead's clamp to the workpiece.

NOTE

If a DC welding machine isn't available, a constant current AC machine will work. E7018 electrodes perform properly in constant current AC mode. In cases where arc blow is a problem, switching from DC to AC may solve the problem.

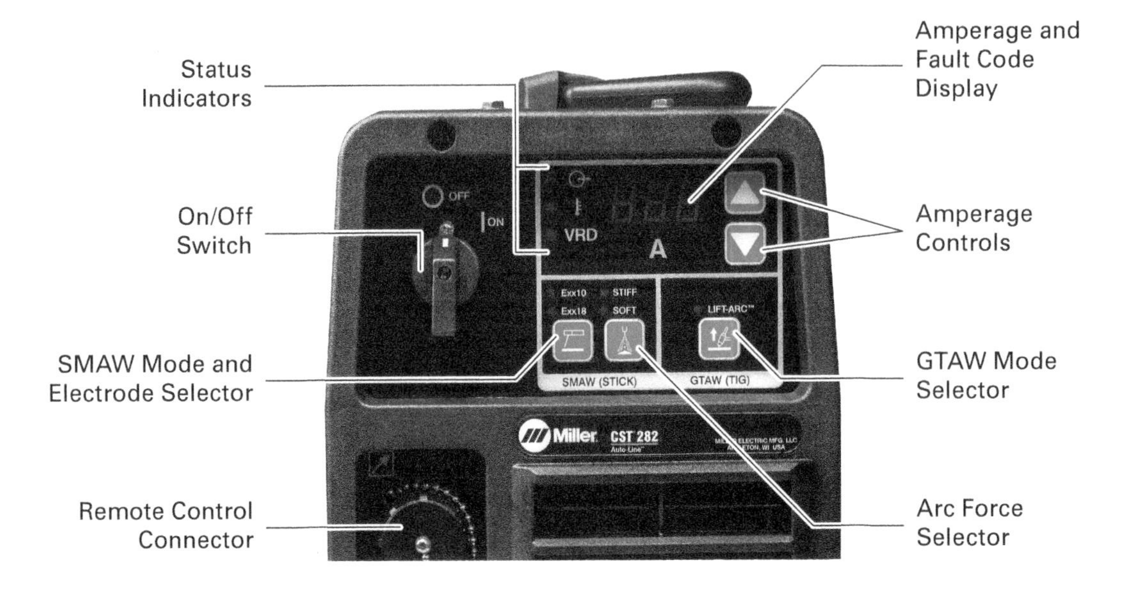

(A) Regular Welding Machine Controls

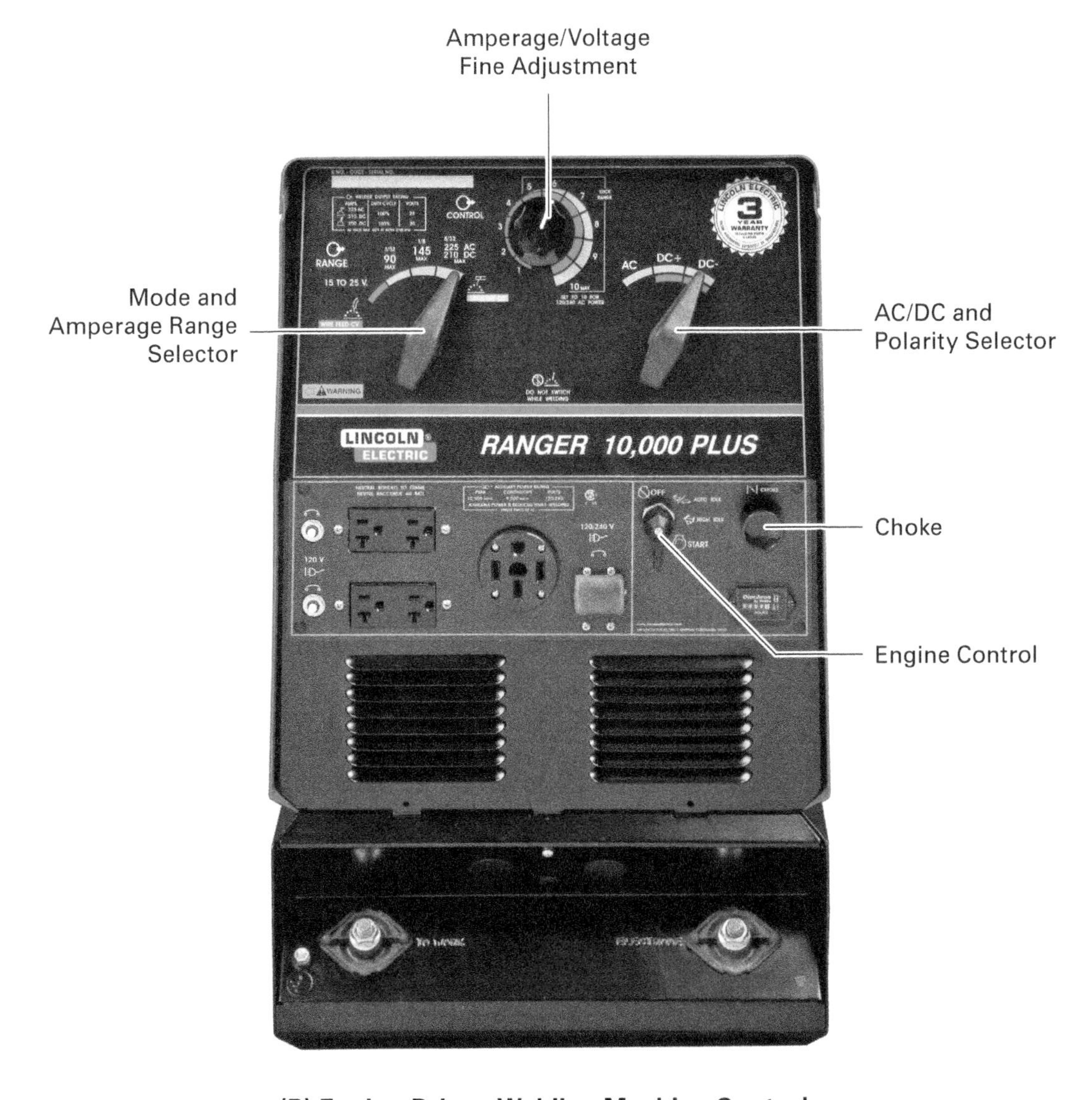

(B) Engine-Driven Welding Machine Controls

Figure 10 Regular and engine-driven welding machines.
Source: © Miller Electric Mfg. LLC (10A), The Lincoln Electric Company, Cleveland, OH, USA (10B)

Step 6 Set the amperage for the electrode type and size. *Table 1* shows typical settings.

Step 7 Confirm that the electrode holder *isn't* resting on a grounded surface.

Step 8 Turn on the welding machine when you're ready to weld.

TABLE 1 Typical Electrode Sizes and Amperages

Electrode	Size	Amperage
E7018	3⁄32" (2.4 mm)	70A to 110A (DCEP) 80A to 120A (AC)
E7018	1⁄8" (3.2 mm)	90A to 160A (DCEP) 100A to 160A (AC)
E7018	5⁄32" (4.0 mm)	130A to 210A (DCEP) 140A to 210A (AC)

NOTE

These values are general guidelines only. Amperage recommendations vary by manufacturer, welding position, current type, and electrode brand. If possible, refer to the WPS for guidance. If a WPS isn't available, follow the electrode manufacturer's guidelines.

If you're using an engine-driven welding machine, you must place it in a safe, properly ventilated space. Startup procedures vary, depending on the machine's engine model. Consult NCCER Module 29107, *SMAW – Equipment and Setup*, for more detail when working with engine-driven welding machines.

WARNING!

Remember that the workpiece lead (ground lead) isn't always grounded. It could have the welding machine's full open-circuit voltage between it and any grounded object. Always be cautious with both welding leads.

WARNING!

When you're finished welding, always remove the electrode from its holder. This prevents arcing, which can cause burns and arc flash.

1.2.8 Preparing the Weld Coupons

When possible, use 3⁄8" (~10 mm) thick carbon steel for welding coupons. This meets the AWS limited-thickness test coupon requirements. Alternatively, use 1⁄4" to 1" (~6 mm to 25 mm) thick steel for practice welds. You'll also need some scrap coupons. One is for practicing stringer beads. Scrap coupons are handy for setting up the welding machine and adjusting its settings. They may be any convenient size or thickness. Clean all coupons with a wire brush or grinder to remove scale and corrosion.

WARNING!

Be extremely careful with powered wire brushes. Verify that the brush's speed rating matches or exceeds the tool's maximum speed. Loose wires can fly off during cleaning. Regardless of the tool type, always wear appropriate PPE when cleaning metal.

NOTE

When preparing test coupons for welder qualifications, use a scrap coupon to check the welding machine's settings. Practicing on a scrap coupon is also a good way to become familiar with the machine before producing the test coupon.

For each weld coupon, you'll need two plates 3" × 7" (7.6 cm × 17.8 cm). You'll also need a backing strip 1⁄4" × 1" × 8" (6.4 mm × 2.5 cm × 20.3 cm). Notice that the backing is 1" (2.5 cm) longer than the coupons. It will extend past them by 1⁄2" (~13 mm) on each end. Bevel the coupons 22½ degrees on one long edge. Don't bevel the backing strip.

The following steps outline preparing the test coupons:

NOTE

This coupon size (3" × 7") meets the minimum AWS size requirement. To save material, your instructor may select a smaller size for practice.

Step 1 Check the bevels. There should be no root face and no dross. The bevel angle should be 22½ degrees. Remove any burrs from the coupon edges.

NOTE

The root opening should be within the tolerances permitted by the applicable code, WPS, or jobsite quality specifications. For these practice exercises, use a tolerance of ±$\frac{1}{16}$" (1.6 mm).

Step 2 Place the backing strip on a smooth, flat surface. Position the coupons on top of the backing strip with the root side down. The bevels should face each other. The root opening should be $\frac{1}{4}$" (6.4 mm) wide (*Figure 11*). Confirm with a spacer or other measuring device.

Step 3 Clamp the coupons to the backing strip. Be sure that the root opening stays $\frac{1}{4}$" (6.4 mm) wide. Confirm that the backing strip is tight against coupons.

Step 4 Flip the assembly over. Tack weld the backing strip where it laps the coupons. Place tack welds in the center and at each end (six welds total). *Figure 12* shows the completed assembly.

Step 5 When you're finished tack welding, *always* remove the electrode from its holder. This prevents arcing. Discard the stub in the proper container.

Step 6 Position the coupon in the required welding position (1, 2, 3, or 4).

Going Green

Welding Coupons

Reuse or recycle welding coupons rather than discarding them. Doing this reduces unnecessary waste and expense. Cut up the larger used coupons into smaller ones. Practice running beads on smaller used coupons.

Backing Strip Length

Using a backing strip longer than the coupons allows the welder to start and stop the bead outside the weld groove. Use at least $\frac{1}{2}$" (~13 mm) extra at each end. *Run-on weld tab* refers to the metal extending past the place where the welding starts. *Run-off weld tab* refers to the metal extending past the place where the welding stops.

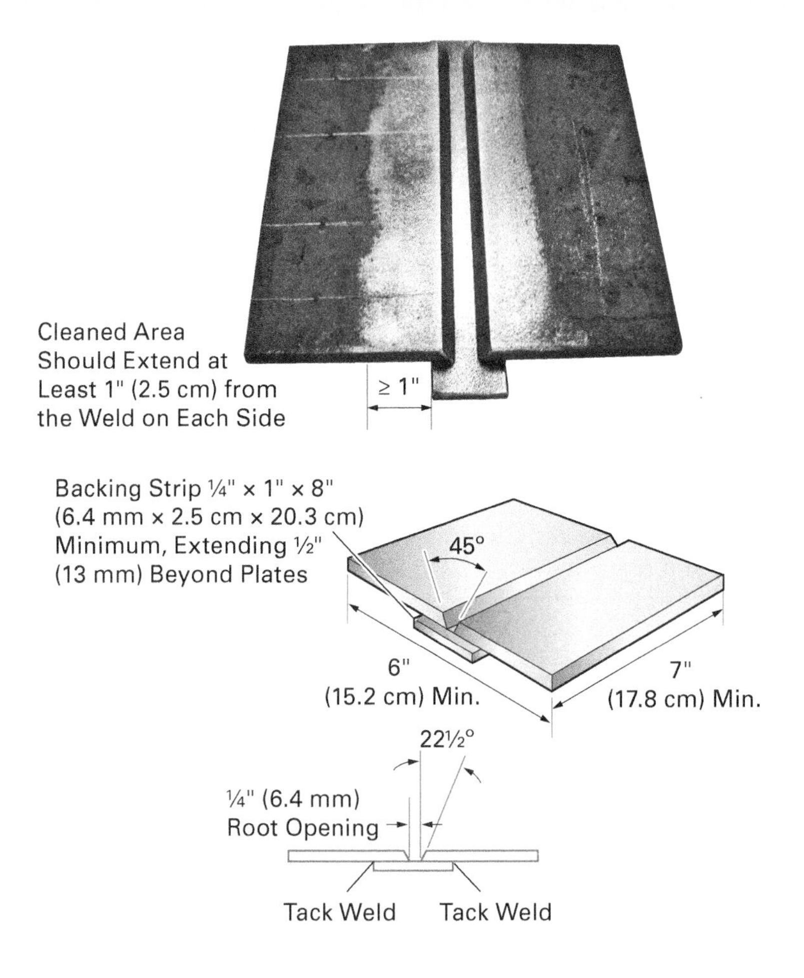

Figure 11 V-groove weld coupon with metal backing.

FLASH

Learning the memory aid **FLASH** helps you produce good welds:

- **F** (fit-up) — Correctly align and tack weld the base material together according to your instructor's specifications.
- **L** (length of arc) — Maintain the right distance between the electrode and the base metal. This distance usually equals the electrode core's diameter.
- **A** (angle) — Two angles are critical: the *work angle* and the *travel angle* (see the diagram).
- **S** (speed) — Travel speed, in inches per minute (IPM), affects the weld's width and fusion.
- **H** (heat) — The amperage setting controls the heat. The proper value depends on the electrode diameter, base metal type, base metal thickness, and welding position.

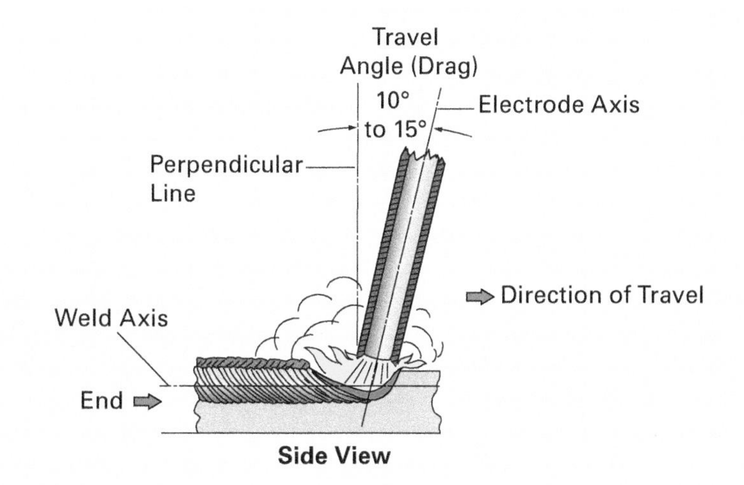

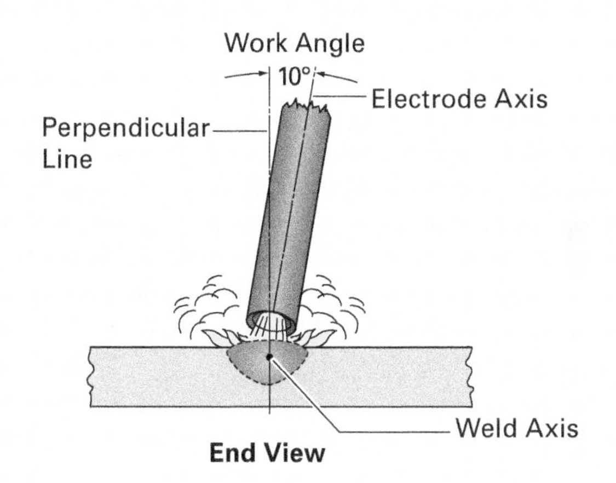

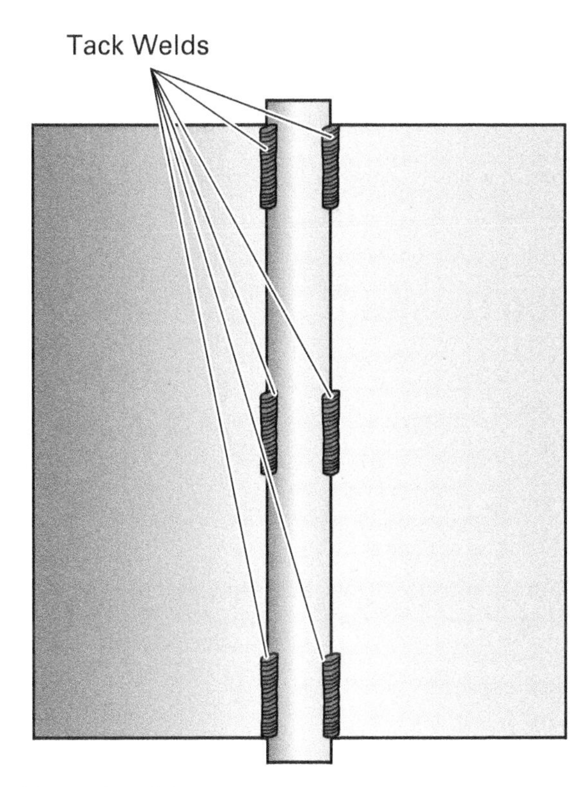

Figure 12 Backing tack welds.

NOTE

When preparing coupons for the horizontal position, ask your instructor which joint preparation to use.

When welding in the horizontal (2G) position, an alternative joint preparation may be better (*Figure 13*). Bevel one plate at approximately 45 degrees. Leave a square edge on the other plate. Place the beveled plate above the square plate. Use a ¼" (6.4 mm) root opening.

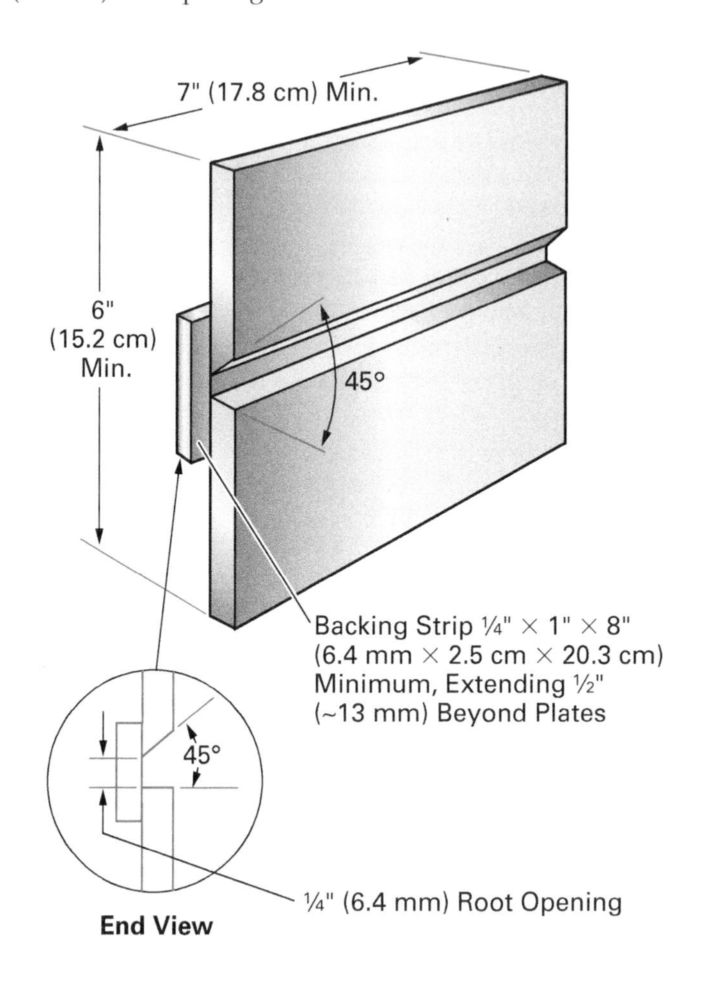

Figure 13 Alternative weld coupons for the horizontal position.

1.0.0 Section Review

1. The angle formed between the bevel face and a line perpendicular to the base metal is the _____.
   a. root angle
   b. bevel angle
   c. groove angle
   d. joint angle

2. Welding operations on zinc or cadmium create _____.
   a. heavy slag
   b. stronger welds
   c. heavy weld spatter
   d. toxic fumes

2.0.0 V-Groove Welds with Backing

Objective

Summarize techniques for making V-groove welds with backing.

a. Describe making V-groove welds with backing in the 1G and 2G positions.

b. Describe making V-groove welds with backing in the 3G and 4G positions.

Performance Tasks

2. Use E7018 electrodes to make flat (1G) V-groove welds with backing.
3. Use E7018 electrodes to make horizontal (2G) V-groove welds with backing.
4. Use E7018 electrodes to make vertical (3G) V-groove welds with backing.
5. Use E7018 electrodes to make overhead (4G) V-groove welds with backing.

Welders frequently produce V-groove welds with backing. They usually use low-hydrogen electrodes. The backing can match the base metal or be a different metal. It can also be a backing weld made with an E7018 electrode.

Some codes and WPSs permit using a low-hydrogen electrode for the root pass of an open-root weld. Doing this, however, requires back gouging to sound metal and back welding. Using backing eliminates these steps. It's a fast and effective way to prepare the joint. Running the root pass is easier with backing too.

A common AWS qualification test requires producing a V-groove weld with steel backing. The most common type is limited thickness with qualification up to $^{3}/_{4}$" (19 mm). It requires $^{3}/_{8}$" (~10 mm) carbon steel plate welded with low-hydrogen electrodes.

Welders can make groove welds in all positions. The weld position identifier for plate depends on the weld type and workpiece orientation (*Figure 14*). All groove welds end in G ("groove"). Groove weld positions for plate are flat (1G), horizontal (2G), vertical (3G), and overhead (4G).

The welder can incline the weld axis up to 15 degrees in the 1G and 2G positions. The AWS standard specifies a permissible inclination in the 3G and 4G positions that depends on the rotational position.

Make groove welds with slight face reinforcement and a gradual transition to the base metal at each toe. Groove welds must not have excessive face reinforcement, underfill, or overlap. These defects make the weld unacceptable. Most codes permit some undercut, provided it's not excessive. *Figure 15* shows acceptable and unacceptable profiles.

NOTE

Welders use the position identifiers (1G, 2G, 3G, and 4G) when referring to qualification tests. When making production welds, welders don't use the identifiers. Instead, they use the terms "flat," "horizontal," "vertical," and "overhead" to identify position.

NOTE

This module provides general guidelines only. Always follow the WPS or jobsite quality specifications. Check with your supervisor if you're unsure about the application's specifications.

Removing Backing

Welders don't automatically remove the backing. In some cases, they can leave it in place. An R in a welding symbol, however, specifies that the welder must remove it.

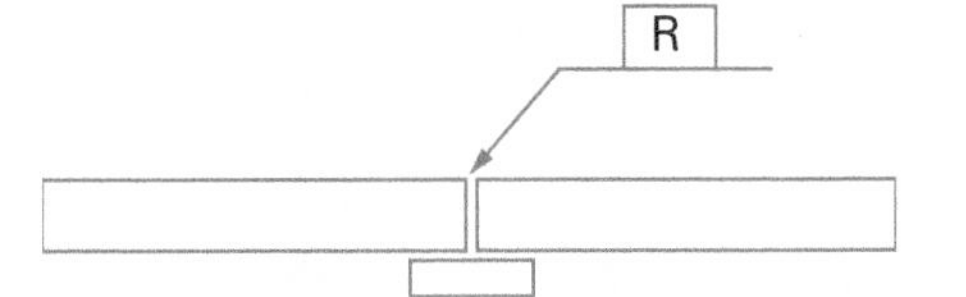

2.1.0 V-Groove Welds with Backing – 1G and 2G Positions

This section introduces producing good V-groove welds with backing in the 1G or 2G position. You'll learn the best technique and the required bead types. These positions are easier than the 3G and 4G positions. Mastering them first makes the harder welds easier.

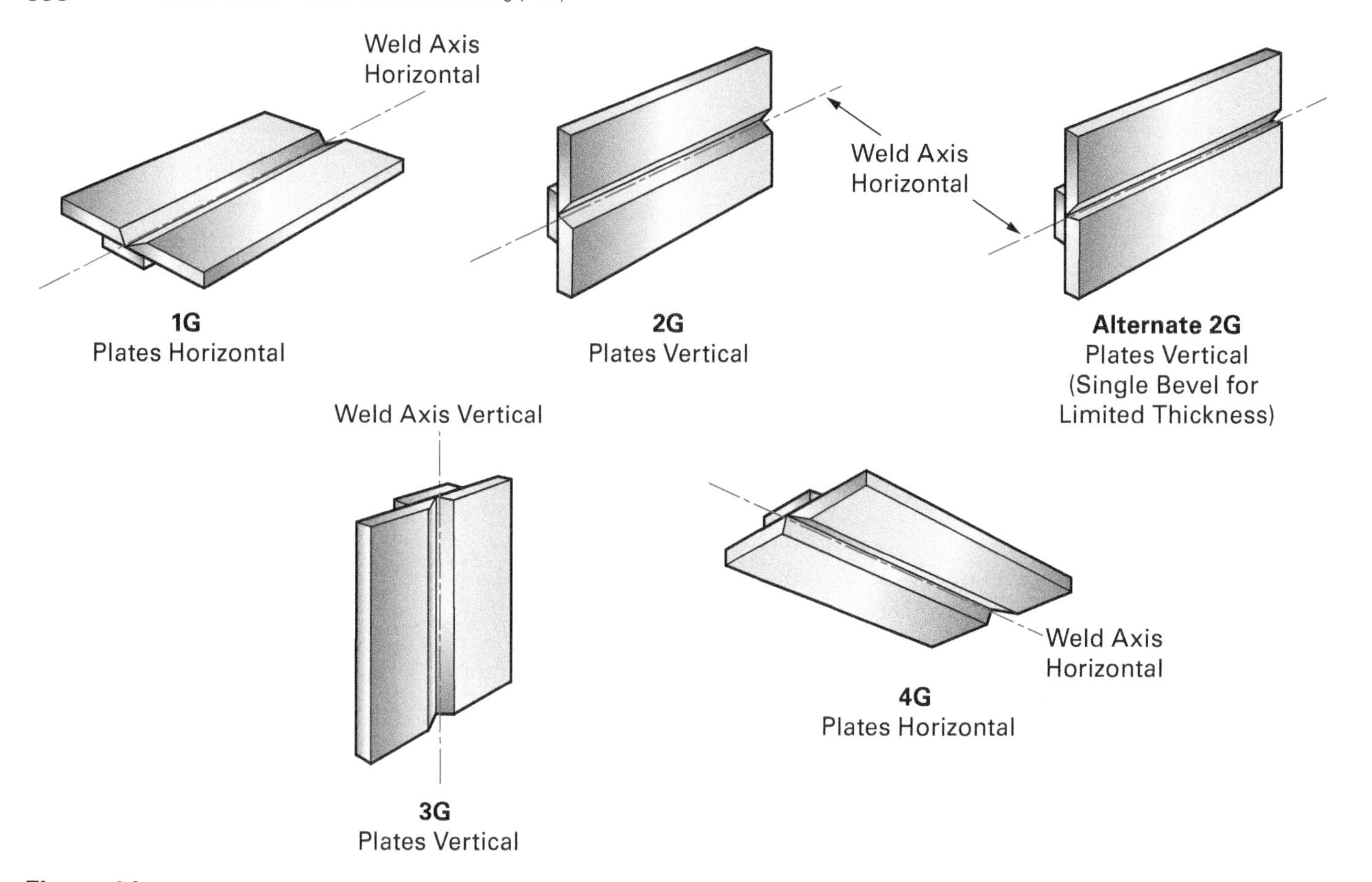

Figure 14 V-groove weld positions for plate.

CAUTION

Never quench functional welds or qualification test coupons. Rapid cooling hardens the metal, which may then crack. It also affects the base metal's mechanical properties. Quenching practice coupons is acceptable. Always quench in a bucket or quench tank rather than a sink or water fountain. Slag can clog drains.

2.1.1 Practicing V-Groove Welds with Backing (1G Position)

Practice V-groove welds with backing in the 1G (flat) position with $^3/_{32}$" (2.4 mm), $^1/_8$" (3.2 mm), or $^5/_{32}$" (4.0 mm) E7018 electrodes. You can use either weave beads or stringer beads. For weave beads, use a 10- to 15-degree drag angle. Keep the electrode at a 0-degree work angle. For stringer beads, use a 10- to 15-degree drag angle too. Adjust the work angle to tie in the weld on one side or the other as required. Be sure to fill the crater at the weld termination. If the coupon gets too hot between passes, quench it in water.

WARNING!

Wear gloves and handle hot coupons with pliers. Be careful of the steam rising from coupons when you quench them. It can burn unprotected skin.

Follow these steps to practice making V-groove welds with metal backing in the 1G position:

Step 1 Tack weld the practice coupon as *Figure 12* shows (six beads).

Step 2 Position the practice coupon flat on the welding table.

Step 3 Run the root pass using E7018 electrodes.

Step 4 Chip, clean, and brush the root weld.

Step 5 Run the fill and cover passes using E7018 electrodes. Chip, clean, and brush each bead. Use stringer or weave beads as your instructor directs. *Figure 16* shows all welding passes with their approximate work angles.

Step 6 When you're finished welding, *always* remove the electrode from its holder. This prevents arcing. Discard the stub in the proper container.

Figure 17 shows the weld's filler and cover passes.

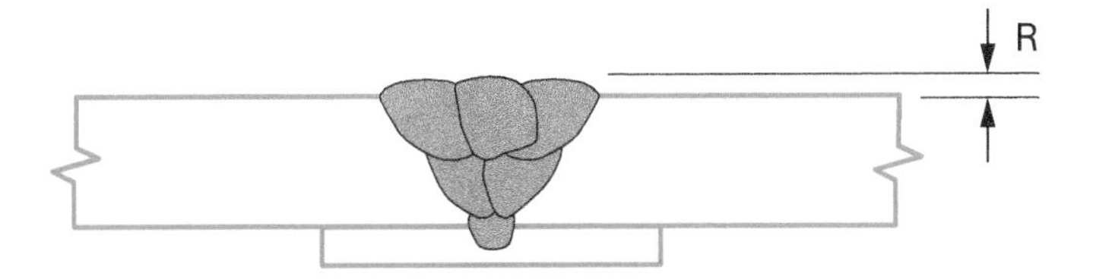

R = Face Reinforcement not to
Exceed the Code-Specified Amount

Acceptable Weld Profile

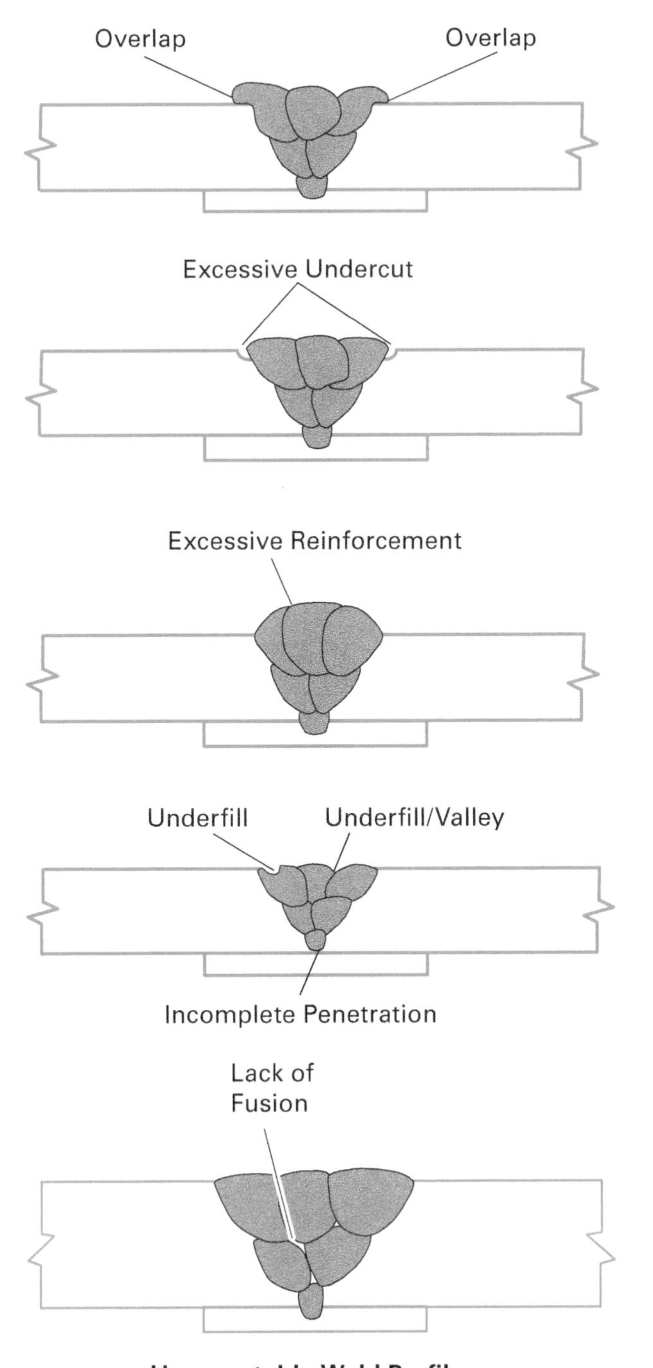

Unacceptable Weld Profiles

Figure 15 Acceptable and unacceptable V-groove weld with backing profiles.

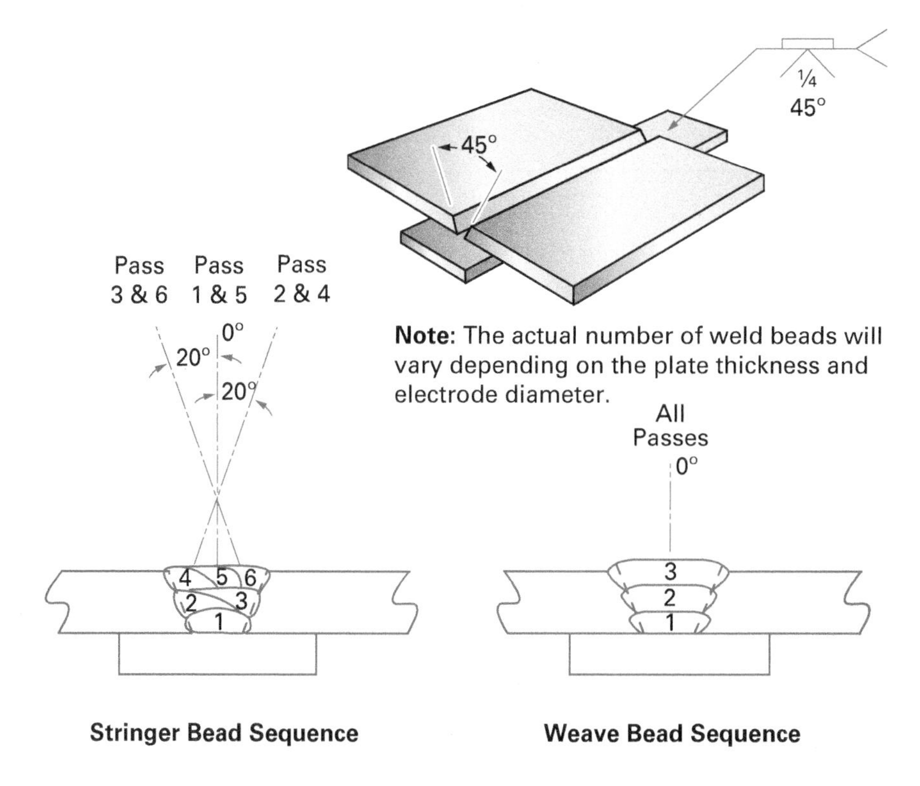

Figure 16 V-groove 1G welding sequence and work angles.

Figure 17 V-groove weld in the 1G position, filler and cover passes.
Source: Zachry Industrial, Inc.

Figure 18 A pad built in the horizontal position.
Source: The Lincoln Electric Company, Cleveland, OH, USA

2.1.2 Practicing V-Groove Welds with Backing (2G Position)

Welders can create horizontal V-groove welds with or without backing. You'll begin by making them with backing. Before you try this weld, however, practice running horizontal stringer beads. Do this by building a pad in the horizontal position (*Figure 18*).

Use $\frac{3}{32}$" (2.4 mm), $\frac{1}{8}$" (3.2 mm), or $\frac{5}{32}$" (4.0 mm) E7018 electrodes. Use a 10- to 15-degree drag angle. Keep the electrode at a 0-degree work angle. To control the weld pool, the work angle may be dropped slightly, but no more than 10 degrees (*Figure 19*).

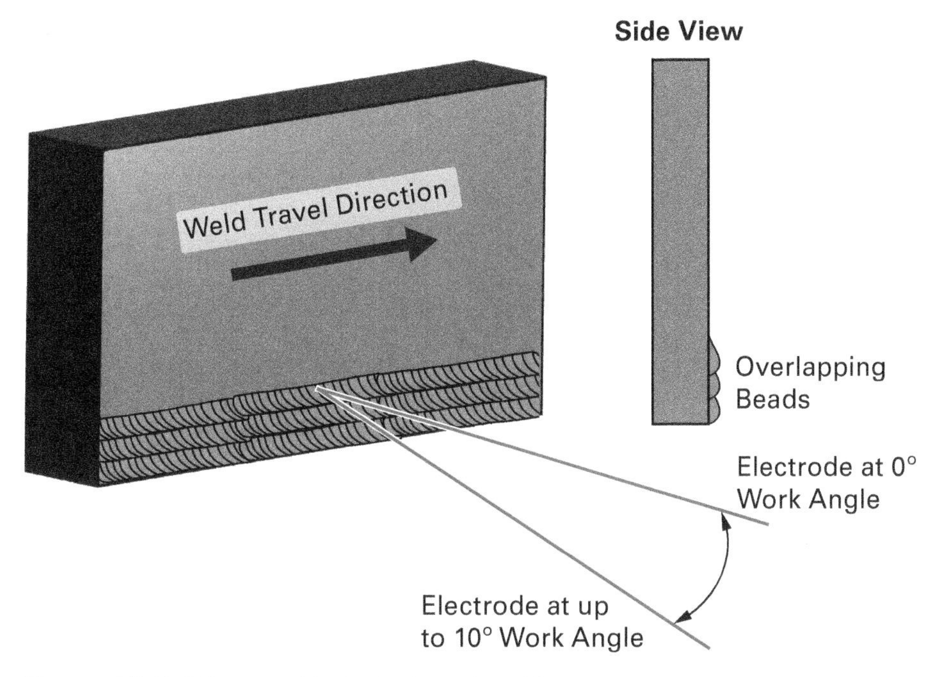

Figure 19 Building a pad in the horizontal position.

Follow these steps to practice making horizontal stringer beads in the 2G position:

Step 1 Tack weld a flat plate coupon in the horizontal position.

Step 2 Run the first pass along the coupon's bottom edge. Use E7018 electrodes.

Step 3 Chip, clean, and brush the weld.

Step 4 Weld the second bead just above the first bead. The beads should overlap as *Figure 19* shows. Chip, clean, and brush it.

Step 5 Continue running beads until you complete the pad.

Step 6 When you're finished welding, *always* remove the electrode from its holder. This prevents arcing. Discard the stub in the proper container.

Once you've produced a satisfactory pad, practice the actual V-groove weld in the 2G (horizontal) position. Use $^{3}/_{32}$" (2.4 mm), $^{1}/_{8}$" (3.2 mm), or $^{5}/_{32}$" (4.0 mm) E7018 electrodes. Create the root pass with a weave bead. Use stringer beads and a 10- to 15-degree drag angle for everything else. Adjust the work angle to tie in the weld on one side or the other as required. Be sure to fill the crater at the weld termination.

Follow these steps to practice making V-groove welds with metal backing in the 2G position:

Step 1 Tack weld the practice coupon as *Figure 12* shows (six beads). Use the standard or alternative coupon preparation as your instructor directs.

Step 2 Tack weld the practice coupon in the horizontal position.

Step 3 Run the root pass using E7018 electrodes. This pass is critical because it ties in the backing strip to both coupons (*Figure 20*). The welder's ability to weave in this bead determines the root pass' quality.

Step 4 Chip, clean, and brush the weld.

Step 5 Run the fill and cover passes using E7018 electrodes. Chip, clean, and brush each bead. *Figure 21* shows all welding passes with their approximate work angles.

Step 6 When you're finished welding, *always* remove the electrode from its holder. This prevents arcing. Discard the stub in the proper container.

NOTE

To make an acceptable weave for the root pass, use a slight zigzag or sawtooth pattern (W-weave). Weave with a longer pause (dwell time) at the top of the weave. This counters gravity's effects on the weld and eliminates excessive weld profile drooping.

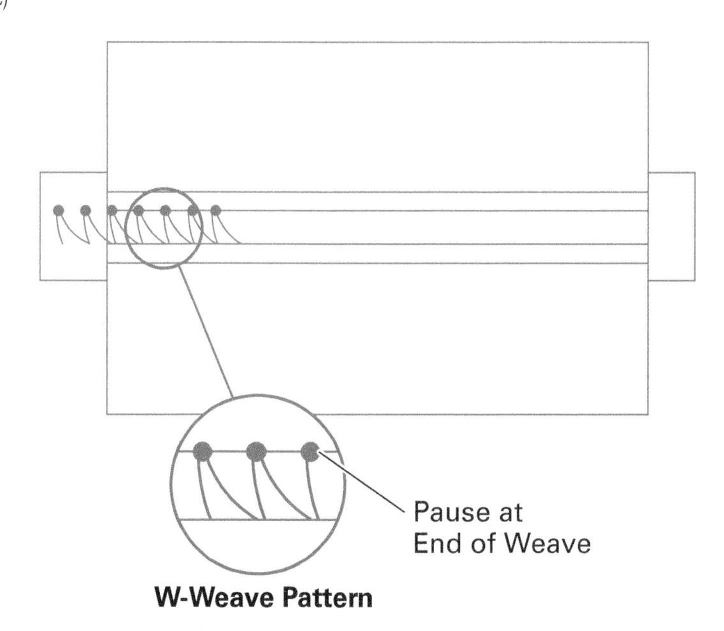

Figure 20 Weaving the root pass.

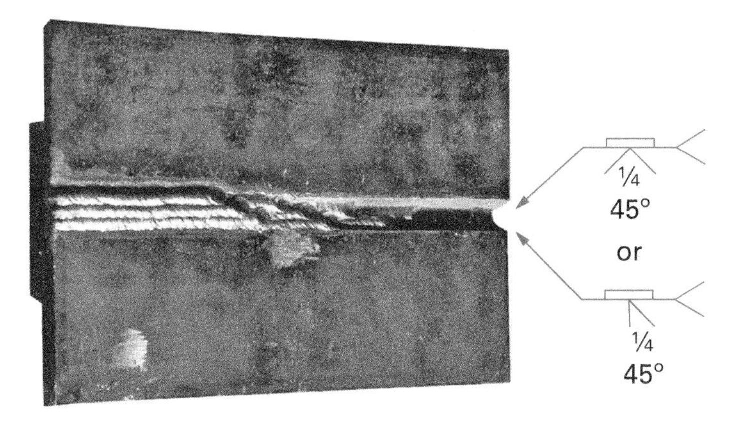

Note: The actual number of weld beads will vary depending on the plate thickness and electrode diameter.

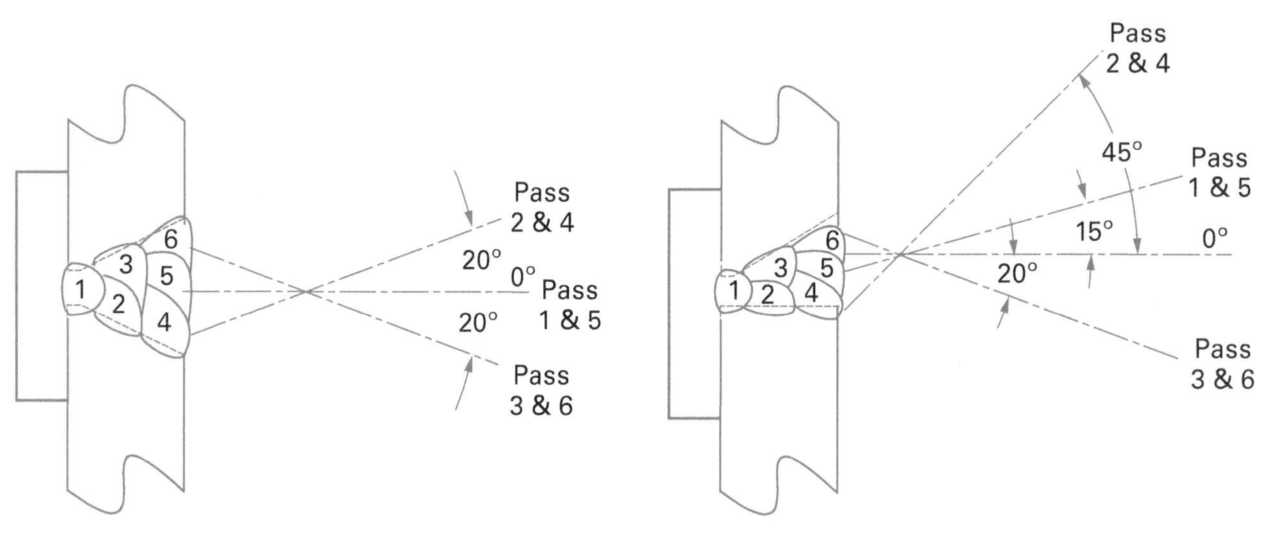

Note: Nonzero work angles are approximate

Figure 21 V-groove 2G welding sequence and work angles.

2.2.0 V-Groove Welds with Backing – 3G and 4G Positions

This section introduces producing good V-groove welds with backing in the 3G or 4G position. You'll learn the best technique and the required bead types. These positions are harder than the 1G and 2G positions.

2.2.1 Practicing V-Groove Welds with Backing (3G Position)

Welders can create vertical V-groove welds with or without backing. You'll begin by making them with backing. Before you try this weld, however, practice running vertical stringer beads. Do this by building a pad in the vertical position.

Use $^{3}/_{32}$" (2.4 mm), $^{1}/_{8}$" (3.2 mm), or $^{5}/_{32}$" (4.0 mm) E7018 electrodes. Set the amperage at the lower end of the correct range. This gives better weld pool control. Use a 0- to 10-degree push angle. Keep the electrode at a 0-degree work angle (*Figure 22*). The width of the stringer beads should be no more than three times the electrode core's diameter.

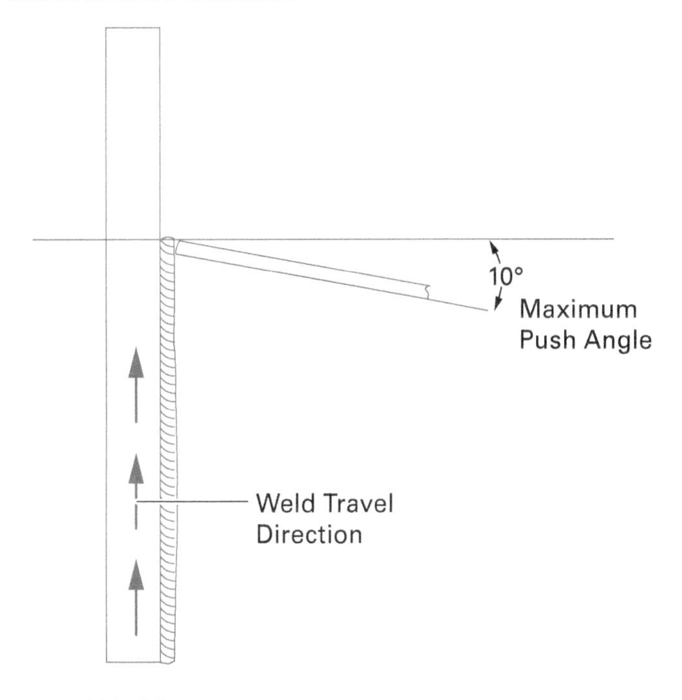

Figure 22 Building a pad in the vertical position (uphill progression).

Follow these steps to practice making vertical stringer beads in the 3G position:

Step 1 Tack weld a flat plate coupon in the vertical position.

Step 2 Run the first pass along one of the coupon's vertical edges. Use E7018 electrodes.

Step 3 Chip, clean, and brush the weld.

Step 4 Weld the second bead next to the first bead. The beads should overlap. Chip, clean, and brush it.

Step 5 Continue running beads until you complete the pad.

Step 6 When you're finished welding, *always* remove the electrode from its holder. This prevents arcing. Discard the stub in the proper container.

Once you've produced a satisfactory pad, practice the actual V-groove weld in the 3G (vertical) position. Use 3⁄32" (2.4 mm), 1⁄8" (3.2 mm), or 5⁄32" (4.0 mm) E7018 electrodes. You can use either weave beads or stringer beads. Set the amperage at the lower end of the correct range. This gives better weld pool control. Use a 0- to 10-degree push angle. Adjust the work angle to tie in the weld on one side or the other as required. When making a weave bead, pause slightly at each toe to penetrate and fill the crater, preventing undercut. Be sure to fill the crater at the weld termination.

Follow these steps to practice making V-groove welds with metal backing in the 3G position:

Step 1 Tack weld the practice coupon as *Figure 12* shows (six beads).

Step 2 Tack weld the practice coupon in the vertical position.

Step 3 Run the root pass using E7018 electrodes from bottom to top (uphill progression). If you're creating a weave bead, weld with a slight oscillation, pausing at each side.

Step 4 Chip, clean, and brush the weld.

Step 5 Run the fill and cover passes using E7018 electrodes uphill. Chip, clean, and brush each bead. *Figure 23* shows all welding passes with their approximate work angles. The figure shows both weave and stringer bead welds.

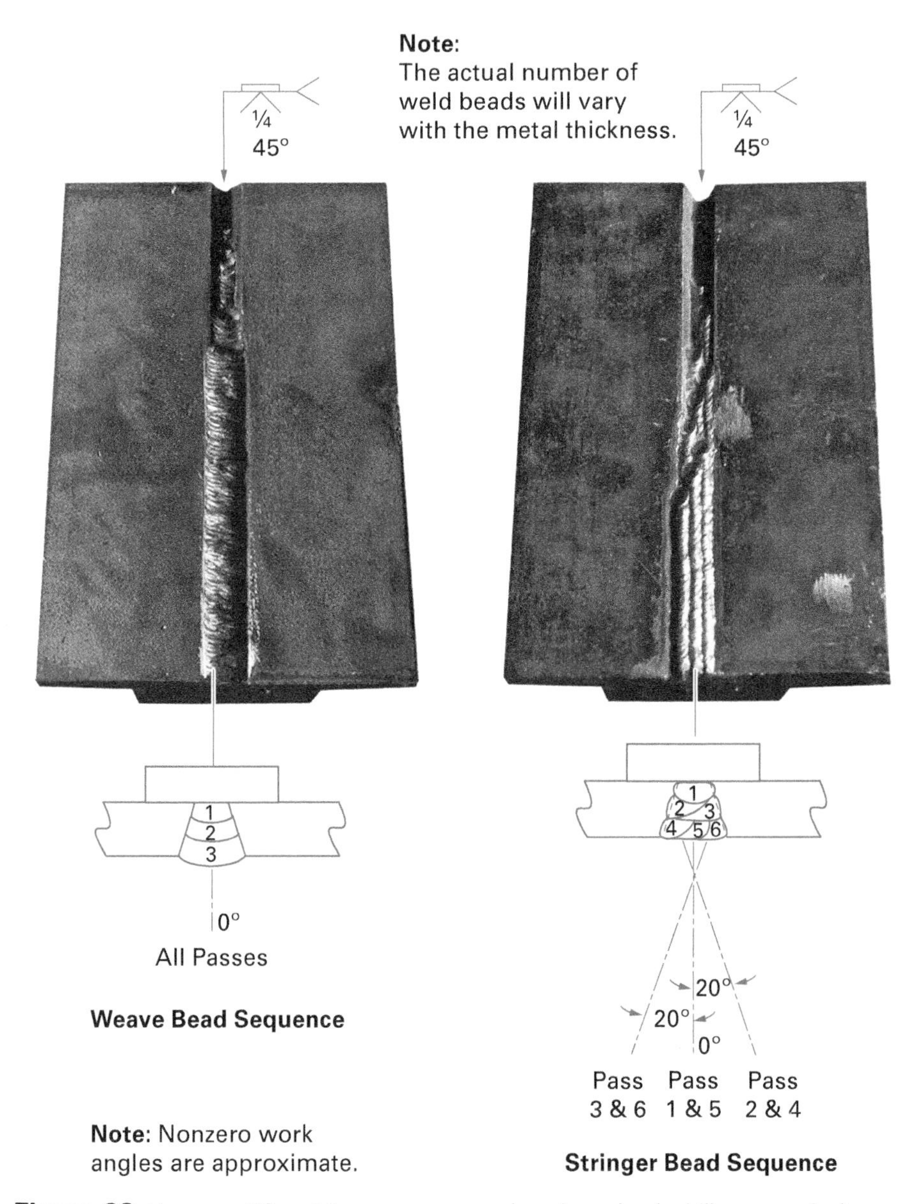

Figure 23 V-groove 3G welding sequence and work angles (uphill progression).

Step 6 When you're finished welding, *always* remove the electrode from its holder. This prevents arcing. Discard the stub in the proper container.

2.2.2 Practicing V-Groove Welds with Backing (4G Position)

Welders can create overhead V-groove welds with or without backing. You'll begin by making them with backing. But before you try this weld, however, practice running overhead stringer beads. Do this by building a pad in the overhead position.

WARNING!

Avoid standing directly under the coupons while welding. Hot slag and molten metal can drop on you, causing severe burns. Besides regular PPE, wear the additional protective clothing required for overhead work.

Use $^3/_{32}$" (2.4 mm), $^1/_8$" (3.2 mm), or $^5/_{32}$" (4.0 mm) E7018 electrodes. Use a 10- to 15-degree drag angle. Keep the electrode at a 0-degree work angle (*Figure 24*).

Follow these steps to practice making overhead stringer beads in the 4G position:

Step 1 Tack weld a flat plate coupon in the overhead position.

Step 2 Run the first pass along one of the coupon's edges. Use E7018 electrodes.

Step 3 Chip, clean, and brush the weld.

Step 4 Weld the second bead next to the first bead. The beads should overlap. Chip, clean, and brush it.

Step 5 Continue running beads until you complete the pad.

Step 6 When you're finished welding, *always* remove the electrode from its holder. This prevents arcing. Discard the stub in the proper container.

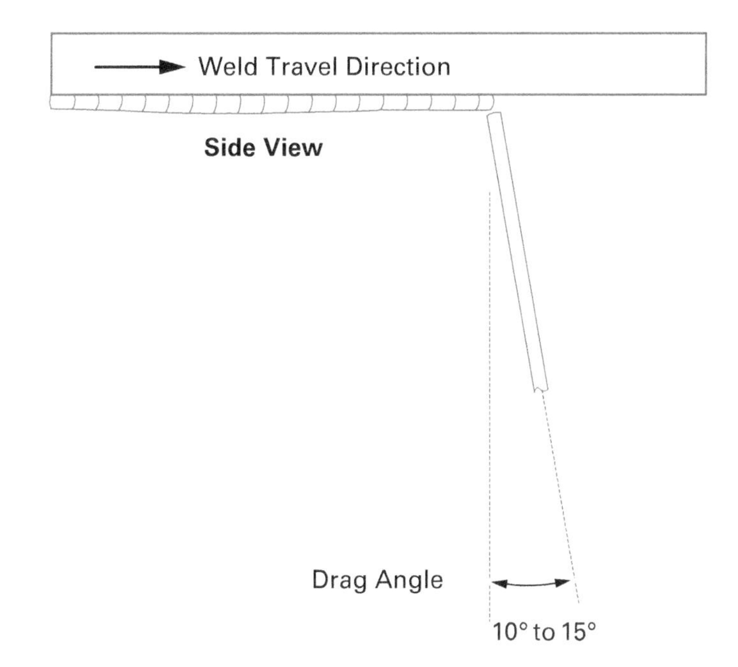

Figure 24 Building a pad in the overhead position.

Once you've produced a satisfactory pad, practice the actual V-groove weld in the 4G (overhead) position. Use 3⁄32" (2.4 mm), 1⁄8" (3.2 mm), or 5⁄32" (4.0 mm) E7018 electrodes. Use stringer beads. Use a 10- to 15-degree drag angle. Adjust the work angle to tie in the weld on one side or the other as required. Be sure to fill the crater at the weld termination.

Follow these steps to practice making V-groove welds with metal backing in the 4G position:

Step 1 Tack weld the practice coupon as *Figure 12* shows (six beads).

Step 2 Tack weld the practice coupon in the overhead position.

Step 3 Run the root pass using E7018 electrodes.

Step 4 Chip, clean, and brush the weld.

Step 5 Run the fill and cover passes using E7018 electrodes. Chip, clean, and brush each bead. *Figure 25* shows all welding passes with their approximate work angles.

Step 6 When you're finished welding, *always* remove the electrode from its holder. This prevents arcing. Discard the stub in the proper container.

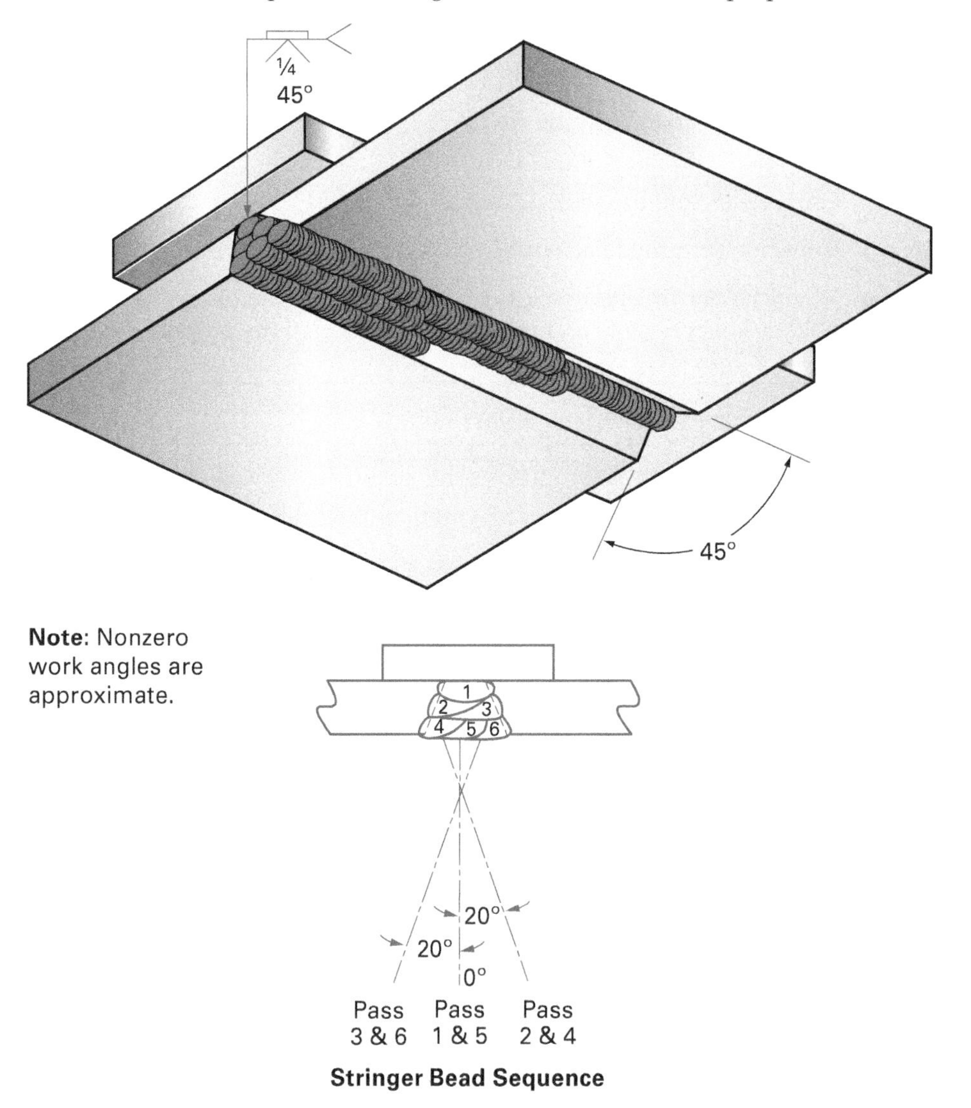

Figure 25 V-groove 4G welding sequence and work angles.

Welders and Multifocal Lenses

In the past, welders who wore multifocal glasses (bifocals, trifocals, and progressives) had a challenging time. Welding helmets and hoods didn't have a lot of extra space inside. The front tinted lens was small, so welders couldn't tilt their heads to bring the right part of their glasses into position. Some welders fitted their helmets with a magnifying lens ("cheater") to help see better. Many still do this.

Today, PPE manufacturers make helmets and hoods designed for welders who wear multifocal glasses. The headgear adjusts to move the front farther away from the face, making room for glasses. The tinted lens is also larger, so welders can look through the different sections of their glasses more easily and see things clearly.

When wearing prescription glasses rather than standard safety glasses, the lenses must be shatterproof. The glasses must also have ANSI-approved side shields.

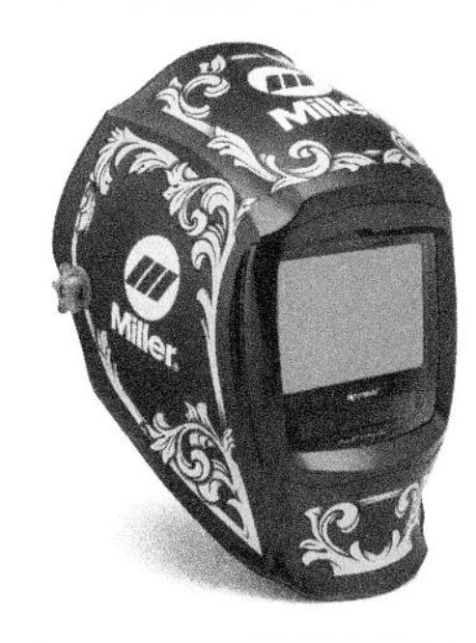

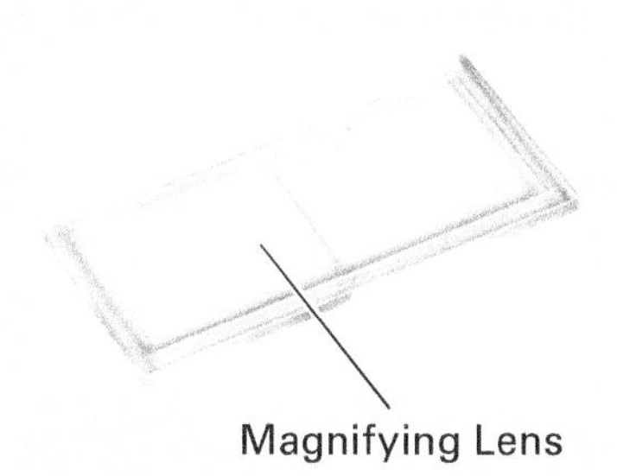

Source: © Miller Electric Mfg. LLC

2.0.0 Section Review

1. Which of the following can be quenched without concern?
   a. Welder qualification test coupons
   b. Functional welds
   c. Practice coupons
   d. Any weld done in the 3G position

2. When making a V-groove weld overhead, you are welding in the _____.
   a. 1G position
   b. 2F position
   c. 3F position
   d. 4G position

Module 29111 Review Questions

1. If a groove for a weld goes through the entire base metal thickness, it makes a(n) _____.
   a. partial joint penetration (PJP) weld
   b. complete joint penetration (CJP) weld
   c. axial joint penetration (AJP) weld
   d. subsurface joint penetration (SJP) weld

Figure RQ01

2. The groove weld in *Figure RQ01* is a _____.
   a. double V-groove
   b. double bevel groove
   c. single V-groove
   d. single bevel groove

3. The place in a welded joint where the two components are closest to each other is the _____.
   a. joint root
   b. bevel depth
   c. groove radius
   d. weld toe

4. Which of the following is *not* true about welding with low-hydrogen electrodes?
   a. You should maintain a short arc.
   b. You should take only as many electrodes from the storage oven as you need.
   c. You should always whip the electrode.
   d. You should never use electrodes with missing flux.

5. When preparing welding coupons for V-groove welds with backing, the backing should extend beyond the coupons by *at least* _____.
   a. ⅛" (~3 mm)
   b. ¼" (~6 mm)
   c. ½" (~13 mm)
   d. ¾" (~19 mm)

6. V-groove welds with backing are popular welds normally made with _____.
   a. fast-fill electrodes
   b. fill-freeze electrodes
   c. fast-freeze electrodes
   d. low-hydrogen electrodes

7. When using weave beads on flat V-groove welds (1G), keep the electrode at a 10- to 15-degree drag angle and a work angle of _____.
   a. 0 degrees
   b. 45 degrees
   c. 60 degrees
   d. 75 degrees

8. When practicing horizontal V-groove welds (2G) using stringer beads, the work angle should be _____.
   a. 5 degrees
   b. 45 degrees
   c. 60 degrees
   d. adjusted as needed to ensure tie-in

9. When using stringer beads on V-groove welds in the vertical position (3G), keep the electrode _____.
   a. always at a 0-degree push angle
   b. at a 0- to 10-degree push angle
   c. at a 15- to 45-degree push angle
   d. always at a 60-degree push angle

10. Before practicing a V-groove weld in the overhead position (4G), you should _____.
    a. change to E6010 electrodes
    b. increase the amperage to the top of the correct range
    c. build a practice pad with stringer beads
    d. change to constant voltage DC

Answers to odd-numbered Module Review Questions are found in *Appendix A*.

Answers to Section Review Questions

Answer	Section Reference	Objective
Section 1.0.0		
1. b	1.1.1	1a
2. d	1.2.4	1b
Section 2.0.0		
1. c	2.1.1	2a
2. d	2.2.2	2b

User Update

Did you find an error? Submit a correction by visiting **https://www.nccer.org/olf** or by scanning the QR code using your mobile device.

SMAW – Open-Root Groove Welds (Plate)

Source: The Lincoln Electric Company, Cleveland, OH, USA

Objectives

Successful completion of this module prepares you to do the following:

1. Summarize groove weld types and preparing to weld them.
   a. Identify groove welds and their properties.
   b. Outline preparing for making open-root single V-groove welds.
2. Summarize techniques for making open-root single V-groove welds.
   a. Describe producing the root pass for open-root single V-groove welds.
   b. Describe making open-root single V-groove welds in the 1G and 2G positions.
   c. Describe making open-root single V-groove welds in the 3G and 4G positions.

Performance Tasks

Under supervision, you should be able to do the following:

1. Safely set up SMAW equipment for groove welding.
2. Use E6010/E6011 and E7018 electrodes to make flat (1G) open-root single V-groove welds.
3. Use E6010/E6011 and E7018 electrodes to make horizontal (2G) open-root single V-groove welds.
4. Use E6010/E6011 and E7018 electrodes to make vertical (3G) open-root single V-groove welds.
5. Use E6010/E6011 and E7018 electrodes to make overhead (4G) open-root single V-groove welds.

Overview

Welders have many choices available when they join plate with groove welds. They can produce them in any position with a variety of electrodes. More importantly, the welds can have an open root or a backing. Open-root welding offers the greater challenge since nothing prevents the weld metal from passing through the groove. This module introduces the fit-up procedures and techniques required to produce open-root single V-groove welds in plate.

Industry Recognized Credentials

If you are training through an NCCER-accredited sponsor, you may be eligible for credentials from NCCER's Registry. The ID number for this module is 29112. Note that this module may have been used in other NCCER curricula and may apply to other level completions. Contact NCCER's Registry at 1.888.622.3720 or go to **www.nccer.org** for more information.

You can also show off your industry-recognized credentials online with NCCER's digital badges. Transform your knowledge, skills, and achievements into badges that you can share across social media platforms, send to your network, and add to your resume. For more information, visit **www.nccer.org**.

NOTE

This module uses US standard and metric units in up to three different ways. This note explains how to interpret them.

Exact Conversions
Exact metric equivalents of US standard units appear in parentheses after the US standard unit. For example: "Measure 18" (45.7 cm) from the end and make a mark."

Approximate Conversions
In some cases, exact metric conversions would be inappropriate or even absurd. In these situations, an approximate metric value appears in parentheses with the ~ symbol in front of the number. For example: "Grip the tool about 3" (~8 cm) from the end."

Parallel but not Equal Values
Certain scenarios include US standard and metric values that are parallel but not equal. In these situations, a slash (/) surrounded by spaces separates the US standard and metric values. For example: "Place the point on the steel rule's 1" / 1 cm mark."

Digital Resources for Welding

Scan this code using the camera on your phone or mobile device to view the digital resources related to this craft.

1.0.0 Open-Root Groove Welds

Performance Task

1. Safely set up SMAW equipment for groove welding.

Objective

Summarize groove weld types and preparing to weld them.

a. Identify groove welds and their properties.
b. Outline preparing for making open-root single V-groove welds.

Welders can join two parts with a groove weld between them. They can also make groove welds in part openings. Groove welds apply to all five basic joint types: butt, lap, T, edge, and corner. Simply fitting the parts together creates some groove welds. Other groove welds require preparation steps. *Complete joint penetration (CJP)* welds go completely through the base metal. *Partial joint penetration (PJP)* welds don't go completely through.

This module introduces open-root single V-groove welds. The open root is a small gap separating the two parts. No backing bridges this gap. Welders often use these welds to join plate. The joint is easy to prepare because it requires single-face beveling only. On the other hand, welding the open root without backing is challenging. Trainee welders must work hard to learn the techniques to create a good root pass.

1.1.0 Groove Weld Styles

A groove weld's name comes from its shape. Welds can have one or two grooves. Single-groove styles include square, bevel, V, U, J, flare V, and flare bevel. Double-groove styles are bevel, V, U, and J. *Figure 1* shows examples of each.

If the parts being joined don't have the right shape (profile), the welder must create it. Most joint configurations start with square (90-degree) edges. A grinder can produce beveled (angled) grooves. Preparing the edges cleans the metal. It also increases the weld size. Welders must produce grooves with good root access. Grooves should never be larger than necessary. The groove preparation's main purpose is to give the welder access to the weld root to achieve the designed weld penetration.

Sometimes, welders combine a fillet weld with a groove weld. This strategy maintains good strength but requires less metal. *Figure 2* shows several combination weld examples.

NOTE

This module includes dimensions with both US standard and metric measurements. Welding symbols in drawings, however, usually show US standard units only.

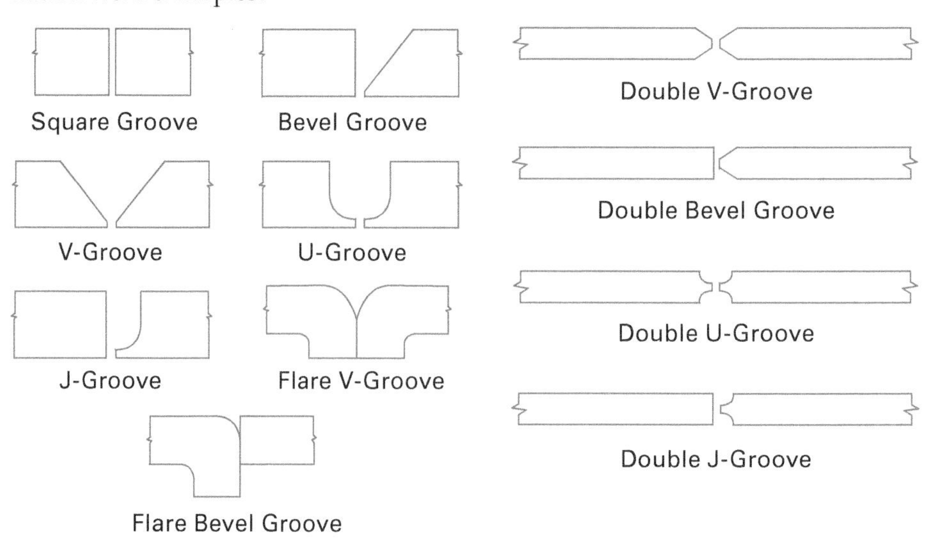

Figure 1 Single- and double-groove welds.

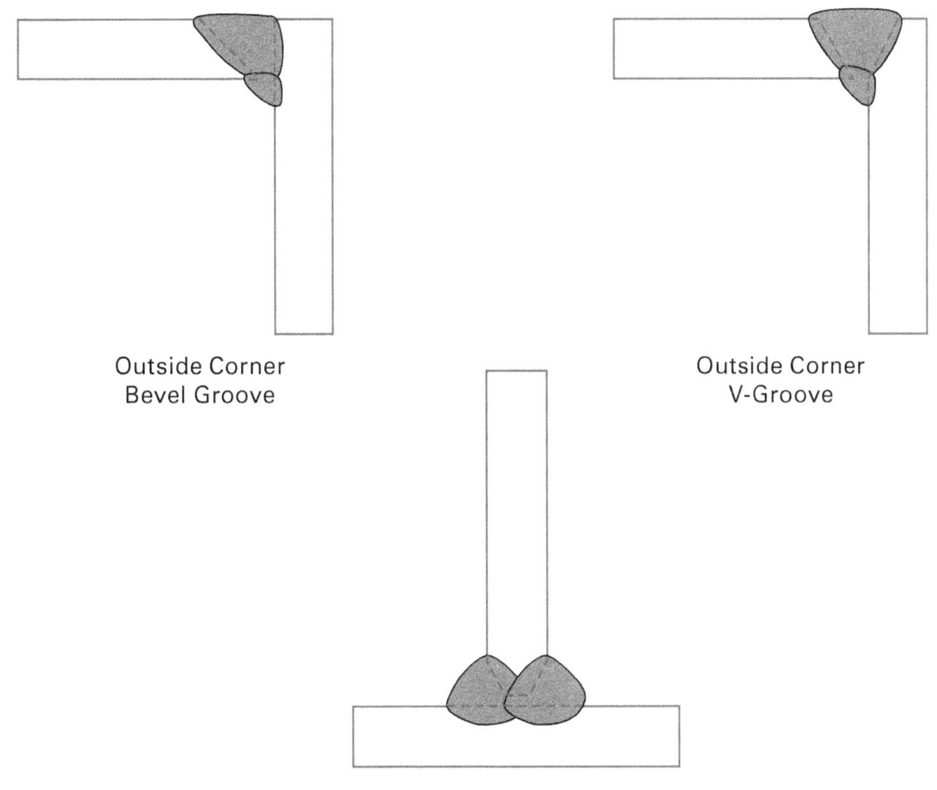

Figure 2 Combination welds.

1.1.1 Groove Weld Terminology

Welders describe weld features and dimensions with a specialized vocabulary. Learning it will help you communicate clearly and accurately with other welders. The following terms describe open-root groove weld features (*Figure 3*):

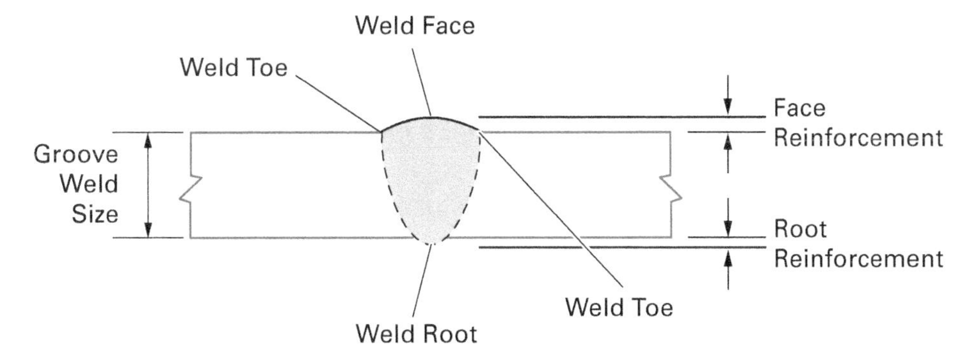

Figure 3 Open-root groove weld features.

Weld face — The weld's exposed surface on the side where the welder produced it.

Weld root — The point where the back of the weld extends farthest into the joint.

Weld toes — The points at the weld face where the weld and base metals meet.

Face reinforcement — Weld metal extending beyond the base metal on a groove weld's face side.

Root reinforcement — Weld metal extending beyond the base metal on a groove weld's root side.

Groove weld size — The depth that the weld penetrates the joint.

The following terms apply to the dimensions of open-root groove welds (*Figure 4*):

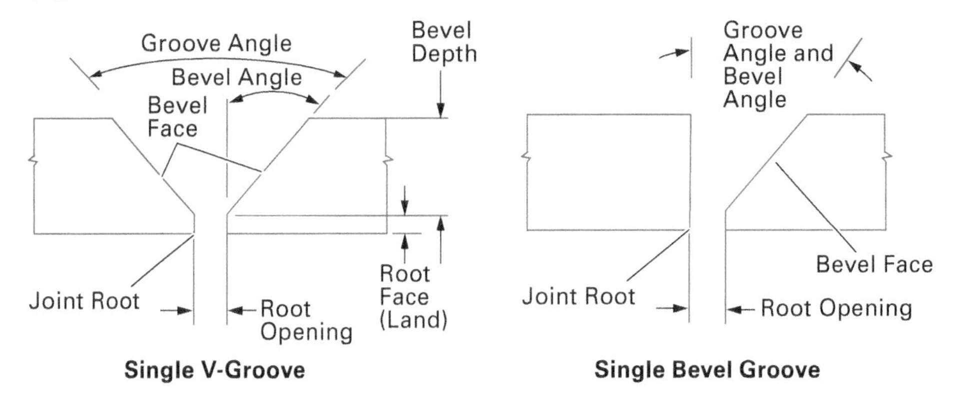

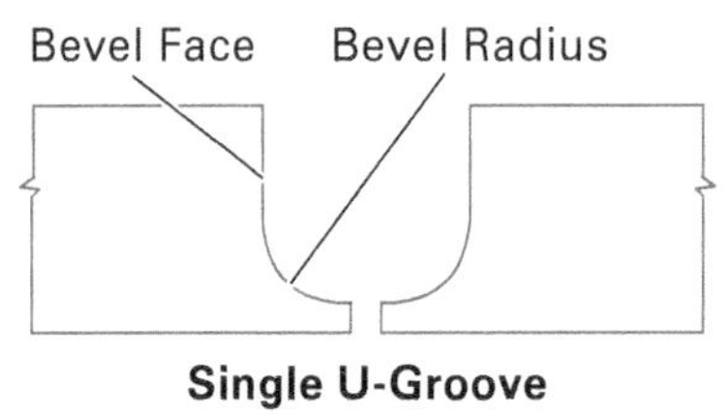

Figure 4 Open-root groove weld dimensions.

Joint preparation — The cut or machined groove shape that penetrates the base metal.

Groove face — Any groove surface before welding.

Bevel face — The groove's angled or curved surface.

Root face — The flattened area at the end of a groove weld's bevel. Also called the *land*.

Groove angle — The included angle between the two groove faces.

Bevel angle — The angle between the bevel face and a line perpendicular to the base metal.

Bevel radius — The radius of the curve forming a J- or U-groove.

Bevel depth — The distance that the joint preparation extends into the base metal.

Joint root — The place in the joint where the weld members are closest together.

Root opening — The distance between the two members at the root.

1.2.0 Welding Preparations

NOTE

The dimensions and specifications in this module represent welding codes in general, not a specific code. Always follow the proper codes for your jobsite. Also follow the applicable Welding Procedure Specification (WPS) and jobsite quality specifications.

This module explains preparing and producing open-root single V-groove welds in all positions. You'll use E6010/E6011 and E7018 (low-hydrogen) electrodes. What you learn here will prepare you for acquiring welder qualifications, as well as producing other groove weld types.

You'll need to prepare before you can weld an open-root single V-groove joint. You must set up the welding area, equipment, and metal. The following sections explain how to do these tasks safely.

1.2.1 Safety Preparations

NOTE

Be sure that you've completed NCCER Module 29101, *Welding Safety*, before continuing.

Cutting and welding equipment produces high temperatures, sparks, and many other hazards. Using equipment safely requires following specific procedures. This section summarizes good practices but isn't a complete safety training. Always complete all training that your workplace requires. Follow all workplace safety guidelines and wear appropriate PPE (*Figure 5*).

1.2.2 Protective Clothing and Equipment

Welding and cutting tasks are dangerous. Unless you wear all required PPE, you're at risk of serious injury. The following guidelines summarize PPE for SMAW work:

- For cutting, always wear close-fitting goggles or safety glasses, plus a full face shield, helmet, or hood. The goggles, glasses, face shield, or helmet/hood lens must have the proper shade value for the type of cutting work (*Figure 6*). Never cut without using the proper lens.
- For welding, always wear safety glasses, plus a helmet or hood. The glasses or helmet/hood lens must have the proper shade value for the welding process (*Figure 6*). Never view the cutting arc directly or indirectly without using the proper lens.

Figure 5 Welding PPE suitable for SMAW work.
Source: © Miller Electric Mfg. LLC

WARNING!

Don't open your welding helmet/hood lens and expose your eyes when grinding or cleaning a weld. Either use a helmet or hood with an auto-darkening lens or a clear lens under a flip-up shaded lens. Alternatively, remove the helmet or hood and use a full face shield.

- Wear protective leather and/or flame-retardant clothing along with welding gloves that will protect you from flying sparks and molten metal.
- Wear high-top safety shoes or boots. Make sure the pant leg covers the tongue and lace area. If necessary, wear leather spats or chaps for protection. Boots without laces or holes are better for welding and cutting.
- Wear a 100 percent cotton cap with no mesh material in its construction. The bill should point to the rear or the side with the most exposed ear. If you must wear a hard hat, use one with a face shield.

WARNING!

Never wear a cap with a button on top. The conductive metal under the fabric is a safety hazard.

- Wear earplugs to protect your ear canals from sparks. Since welding and cutting can be noisy, always wear hearing protection.

1.2.3 Fire/Explosion Prevention

All processes in this module can produce high temperatures. Welding or cutting a vessel or container that once contained combustible, flammable, or explosive materials is hazardous. Residues can catch fire or explode. Before welding or cutting vessels, check whether they contained explosive, hazardous, or flammable materials. These include petroleum products, citrus products, or chemicals that release toxic fumes when heated.

American Welding Society (AWS) F4.1, Safe Practices for the Preparation of Containers and Piping for Welding and Cutting, and *ANSI/AWS Z49.1* describe safe practices for these situations. Begin by cleaning the container to remove any residue. Steam cleaning, washing with detergent, or flushing with water are possible methods. Sometimes you must combine these to get good results.

WARNING!

Clean containers only in well-ventilated areas. Vapors can accumulate in a confined space during cleaning, causing explosions or toxic substance exposure.

Arc Welding Processes

Process	Electrode	Amperage	Lens Shade Numbers (2 3 4 5 6 7 8 9 10 11 12 13 14)
Shielded Metal Arc Welding (SMAW)	< 3/32" (2.4 mm)	< 60A	7
	3/32" – 5/32" (2.4 mm – 4.0 mm)	60A – 160A	8 9 10
	5/32" – 1/4" (4.0 mm – 6.4 mm)	160A – 250A	10 11 12
	> 1/4" (6.4 mm)	250A – 550A	11 12 13 14
Gas Metal Arc Welding (GMAW) and Flux Cored Arc Welding (FCAW)		< 60A	7
		60A – 160A	10 11
		160A – 250A	10 11 12
		250A – 500A	10 11 12 13 14
Gas Tungsten Arc Welding (GTAW)		< 50A	8 9 10
		50A – 150A	8 9 10 11 12
		150A – 500A	10 11 12 13 14
Plasma Arc Welding (PAW)		< 20A	6 7 8
		20A – 100A	8 9 10
		100A – 400A	10 11 12
		400A – 800A	11 12 13 14
Carbon Arc Welding (CAW)			14

Arc Cutting Processes

Process	Electrode	Amperage	Lens Shade Numbers (2 3 4 5 6 7 8 9 10 11 12 13 14)
Plasma Arc Cutting (PAC)		< 20A	4
		20A – 40A	5
		40A – 60A	6
		60A – 80A	8
		80A – 300A	8 9
		300A – 400A	9 10 11 12
		400A – 800A	10 11 12 13 14
Air-Carbon Arc Cutting (A-CAC)	Light	< 500A	10 11 12
	Heavy	500A – 1,000A	11 12 13 14

Gas Processes

Process	Type	Plate Thickness	Lens Shade Numbers (2 3 4 5 6 7 8 9 10 11 12 13 14)
Oxyfuel Welding (OFW)	Light	< 1/8" (3 mm)	4 5
	Medium	1/8" – 1/2" (3 mm – 13 mm)	5 6
	Heavy	> 1/2" (13 mm)	6 7 8
Oxyfuel Cutting (OC)	Light	< 1" (25 mm)	3 4
	Medium	1" – 6" (25 mm – 150 mm)	4 5
	Heavy	> 6" (150 mm)	5 6
Torch Brazing (TB)			3 4
Torch Soldering (TS)			2

Lens shade values are based on OSHA minimum values and ANSI/AWS suggested values. Choose a lens shade that's too dark to see the weld zone. Without going below the required minimum value, reduce the shade value until it's visible. If possible, use a lens that absorbs yellow light when working with fuel gas processes.

Figure 6 Guide to lens shade numbers.

After cleaning the container, you must formally confirm that it's safe for welding or cutting. *American Welding Society (AWS) F4.1, Safe Practices for the Preparation of Containers and Piping for Welding and Cutting*, outlines the proper procedure. The following three paragraphs summarize the process.

Immediately before work begins, a qualified person must check the container with an appropriate test instrument and document that it's safe for welding or cutting. Tests should check for relevant hazards (flammability, toxicity, etc.). During work, repeated tests must confirm that the container and its surroundings remain safe.

Alternatively, fill the container with an inert material ("inerting") to drive out any hazardous vapors. Water, sand, or an inert gas like argon meets this requirement. Water must fill the container to within 3" (~7.5 cm) or less of the work location. The container must also have a vent above the water so air can escape as it expands (*Figure 7*). Sand must completely fill the container.

When using an inert gas, a qualified person must supervise, confirming that the correct amount of gas keeps the container safe throughout the work. Using an inert gas also requires additional safety procedures to avoid accidental suffocation.

NOTE

A *lower explosive limit (LEL)* gas detector can check for flammable or explosive gases in a container or its surroundings. Gas monitoring equipment checks for specific substances, so be sure to use a suitable instrument. Test equipment must be checked and calibrated regularly. Always follow the manufacturer's guidelines.

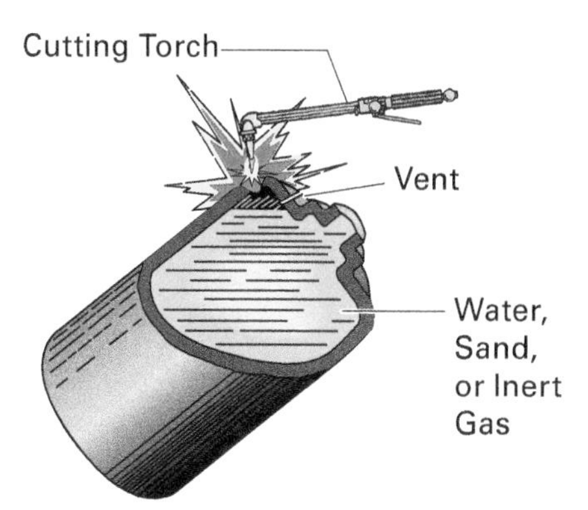

Note: *ANSI Z49.1* and AWS standards should be followed.

Figure 7 "Inerting" a container.

WARNING!

Never weld or cut drums, barrels, tanks, vessels, or other containers until they have been emptied and cleaned thoroughly. Residues, such as detergents, solvents, greases, tars, or corrosive/caustic materials, can produce flammable, toxic, or explosive vapors when heated. Never assume that a container is clean and safe until a qualified person has checked it with a suitable test instrument. Never weld or cut in places with explosive vapors, dust, or combustible products in the air.

The following general fire- and explosion-prevention guidelines apply to cutting and SMAW work:

- Never carry matches or gas-filled lighters in your pockets. Sparks can ignite the matches or cause the lighter to explode.
- Always comply with all site and employer requirements for hot work permits and fire watches.
- Never use oxygen in place of compressed air to blow off the workpiece or your clothing. Never release large amounts of oxygen. Oxygen makes fires start readily and burn intensely. Oxygen trapped in fabric makes it ignite easily if a spark falls on it. Keep oxygen away from oil, grease, and other petroleum products.
- Remove all flammable materials from the work area or cover them with a fire-resistant blanket.
- Before welding, heating, or cutting, confirm that an appropriate fire extinguisher is available. It must have a valid inspection tag and be in good condition. All workers in the area must know how to use it.
- Never release a large amount of fuel gas into the work environment. Some gases (propane and propylene) accumulate on the floor, while others (acetylene and natural gas) accumulate near the ceiling. Either can ignite far from the release point. Acetylene will explode at lower concentrations than other fuel gases.
- Prevent fires by maintaining a neat and clean work area. Confirm that metal scrap and slag are cold before disposing of them.

1.2.4 Work Area Ventilation

Vapors and fumes fill the air around their sources. Welding and cutting fumes can be hazardous. Welders often work above the area from which fumes come. Good ventilation and source extraction equipment help remove the vapors and protect the welder. The following lists good ventilation practices to use when welding and cutting:

- Always weld or cut in a well-ventilated area. Operations involving hazardous materials, such as metal coatings and toxic metals, produce dangerous fumes. You must wear a suitable respirator when working with them. Follow your workplace's safety protocols and confirm that your respirator is in good condition.

WARNING!

Cadmium, mercury, lead, zinc, chromium, and beryllium produce toxic fumes when heated. When welding or cutting these materials, wear appropriate respiratory PPE. Supplied-air respirators (SARs) work best for long-term work.

- Never weld or cut in a confined space without a confined space permit. Prepare for working in the confined space by following all safety protocols.
- Set up a confined space ventilation system before beginning work.
- Never ventilate with oxygen or breathe pure oxygen while working. Both actions are very dangerous.

WARNING!

When working around other craftworkers, always be aware of what they're doing. Take precautions to keep your work from endangering them.

Going Green

Vapor Extraction Systems

Protect the environment by using a vapor extraction system that filters out the welding fumes. This strategy is superior to simply releasing the fumes into the environment.

1.2.5 Preparing the Welding Area

To practice open-root single V-groove welding, you'll need a welding table, bench, or stand (*Figure 8*). It should have a way to place welding coupons out of position.

Figure 8 Welding station with source extraction.
Source: The Lincoln Electric Company, Cleveland, OH, USA

The following steps outline setting up the welding area:

Step 1 Check the work area ventilation. If necessary, use fans, windows, or source extraction equipment.

Step 2 Check the area for fire hazards. Remove any flammable materials. Cover with a welding blanket anything flammable that you can't move.

Step 3 Locate the nearest fire extinguisher. Confirm that it's charged and that you know how to use it.

Step 4 Position the welding machine near the welding table.

Step 5 Set up welding screens around the welding area.

1.2.6 Electrodes

These welding exercises require E6010/E6011 and E7018 electrodes. Use a diameter of $\frac{3}{32}$" (2.4 mm), $\frac{1}{8}$" (3.2 mm), or $\frac{5}{32}$" (4.0 mm). Remember that E7018 electrodes must stay dry. Leave them in their storage oven until you're ready to weld. Remove only as many as you'll use during the session.

Keep electrodes in a pouch or holder so they don't get damaged. Never leave them loose on the table. They can fall on the floor and be damaged or create a slip hazard. Have a suitable metal container nearby to hold discarded electrode stubs.

NOTE

The electrode sizes are recommendations only. Check with your instructor for the correct size to use.

WARNING!

Don't throw electrode stubs on the floor. They can roll under your feet, causing accidents.

Electrode Housekeeping

Electrode stubs on the floor can roll under your feet, creating a serious hazard. Many employers discipline employees who throw stubs on the floor. Some even make this practice a dismissible offense. Always place electrode stubs in a suitable container. Good welders keep their workspaces clean and tidy.

During critical welding jobs, some companies require welders to turn in their electrode stubs at the end of the shift. The number of stubs must match the number of electrodes they received at the start of the shift.

Use a whipping motion with E6010/E6011 electrodes when depositing a stringer bead. This controls the weld pool better. With E7018 electrodes, *don't* whip the electrode. Maintain a short arc. Low-hydrogen electrodes rely on the slag to protect the molten metal. They don't produce a heavy gaseous shield.

The following general guidelines apply to welding with either E6010/E6011 or E7018 electrodes:

- Remove all slag between passes.
- Restart by striking the arc ahead of the crater. Move quickly back into the crater and then proceed as usual. This technique welds over the arc strike. It eliminates porosity and gives a smoother restart.
- When running weave beads, use a slight side-to-side motion. Pause briefly at the weld toe to tie in and flatten the bead.
- Never use electrodes with chipped or missing flux.
- When welding vertically, set the amperage at the lower end of the recommended range. This gives better weld pool control.

Running E7018 Electrodes

Moving the electrode from side to side makes a weave bead. Many different patterns will produce a weave bead, including circles, crescents, and zigzags. Pausing at the edges flattens out the weld, giving it the proper profile.

1.2.7 Preparing the Welding Machine

Welders will use many different welding machine brands and models. Not all will have manuals with them. Learn to recognize welding machine types and their controls, so you can use them safely and correctly (*Figure 9*).

The following steps outline setting up a regular welding machine:

Step 1 Locate the nearest power disconnect so you can shut down the machine quickly.

Step 2 Plug in the machine.

Step 3 Select SMAW mode (constant current DC operation).

Step 4 If the machine has a polarity selector, set it to DCEP. If not, connect the leads for DCEP.

Step 5 Connect the workpiece lead's clamp to the workpiece.

Step 6 Set the amperage for the electrode type and size. *Table 1* shows typical settings.

NOTE

If a DC welding machine isn't available, a constant current AC machine will work. You'll need to use E6011 electrodes instead of E6010. E7018 electrodes perform properly in constant current AC mode. In cases where arc blow is a problem, switching from DC to AC may solve the problem.

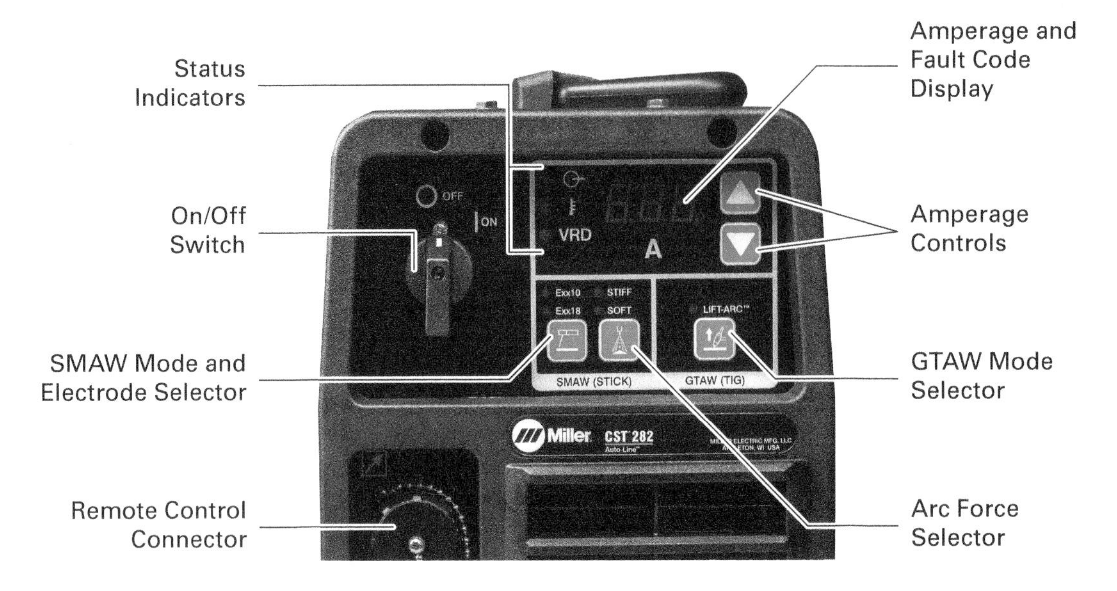

(A) Regular Welding Machine Controls

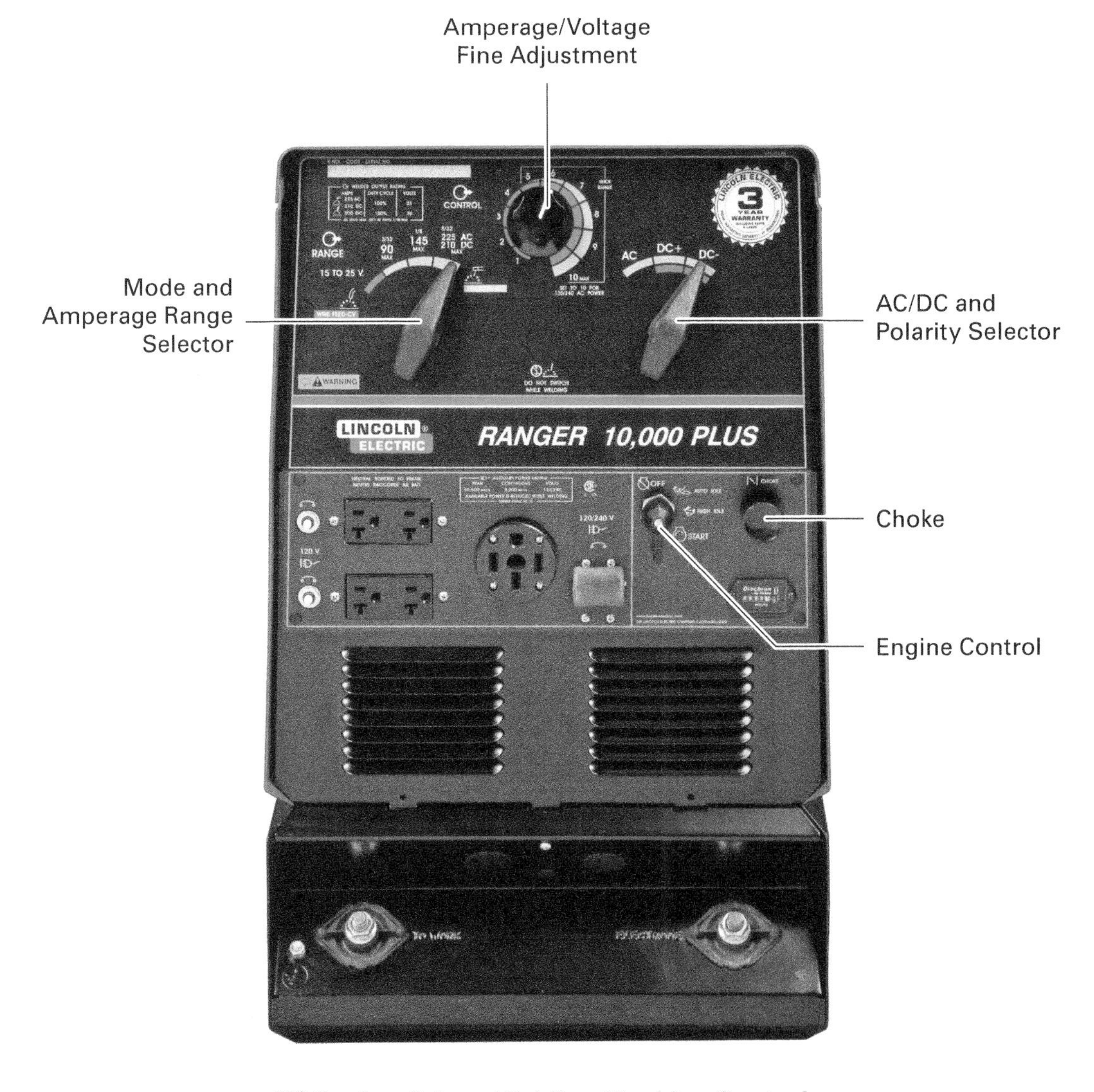

(B) Engine-Driven Welding Machine Controls

Figure 9 Regular and engine-driven welding machines.

Source: © Miller Electric Mfg. LLC (9A), The Lincoln Electric Company, Cleveland, OH, USA (9B)

TABLE 1 Typical Electrode Sizes and Amperages

Electrode	Size	Amperage
E6010	$\frac{3}{32}$" (2.4 mm)	40A to 80A (DCEP)
E6010	$\frac{1}{8}$" (3.2 mm)	70A to 130A (DCEP)
E6010	$\frac{5}{32}$" (4.0 mm)	90A to 165A (DCEP)
E6011	$\frac{3}{32}$" (2.4 mm)	40A to 90A (AC)
E6011	$\frac{1}{8}$" (3.2 mm)	65A to 120A (AC)
E6011	$\frac{5}{32}$" (4.0 mm)	115A to 150A (AC)
E7018	$\frac{3}{32}$" (2.4 mm)	70A to 110A (DCEP) 80A to 120A (AC)
E7018	$\frac{1}{8}$" (3.2 mm)	90A to 160A (DCEP) 100A to 160A (AC)
E7018	$\frac{5}{32}$" (4.0 mm)	130A to 210A (DCEP) 140A to 210A (AC)

NOTE

These values are general guidelines only. Amperage recommendations vary by manufacturer, welding position, current type, and electrode brand. If possible, refer to the WPS for guidance. If a WPS isn't available, follow the electrode manufacturer's guidelines.

Step 7 Confirm that the electrode holder *isn't* resting on a grounded surface.

Step 8 Turn on the welding machine when you're ready to weld.

If you're using an engine-driven welding machine, you must place it in a safe, properly ventilated space. Startup procedures vary, depending on the machine's engine model. Consult NCCER Module 29107, *SMAW – Equipment and Setup*, for more detail when working with engine-driven welding machines.

WARNING!

Remember that the workpiece lead (ground lead) isn't always grounded. It could have the welding machine's full open-circuit voltage between it and any grounded object. Always be cautious with both welding leads.

WARNING!

When you're finished welding, always remove the electrode from its holder. This prevents arcing, which can cause burns and arc flash.

1.2.8 Preparing the Weld Coupons

When possible, use $\frac{3}{8}$" (~10 mm) thick carbon steel for welding coupons. This meets the AWS limited-thickness test coupon requirements. Alternatively, use $\frac{1}{4}$" to 1" (~6 mm to 25 mm) thick steel for practice welds. You'll also need some scrap coupons. One is for practicing stringer beads. Scrap coupons are handy for setting up the welding machine and adjusting its settings. They may be any convenient size or thickness. Clean all coupons with a wire brush or grinder to remove scale and corrosion.

WARNING!

Be extremely careful with powered wire brushes. Verify that the brush's speed rating matches or exceeds the tool's maximum speed. Loose wires can fly off during cleaning. Regardless of the tool type, always wear appropriate PPE when cleaning metal.

NOTE

When preparing test coupons for welder qualifications, use a scrap coupon to check the welding machine's settings. Practicing on a scrap coupon is also a good way to become familiar with the machine before producing the test coupon.

For each weld coupon, you'll need two plates 3" × 7" (7.6 cm × 17.8 cm). Bevel the coupons 30 to $37\frac{1}{2}$ degrees on one long edge.

The following steps outline preparing the test coupons:

NOTE

This coupon size (3" × 7") meets the minimum AWS size requirement. To save material, your instructor may select a smaller size for practice.

Step 1 Check the bevels. There should be no dross. The root face should be between $\frac{1}{16}$" and $\frac{1}{8}$" (1.6 mm and 3.2 mm). The bevel angle should be 30 to $37\frac{1}{2}$ degrees. Remove any burrs from the coupon edges.

NOTE

The root opening should be within the tolerances permitted by the applicable code, WPS, or jobsite quality specifications. For these practice exercises, the root opening should be no larger than 3⁄16" (4.8 mm) for a 3⁄8" (~10 mm) plate.

NOTE

Most welding codes allow the root face and root opening on open-root welds to be between 0" and 1⁄8" (0 mm and 3.2 mm). Adjust the root face and root opening as needed. Always follow the applicable code, WPS, or jobsite quality specifications.

Step 2 Position the coupons on a smooth, flat surface with the root side up. The bevels should face each other. The root opening should be 1⁄8" (3.2 mm) wide (*Figure 10*). Confirm with a spacer or other measuring device.

Step 3 Tack weld the coupons on each end (*Figure 10*).

Step 4 When you're finished tack welding, *always* remove the electrode from its holder. This prevents arcing. Discard the stub in the proper container.

Step 5 Position the coupon in the required welding position (1, 2, 3, or 4).

Can You Spare Some Change?

Nickels and dimes make excellent thickness gauges for fitting up open-root single V-groove weld coupons. Check the root face (land) with the nickel's thickness. Check the root opening with the dime's thickness.

Going Green

Welding Coupons

Reuse or recycle welding coupons rather than discarding them. Doing this reduces unnecessary waste and expense. Cut up the larger used coupons into smaller ones. Practice running beads on smaller used coupons.

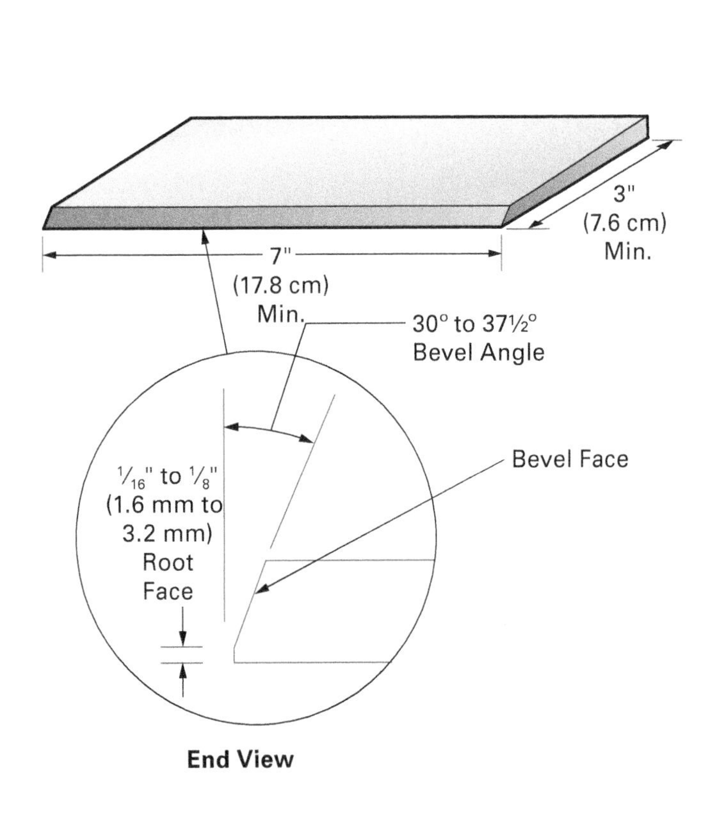

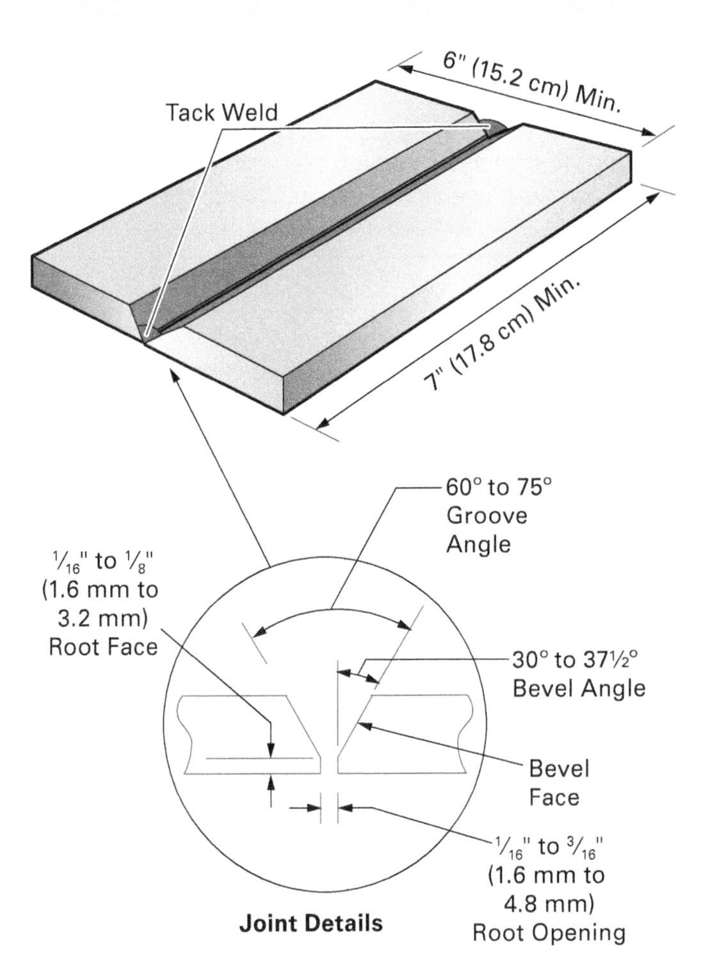

Figure 10 Open-root single V-groove weld coupon.

FLASH

Learning the memory aid **FLASH** helps you produce good welds:

- **F** (fit-up) — Correctly align and tack weld the base material together according to your instructor's specifications.
- **L** (length of arc) — Maintain the right distance between the electrode and the base metal. This distance usually equals the electrode core's diameter.
- **A** (angle) — Two angles are critical: the *work angle* and the *travel angle* (see the diagram).
- **S** (speed) — Travel speed, in inches per minute (IPM), affects the weld's width and fusion.
- **H** (heat) — The amperage setting controls the heat. The proper value depends on the electrode diameter, base metal type, base metal thickness, and welding position.

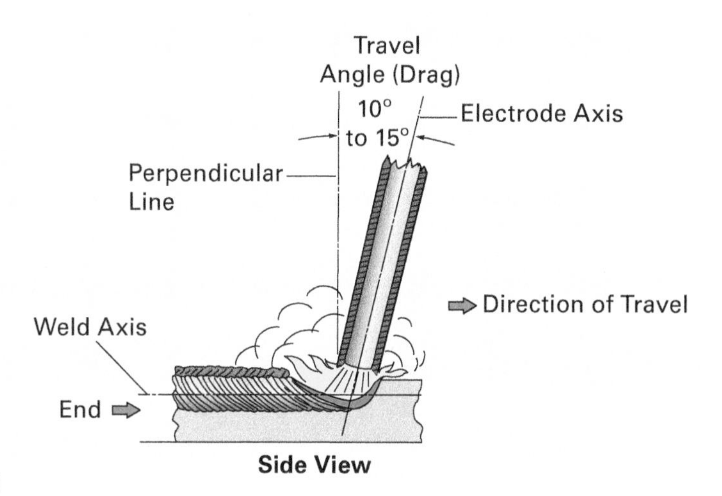

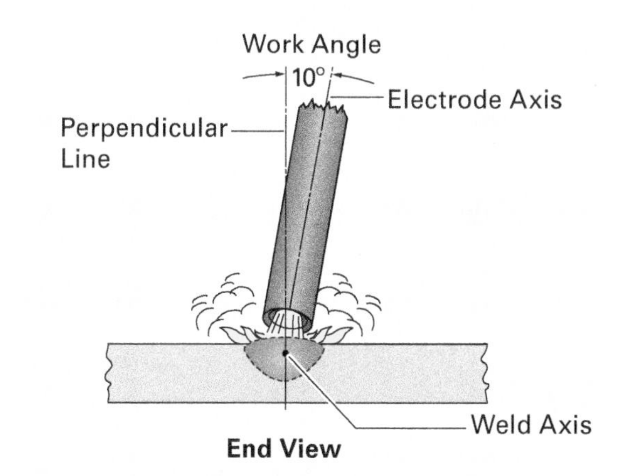

1.0.0 Section Review

1. The groove's angled or curved surface is the _____.
   a. bevel face
   b. groove radius
   c. root face
   d. joint root

2. Which of the following is a common powered wire brush hazard?
   a. The removed scale and corrosion can ignite.
   b. The wires can overheat and release toxic fumes.
   c. The wires can fly off during cleaning.
   d. The brush can cause arc strikes on the base metal.

2.0.0 Open-Root Single V-Groove Welds

Performance Tasks

2. Use E6010/E6011 and E7018 electrodes to make flat (1G) open-root single V-groove welds.
3. Use E6010/E6011 and E7018 electrodes to make horizontal (2G) open-root single V-groove welds.
4. Use E6010/E6011 and E7018 electrodes to make vertical (3G) open-root single V-groove welds.
5. Use E6010/E6011 and E7018 electrodes to make overhead (4G) open-root single V-groove welds.

Objective

Summarize techniques for making open-root single V-groove welds.

a. Describe producing the root pass for open-root single V-groove welds.
b. Describe making open-root single V-groove welds in the 1G and 2G positions.
c. Describe making open-root single V-groove welds in the 3G and 4G positions.

Welders frequently produce open-root single V-groove welds on plate and pipe. Learning to weld plate prepares you to tackle pipe later. This module's procedures align with the *ASME Boiler and Pressure Vessel Code, Section IX, Welding and Brazing Qualifications*. Some welding jobs may require qualifications from other standards, such as *AWS D1.1*.

Welders can make open-root single V-groove welds in all positions. The weld position identifier for plate depends on the weld type and workpiece orientation (*Figure 11*). All groove welds end in G ("groove"). Groove weld positions for plate are flat (1G), horizontal (2G), vertical (3G), and overhead (4G).

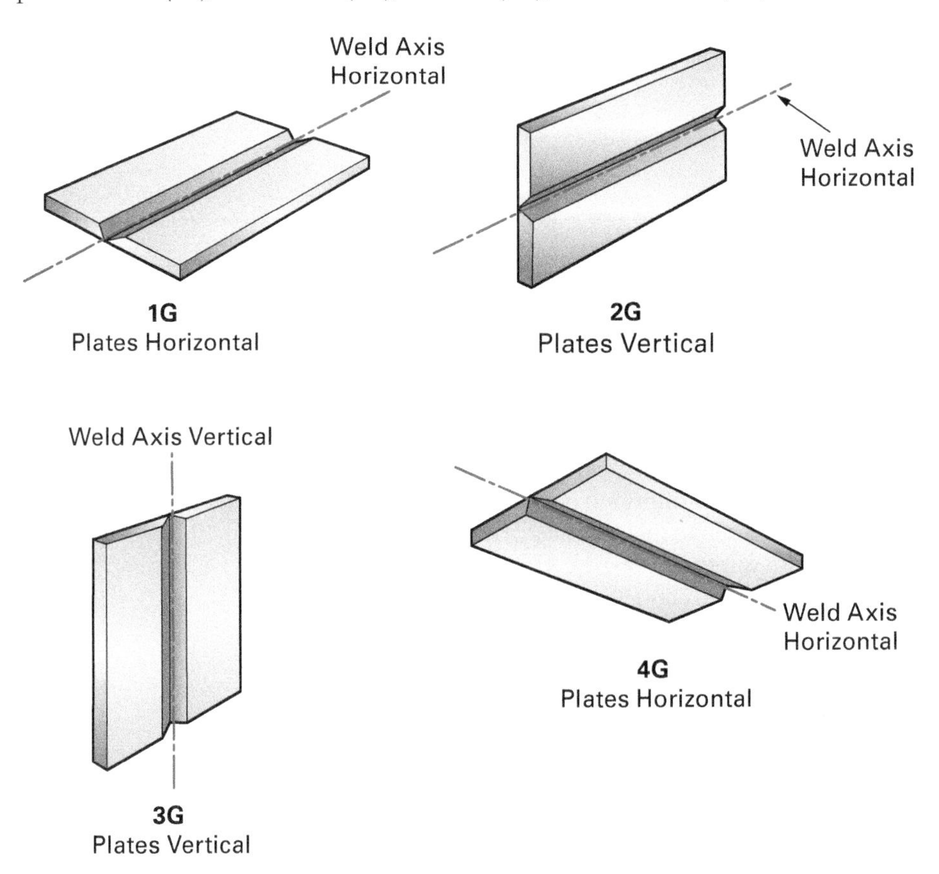

Figure 11 V-groove weld positions for plate.

When they use low-hydrogen electrodes for filler passes, welders often must grind first. Low-hydrogen electrodes have shallow penetration qualities, so inclusions can be a problem. Always clean the weld between passes. Tight slag usually indicates poor welding technique.

The welder can incline the weld axis up to 15 degrees in the 1G and 2G positions. The AWS standard specifies a permissible inclination in the 3G and 4G positions that depends on the rotational position.

Make groove welds with slight face reinforcement and a gradual transition to the base metal at each toe. Groove welds must not have excessive reinforcement (face or root), underfill, or overlap. These defects make the weld unacceptable. Most codes permit some undercut, provided it's not excessive. *Figure 12* shows acceptable and unacceptable profiles.

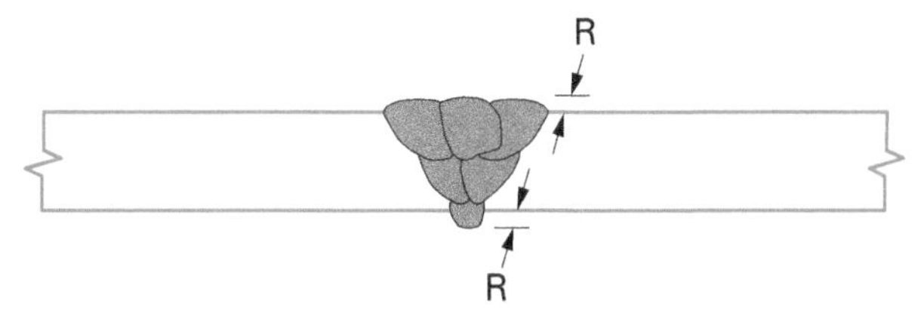

R = Face and Root Reinforcement not to Exceed the Code-Specified Amount

Acceptable Weld Profile

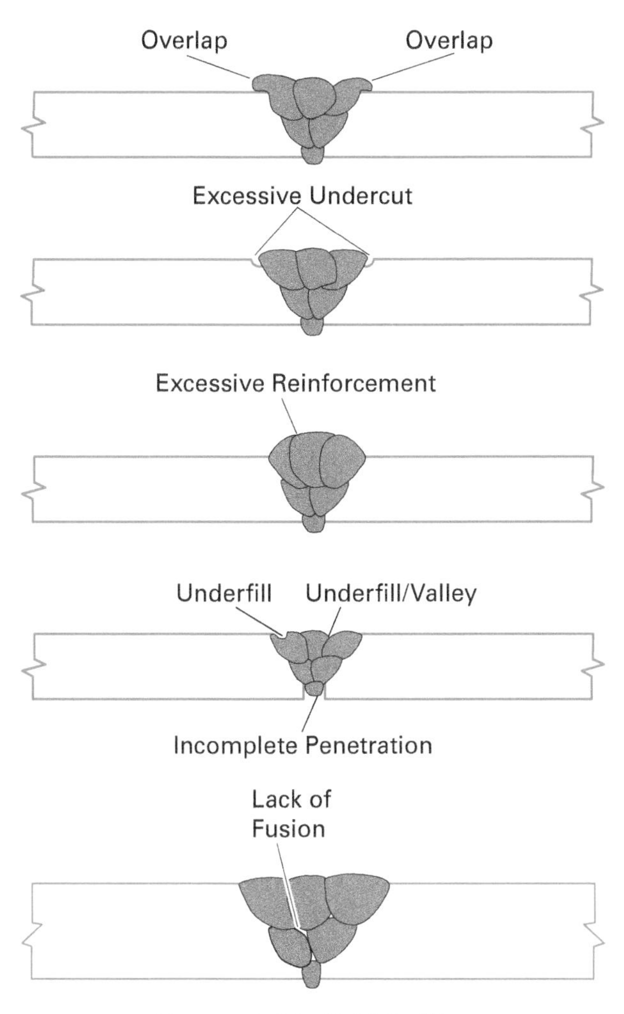

Unacceptable Weld Profiles

Figure 12 Acceptable and unacceptable open-root single V-groove weld profiles.

NOTE

Sometimes, the code, WPS, or jobsite quality specifications don't permit grinding welds.

NOTE

Welders use the position identifiers (1G, 2G, 3G, and 4G) when referring to qualification tests. When making production welds, welders don't use the identifiers. Instead, they use the terms "flat," "horizontal," "vertical," and "overhead" to identify position.

NOTE

This module provides general guidelines only. Always follow the WPS or jobsite quality specifications. Check with your supervisor if you're unsure about the application's specifications.

2.1.0 Producing the Root Pass

Producing the root pass is the most challenging part of making an open-root single V-groove weld. The welder makes the root pass from the V-groove side. It must completely penetrate but not have excessive root reinforcement. Welders use a technique called "running a keyhole" to control penetration.

The arc makes the keyhole by melting away the root faces (lands) of the plates. The molten metal flows to the back side of the keyhole, forming the weld (*Figure 13*). The keyhole should be centered and about $^{1}/_{8}$" to $^{3}/_{16}$" (~3 mm to 5 mm) in diameter. A larger keyhole will produce excess root reinforcement. If the keyhole closes, there won't be adequate penetration. Root reinforcement should be either flush or no more than the code, WPS, or jobsite quality specifications permit. A maximum of $^{1}/_{8}$" (3.2 mm) is common under some codes.

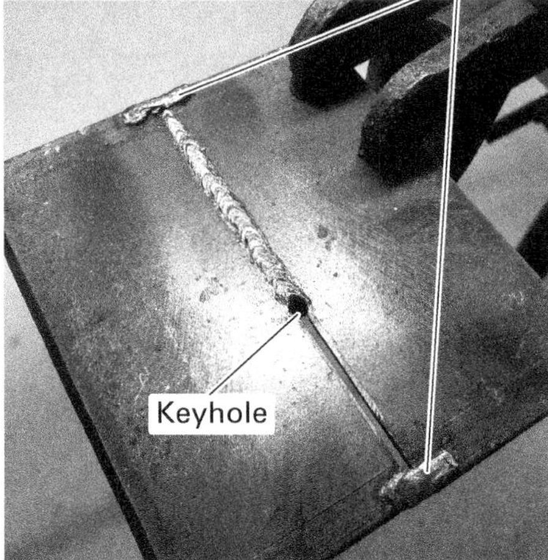

Top (V-Groove) **Bottom (Root Opening)**

Figure 13 The root pass.
Source: Courtesy John Elliott, Lee College

Control the keyhole by using a fast-freeze E6010/E6011 electrode and whipping it. Move the electrode forward about $^{3}/_{16}$" to $^{1}/_{4}$" (~5 mm to 6 mm). Move back about $^{1}/_{8}$" (~3 mm) and pause. Keep the arc length consistent as you move.

Carefully watch the keyhole's back edge to ensure good fusion. This area should be round. If it becomes V-shaped, proper fusion isn't happening. If the keyhole starts to grow, reduce the pause time and increase the whip's forward length. If the keyhole starts to close, increase the pause time and decrease the whip's forward length. Amperage, root face size, and the root opening can also affect the keyhole. Adjust these factors within what the code, WPS, or jobsite quality specifications permit.

After running the root pass, clean and inspect it. Look for excessive buildup or undercut. These could trap slag during later passes. Remove excessive buildup or undercut with a hand grinder. Grind the root pass face with the grinding disc's edge (*Figure 14*). Don't grind through the root pass or widen the groove.

Hold It Open

Before beginning the root pass, tack weld the end opposite to the starting point with a bead about $^{1}/_{2}$" (~13 mm) long. This will prevent the root opening from closing at the opposite end as you begin welding.

2.2.0 Open-Root Single V-Groove Welds – 1G and 2G Positions

This section introduces producing good open-root single V-groove welds in the 1G or 2G position. You'll learn the best technique and the required bead types. These positions are easier than the 3G and 4G positions. Mastering them first makes the harder welds easier.

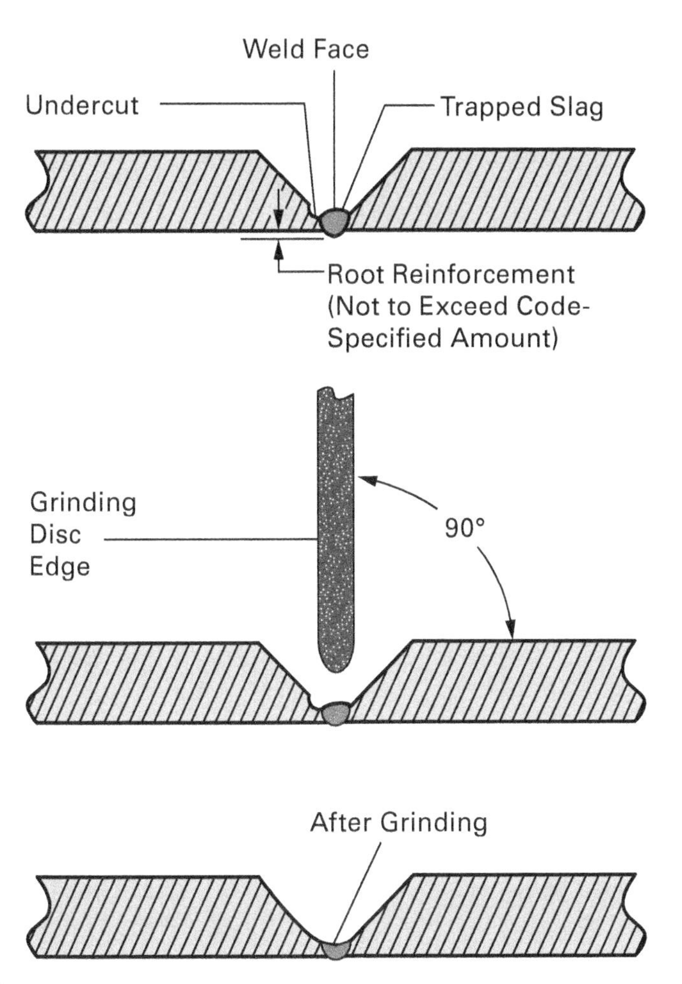

Figure 14 Grinding the root pass.

2.2.1 Practicing Open-Root Single V-Groove Welds (1G Position)

Practice open-root single V-groove welds in the 1G (flat) position with $^3\!/_{32}$" (2.4 mm), $^1\!/_8$" (3.2 mm), or $^5\!/_{32}$" (4.0 mm) E6010/E6011 and E7018 electrodes. You can use either weave beads or stringer beads. For weave beads, use a 10- to 15-degree drag angle. Keep the electrode at a 0-degree work angle. For stringer beads, use a 10- to 15-degree drag angle too. Adjust the work angle to tie in the weld on one side or the other as required. Be sure to fill the crater at the weld termination. If the coupon gets too hot between passes, quench it in water.

CAUTION

Never quench functional welds or qualification test coupons. Rapid cooling hardens the metal, which may then crack. It also affects the base metal's mechanical properties. Quenching practice coupons is acceptable. Always quench in a bucket or quench tank rather than a sink or water fountain. Slag can clog drains.

WARNING!

Wear gloves and handle hot coupons with pliers. Be careful of the steam rising from coupons when you quench them. It can burn unprotected skin.

Follow these steps to practice making open-root single V-groove welds in the 1G position:

Step 1 Tack weld the practice coupon as *Figure 10* shows.

Step 2 Clean and feather (taper) the tack welds with the edge of a grinding disc. Feathering the tack welds helps fuse them into the root pass.

Step 3 Position the practice coupon flat on the welding table.

Step 4 Run the root pass using E6010/E6011 electrodes.

Step 5 Chip, clean, and brush the root weld. Grind if required.

Step 6 Run the fill and cover passes using E7018 electrodes. Chip, clean, and brush each bead. Use stringer or weave beads as your instructor directs. *Figure 15* shows all welding passes with their approximate work angles.

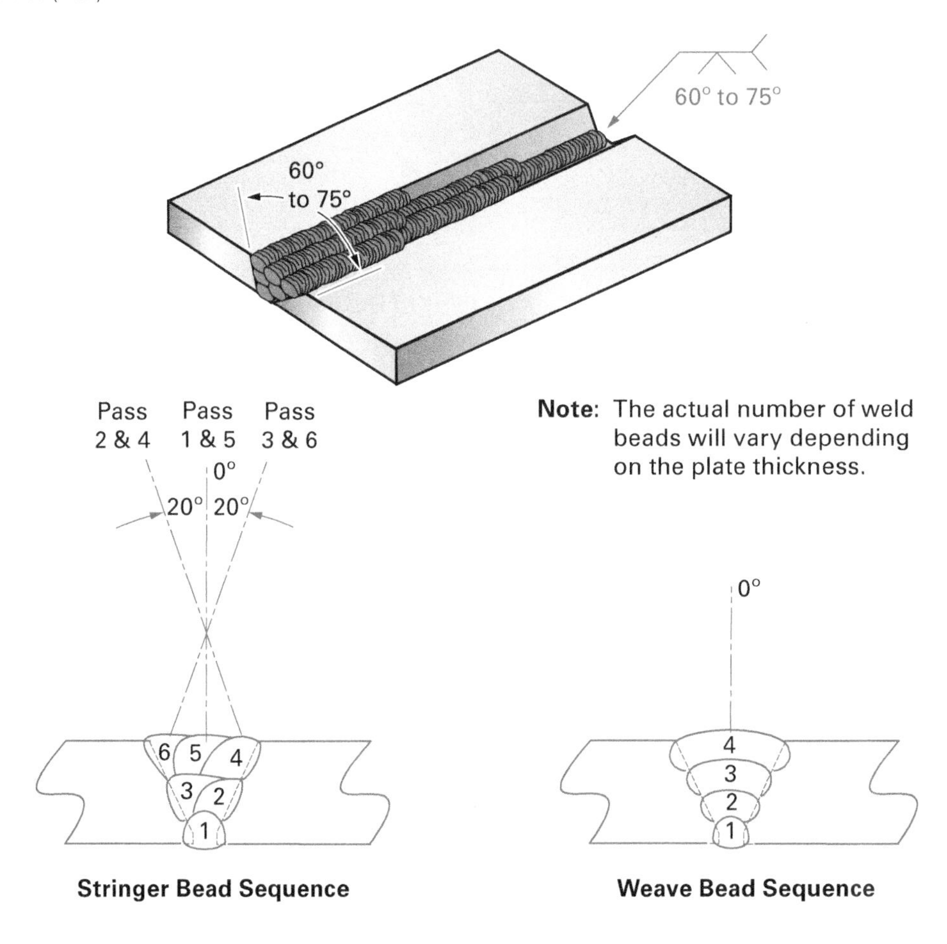

Figure 15 Open-root single V-groove 1G welding sequence and work angles.

Step 7 When you're finished welding, *always* remove the electrode from its holder. This prevents arcing. Discard the stub in the proper container.

2.2.2 Practicing Open-Root Single V-Groove Welds (2G Position)

Welders can create horizontal open-root single V-groove welds. Before you try this weld, however, practice running horizontal stringer beads. Do this by building a pad in the horizontal position (*Figure 16*).

Figure 16 A pad built in the horizontal position.
Source: The Lincoln Electric Company, Cleveland, OH, USA

Use $^3/_{32}$" (2.4 mm), $^1/_8$" (3.2 mm), or $^5/_{32}$" (4.0 mm) E6010/E6011 electrodes. Use a 10- to 15-degree drag angle. Keep the electrode at a 0-degree work angle. To control the weld pool, the work angle may be dropped slightly, but no more than 10 degrees (*Figure 17*).

Follow these steps to practice making horizontal stringer beads in the 2G position:

Step 1 Tack weld a flat plate coupon in the horizontal position.

Step 2 Run the first pass along the coupon's bottom edge. Use E6010/E6011 electrodes.

Step 3 Chip, clean, and brush the weld.

Step 4 Weld the second bead just above the first bead. The beads should overlap as *Figure 17* shows. Chip, clean, and brush it.

Step 5 Continue running beads until you complete the pad.

Step 6 When you're finished welding, *always* remove the electrode from its holder. This prevents arcing. Discard the stub in the proper container.

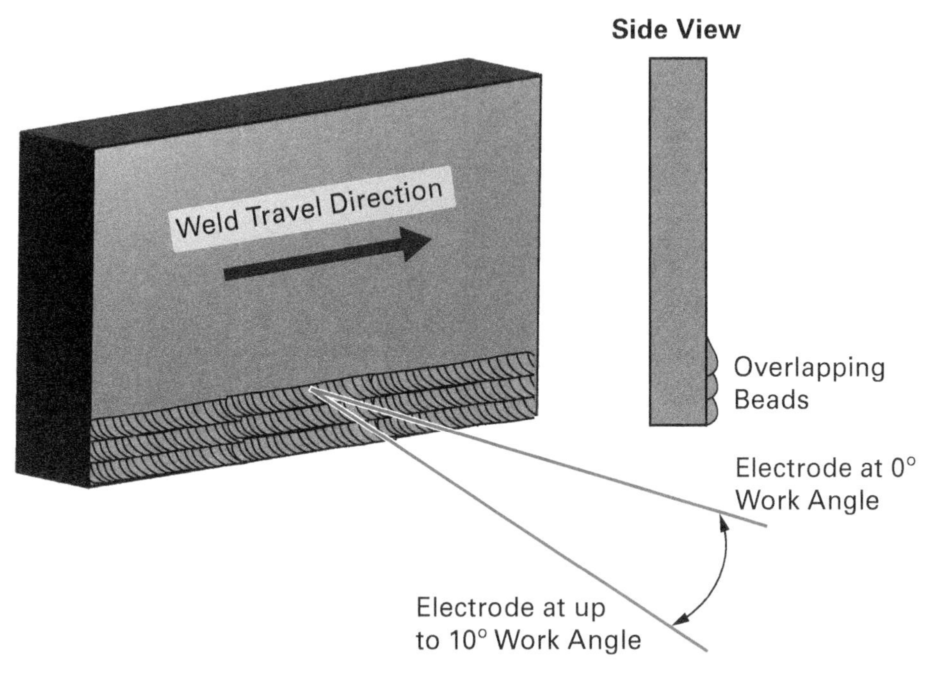

Figure 17 Building a pad in the horizontal position.

Once you've produced a satisfactory pad, practice the actual open-root single V-groove weld in the 2G (horizontal) position. Use $^3/_{32}$" (2.4 mm), $^1/_8$" (3.2 mm), or $^5/_{32}$" (4.0 mm) E6010/E6011 and E7018 electrodes. Use stringer beads and a 10- to 15-degree drag angle. Adjust the work angle to tie in the weld on one side or the other as required. Be sure to fill the crater at the weld termination.

Follow these steps to practice making open-root single V-groove welds in the 2G position:

Step 1 Tack weld the practice coupon as *Figure 10* shows. Use the standard or alternative coupon preparation as your instructor directs.

Step 2 Clean and feather (taper) the tack welds with the edge of a grinding disc. Feathering the tack welds helps fuse them into the root pass.

Step 3 Tack weld the practice coupon in the horizontal position.

Step 4 Run the root pass using E6010/E6011 electrodes.

Step 5 Chip, clean, and brush the weld.

Step 6 Run the fill and cover passes using E7018 electrodes. Chip, clean, and brush each bead. *Figure 18* shows all welding passes with their approximate work angles.

Step 7 When you're finished welding, *always* remove the electrode from its holder. This prevents arcing. Discard the stub in the proper container.

2.3.0 Open-Root Single V-Groove Welds – 3G and 4G Positions

This section introduces producing good open-root single V-groove welds in the 3G or 4G position. You'll learn the best technique and the required bead types. These positions are harder than the 1G and 2G positions.

2.3.1 Practicing Open-Root Single V-Groove Welds (3G Position)

Welders can create open-root single V-groove welds in the vertical position. Before you try this weld, however, practice running vertical weave beads. Do this by building a pad in the vertical position.

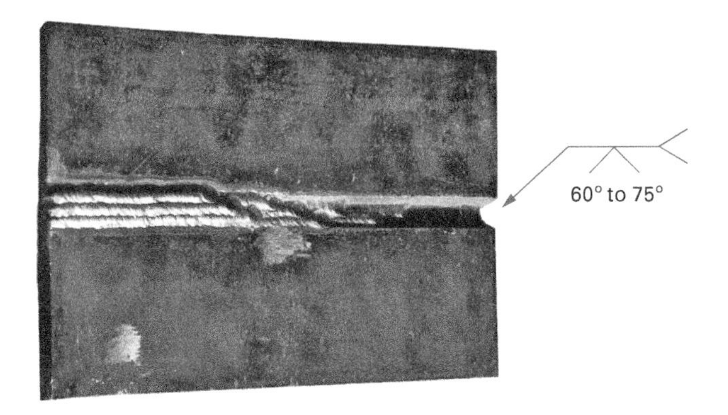

Note: The actual number of weld beads will vary depending on the plate thickness and electrode diameter.

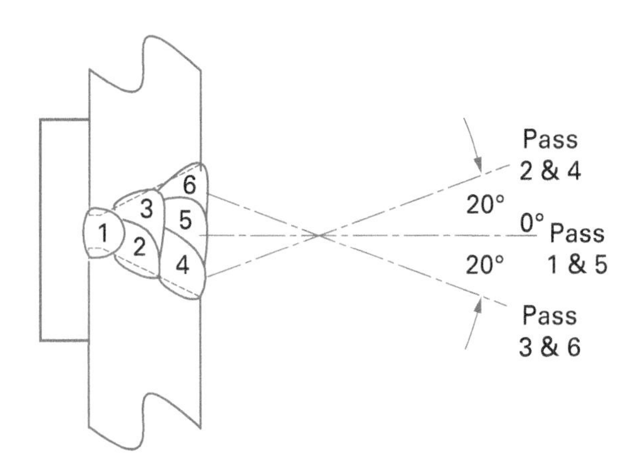

Note: Nonzero work angles are approximate.

Figure 18 Open-root single V-groove 2G welding sequence and work angles.

Use $\frac{3}{32}$" (2.4 mm), $\frac{1}{8}$" (3.2 mm), or $\frac{5}{32}$" (4.0 mm) E6010/E6011 electrodes. Set the amperage at the lower end of the correct range. This gives better weld pool control. Use a 0- to 10-degree push angle. Keep the electrode at a 0-degree work angle for the first pass. Use a 0- to 10-degree work angle toward the previous bead for subsequent beads (*Figure 19*). Use a slight side-to-side motion with a brief pause at each weld toe. This ties in better and controls the weld pool. The width of the weave beads should be no more than five times the electrode core's diameter.

Follow these steps to practice making vertical weave beads in the 3G position:

Step 1 Tack weld a flat plate coupon in the vertical position.

Step 2 Using a whipping motion, run the first pass along one of the coupon's vertical edges from bottom to top. Use E6010/E6011 electrodes.

Step 3 Chip, clean, and brush the weld.

Step 4 Weld the second bead next to the first bead. The beads should overlap. Chip, clean, and brush it.

Step 5 Continue running beads until you complete the pad.

Step 6 When you're finished welding, *always* remove the electrode from its holder. This prevents arcing. Discard the stub in the proper container.

NOTE

Some codes don't allow weave beads for this weld. Always consult the code, WPS, or jobsite quality specifications.

NOTE

Some codes, such as the API, allow making vertical welds in a downhill progression. Usually, this requires that the entire joint be welded with fast-freeze electrodes like E6010, E6011, E7010, or E8010. Low-hydrogen electrodes like E7018 are not suitable for downhill progression.

Once you've produced a satisfactory pad, practice the actual open-root single V-groove weld in the 3G (vertical) position. Use $\frac{3}{32}$" (2.4 mm), $\frac{1}{8}$" (3.2 mm), or $\frac{5}{32}$" (4.0 mm) E6010/E6011 and E7018 electrodes. Use weave beads for everything but the root pass. Set the amperage at the lower end of the correct range. This gives better weld pool control. Use a 0- to 10-degree push angle. Adjust the work angle to tie in the weld on one side or the other as required. Pause slightly at each toe to penetrate and fill the crater, preventing undercut. Be sure to fill the crater at the weld termination.

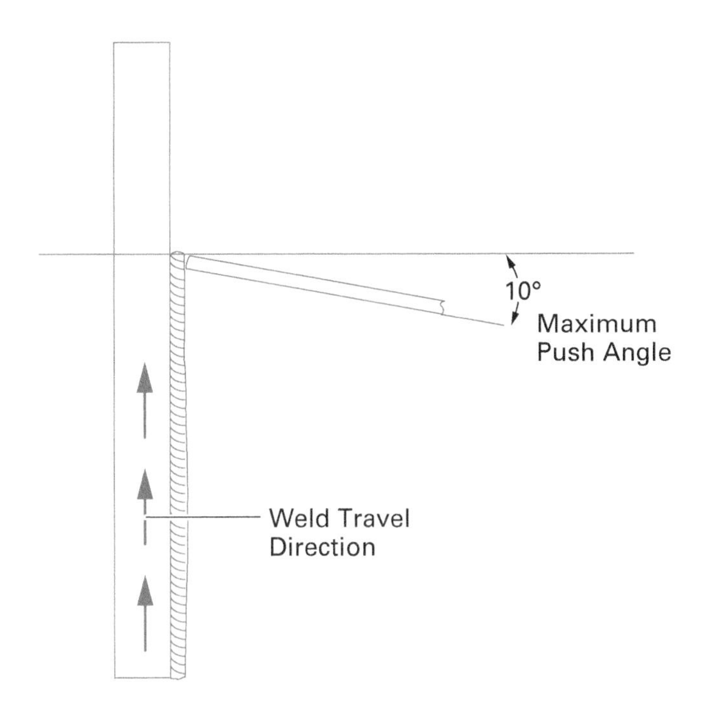

Figure 19 Building a pad in the vertical position (uphill progression).

Follow these steps to practice making open-root single V-groove welds in the 3G position:

Step 1 Tack weld the practice coupon as *Figure 10* shows.

Step 2 Clean and feather (taper) the tack welds with the edge of a grinding disc. Feathering the tack welds helps fuse them into the root pass.

Step 3 Tack weld the practice coupon in the vertical position.

Step 4 Run the root pass as a stringer bead using E6010/E6011 electrodes from bottom to top (uphill progression).

Step 5 Chip, clean, and brush the weld.

Step 6 Run the fill and cover passes uphill as weave beads using E7018 electrodes. Chip, clean, and brush each bead. *Figure 20* shows all welding passes with their approximate work angles.

Step 7 When you're finished welding, *always* remove the electrode from its holder. This prevents arcing. Discard the stub in the proper container.

2.3.2 Practicing Open-Root Single V-Groove Welds (4G Position)

Welders can create open-root single V-groove welds in the overhead position. Before you try this weld, however, practice running overhead stringer beads. Do this by building a pad in the overhead position.

WARNING!

Avoid standing directly under the coupons while welding. Hot slag and molten metal can drop on you, causing severe burns. Besides regular PPE, wear the additional protective clothing required for overhead work.

Use $^{3}/_{32}$" (2.4 mm), $^{1}/_{8}$" (3.2 mm), or $^{5}/_{32}$" (4.0 mm) E6010/E6011 electrodes. Use a 10- to 15-degree drag angle. Keep the electrode at a 0-degree work angle for the first pass. Use a 10-degree work angle toward the previous bead for subsequent beads (*Figure 21*).

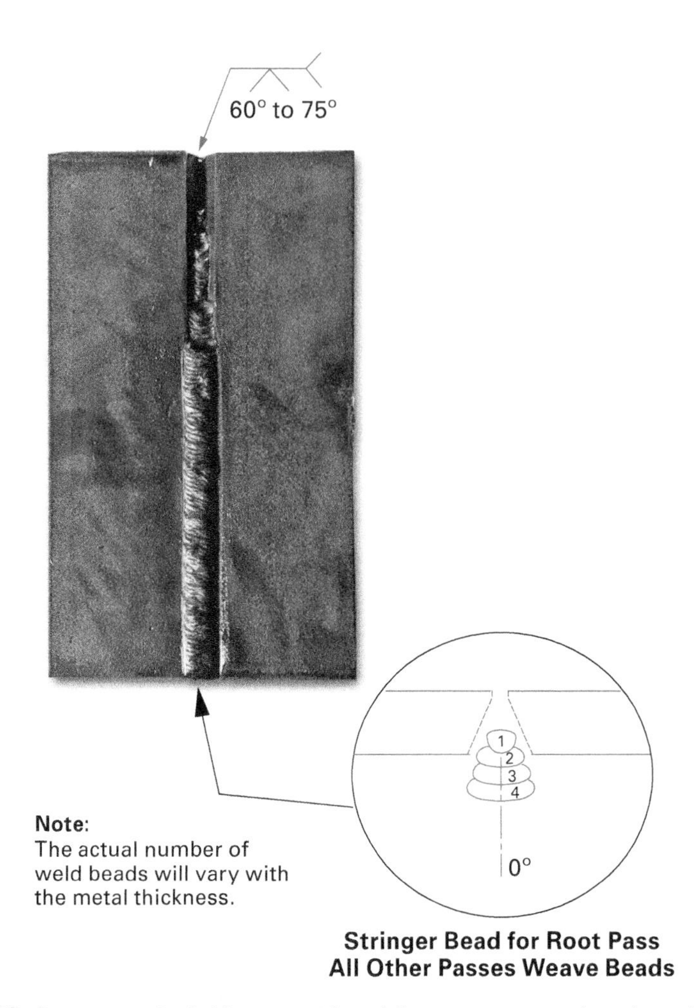

Figure 20 Open-root single V-groove 3G welding sequence and work angles (uphill progression).

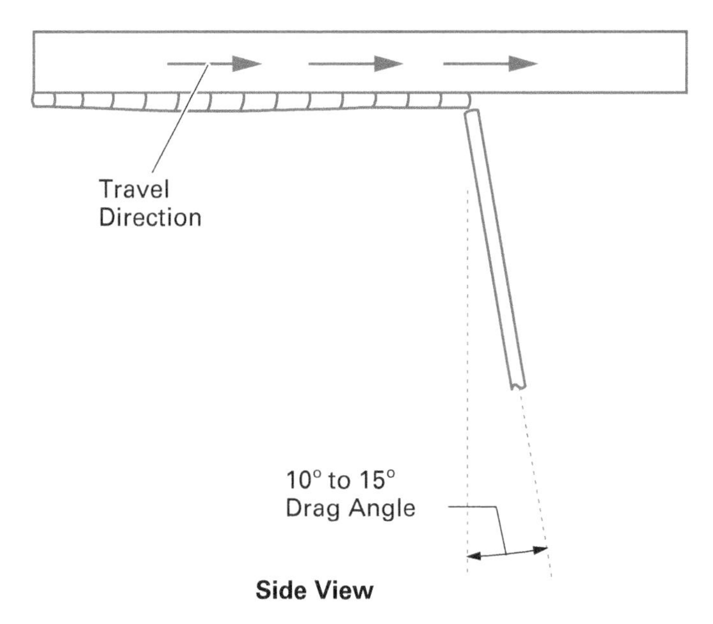

Figure 21 Building a pad in the overhead position.

Follow these steps to practice making overhead stringer beads in the 4G position:

Step 1 Tack weld a flat plate coupon in the overhead position.

Step 2 Run the first pass along one of the coupon's edges. Use E6010/E6011 electrodes.

Step 3 Chip, clean, and brush the weld.

Step 4 Weld the second bead next to the first bead. The beads should overlap. Chip, clean, and brush it.

Step 5 Continue running beads until you complete the pad.

Step 6 When you're finished welding, *always* remove the electrode from its holder. This prevents arcing. Discard the stub in the proper container.

Once you've produced a satisfactory pad, practice the actual open-root single V-groove weld in the 4G (overhead) position. Use $^{3}/_{32}$" (2.4 mm), $^{1}/_{8}$" (3.2 mm), or $^{5}/_{32}$" (4.0 mm) E6010/E6011 and E7018 electrodes. Use stringer beads. Use a 10- to 15-degree drag angle. Adjust the work angle to tie in the weld on one side or the other as required. Be sure to fill the crater at the weld termination.

Follow these steps to practice making open-root single V-groove welds in the 4G position:

Step 1 Tack weld the practice coupon as *Figure 10* shows.

Step 2 Clean and feather (taper) the tack welds with the edge of a grinding disc. Feathering the tack welds helps fuse them into the root pass.

Step 3 Tack weld the practice coupon in the overhead position.

Step 4 Run the root pass using E6010/E6011 electrodes.

Step 5 Chip, clean, and brush the weld.

Step 6 Run the fill and cover passes using E7018 electrodes. Chip, clean, and brush each bead. *Figure 22* shows all welding passes with their approximate work angles.

Step 7 When you're finished welding, *always* remove the electrode from its holder. This prevents arcing. Discard the stub in the proper container.

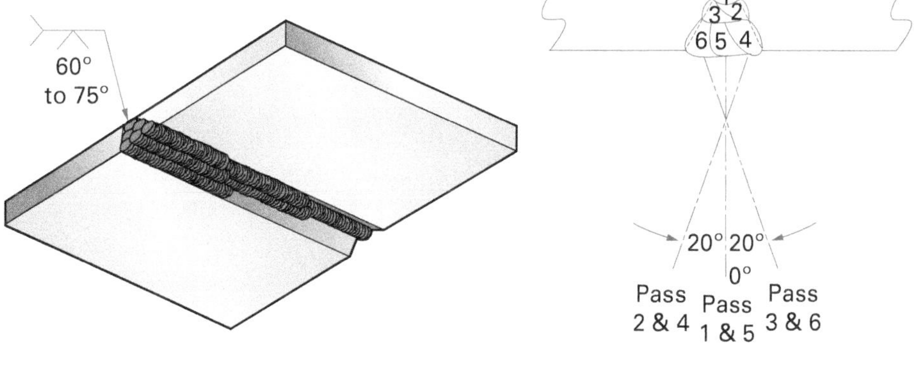

Figure 22 Open-root single V-groove 4G welding sequence and work angles.

Welders and Multifocal Lenses

In the past, welders who wore multifocal glasses (bifocals, trifocals, and progressives) had a challenging time. Welding helmets and hoods didn't have a lot of extra space inside. The front tinted lens was small, so welders couldn't tilt their heads to bring the right part of their glasses into position. Some welders fitted their helmets with a magnifying lens ("cheater") to help see better. Many still do this.

Today, PPE manufacturers make helmets and hoods designed for welders who wear multifocal glasses. The headgear adjusts to move the front farther away from the face, making room for glasses. The tinted lens is also larger, so welders can look through the different sections of their glasses more easily and see things clearly.

When wearing prescription glasses rather than standard safety glasses, the lenses must be shatterproof. The glasses must also have ANSI-approved side shields.

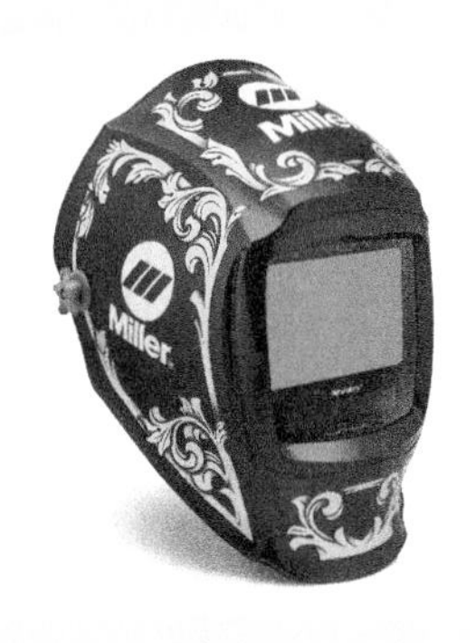

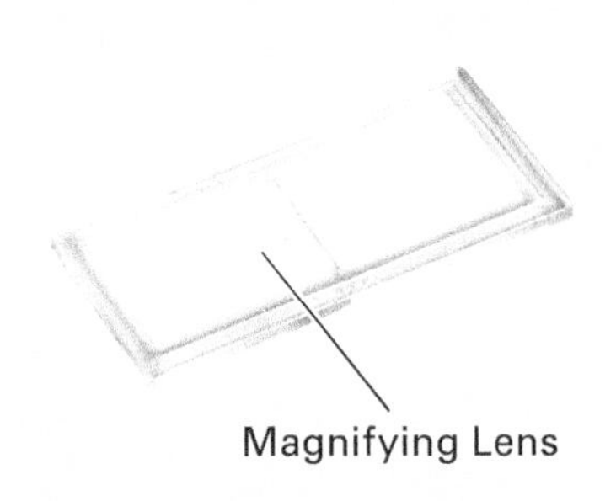

Source: © Miller Electric Mfg. LLC

2.0.0 Section Review

1. The key to a successful root pass when welding an open-root single V-groove weld is controlling the _____.
   a. weave
   b. groove angle
   c. keyhole
   d. bevel angle

2. When making an open-root single V-groove weld horizontally, you are welding in the _____.
   a. 1G position
   b. 2G position
   c. 3G position
   d. 4G position

3. You're producing an open-root single V-groove weld in the 3G position. Unless the applicable code, WPS, or jobsite quality specifications forbid it, what bead style should you use?
   a. Weave beads
   b. Root face beads
   c. Stringer beads
   d. Keyhole beads

Module 29112 Review Questions

1. Which of the following is an advantage of welds that combine a fillet weld with a groove weld?
   a. The weld resists corrosion very well in damp environments.
   b. The weld maintains good strength but uses less weld metal.
   c. The weld doesn't have porosity problems.
   d. The weld is unusually ductile.

2. *Land* is another name for _____.
   a. bevel face
   b. bevel angle
   c. root face
   d. groove preparation

3. When depositing a stringer bead with an E6010/E6011 electrode, it's best to _____.
   a. use a whipping motion
   b. maintain a very long arc
   c. preheat the electrode first
   d. quench the finished bead

4. An open-root single V-groove weld usually has a bevel angle _____.
   a. less than 20 degrees
   b. between 20 and $22^1/_2$ degrees
   c. between 30 and $37^1/_2$ degrees
   d. larger than 45 degrees

5. When using low-hydrogen electrodes for filler passes, welders need to _____.
   a. use a special fume extraction system
   b. weld out of position
   c. use a work angle of 45 degrees
   d. be especially alert for inclusions

6. When making an open-root single V-groove weld, melting the root faces produces an opening called a(n) _____.
   a. keyhole
   b. port
   c. overlap
   d. root preparation

7. After producing an open-root single V-groove weld's root pass, you should _____.
   a. preheat the finished root pass
   b. inspect for excessive buildup or undercut
   c. leave a thin layer of slag in place
   d. inspect for ingrown weld toe or overpass

8. When producing an open-root single V-groove weld in the flat (1G) position, you should _____.
   a. always use keyhole beads
   b. never use stringer beads
   c. use either weave or stringer beads
   d. never use weave beads

9. When producing an open-root single V-groove weld in the horizontal (2G) position, you should use _____.
   a. reverse root pass beads
   b. stringer beads
   c. weave and stringer beads
   d. weave beads

10. When welding in the vertical (3G) position, what can you do to improve weld pool control?
    a. Switch to an E7020 electrode.
    b. Use a fan to disperse the shielding gases.
    c. Maintain a very long arc.
    d. Set the amperage at the lower end of the correct range.

Answers to odd-numbered Module Review Questions are found in *Appendix A*.

Answers to Section Review Questions

Answer	Section Reference	Objective
Section 1.0.0		
1. a	1.1.1	1a
2. c	1.2.8	1b
Section 2.0.0		
1. c	2.1.0	2a
2. b	2.2.2	2b
3. a	2.3.1	2c

User Update

Did you find an error? Submit a correction by visiting **https://www.nccer.org/olf** or by scanning the QR code using your mobile device.

APPENDIX A Review Question Answer Keys

MODULE 29101

Answer	Section Reference
1. d	1.1.1
3. b	1.2.3
5. c	2.2.1
7. d	3.2.2
9. c	3.3.1
11. a	3.4.1
13. a	3.6.0
15. b	3.7.0

MODULE 29102

Answer	Section Reference
1. b	1.1.0
3. b	1.2.1; *Figure 3*
5. c	2.1.3
7. c	2.2.1
9. c	2.3.1
11. c	2.4.5
13. c	3.1.2
15. d	3.3.1
17. d	4.1.2
19. c	4.2.1

MODULE 29103

Answer	Section Reference
1. d	1.1.0
3. b	1.1.1
5. c	1.2.1
7. a	2.1.1
9. c	2.1.2
11. d	2.3.0
13. a	3.1.2
15. b	3.3.2

MODULE 29104

Answer	Section Reference
1. b	1.0.0
3. d	1.2.1
5. b	2.1.1
7. d	2.3.1
9. b	3.1.2; 3.2.1
11. a	3.2.1
13. a	3.2.2
15. d	3.3.2

MODULE 29105

Answer	Section Reference
1. d	1.1.1
3. a	1.2.5
5. a	1.3.1
7. d	2.2.0
9. b	2.2.8
11. a	2.2.9
13. b	2.3.0
15. d	3.2.0

MODULE 29106

Answer	Section Reference
1. b	1.1.1
3. c	1.1.5
5. c	1.2.1
7. b	2.1.1
9. d	2.2.1
11. a	2.2.3
13. b	3.1.2
15. c	3.1.4
17. c	3.3.4
19. a	4.2.1; *Table 1*

MODULE 29107

Answer	Section Reference
1. b	1.0.0
3. c	1.1.3
5. b	1.1.4
7. a	1.1.5
9. b	1.2.5
11. b	2.1.3
13. d	2.2.1; *Table 1*
15. c	3.2.1

MODULE 29108

Answer	Section Reference
1. d	1.1.0
3. a	1.1.1
5. c	1.1.2
7. d	1.2.4
9. b	2.2.1

MODULE 29109

Answer	Section Reference
1. c	1.1.0
3. c	2.1.1
5. a	2.2.1
7. a	2.2.6
9. c	2.3.2

MODULE 29110

Answer	Section Reference
1. c	1.1.2
3. b	1.3.2
5. a	1.4.3
7. c	2.1.0
9. c	2.2.3

MODULE 29111

Answer	Section Reference
1. b	1.0.0
3. a	1.1.1
5. c	1.2.8
7. a	2.1.1
9. b	2.2.1

MODULE 29112

Answer	Section Reference
1. b	1.1.0
3. a	1.2.6
5. d	2.0.0
7. b	2.1.0
9. b	2.2.2

APPENDIX 29102A

Performance Accreditation Tasks

The American Welding Society (AWS) School Excelling through National Skills Standards Education (SENSE) program is a comprehensive set of minimum Standards and Guidelines for Welding Education programs. The following Performance Accreditation Tasks (PATs) are aligned with and designed around the SENSE program.

PATs correspond to and support the learning objectives in *AWS EG2.0, Guide for the Training and Qualification of Welding Personnel: Entry-Level Welder*.

Note that to satisfy all learning objectives in *AWS EG2.0*, the instructor must also use the PATs contained in the second level of the NCCER *Welding* curriculum.

PATs 1 and 2 correspond to *AWS EG2.0, Module 8 – Thermal Cutting Processes, Unit 1 – Manual OFC Principles*, Key Indicators 3, 4, 5, 6, and 7.

PATs provide specific acceptable criteria for performance and help to ensure a true competency-based welding program for students.

The following tasks test your competency with an oxyfuel cutting torch. Don't perform these tasks until your instructor directs you to do so. Practice the tasks until you're thoroughly familiar with the procedures.

After you complete each task, take it to your instructor for evaluation.

Performance Accreditation Tasks

Module 29102

SETTING UP, IGNITING, ADJUSTING, AND SHUTTING DOWN OXYFUEL EQUIPMENT

Using oxyfuel equipment that has been completely disassembled, demonstrate how to:

- Set up oxyfuel equipment
- Ignite and adjust the flame
  - Carburizing
  - Neutral
  - Oxidizing
- Shut off the torch
- Shut down the oxyfuel equipment

Criteria for Acceptance:

- Set up the oxyfuel equipment in the correct sequence ____________
- Demonstrate that there are no leaks ____________
- Properly adjust all three flames ____________
- Shut off the torch in the correct sequence ____________
- Shut down the oxyfuel equipment ____________

Performance Accreditation Tasks

Module 29102

CUTTING A SHAPE

Using a carbon steel plate, lay out and cut in the flat position the shape shown in the figure. If available, use a machine track cutter to straight cut the longer dimension.

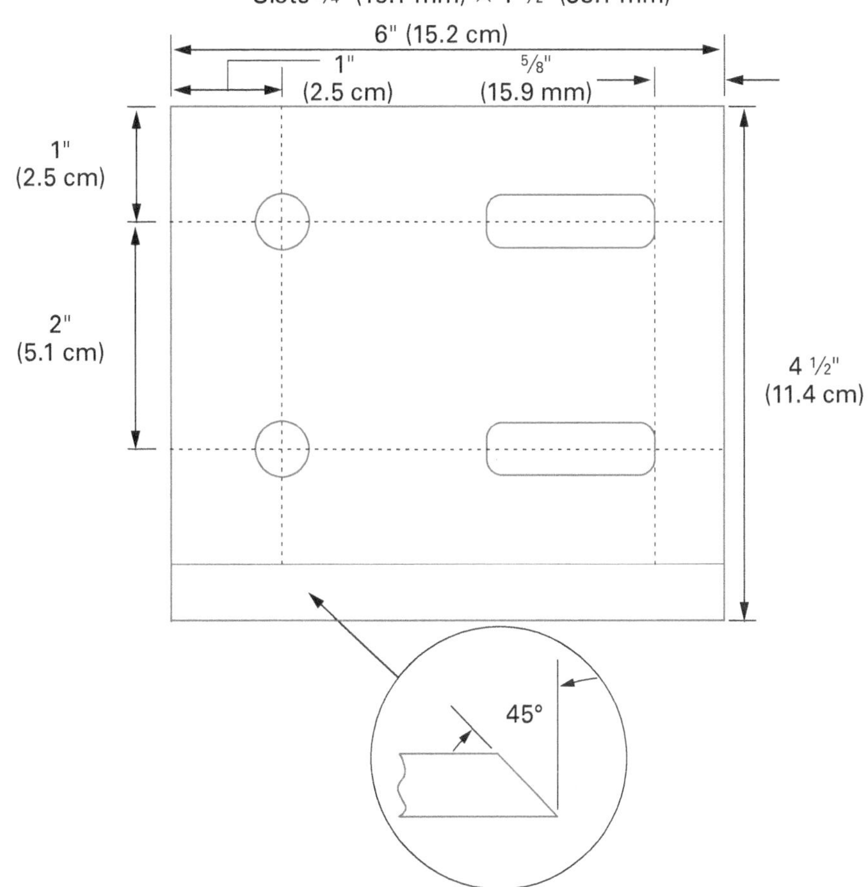

Criteria for Acceptance:

- Outside dimensions ±⅛" (3.2 mm) ____________
- Inside dimensions (holes and slots) ±⅛" (3.2 mm) ____________
- Square ±5° ____________
- Bevel ±5° ____________
- Minimal amount of dross sticking to plate which can be easily removed ____________
- Square kerf face with minimal notching not exceeding ⅛" (3.2 mm) deep ____________
- Confirm acceptable surface roughness with a surface roughness tool ____________

APPENDIX 29103A

Performance Accreditation Task

The American Welding Society (AWS) School Excelling through National Skills Standards Education (SENSE) program is a comprehensive set of minimum Standards and Guidelines for Welding Education programs. The following Performance Accreditation Task (PAT) is aligned with and designed around the SENSE program.

PATs correspond to and support the learning objectives in *AWS EG2.0, Guide for the Training and Qualification of Welding Personnel: Entry-Level Welder*.

Note that to satisfy all learning objectives in *AWS EG2.0*, the instructor must also use the PATs contained in the second level of the NCCER *Welding* curriculum.

PAT 1 corresponds to *AWS EG2.0, Module 8 – Thermal Cutting Processes, Unit 3 – Manual Plasma Arc Cutting (PAC)*, Key Indicators 3, 4, 5, and 6.

PATs provide specific acceptable criteria for performance and help to ensure a true competency-based welding program for students.

The following task tests your competency with a plasma arc cutting torch. Don't perform this task until your instructor directs you to do so. Practice the task until you're thoroughly familiar with the procedure.

After you complete the task, take it to your instructor for evaluation.

Performance Accreditation Task **Module 29103**

PLASMA ARC CUTTING

Using electrically conductive material, lay out and cut in the flat position the shape shown in the figure.

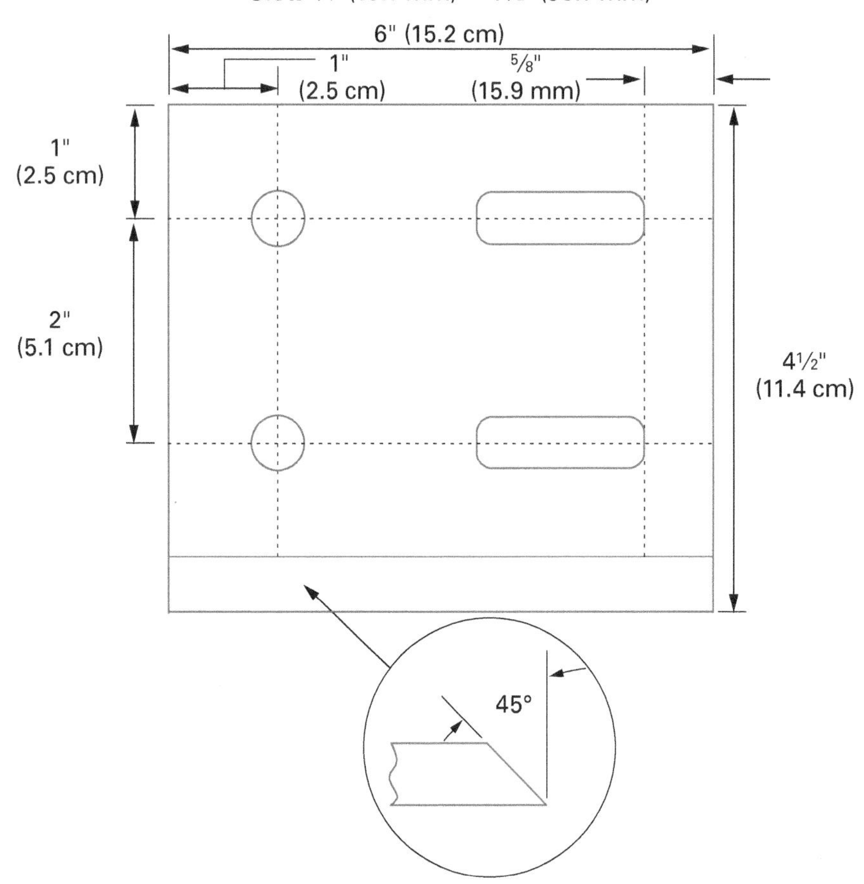

Criteria for Acceptance:

- Outside dimensions ±⅛" (3.2 mm) ____________
- Inside (holes and slots) dimensions ±⅛" (3.2 mm) ____________
- Square ±5° ____________
- Bevel ±5° ____________
- Minimal amount of dross sticking to plate which can be easily removed ____________
- Square kerf face with minimal notching not exceeding ±⅛" (3.2 mm) deep ____________

APPENDIX 29104A

Performance Accreditation Task

The American Welding Society (AWS) School Excelling through National Skills Standards Education (SENSE) program is a comprehensive set of minimum Standards and Guidelines for Welding Education programs. The following Performance Accreditation Task (PAT) is aligned with and designed around the SENSE program.

PATs correspond to and support the learning objectives in *AWS EG2.0, Guide for the Training and Qualification of Welding Personnel: Entry-Level Welder*.

Note that to satisfy all learning objectives in *AWS EG2.0*, the instructor must also use the PATs contained in the second level of the NCCER *Welding* curriculum.

PAT 1 corresponds to *AWS EG2.0, Module 8 – Thermal Cutting Processes, Unit 4 – Air-Carbon Arc Cutting*, Key Indicators 3, 4, and 5.

PATs provide specific acceptable criteria for performance and help to ensure a true competency-based welding program for students.

The following task tests your competency with A-CAC equipment. Don't perform this task until your instructor directs you to do so. Practice the task until you're thoroughly familiar with the procedure.

After you complete the task, take it to your instructor for evaluation.

Performance Accreditation Task

Module 29104

A-CAC WASHING AND GOUGING

Perform A-CAC Washing

Using any of the materials identified below, perform A-CAC washing to remove the portion identified by the instructor. Materials that can be used for this task include:

- Steel backing strip on a butt weld
- Excess buildup on the face of a weld
- Rivets or bolts in a plate
- Blocks, angles, clips, eyes, D-rings, or items welded to a plate

Criteria for Acceptance

- Material removed flush with the base metal surface ____________
- No notching in the surface of the base metal ____________

Perform A-CAC gouging

Using carbon steel plate ½" (~13 mm) thick or thicker, gouge a U-groove at least 8" (~20 cm) long, as shown in the figure, in the 1G and 2F positions.

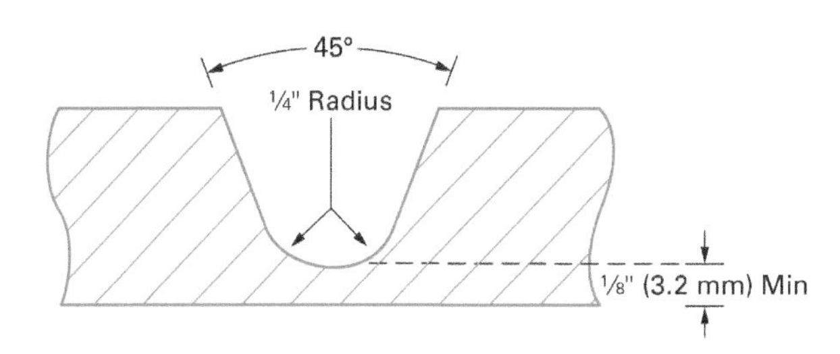

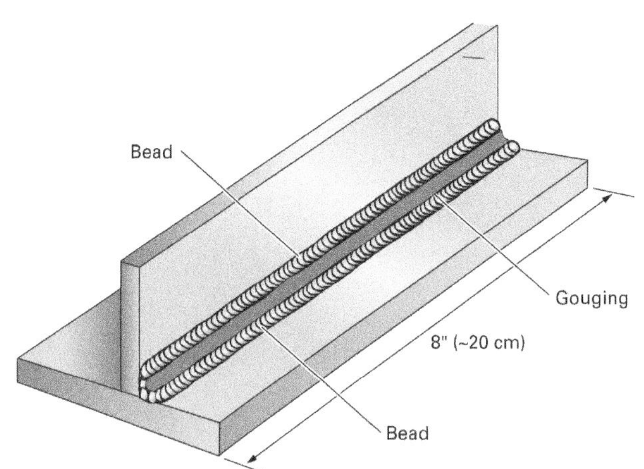

Criteria for Acceptance

- Groove width and depth are uniform ____________
- Groove walls are smooth and uniform ____________
- No dross within the groove ____________

APPENDIX 29105A

Performance Accreditation Tasks

The American Welding Society (AWS) School Excelling through National Skills Standards Education (SENSE) program is a comprehensive set of minimum Standards and Guidelines for Welding Education programs. The following Performance Accreditation Tasks (PATs) are aligned with and designed around the SENSE program.

PATs correspond to and support the learning objectives in *AWS EG2.0, Guide for the Training and Qualification of Welding Personnel: Entry-Level Welder*.

Note that to satisfy all learning objectives in *AWS EG2.0*, the instructor must also use the PATs contained in the second level of the NCCER *Welding* curriculum.

PAT 1 corresponds to no *AWS EG2.0* reference.

PAT 2 corresponds to *AWS EG2.0, Module 8 – Thermal Cutting Processes, Unit 1 – Manual OFC*, Key Indicators 3, 4, 5, and 7.

PATs provide specific acceptable criteria for performance and help to ensure a true competency-based welding program for students.

The following tasks test your competency preparing base metal. Don't perform these tasks until your instructor directs you to do so. Practice the tasks until you're thoroughly familiar with the procedures.

After you complete each task, take it to your instructor for evaluation.

Performance Accreditation Tasks

Module 29105

PREPARE PLATE JOINTS MECHANICALLY

Using a nibbler, cutter, or grinder, mechanically prepare the edge of a ¼" to ¾" (~6 mm to 20 mm) thick carbon steel plate with a bevel angle of 30 to 37½ degrees, at the discretion of the instructor.

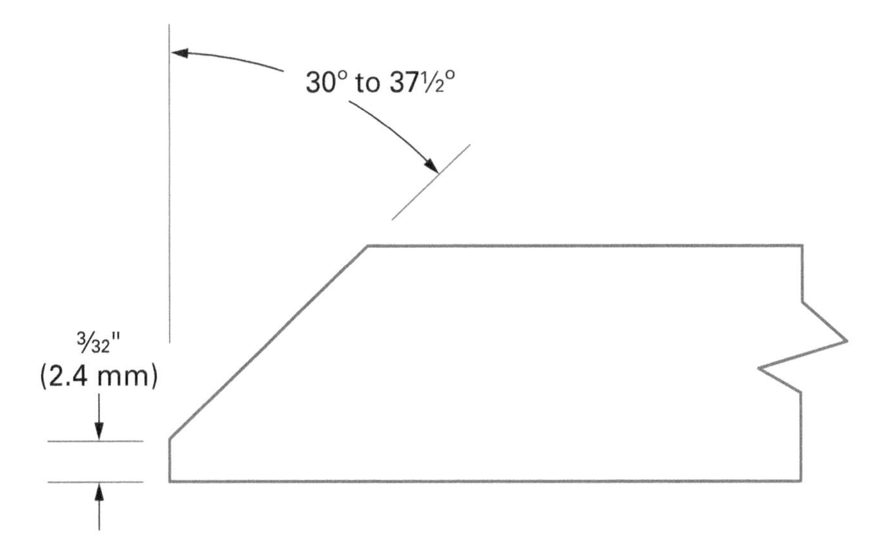

Criteria for Acceptance:

- Bevel angle ±2½° __________
- Bevel face smooth and uniform to 1⁄16" (1.6 mm) __________
- Root face ±1⁄32" (0.8 mm) __________

Performance Accreditation Tasks

Module 29105

PREPARE PLATE JOINTS THERMALLY

Using oxyfuel or plasma arc cutting equipment, thermally prepare the edge of a ¼" to ¾" (~6 mm to 20 mm) thick carbon steel plate with a bevel angle of 22½ degrees, at the discretion of the instructor.

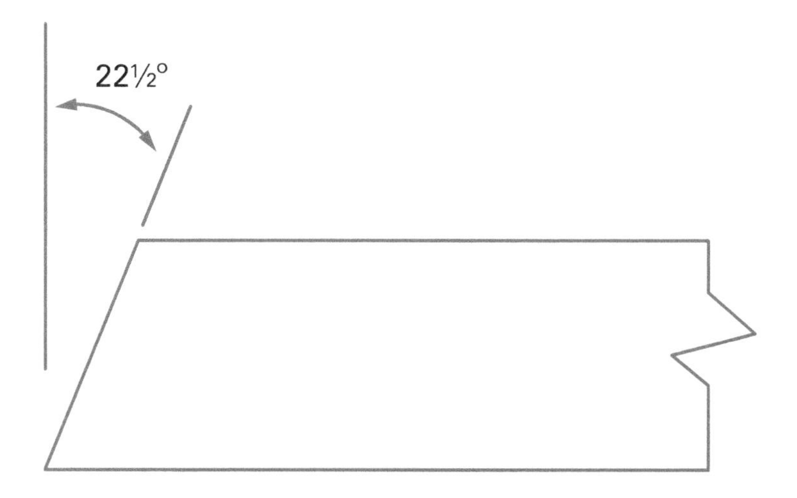

Criteria for Acceptance:

- Bevel angle ±2½° __________
- No dross __________
- Minimal notching, not exceeding 1⁄16" (1.6 mm) deep on the kerf face __________

APPENDIX 29106A

Performance Accreditation Task

The American Welding Society (AWS) School Excelling through National Skills Standards Education (SENSE) program is a comprehensive set of minimum Standards and Guidelines for Welding Education programs. The following Performance Accreditation Task (PAT) is aligned with and designed around the SENSE program.

PATs correspond to and support the learning objectives in AWS EG2.0, Guide for the Training and Qualification of Welding Personnel: Entry-Level Welder.

Note that to satisfy all learning objectives in AWS EG2.0, the instructor must also use the PATs contained in the second level of the NCCER Welding curriculum.

PAT 1 corresponds to AWS EG2.0, Module 9 – Welding Inspection and Testing, Key Indicators 1 and 2.

PATs provide specific acceptable criteria for performance and help to ensure a true competency-based welding program for students.

The following task tests your competency in performing a visual weld test inspection. Don't perform this task until your instructor directs you to do so. Practice the task until you're thoroughly familiar with the procedure.

After you complete the task, take it to your instructor for evaluation.

Performance Accreditation Task Module 29106

VISUAL WELD TEST INSPECTION

Obtain a completed fillet weld on ¼" (3.2 mm) plate a minimum of 6" (15.2 cm) long. The plate should also include a complete joint penetration groove weld a minimum of 6" (15.2 cm) long. Finally, the plate should have an unwelded but properly prepared edge. Perform a visual test inspection (VT) on the prepared edge and on each weld using the following guide and acceptance criteria.

Visual Test Inspection Report

Trainee Inspector ______________________ **Instructor** ______________________

Discontinuity Category	Acceptance Criteria	Base Metal Preparation
Examine any cut surfaces and edges of prepared base metal parts	Surface free from contaminants, obvious notches, roughness, unevenness, and complying with the drawing	Acceptable __________ or Reason for rejection __________

Discontinuity Category	Acceptance Criteria	¼" Fillet Weld	CJP Groove Weld
Crack Prohibition	Any crack shall be unacceptable regardless of size or location	Acceptable __________ or Reason for rejection __________	Acceptable __________ or Reason for rejection __________
Weld/Base Metal Fusion	Complete fusion shall exist between adjacent layers of weld metal and base metal	Acceptable __________ or Reason for rejection __________	Acceptable __________ or Reason for rejection __________
Crater Cross Section	All craters shall be filled to provide the specified weld size	Acceptable __________ or Reason for rejection __________	Acceptable __________ or Reason for rejection __________
Weld Profiles	Weld profiles shall be in accordance with Figure 7 and Figure 8 in Module 29106	Acceptable __________ or Reason for rejection __________	Acceptable __________ or Reason for rejection __________
Fillet Weld Size	The specified nominal fillet weld size tolerance is ± $\frac{1}{16}$"	Acceptable __________ or Reason for rejection __________	N/A
Undercut	Undercut shall not exceed $\frac{1}{32}$" for any accumulated length up to 2 inches	Acceptable __________ or Reason for rejection __________	Acceptable __________ or Reason for rejection __________
Porosity	The sum of visible porosity $\frac{1}{32}$" or greater shall not exceed ⅜" in any linear inch and shall not exceed ¾" in any linear 12 inches of weld	Acceptable __________ or Reason for rejection __________	Acceptable __________ or Reason for rejection __________
Complete Weld	Circle Accept or Reject for fillet and groove final inspection	Accept or Reject	Accept or Reject

Trainee Inspector's Signature ______________________ **Date** __________

Instructor's Verification: PASS/FAIL Signature ______________________

APPENDIX 29108A

SDS NO: 415884-B-EN-NA
REVISED: October 29, 2018
C5085
Page 1 of 6

B

SAFETY DATA SHEET

This Safety Data Sheet (SDS) is for welding consumables and related products and may be used to comply with OSHA's Hazard Communication standard, 29 CFR 1910.1200, and Superfund Amendments and Reauthorization Act (SARA) of 1986 Public Law 99-499 and Canadian Workplace Hazardous Materials Information System (WHMIS) per Health Canada administrative policy. The OSHA standard must be consulted for specific requirements. This Safety Data Sheet complies with ISO 11014-1 and ANSI Z400.1. This document is translated in several languages and is available on our website at www.hobartbrothers.com, from your sales representative or by calling customer service at 1 (937) 332-4000.

SECTION 1 – IDENTIFICATION

Manufacturer/Supplier
Name: HOBART BROTHERS LLC
Address: 101 TRADE SQUARE EAST, TROY, OH 45373
Canadian Address: 2570 NORTH TALBOT ROAD, OLDCASTLE, ONTARIO, CANADA N0R1L0
Website: www.hobartbrothers.com

Telephone No: +1 (937) 332-4000
Emergency No: +1 (800) 424-9300
Canada: +1 (519) 737-3053

Product Type: SHIELDED METAL ARC WELDING (SMAW) ELECTRODES

GROUP A: Product For: Carbon Steel
Trade Name: **DECKMASTER** 1139; **HOBART** 12, 14A, 24, 335A, 335C, 447A, 447C, 610; **PIPEMASTER** 60, PRO-60; **ROCKET** 24

GROUP B: Product For: Low Hydrogen Carbon Steel
Trade Name: **BOILERMAKER** 18; **HOBART** 16, 18AC, 418, 718MC, 718 TSR, 7018XLM; **SOFT-ARC** 7018-1

GROUP C: Product For: Low Hydrogen Low-Alloy Steel
Trade Name: **BOILERMAKER** 18A1, B2, B3; **HOBALLOY** 7018A1, 8018B2, 8018B2L, 8018B6, 8018B8, 8018C1, 8018C2, 8018C3, 8018G, 9015B91, 9018B3, 9018B3L, 9018M, E10018D2, 10018M, 10518M, 11018M, 12018M

GROUP D: Product For: Cellulosic Low Alloy Steel
Trade Name: **PIPEMASTER** 70, 80, 90

AWS Specification: AWS A5.1 & A5.5: E6010, E6011, E6012, E6013, E6022, E7014, E7024, E7024-1, E7016, E7018, E7018-1, E7018-M, E7018-A1, E7018-G, E8018-B2, E8018-B2L, E8018-B6, E8018-B8, E8018-C1, E8018-C2, E8018-C3, E8018-G, E9015-B9, E9018-B3, E9018-B3L, E9018-M, E10018-D2, E10018-M, E10518-M, E11018-M, E12018-M, E7010-P1, E8010-P1, E9010-G, E9010-P1

Recommended Use: SHIELDED METAL ARC WELDING (SMAW) ELECTRODES
Restrictions on Use: Use only as indicated for welding operations

SECTION 2 – IDENTIFICATION OF HAZARDS

HAZARD CLASSIFICATION – The products described in Section 1 are not classified as hazardous according to applicable GHS hazard classification criteria as required and defined in OSHA Hazard Communication Standard (29 CFR Part 1910.1200).

LABEL ELEMENTS: **Hazard Symbol** – No symbol required
Hazard Statement – Not applicable
Signal Word – No signal word required
Precautionary Statement – Not Applicable

HAZARDS NOT OTHERWISE CLASSIFIED

WARNING! - Avoid breathing welding fumes and gases, they may be dangerous to your health. Always use adequate ventilation. Always use appropriate personal protective equipment.
PRIMARY ROUTES OF ENTRY: Respiratory System, Eyes and/or Skin.
ELECTRIC SHOCK: Arc welding and associated processes can kill. See Section 8.
ARC RAYS: The welding arc can injure eyes and burn skin.
FUMES AND GASES: Can be dangerous to your health.

Welding fumes and gases cannot be classified simply. The composition and quantity of both are dependent upon the metal being welded, the process, procedures and electrodes used. Most fume ingredients are present as complex oxides and compounds and not as pure metals. When the electrode is consumed, the fume and gas decomposition products generated are different in percent and form from the ingredients listed in Section 3. Decomposition products of normal operation include those originating from the volatilization, reaction or oxidation, plus those from the base metal and coating, etc., of the materials shown in Section 3 of this Safety Data Sheet. Monitor for the component materials identified in the list in Section 3.

Fumes from the use of this product may contain complex oxides or compounds of the following elements and molecules: amorphous silica fume, calcium oxide, chromium, fluorspar or fluorides, manganese, nickel, silica, strontium and vanadium. Other reasonably expected constituents of the fume would also include complex oxides of iron, titanium, silicon and molybdenum. Gaseous reaction products may include carbon monoxide and carbon dioxide. Ozone and nitrogen oxides may be formed by the radiation from the arc. Other conditions which also influence the composition and quantity of the fumes and gases to which workers may be exposed include: coatings on the metal being welded (such as paint, plating or galvanizing), the number of welders and the volume of the work area, the quality and amount of ventilation, the position of the welder's head with respect to the fume plume, as well as the presence of contaminants in the atmosphere (such as chlorinated hydrocarbon vapors from cleaning and degreasing activities). One recommended way to determine the composition and quantity of fumes and gases to which workers are exposed is to take an air sample inside the welder's helmet if worn or in the worker's breathing zone. See ANSI/AWS F1.1 and F1.3, available from the "American Welding Society", 8669 NW 36 Street, # 130, Miami, Florida 33166-6672, Phone: 800-443-9353 or 305-443-9353.

SECTION 3 – COMPOSITION/INFORMATION ON INGREDIENTS

HAZARDOUS INGREDIENTS

IMPORTANT - This section covers the hazardous materials from which this product is manufactured. This data has been classified according to the criteria of the Globally Harmonized System of Classification and Labeling of Chemicals (GHS) as required and defined in OSHA Hazard Communication Standard (29 CFR Part 1910.1200). The fumes and gases produced during welding with normal use of this product are addressed in Section 8.

Source: Courtesy of Hobart Brothers LLC

SDS NO: 415884-B-EN-NA
REVISED: October 29, 2018
C5085
Page 2 of 6

B

SAFETY DATA SHEET

INGREDIENT	CAS NO.	EINECS[r]	GROUP AND %WEIGHT				GHS Classification(s)	GHS HAZARD STATEMENTS (See Section 16 for Complete Phrases)
			A	B	C	D		
ALUMINUM OXIDE	1344-28-1	215-691-6	<5	<1	<1	---	NONE	
CALCIUM CARBONATE	1317-65-3	215-279-6	---	2-10	2-10	---	NONE	
CELLULOSE	9004-34-6	232-674-9	<5	---	---	<5	NONE	
CHROMIUM (metal)	7440-47-3	231-157-5	---	---	<9	---	NONE	
FLUORSPAR	7789-75-5	232-188-7	---	1-12	4-15	---	NONE	
IRON	7439-89-6	231-096-4	70-90	60-80	60-90	70-90	NONE	
MAGNESIUM CARBONATE	546-93-0	208-915-9	<2	<5	<1	<1	NONE	
MANGANESE	7439-96-5	231-105-1	1-5	1-5	1-5	1-5	- Acute Tox. 4 (Inhalation)[(1)] - Acute Tox. 4 (Oral)[(1)] - STOT RE 1[(2)]	H332 H302 H372
MICA	12001-26-2	None	<5	---	---	---	NONE	
MOLYBDENUM	7439-98-7	231-107-2	---	---	<2	<1	- STOT RE 2[(2)] - Eye Irrit. 2[(3)] - STOT SE 3[(4)]	H373 H319 H335
NICKEL	7440-02-0	231-111-4	---	---	<5	<2	Powder/Element: - Carc. 2[(5)] - Skin Sens. 1[(6)] - STOT RE 1[(2)] - Aquatic Chronic 3	H351 H317 H372 H412
POTASSIUM SILICATE	1312-76-1	215-199-1	<2	<2	<2	<2	NONE	
SILICA	14808-60-7	238-878-4	<7	<8	<7	<7	- STOT RE 2[(2)] - Carc. 2[(5)] - Acute Tox. 4 (Inhalation)[(1)]	H373 H351 H332
(Amorphous Silica Fume)	69012-64-2	273-761-1	---	---	---	---	NONE	
SILICON	7440-21-3	231-130-8	<2	<2	<5	<2	NONE	
SODIUM SILICATE	1344-09-8	215-687-4	<2	<2	<2	<2	NONE	
STRONTIUM CARBONATE	1633-05-2	216-643-7	---	<2	<2	---	NONE	
TITANIUM DIOXIDE	13463-67-7	236-675-5	<14	<10	<5	<5	- Carc. 2[(5)]	H351
VANADIUM	7440-62-2	231-171-1	---	---	<1	---	- Acute Tox. 4 (Inhalation)[(6)] - STOT RE 2[(7)] - Eye Irrit. 1[(8)] - Aquatic Chronic 2	H332 H373 H318 H411
HEXAVALENT CHROMIUM [CHROMIUM (VI) TRIOXIDE] (Fume constituent)	1333-82-0	215-607-8	Varies	Varies	Varies	Varies	- Ox. Sol. 1[(7)] - Carc. 1A[(5)] - Muta. 1B[(8)] - Repr. Tox 2[(9)] - Acute Tox. 2 (Inhalation)[(1)] - Acute Tox. 3 (Skin & Oral)[(1)] - STOT RE 1[(2)] - Skin Corr. 1A[(10)] - Skin Sens. 1[(6)] - Resp. Sens. 1[(11)] - Aquatic Acute 1 - Aquatic Chronic 1	H271 H350 H340 H361f H330 H311, H301 H372 H314 H317 H334, H317 H400 H410

--- Dashes indicate the ingredient is not present within the group of products **Γ** – European Inventory of Existing Chemical Commercial Substance Number **(1)** Acute toxicity (Cat. 1, 2, 3 and 4) **(2)** Specific target organ toxicity (STOT) – repeated exposure (Cat. 1 and 2) **(3)** Serious eye damage/eye irritation (Cat. 1 and 2) **(4)** Specific target organ toxicity (STOT) – single exposure ((Cat. 1, 2) and Cat. 3 for narcotic effects and respiratory tract irritation, only) **(5)** Carcinogenicity (Cat. 1A, 1B and 2) **(6)** Skin sensitization (Cat. 1, Sub-cat. 1A and 1B) **(7)** Oxidizing solid (Cat. 1, 2 and 3) **(8)** Germ cell mutagenicity (Cat. 1A, 1B and 2) **(9)** Reproductive toxicity (Cat. 1A, 1B and 2) **(10)** Skin corrosion/irritation (Cat. 1, 1A, 1B, 1C and 2) **(11)** Respiratory sensitization (Cat. 1, Sub-cat. 1A and 1B)

Source: Courtesy of Hobart Brothers LLC

SDS NO: 415884-B-EN-NA
REVISED: October 29, 2018
C5085
Page 3 of 6

B

SAFETY DATA SHEET

SECTION 4 – FIRST AID MEASURES

INGESTION: Not an expected route of exposure. Do not eat, drink, or smoke while welding; wash hands thoroughly before performing these activities. If symptoms develop, seek medical attention at once.
INHALATION during welding: If breathing is difficult, provide fresh air and contact physician. If breathing has stopped, perform artificial respiration and obtain medical assistance at once.
SKIN CONTACT during welding: Remove contaminated clothing and wash the skin thoroughly with soap and water. If symptoms develop, seek medical attention at once.
EYE CONTACT during welding: Dust or fume from this product should be flushed from the eyes with copious amounts of clean, tepid water until victim is transported to an emergency medical facility. Do not allow victim to rub or keep eyes tightly closed. Obtain medical assistance at once.
Arc rays can injure eyes. If exposed to arc rays, move victim to dark room, remove contact lenses as necessary for treatment, cover eyes with a padded dressing and rest. Obtain medical assistance if symptoms persist.

Section 11 of this SDS covers the acute effects of overexposure to the various ingredients within the welding consumable. Section 8 of this SDS lists the exposure limits and covers methods for protecting yourself and your co-workers.

SECTION 5 – FIRE-FIGHTING MEASURES

Fire Hazards: Welding consumables applicable to this sheet as shipped are nonreactive, nonflammable, non-explosive and essentially nonhazardous until welded.

Welding arcs and sparks can ignite combustibles and flammable products. If there are flammable materials, including fuel or hydraulic lines, in the work area and the worker cannot move the work or the flammable material, a fire-resistant shield such as a piece of sheet metal or fire resistant blanket should be placed over the flammable material. If welding work is conducted within 35 feet or so of flammable materials, station a responsible person in the work zone to act as fire watcher to observe where sparks are flying and to grab an extinguisher or sound the alarm if needed.
Unused welding consumables may remain hot for a period of time after completion of a welding process. See American National Standard Institute (ANSI) Z49.1 for further general safety information on the use and handling of welding consumables and associated procedures.
Suitable Extinguishing Media: This product is essentially nonflammable until welded; therefore, use a suitable extinguishing agent for a surrounding fire.
Unsuitable Extinguishing Media: None known.

SECTION 6 - ACCIDENTAL RELEASE MEASURES

In the case of a release of solid welding consumable products, solid objects can be picked up and placed into a disposal container. If airborne dust and/or fume is present, use adequate engineering controls and, if needed, personal protection to prevent overexposure. Refer to recommendations in Section 8. Wear proper personal protective equipment while handling. Do not discard as general trash.

SECTION 7 - HANDLING AND STORAGE

HANDLING: No specific requirements in the form supplied. Handle with care to avoid cuts. Wear gloves when handling welding consumables. Avoid exposure to dust. Do not ingest. Some individuals can develop an allergic reaction to certain materials. Retain all warning and product labels.
STORAGE: Keep separate from acids and strong bases to prevent possible chemical reactions.

SECTION 8 - EXPOSURE CONTROLS AND PERSONAL PROTECTION

Read and understand the instructions and the labels on the packaging. Welding fumes do not have a specific OSHA PEL (Permissible Exposure Limit) or ACGIH TLV (Threshold Limit Value). The OSHA PEL for Particulates – Not Otherwise Regulated (PNOR) is 5 mg/m³ – Respirable Fraction, 15 mg/m³ – Total Dust. The ACGIH TLV for Particles – Not Otherwise Specified (PNOS) is 3 mg/m³ – Respirable Particles, 10 mg/m³ – Inhalable Particles. The individual complex compounds within the fume may have a lower OSHA PEL or ACGIH TLV than the OSHA PNOR and ACGIH PNOS. An Industrial Hygienist, the OSHA PELs for Air Contaminants (29 CFR 1910.1000), and the ACGIH TLVs should be consulted to determine the specific fume constituents present and their respective exposure limits. All exposure limits are in milligrams per cubic meter (mg/m³).

INGREDIENT	CAS	EINECS	OSHA PEL	ACGIH TLV
ALUMINUM###	7429-90-5	231-072-3	5 R*, 15 (Dust)	1 R* {A4} 5 (Welding fumes, as Al)
CALCIUM CARBONATE	1317-65-3	215-279-6	5 R*, 5 (as CaO)	3 R*, 2 (as CaO)
CELLULOSE	9004-34-6	232-674-9	5 R*	10 Dust
CHROMIUM#	7440-47-3	231-157-5	1 (Metal) 0.5 (Cr II & Cr III Cpnds) 0.005 (Cr VI Cpnds, Calif. OSHA PEL)	0.5 (Metal) 0.003 (Cr III Cpnds) {A4; DSEN; RSEN} 0.0002 (Cr VI Sol Cpnds) {A1; Skin; DSEN; RSEN} 0.0005 (Cr VI STEL)
FLUORSPAR	7789-75-5	232-188-7	2.5 (as F)	2.5 (as F) {A4}
IRON+	7439-89-6	231-096-4	5 R*	5 R* (Fe_2O_3) {A4}
IRON OXIDE	1309-37-1	215-168-2	10 (Oxide Fume)	5 R* (Fe_2O_3) {A4}
MAGNESIUM CARBONATE+	546-93-0	208-915-9	5 R*	3 R*
MANGANESE#	7439-96-5	231-105-1	5 CL ** (Fume) 1, 3 STEL*** ■	0.1 I* {A4} ♦ 0.02 R* ♦♦
MICA	12001-26-2	None	3 R*■	3 R*
MOLYBDENUM	7439-98-7	231-107-2	5 R*	3 R*; 10 I* (Ele and Insol) 0.5 R* (Sol Cpnds) {A3}
NICKEL#	7440-02-0	231-111-4	1 (Metal) 1 (Sol Cpnds) 1 (Insol Cpnds)	1.5 I* (Ele) {A5} 0.1 I* (Sol Cpnds) {A4} 0.2 I* (Insol Cpnds) {A1}
POTASSIUM SILICATE	1312-76-1	215-199-1	Not established	Not established
SILICA++	14808-60-7	238-878-4	0.05 R*	0.025 R* {A2}
(Amorphous Silica Fume)	69012-64-2	273-761-1	0.8	2 R*
SILICON+	7440-21-3	231-130-8	5 R*	3 R*
SODIUM SILICATE	1344-09-8	215-687-4	Not established	Not established
STRONTIUM CARBONATE+	1633-05-2	216-643-7	5 R*	3 R*
TITANIUM DIOXIDE	13463-67-7	236-675-5	15 (Dust)	10 {A4}
VANADIUM	7440-62-2	231-171-1	Not Established (Ele) 1 TWA, 3 STEL***■ (Ele) 0.1 CL** (Fume as V_2O_5) 0.5 R* CL** (Dust as V_2O_5)	Not Established (Ele) 0.05 R* (Dust as V_2O_5) {A4} 0.05 I* (Fume as V_2O_5) {A3}

R* - Respirable Fraction I* - Inhalable Fraction ** - Ceiling Limit *** - Short Term Exposure Limit + - As a nuisance particulate covered under "Particulates Not Otherwise Regulated" by OSHA or "Particulates Not Otherwise Classified" by ACGIH ++ - Crystalline silica is bound within the product as it exists in the package. However, research indicates silica is present in welding fume in the amorphous (noncrystalline) form #- Reportable material under Section 313 of SARA ## - Reportable material under Section 313 of SARA only in fibrous form ■ - NIOSH REL TWA and STEL ■■ - AIHA Ceiling Limit of 1 mg/m³ ♦ - Limit of 0.1 mg/m³ is for Inhalable Mn in 2015 by ACGIH ♦♦ - Limit of 0.02 mg/m³ is for Respirable Mn in 2015 by ACGIH Ele – Element Sol – Soluble Insol – Insoluble Inorg – Inorganic Cpnds – Compounds NOS – Not Otherwise Specified

Source: Courtesy of Hobart Brothers LLC

SAFETY DATA SHEET

{A1} - Confirmed Human Carcinogen per ACGIH {A2} - Suspected Human Carcinogen per ACGIH {A3} - Confirmed Animal Carcinogen with Unknown Relevance to Humans per ACGIH {A4} - Not Classifiable as a Human Carcinogen per ACGIH {A5} - Not Suspected as a Human Carcinogen per ACGIH (noncrystalline form) DSEN – Dermal Sensitization RSEN – Respiratory Sensitization EINECS – European Inventory of Existing Commercial Chemical Substances OSHA – U.S. Occupational Safety and Health Administration ACGIH – American Conference of Governmental Industrial Hygienists

VENTILATION: Use enough ventilation or local exhaust at the arc or both to keep the fumes and gases below the PEL/TLV in the worker's breathing zone and the general area. Train the welder to keep his head out of the fumes.
RESPIRATORY PROTECTION: Use NIOSH-approved or equivalent fume respirator or air supplied respirator when welding in confined space or where local exhaust or ventilation does not keep exposure below the regulatory limits.
EYE PROTECTION: Wear helmet or use face shield with filter lens for open arc welding processes. As a rule of thumb begin with Shade Number 14. Adjust if needed by selecting the next lighter and/or darker shade number. Provide protective screens and flash goggles, if necessary, to shield others from the weld arc flash.
PROTECTIVE CLOTHING: Wear hand, head and body protection which help to prevent injury from radiation, sparks and electrical shock. See ANSI Z49.1. At a minimum this includes welder's gloves and a protective face shield, and may include arm protectors, aprons, hats, shoulder protection as well as dark non-synthetic clothing. Train the welder not to touch live electrical parts and to insulate himself from work and ground.
PROCEDURE FOR CLEANUP OF SPILLS OR LEAKS: Not applicable
SPECIAL PRECAUTIONS (IMPORTANT): When welding with electrodes that require special ventilation (such as stainless or hardfacing, or other products which require special ventilation, or on lead- or cadmium-plated steel and other metals or coatings like galvanized steel, which produce hazardous fumes) maintain exposure below the PEL/TLV. Use industrial hygiene monitoring to ensure that your use of this material does not create exposures which exceed PEL/TLV. Always use exhaust ventilation. Refer to the following sources for important additional information: American National Standard Institute (ANSI) Z49.1; Safety in Welding and Cutting published by the American Welding Society, 8669 NW 36 Street, # 130, Miami, Florida 33166-6672, Phone: 800-443-9353 or 305-443-9353; and OSHA Publication 2206 (29 CFR 1910), U.S. Government Printing Office, Washington, DC 20402.

SECTION 9 – PHYSICAL AND CHEMICAL PROPERTIES

Welding consumables applicable to this sheet as shipped are nonreactive, nonflammable, non-explosive and essentially nonhazardous until welded.
PHYSICAL STATE: Solid
APPEARANCE: Cored Wire/Coated Rod
COLOR: Gray
ODOR: Not Applicable
ODOR THRESHOLD: Not Applicable
pH: Not Applicable
MELTING POINT/FREEZING POINT: Not Available
INITIAL BOILING POINT AND BOILING RANGE: Not Available
FLASH POINT: Not Available
EVAPORATION RATE: Not Applicable
FLAMMABILITY (SOLID, GAS): Not Available
UPPER/LOWER FLAMMABILITY OR EXPLOSIVE LIMITS: Not Available
VAPOR PRESSURE: Not Applicable
VAPOR DENSITY: Not Applicable
RELATIVE DENSITY: Not Available
SOLUBILITY(IES): Not Available
PARTITION COEFFICIENT: N-OCTANOL/WATER: Not Applicable
AUTO-IGNITION TEMPERATURE: Not Available
DECOMPOSITION TEMPERATURE: Not Available
VISCOSITY: Not Applicable

SECTION 10 – STABILITY AND REACTIVITY

GENERAL: Welding consumables applicable to this sheet are solid and nonvolatile as shipped. This product is only intended for use per the welding parameters it was designed for. When this product is used for welding, hazardous fumes may be created. Other factors to consider include the base metal, base metal preparation and base metal coatings. All of these factors can contribute to the fume and gases generated during welding. The amount of fume varies with the welding parameters.
STABILITY: This product is stable under normal conditions.
REACTIVITY: Contact with acids or strong bases may cause generation of gas.

SECTION 11 – TOXICOLOGICAL INFORMATION

SHORT-TERM (ACUTE) OVEREXPOSURE EFFECTS: Welding Fumes - May result in discomfort such as dizziness, nausea or dryness or irritation of nose, throat or eyes. **Aluminum Oxide** - Irritation of the respiratory system. **Calcium Oxide** - Dust or fumes may cause irritation of the respiratory system, skin and eyes. **Chromium** - Inhalation of fume with chromium (VI) compounds can cause irritation of the respiratory tract, lung damage and asthma-like symptoms. Swallowing chromium (VI) salts can cause severe injury or death. Dust on skin can form ulcers. Eyes may be burned by chromium (VI) compounds. Allergic reactions may occur in some people. **Fluorides** - Fluoride compounds evolved may cause skin and eye burns, pulmonary edema and bronchitis. **Iron, Iron Oxide** - None are known. Treat as nuisance dust or fume. **Magnesium, Magnesium Oxide** - Overexposure to the oxide may cause metal fume fever characterized by metallic taste, tightness of chest and fever. Symptoms may last 24 to 48 hours following overexposure. **Manganese** - Metal fume fever characterized by chills, fever, upset stomach, vomiting, irritation of the throat and aching of body. Recovery is generally complete within 48 hours of the overexposure. **Mica** - Dust may cause irritation of the respiratory system, skin and eyes. **Molybdenum** - Irritation of the eyes, nose and throat. **Nickel, Nickel Compounds** - Metallic taste, nausea, tightness in chest, metal fume fever, allergic reaction. **Potassium Silicate** - Dust or fumes may cause irritation of the respiratory system, skin and eyes. **Silica (Amorphous)** - Dust and fumes may cause irritation of the respiratory system, skin and eyes. **Sodium Silicate** - Dust or fumes may cause irritation of the respiratory system, skin and eyes. **Strontium Compounds** - Strontium salts are generally non-toxic and are normally present in the human body. In large oral doses, they may cause gastrointestinal disorders, vomiting and diarrhea. **Titanium Dioxide** - Irritation of respiratory system. **Vanadium** - Overexposure to the oxide causes green tongue, cough, metallic taste, throat irritation and eczema.

LONG-TERM (CHRONIC) OVEREXPOSURE EFFECTS: Welding Fumes - Excess levels may cause bronchial asthma, lung fibrosis, pneumoconiosis or "siderosis." Studies have concluded that there is sufficient evidence for ocular melanoma in welders. **Aluminum Oxide** - Pulmonary fibrosis and emphysema. **Calcium Oxide** - Prolonged overexposure may cause ulceration of the skin and perforation of the nasal septum, dermatitis and pneumonia. **Chromium** - Ulceration and perforation of nasal septum. Respiratory irritation may occur with symptoms resembling asthma. Studies have shown that chromate production workers exposed to hexavalent chromium compounds have an excess of lung cancers. Chromium (VI) compounds are more readily absorbed through the skin than chromium (III) compounds. Good practice requires the reduction of employee exposure to chromium (III) and (VI) compounds. **Fluorides** - Serious bone erosion (Osteoporosis) and mottling of teeth. **Iron, Iron Oxide Fumes** - Can cause siderosis (deposits of iron in lungs) which some researchers believe may affect pulmonary function. Lungs will clear in time when exposure to iron and its compounds ceases. Iron and magnetite (Fe_3O_4) are not regarded as fibrogenic materials. **Magnesium, Magnesium Oxide** - No adverse long term health effects have been reported in the literature. **Manganese** - Long-term overexposure to manganese compounds may affect the central nervous system. Symptoms may be similar to Parkinson's disease and can include slowness, changes in handwriting, gait impairment, muscle spasms and cramps and less commonly, tremor and behavioral changes. Employees who are overexposed to manganese compounds should be seen by a physician for early detection of neurologic problems. Overexposure to manganese and manganese compounds above safe exposure limits can cause irreversible damage to the central nervous system, including the brain, symptoms of which may include slurred speech, lethargy, tremor, muscular weakness, psychological disturbances and spastic gait. **Mica** - Prolonged overexposure may cause scarring of the lungs and pneumoconiosis characterized by cough, shortness of breath, weakness and weight loss. **Molybdenum** - Prolonged overexposure may result in loss of appetite, weight loss, loss of muscle coordination, difficulty in breathing and anemia. **Nickel, Nickel Compounds** - Lung fibrosis or pneumoconiosis. Studies of nickel refinery workers indicated a higher incidence of lung and nasal cancers. **Potassium Silicate** - Prolonged overexposure may cause ulceration of the skin and perforation of the nasal septum, dermatitis and pneumonia. **Silica (Amorphous)** - Research indicates that silica is present in welding fume in the amorphous form. Long term overexposure may cause

Source: Courtesy of Hobart Brothers LLC

SAFETY DATA SHEET

pneumoconiosis. Noncrystalline forms of silica (amorphous silica) are considered to have little fibrotic potential. **Sodium Silicate** - Prolonged overexposure may cause ulceration of the skin and perforation of the nasal septum, dermatitis and pneumonia. **Strontium Compounds** - Strontium at high doses is known to concentrate in bone. Major signs of chronic toxicity, which involve the skeleton, have been labeled as "strontium rickets". **Titanium Dioxide** - Pulmonary irritation and slight fibrosis. **Vanadium** - Prolonged overexposure to vanadium pentoxide can cause nasal catarrh or nose bleeds and chronic respiratory problems.

MEDICAL CONDITIONS AGGRAVATED BY EXPOSURE: Persons with pre-existing impaired lung functions (asthma-like conditions). Persons with a pacemaker should not go near welding and cutting operations until they have consulted their doctor and obtained information from the manufacturer of the device. Respirators are to be worn only after being medically cleared by your company-designated physician.

EMERGENCY AND FIRST AID PROCEDURES: Call for medical aid. Employ first aid techniques recommended by the American Red Cross. If irritation or flash burns develop after exposure, consult a physician.

CARCINOGENICITY: Chromium VI compounds, nickel compounds and silica (crystalline quartz) are classified as IARC Group 1 and NTP Group K carcinogens. Titanium dioxide, nickel metal/alloys, vanadium (V_2O_5) and welding fumes are classified as IARC Group 2B carcinogens.

CALIFORNIA PROPOSITION 65:
⚠**WARNING:** These products can expose you to chemicals, including titanium dioxide and/or chromium and/or nickel, which are known to the State of California to cause cancer, and to carbon monoxide, which is known to the State of California to cause birth defects or other reproductive harm. For more information, go to www.P65Warnings.ca.gov.

INGREDIENT	CAS	IARC^E	NTP^Z	OSHA^H	65$^\Theta$
ALUMINUM OXIDE	1344-28-1	---	---	---	---
CALCIUM CARBONATE	1317-65-3	---	---	---	---
CELLULOSE	9004-34-6	---	---	---	---
CHROMIUM	7440-47-3	3^{Σ}, $1^{\Sigma\Sigma}$	$K^{\Sigma\Sigma}$	$X^{\Sigma\Sigma}$	$X^{\Sigma\Sigma}$
FLUORSPAR	7789-75-5	---	---	---	---
IRON	7439-89-6	---	---	---	---
IRON OXIDE	1309-37-1	3	---	---	---
MAGNESIUM CARBONATE	546-93-0	---	---	---	---
MANGANESE	7439-96-5	---	---	---	---
MICA	12001-26-2	---	---	---	---
MOLYBDENUM	7439-98-7	---	---	---	---
NICKEL	7440-02-0	$2B^{\beta}$, $1^{\beta\beta}$	S^{β}, $K^{\beta\beta}$	---	X^{β}, $X^{\beta\beta}$
POTASSIUM SILICATE	1312-76-1	---	---	---	---
SILICA	14808-60-7	1^{Ψ}	K	---	X
(Amorphous Silica fume)	69012-64-2	3	--	---	---
SILICON	7440-21-3	---	---	---	---
SODIUM SILICATE	1344-09-8	---	---	---	---
STRONTIUM CARBONATE	1633-05-2	---	---	---	---
TITANIUM DIOXIDE	13463-67-7	2B	---	---	X
Ultraviolet Radiation	---	1	---	---	---
VANADIUM	7440-62-2	$2B^{\Omega}$	---	---	X^{Ω}
Welding Fumes	--	1	--	---	--

E – International Agency for Research on Cancer (1 – Carcinogenic to Humans, 2A – Probably Carcinogenic to Humans, 2B – Possibly Carcinogenic to Humans, 3 – Not Classifiable as to its Carcinogenicity to Humans, 4 --- Probably Not Carcinogenic to Humans) Z – US National Toxicology Program (K – Known Carcinogen, S – Suspected Carcinogen) H – OSHA Designated Carcinogen List Θ – California Proposition 65 (X – On Proposition 65 list) Σ – Chromium Metal and Chromium III Compounds ΣΣ – Chromium VI β – Nickel metal and alloys ββ -- Nickel compounds Ψ – Silica Crystalline α-Quartz Ω – Vanadium pentoxide --- Dashes indicate the ingredient is not listed with the IARC, NTP, OSHA or Proposition 65

SECTION 12 – ECOLOGICAL INFORMATION

Welding processes can release fumes directly to the environment. Welding wire can degrade if left outside and unprotected. Residues from welding consumables and processes could degrade and accumulate in the soil and groundwater.

SECTION 13 – DISPOSAL CONSIDERATIONS

Use recycling procedures if available. Discard any product, residue, packaging, disposable container or liner in an environmentally acceptable manner, in full compliance with federal, state and local regulations.

SECTION 14 – TRANSPORT INFORMATION

No international regulations or restrictions are applicable. No special precautions are necessary.

SECTION 15 – REGULATORY INFORMATION

Read and understand the manufacturer's instructions, your employer's safety practices and the health and safety instructions on the label and the safety data sheet. Observe all local and federal rules and regulations. Take all necessary precautions to protect yourself and others.
United States EPA Toxic Substance Control Act: All constituents of these products are on the TSCA inventory list or are excluded from listing.
CERCLA/SARA TITLE III: Reportable Quantities (RQs) and/or Threshold Planning Quantities (TPQs):

Ingredient name	**RQ(lb)**	**TPQ (lb)**
Products on this SDS are a solid solution in the form of a solid article.	--	--

Spills or releases resulting in the loss of any ingredient at or above its RQ require immediate notification to the National Response Center and to your Local Emergency Planning Committee.
Section 311 Hazard Class
As shipped: Immediate In use: Immediate delayed
EPCRA/SARA TITLE III 313 TOXIC CHEMICALS: The following metallic components are listed as SARA 313 "Toxic Chemicals" and potentially subject to annual SARA 312 reporting: Aluminum Oxide (fibrous forms),Chromium, Manganese, Nickel and Vanadium. See Section 3 for weight percentage.
CANADIAN WHMIS CLASSIFICATION: Class D; Division 2, Subdivision A
CANADIAN CONTROLLED PRODUCTS REGULATION: This product has been classified in accordance with the hazard criteria of the CPR and the SDS contains all of the information required by the CPR.
CANADIAN ENVIRONMENTAL PROTECTION ACT (CEPA): All constituents of these products are on the Domestic Substance List (DSL).

Source: Courtesy of Hobart Brothers LLC

SDS NO: 415884-B-EN-NA
REVISED: October 29, 2018
C5085
Page 6 of 6

B

SAFETY DATA SHEET

SECTION 16 – OTHER INFORMATION

The following Hazard Statements, provided in the OSHA Hazard Communication Standard (29 CFR Part 1910.1200) correspond to the columns labeled 'GHS Hazard Statements' within Section 3 of this safety data sheet. Take appropriate precautions and protective measures to eliminate or limit the associated hazard.

H271: May cause fire or explosion; strong oxidizer
H301: Toxic if swallowed
H302: Harmful if swallowed
H311: Toxic in contact with skin
H314: Causes severe skin burns and eye damage
H317: May cause an allergic skin reaction
H318: Causes serious eye damage
H319: Causes serious eye irritation
H330: Fatal if inhaled
H332: Harmful if inhaled
H334: May cause allergy or asthma symptoms or breathing difficulties if inhaled
H335: May cause respiratory irritation
H340: May cause genetic defects
H350: May cause cancer
H351: Suspected of causing cancer
H361f: Suspected of damaging fertility or the unborn child
H372: Causes damage to organs through prolonged or repeated exposure
H373: May cause damage to organs through prolonged or repeated exposure
H400: Very toxic to aquatic life.
H410: Very toxic to aquatic life with long lasting effects
H411: Toxic to aquatic life with long lasting effects
H412: Harmful to aquatic life with long lasting effects.

For additional information please refer to the following sources:

USA: **American National Standard Institute (ANSI) Z49.1** "Safety in Welding and Cutting", **ANSI/American Welding Society (AWS) F1.5** "Methods for Sampling and Analyzing Gases from Welding and Allied Processes", **ANSI/AWS F1.1** "Method for Sampling Airborne Particles Generated by Welding and Allied Processes", **AWSF3.2M/F3.2** "Ventilation Guide for Weld Fume", American Welding Society, 8669 NW 36 Street, # 130, Miami, Florida 33166-6672, Phone: 800-443-9353 or 305-443-9353. Safety and Health Fact Sheets available from AWS at www.aws.org.

OSHA Publication 2206 (29 C.F.R. 1910), U.S. Government Printing Office, Superintendent of Documents, P.O. Box 371954, Pittsburgh, PA 15250-7954.

Threshold Limit Values and Biological Exposure Indices, American Conference of Governmental Industrial Hygienists (ACGIH), 6500 Glenway Ave., Cincinnati, Ohio 45211, USA.

NFPA 51B "Standard for Fire Prevention During Welding, Cutting and Other Hot Work" published by the National Fire Protection Association, 1 Batterymarch Park, Quincy, MA 02169.

Canada: **CSA Standard CAN/CSA-W117.2-01** "Safety in Welding, Cutting and Allied Processes".

Hobart Brothers LLC strongly recommends the users of this product study this SDS, the product label information and become aware of all hazards associated with welding. Hobart Brothers LLC believes this data to be accurate and to reflect qualified expert opinion regarding current research. However, Hobart Brothers LLC cannot make any expressed or implied warranty as to this information.

Source: Courtesy of Hobart Brothers LLC

APPENDIX 29109A

Performance Accreditation Tasks

The American Welding Society (AWS) School Excelling through National Skills Standards Education (SENSE) program is a comprehensive set of minimum Standards and Guidelines for Welding Education programs. The following Performance Accreditation Tasks (PATs) are aligned with and designed around the SENSE program.

PATs correspond to and support the learning objectives in *AWS EG2.0, Guide for the Training and Qualification of Welding Personnel: Entry-Level Welder*.

Note that to satisfy all learning objectives in *AWS EG2.0*, the instructor must also use the PATs contained in the second level of the NCCER *Welding* curriculum.

PATs 1 and 2 correspond to *AWS EG2.0, Module 4 – Shielded Metal Arc Welding*, Key Indicators 3 and 4.

PATs 3 through 10 correspond to *AWS EG2.0, Module 4 – Shielded Metal Arc Welding*, Key Indicators 3, 4, and 5.

PATs provide specific acceptable criteria for performance and help to ensure a true competency-based welding program for students.

The following tasks test your competency running beads and producing fillet welds with SMAW equipment and techniques. Don't perform these tasks until your instructor directs you to do so. Practice the tasks until you're thoroughly familiar with the procedures.

After you complete each task, take it to your instructor for evaluation.

Performance Accreditation Tasks

Module 29109

Surfacing Welds with E6010/E6011 Electrodes in the Flat Position

Using $\frac{3}{32}$" to $\frac{5}{32}$" (2.4 mm to 4.0 mm) E6010/E6011 electrodes, build up a pad with surfacing welds on carbon steel plate in the flat position as indicated.

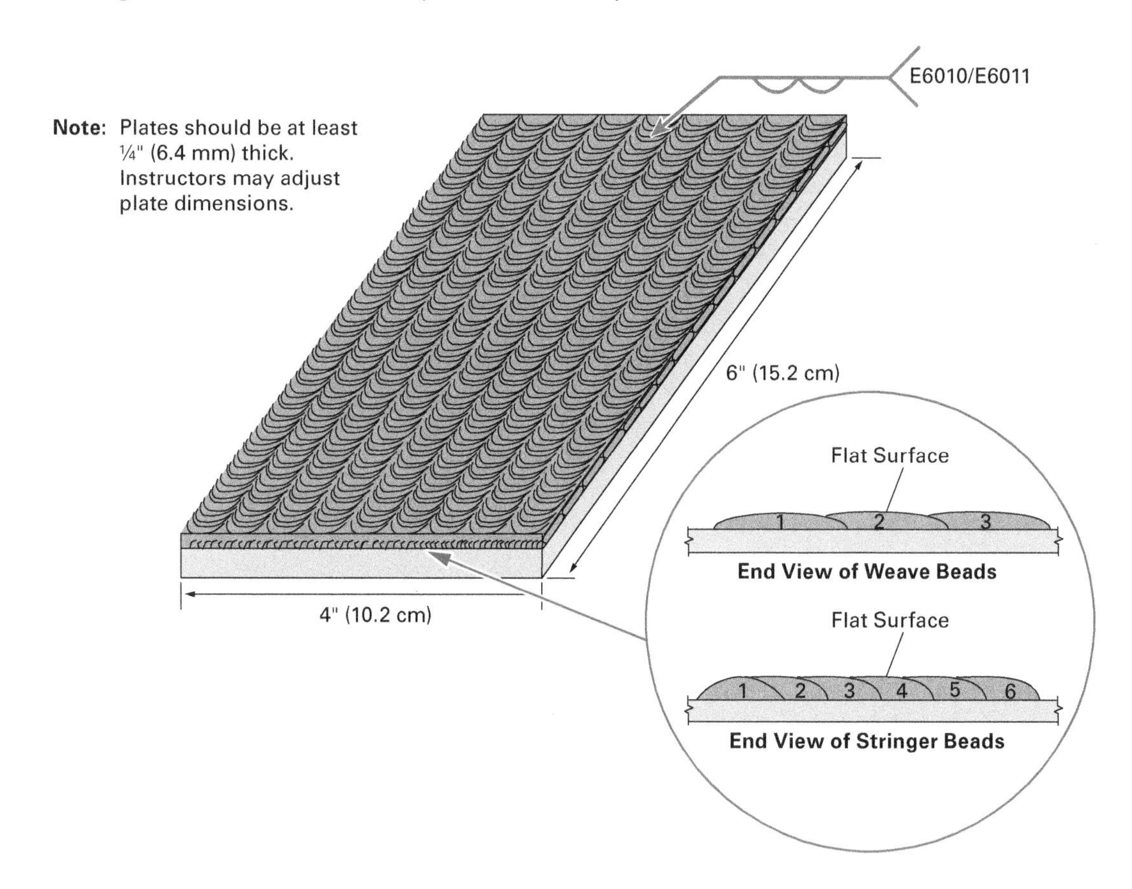

Criteria for Acceptance:

- No arc strikes outside the weld area ________
- Weld beads straight to within $\frac{1}{8}$" (3.2 mm) ________
- Uniform rippled appearance on the bead face with no valley between the beads and acceptable tie-in ________
- Craters and restarts filled to the full cross section of the weld ________
- Face of the pad flat to within $\frac{1}{8}$" (3.2 mm) ________
- Smooth transition with complete fusion at the toes of the weld ________
- No pores larger than $\frac{3}{32}$" (2.4 mm) ________
- No undercut greater than $\frac{1}{32}$" (0.8 mm) deep or 10% of the base metal thickness, whichever is less ________
- No overlap ________
- No inclusions ________
- No cracks ________

Performance Accreditation Tasks

Module 29109

Surfacing Welds with E7018 Electrodes in the Flat Position

Using 3⁄32" to 5⁄32" (2.4 mm to 4.0 mm) E7018 electrodes, build up a pad with surfacing welds on carbon steel plate in the flat position as indicated.

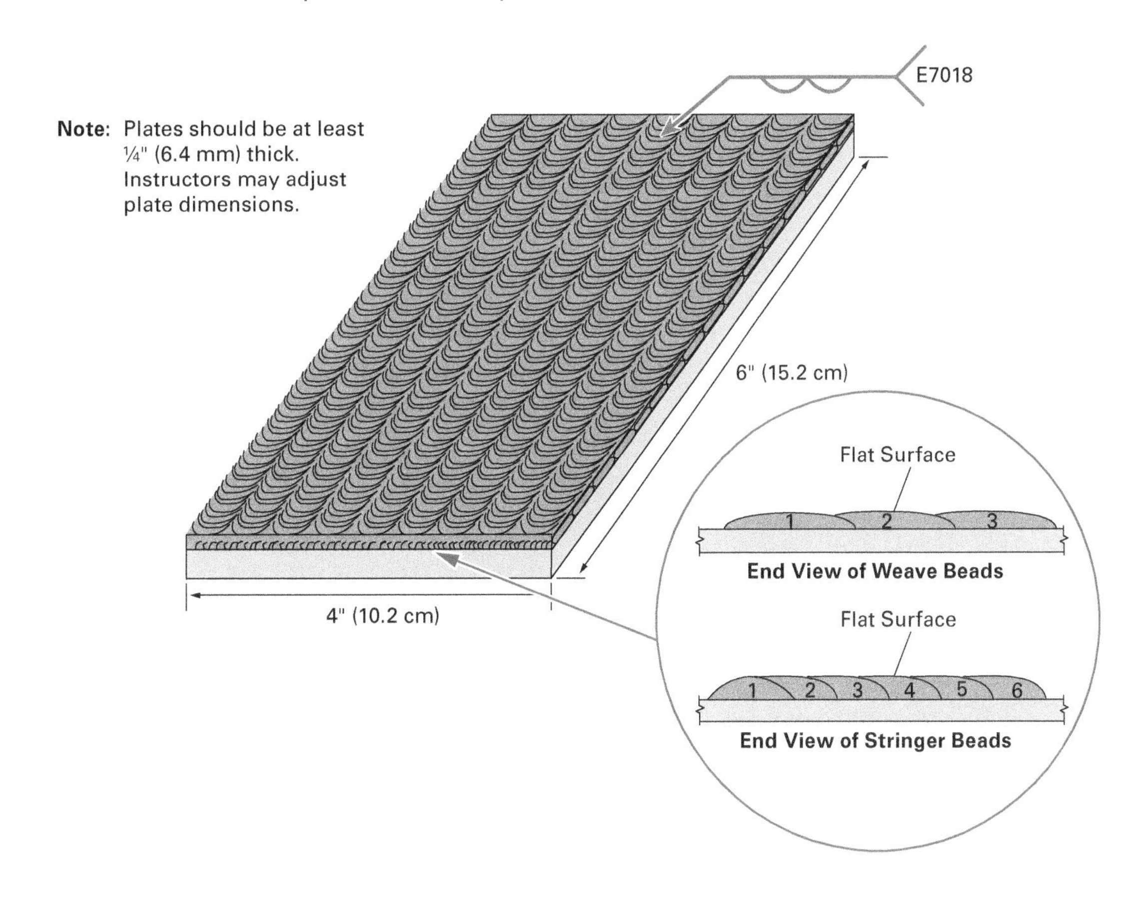

Criteria for Acceptance:

- No arc strikes outside the weld area ______
- Weld beads straight to within 1⁄8" (3.2 mm) ______
- Uniform rippled appearance on the bead face with no valley between the beads and acceptable tie-in ______
- Craters and restarts filled to the full cross section of the weld ______
- Face of the pad flat to within 1⁄8" (3.2 mm) ______
- Smooth transition with complete fusion at the toes of the weld ______
- No pores larger than 3⁄32" (2.4 mm) ______
- No undercut greater than 1⁄32" (0.8 mm) deep or 10% of the base metal thickness, whichever is less ______
- No overlap ______
- No inclusions ______
- No cracks ______

Performance Accreditation Tasks

Module 29109

Fillet Weld with E6010/E6011 Electrodes in the Flat (1F) Position

Using $^3/_{32}$" to $^5/_{32}$" (2.4 mm to 4.0 mm) E6010/E6011 electrodes, make a fillet weld on carbon steel plate in the flat position as indicated.

Note: Plates should be at least ¼" (6.4 mm) thick.

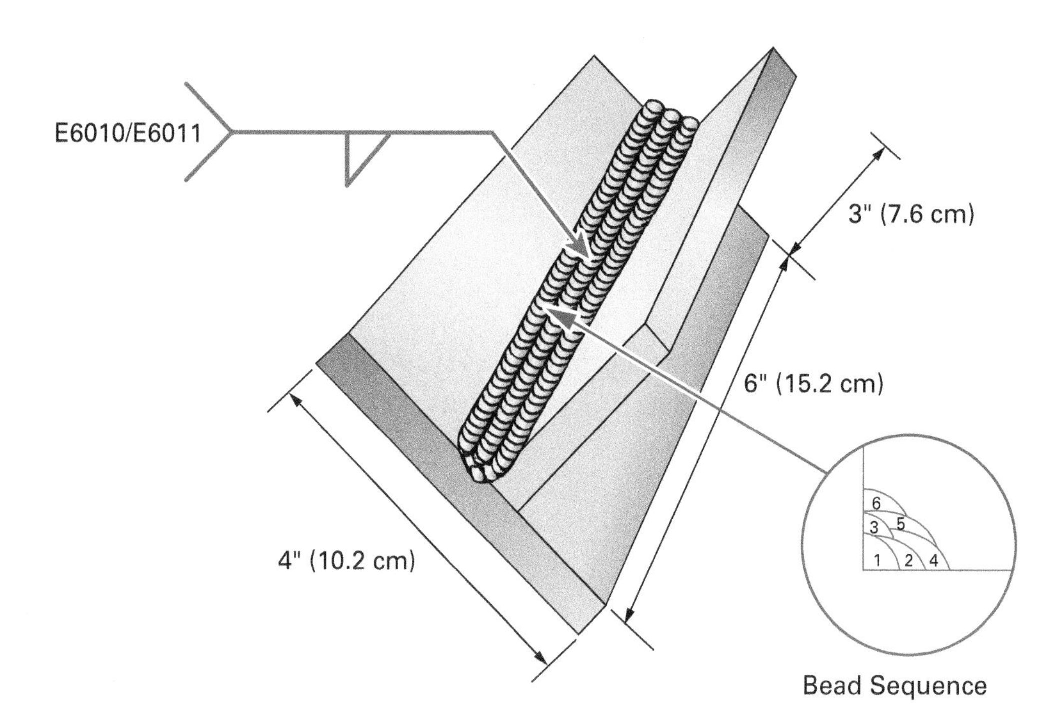

Criteria for Acceptance:

- No arc strikes outside the weld area ________
- Uniform rippled appearance on the bead face with no valley between the beads and acceptable tie-in ________
- Craters and restarts filled to the full cross section of the weld ________
- Uniform weld size ±$^1/_{16}$" (1.6 mm) ________
- Acceptable weld profile in accordance with Module 29109 *Figure 3* ________
- Smooth transition with complete fusion at the toes of the weld ________
- No pores larger than $^3/_{32}$" (2.4 mm) ________
- No undercut greater than $^1/_{32}$" (0.8 mm) deep or 10% of the base metal thickness, whichever is less ________
- No overlap ________
- No inclusions ________
- No cracks ________

Fillet Weld with E7018 Electrodes in the Flat (1F) Position

Using 3/32" to 5/32" (2.4 mm to 4.0 mm) E7018 electrodes, make a fillet weld on carbon steel plate in the flat position as indicated.

Note: Plates should be at least ¼" (6.4 mm) thick.

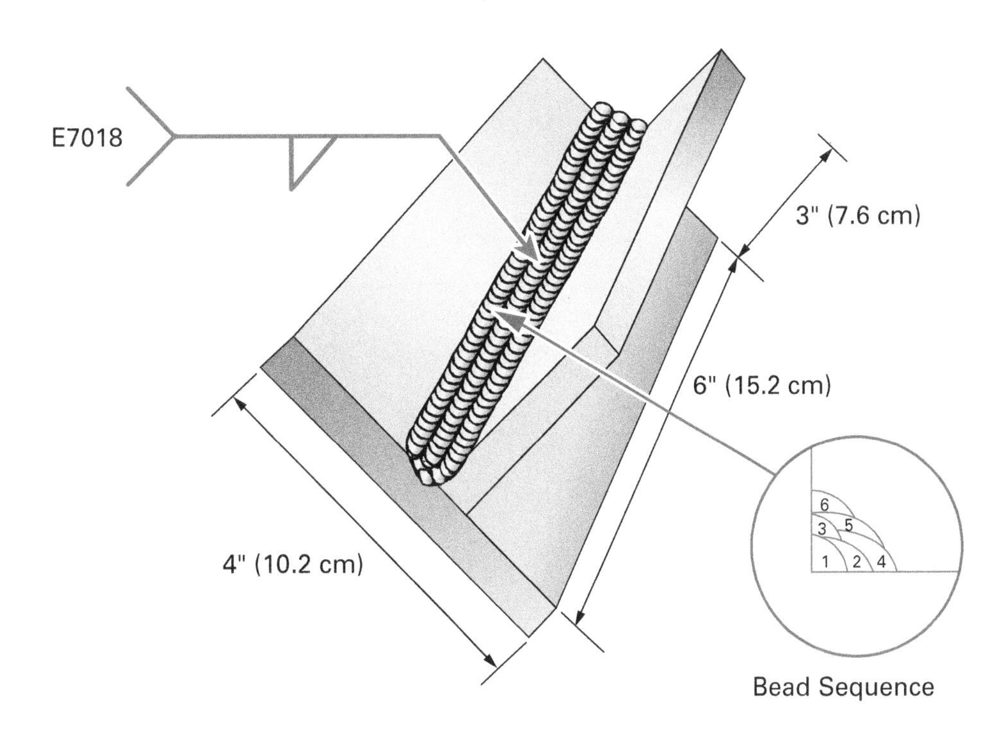

Criteria for Acceptance:

- No arc strikes outside the weld area ____________
- Uniform rippled appearance on the bead face with no valley between the beads and acceptable tie-in ____________
- Craters and restarts filled to the full cross section of the weld ____________
- Uniform weld size ±1/16" (1.6 mm) ____________
- Acceptable weld profile in accordance with Module 29109 *Figure 3* ____________
- Smooth transition with complete fusion at the toes of the weld ____________
- No pores larger than 3/32" (2.4 mm) ____________
- No undercut greater than 1/32" (0.8 mm) deep or 10% of the base metal thickness, whichever is less ____________
- No overlap ____________
- No inclusions ____________
- No cracks ____________

Performance Accreditation Tasks — Module 29109

Fillet Weld with E6010/E6011 Electrodes in the Horizontal (2F) Position

Using $^3/_{32}$" to $^5/_{32}$" (2.4 mm to 4.0 mm) E6010/E6011 electrodes, make a fillet weld on carbon steel plate in the horizontal position as indicated.

Note: Plates should be at least ¼" (6.4 mm) thick.

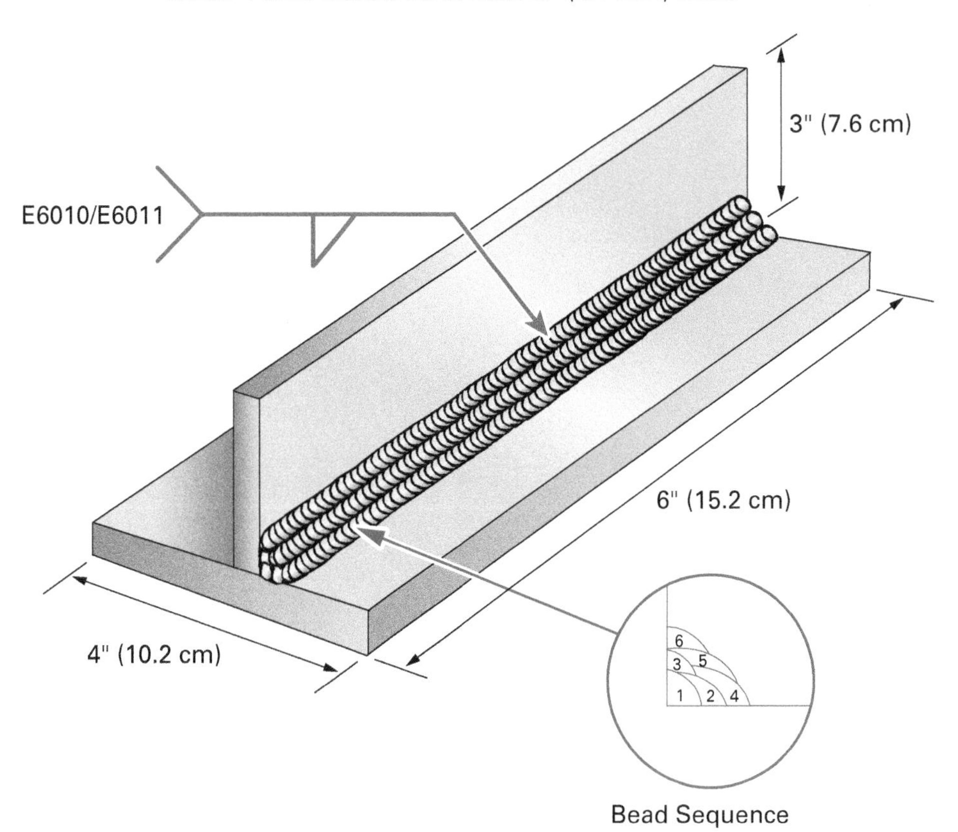

Criteria for Acceptance:

- No arc strikes outside the weld area ______
- Uniform rippled appearance on the bead face with no valley between the beads and acceptable tie-in ______
- Craters and restarts filled to the full cross section of the weld ______
- Uniform weld size $\pm ^1/_{16}$" (1.6 mm) ______
- Acceptable weld profile in accordance with Module 29109 *Figure 3* ______
- Smooth transition with complete fusion at the toes of the weld ______
- No pores larger than $^3/_{32}$" (2.4 mm) ______
- No undercut greater than $^1/_{32}$" (0.8 mm) deep or 10% of the base metal thickness, whichever is less ______
- No overlap ______
- No inclusions ______
- No cracks ______

Performance Accreditation Tasks **Module 29109**

Fillet Weld with E7018 Electrodes in the Horizontal (2F) Position

Using $^{3}/_{32}$" to $^{5}/_{32}$" (2.4 mm to 4.0 mm) E7018 electrodes, make a fillet weld on carbon steel plate in the horizontal position as indicated.

Note: Plates should be at least ¼" (6.4 mm) thick.

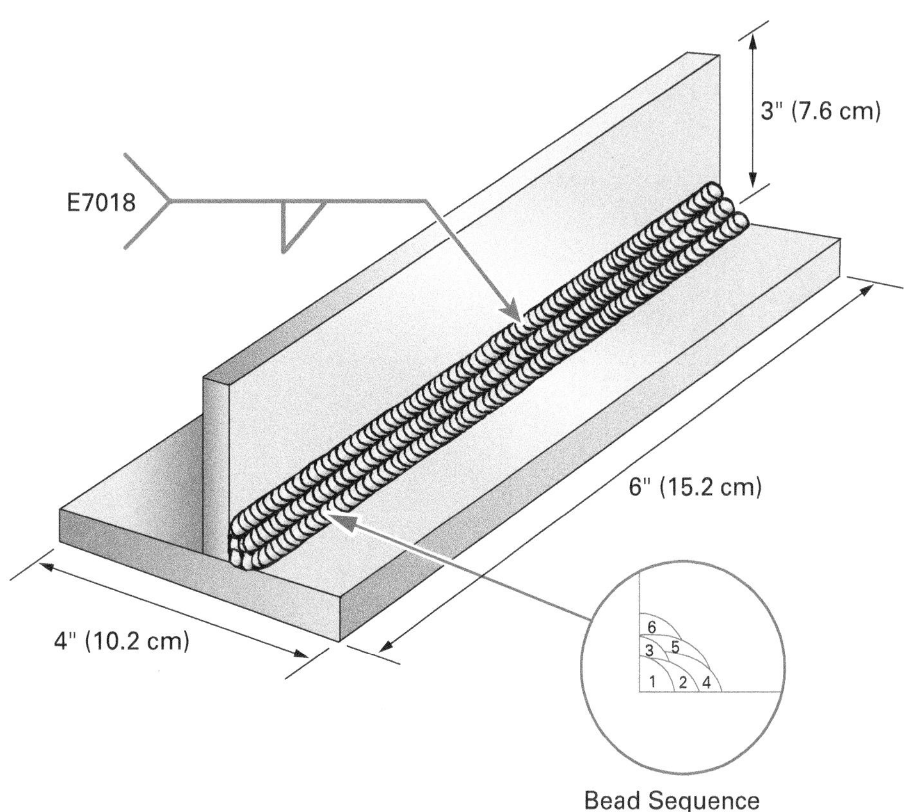

Criteria for Acceptance:

- No arc strikes outside the weld area ____________
- Uniform rippled appearance on the bead face with no valley between the beads and acceptable tie-in ____________
- Craters and restarts filled to the full cross section of the weld ____________
- Uniform weld size ±$^{1}/_{16}$" (1.6 mm) ____________
- Acceptable weld profile in accordance with Module 29109 *Figure 3* ____________
- Smooth transition with complete fusion at the toes of the weld ____________
- No pores larger than $^{3}/_{32}$" (2.4 mm) ____________
- No undercut greater than $^{1}/_{32}$" (0.8 mm) deep or 10% of the base metal thickness, whichever is less ____________
- No overlap ____________
- No inclusions ____________
- No cracks ____________

Performance Accreditation Tasks

Module 29109

Fillet Weld with E6010/E6011 Electrodes in the Vertical (3F) Position

Using $^{3}/_{32}$" to $^{5}/_{32}$" (2.4 mm to 4.0 mm) E6010/E6011 electrodes, make a fillet weld on carbon steel plate in the vertical position as indicated.

Note: Plates should be at least ¼" (6.4 mm) thick.

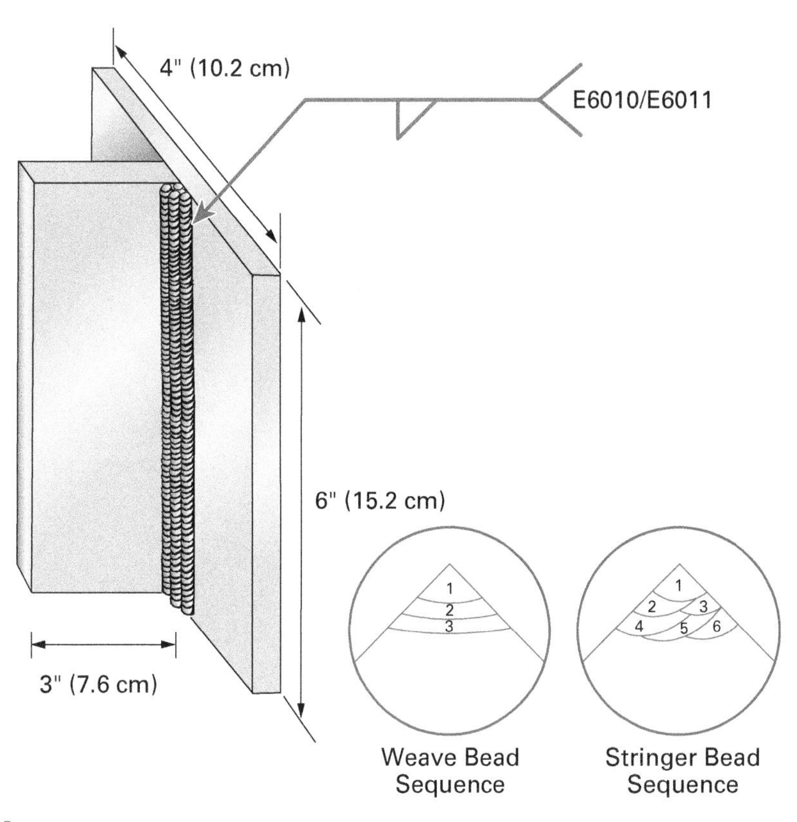

Criteria for Acceptance:

- No arc strikes outside the weld area ______
- Uniform rippled appearance on the bead face with no valley between the beads and acceptable tie-in ______
- Craters and restarts filled to the full cross section of the weld ______
- Uniform weld size $\pm^{1}/_{16}$" (1.6 mm) ______
- Acceptable weld profile in accordance with Module 29109 *Figure 3* ______
- Smooth transition with complete fusion at the toes of the weld ______
- No pores larger than $^{3}/_{32}$" (2.4 mm) ______
- No undercut greater than $^{1}/_{32}$" (0.8 mm) deep or 10% of the base metal thickness, whichever is less ______
- No overlap ______
- No inclusions ______
- No cracks ______

Performance Accreditation Tasks **Module 29109**

Fillet Weld with E7018 Electrodes in the Vertical (3F) Position

Using $^{3}/_{32}$" to $^{5}/_{32}$" (2.4 mm to 4.0 mm) E7018 electrodes, make a fillet weld on carbon steel plate in the vertical position as indicated.

Note: Plates should be at least ¼" (6.4 mm) thick.

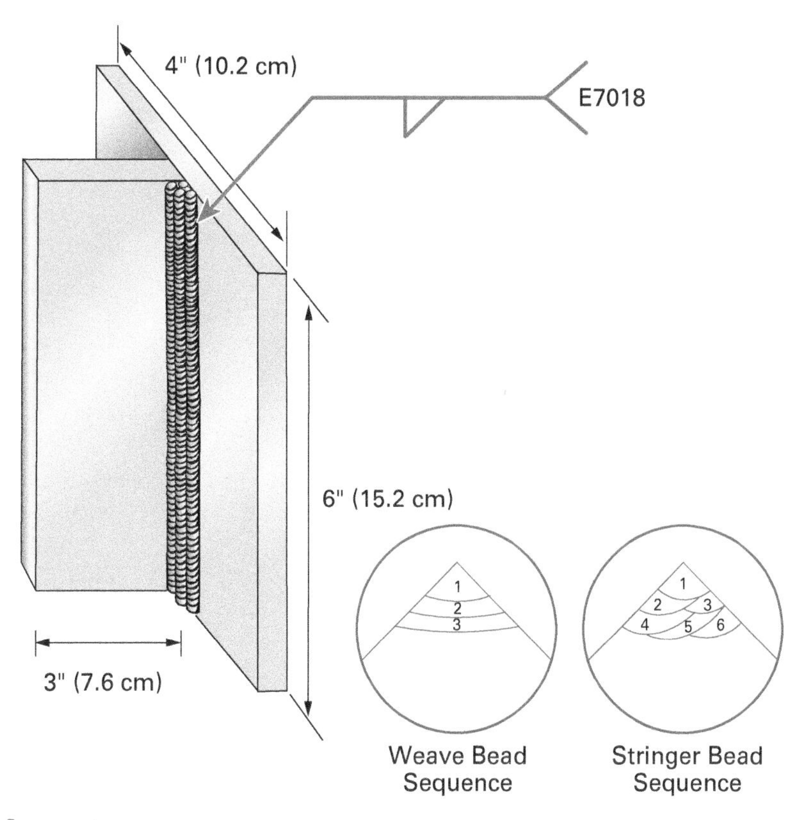

Criteria for Acceptance:

- No arc strikes outside the weld area ______
- Uniform rippled appearance on the bead face with no valley between the beads and acceptable tie-in ______
- Craters and restarts filled to the full cross section of the weld ______
- Uniform weld size ±$^{1}/_{16}$" (1.6 mm) ______
- Acceptable weld profile in accordance with Module 29109 *Figure 3* ______
- Smooth transition with complete fusion at the toes of the weld ______
- No pores larger than $^{3}/_{32}$" (2.4 mm) ______
- No undercut greater than $^{1}/_{32}$" (0.8 mm) deep or 10% of the base metal thickness, whichever is less ______
- No overlap ______
- No inclusions ______
- No cracks ______

Performance Accreditation Tasks

Module 29109

Fillet Weld with E6010/E6011 Electrodes in the Overhead (4F) Position

Using $^3/_{32}$" to $^5/_{32}$" (2.4 mm to 4.0 mm) E6010/E6011 electrodes, make a fillet weld on carbon steel plate in the overhead position as indicated.

Note: Plates should be at least ¼" (6.4 mm) thick.

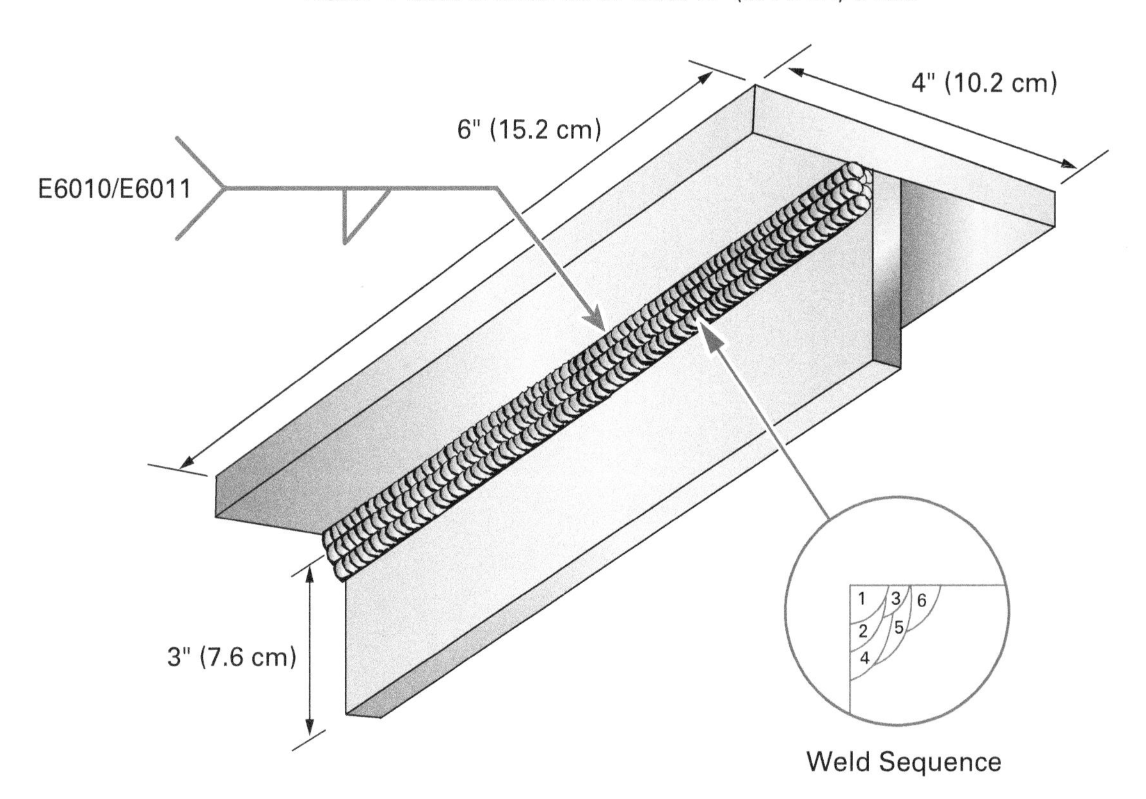

Criteria for Acceptance:

- No arc strikes outside the weld area ________
- Uniform rippled appearance on the bead face with no valley between the beads and acceptable tie-in ________
- Craters and restarts filled to the full cross section of the weld ________
- Uniform weld size ±$^1/_{16}$" (1.6 mm) ________
- Acceptable weld profile in accordance with Module 29109 *Figure 3* ________
- Smooth transition with complete fusion at the toes of the weld ________
- No pores larger than $^3/_{32}$" (2.4 mm) ________
- No undercut greater than $^1/_{32}$" (0.8 mm) deep or 10% of the base metal thickness, whichever is less ________
- No overlap ________
- No inclusions ________
- No cracks ________

Performance Accreditation Tasks

Module 29109

Fillet Weld with E7018 Electrodes in the Overhead (4F) Position

Using 3/32" to 5/32" (2.4 mm to 4.0 mm) E7018 electrodes, make a fillet weld on carbon steel plate in the overhead position as indicated.

Note: Plates should be at least ¼" (6.4 mm) thick.

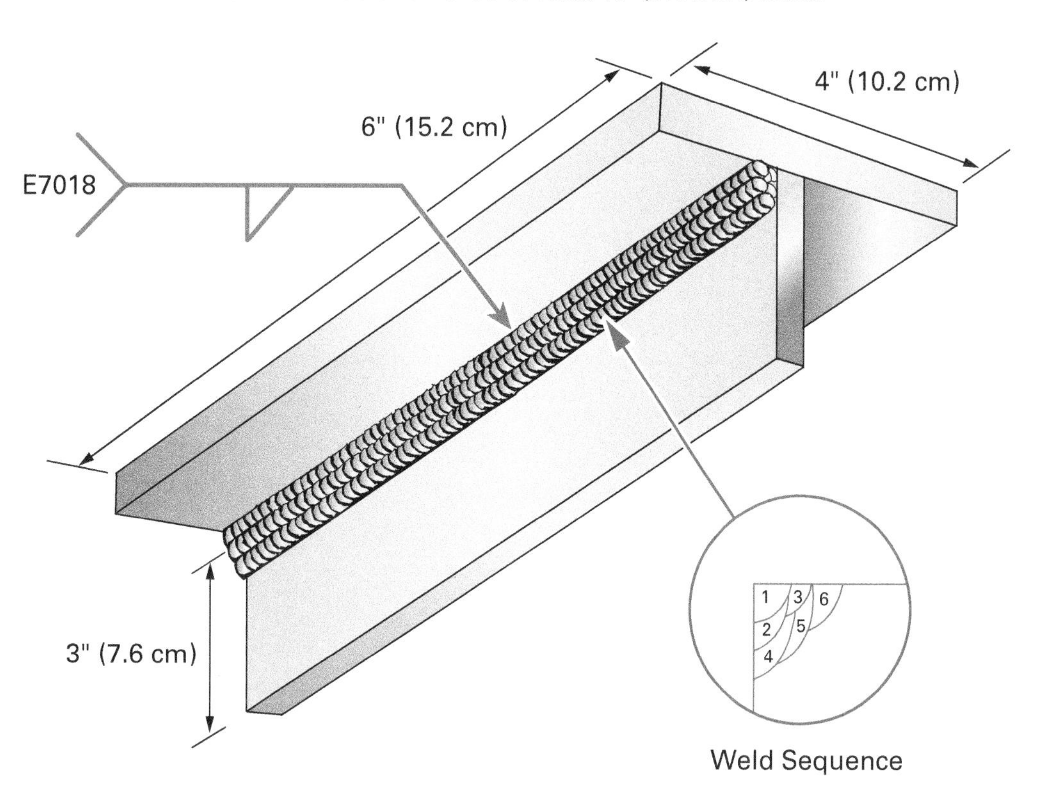

Criteria for Acceptance:

- No arc strikes outside the weld area ______
- Uniform rippled appearance on the bead face with no valley between the beads and acceptable tie-in ______
- Craters and restarts filled to the full cross section of the weld ______
- Uniform weld size ±1/16" (1.6 mm) ______
- Acceptable weld profile in accordance with Module 29109 *Figure 3* ______
- Smooth transition with complete fusion at the toes of the weld ______
- No pores larger than 3/32" (2.4 mm) ______
- No undercut greater than 1/32" (0.8 mm) deep or 10% of the base metal thickness, whichever is less ______
- No overlap ______
- No inclusions ______
- No cracks ______

APPENDIX 29111A

Performance Accreditation Tasks

The American Welding Society (AWS) School Excelling through National Skills Standards Education (SENSE) program is a comprehensive set of minimum Standards and Guidelines for Welding Education programs. The following Performance Accreditation Tasks (PATs) are aligned with and designed around the SENSE program.

PATs correspond to and support the learning objectives in *AWS EG2.0, Guide for the Training and Qualification of Welding Personnel: Entry-Level Welder*.

Note that to satisfy all learning objectives in *AWS EG2.0*, the instructor must also use the PATs contained in the second level of the NCCER *Welding* curriculum.

PATs 1 through 4 correspond to *AWS EG2.0, Module 4 – Shielded Metal Arc Welding*, Key Indicators 3, 4, and 6. PATs 2 and 3 also correspond to *AWS EG2.0, Module 4 – Shielded Metal Arc Welding*, Key Indicator 7, if the instructor chooses to perform the guided bend test. Refer to NCCER Module 29106, *Weld Quality*, for guided bend acceptance criteria.

PATs provide specific acceptable criteria for performance and help to ensure a true competency-based welding program for students.

The following tasks test your competency in welding V-groove welds in the 1G, 2G, 3G, and 4G positions. Don't perform these tasks until your instructor directs you to do so. Practice the tasks until you're thoroughly familiar with the procedures.

After you complete each task, take it to your instructor for evaluation.

Performance Accreditation Tasks Module 29111

V-Groove Welds with Backing in the Flat (1G) Position

Using $\frac{3}{32}$", $\frac{1}{8}$", or $\frac{5}{32}$" (2.4 mm, 3.2 mm, or 4.0 mm) E7018 electrodes, make a V-groove weld with steel backing on carbon steel plate in the flat position as indicated.

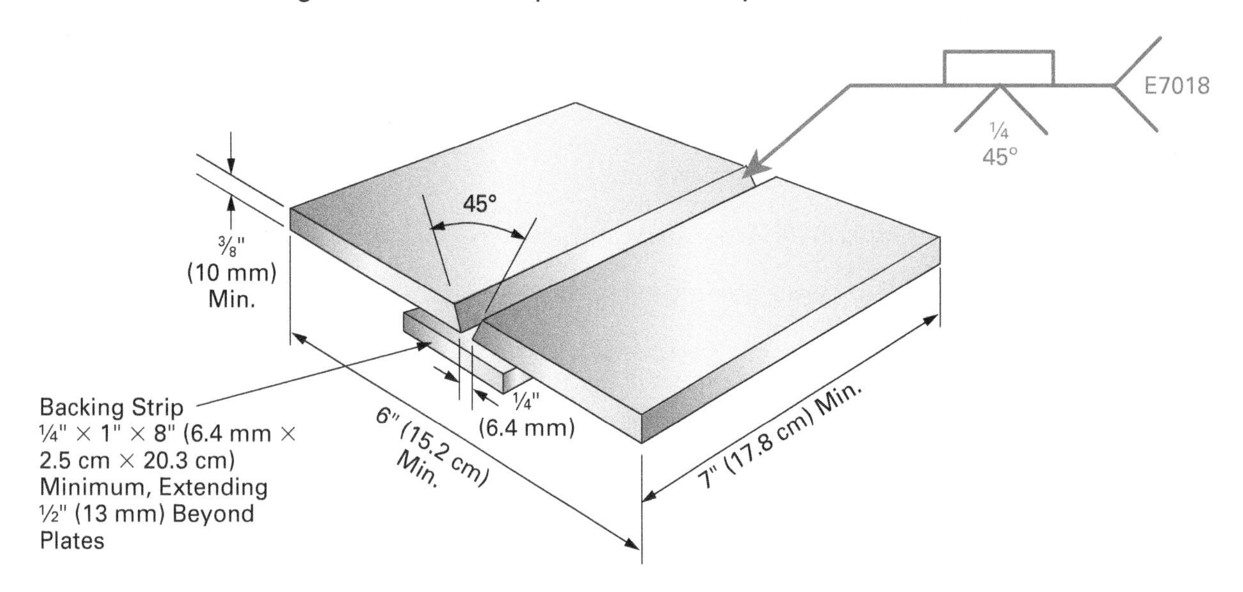

Inspection Hold Points:

- Fit-up __________
- Root pass __________
- Final __________

Criteria for Acceptance:

- No arc strikes outside the weld area __________
- Uniform rippled appearance on the bead face with no valley between the beads and acceptable tie-in __________
- Craters and restarts filled to the full cross section of the weld __________
- Uniform weld size $\pm\frac{1}{16}$" (1.6 mm) __________
- Acceptable weld profile in accordance with Module 29111 *Figure 15* __________
- Smooth transition with complete fusion at the toes of the weld __________
- Face reinforcement no greater than $\frac{1}{8}$" (3.2 mm) __________
- No pores larger than $\frac{3}{32}$" (2.4 mm) __________
- No undercut greater than $\frac{1}{32}$" (0.8 mm) deep or 10% of the base metal thickness, whichever is less __________
- No overlap __________
- No inclusions __________
- No cracks __________
- Acceptable guided bend test results (instructor option) __________

Performance Accreditation Tasks

Module 29111

V-Groove Welds with Backing in the Horizontal (2G) Position

Using ³⁄₃₂", ⅛", or ⁵⁄₃₂" (2.4 mm, 3.2 mm, or 4.0 mm) E7018 electrodes, make a V-groove weld with steel backing on carbon steel plate in the horizontal position as indicated.

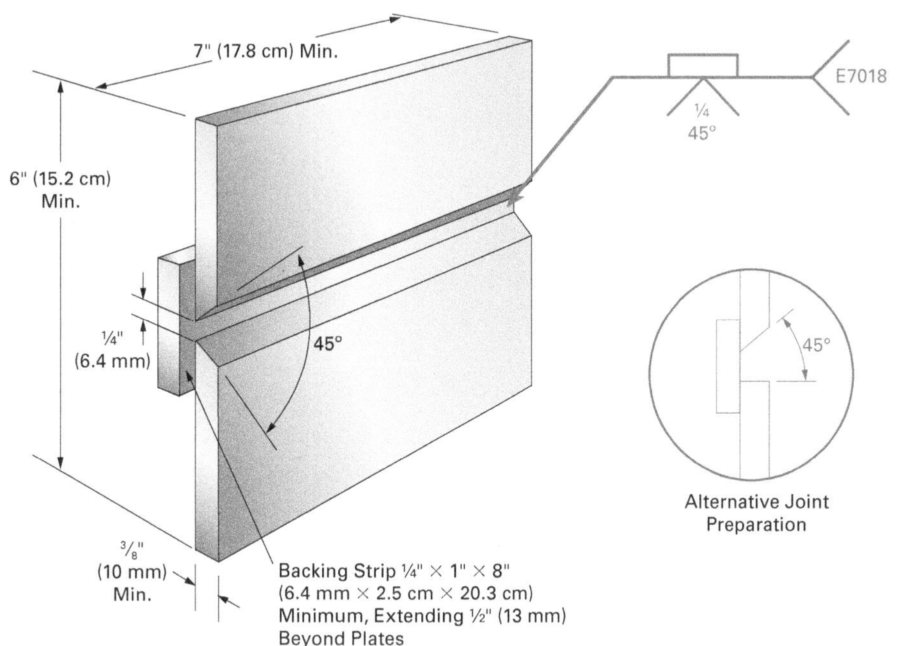

Inspection Hold Points:

- Fit-up ____________
- Root pass ____________
- Final ____________

Criteria for Acceptance:

- No arc strikes outside the weld area ____________
- Uniform rippled appearance on the bead face with no valley between the beads and acceptable tie-in ____________
- Craters and restarts filled to the full cross section of the weld ____________
- Uniform weld size ±¹⁄₁₆" (1.6 mm) ____________
- Acceptable weld profile in accordance with Module 29111 *Figure 15* ____________
- Smooth transition with complete fusion at the toes of the weld ____________
- Face reinforcement no greater than ⅛" (3.2 mm) ____________
- No pores larger than ³⁄₃₂" (2.4 mm) ____________
- No undercut greater than ¹⁄₃₂" (0.8 mm) deep or 10% of the base metal thickness, whichever is less ____________
- No overlap ____________
- No inclusions ____________
- No cracks ____________
- Acceptable guided bend test results (instructor option) ____________

Performance Accreditation Tasks

Module 29111

V-Groove Welds with Backing in the Vertical (3G) Position

Using 3⁄32", 1⁄8", or 5⁄32" (2.4 mm, 3.2 mm, or 4.0 mm) E7018 electrodes, make a V-groove weld with steel backing on carbon steel plate in the vertical position as indicated.

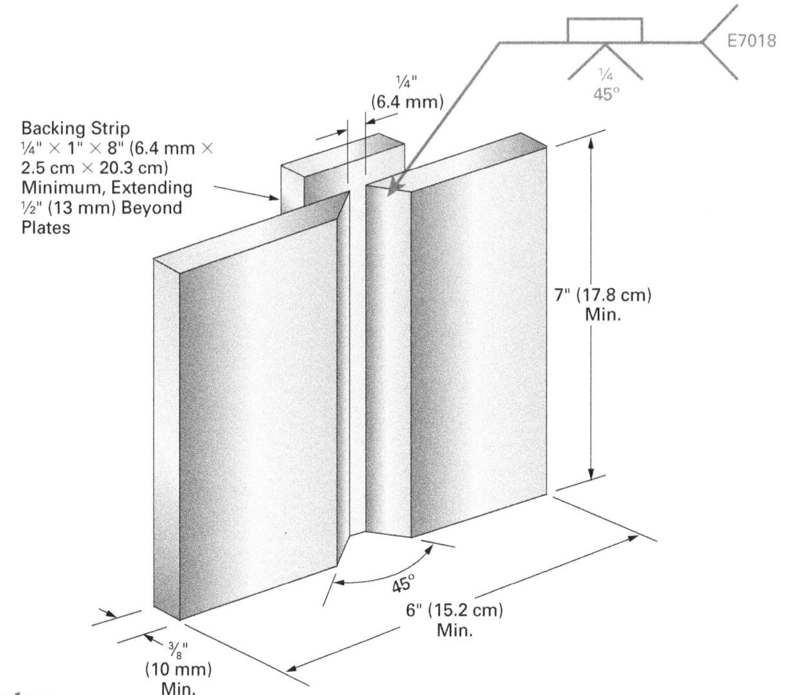

Inspection Hold Points:

- Fit-up ____________
- Root pass ____________
- Final ____________

Criteria for Acceptance:

- No arc strikes outside the weld area ____________
- Uniform rippled appearance on the bead face with no valley between the beads and acceptable tie-in ____________
- Craters and restarts filled to the full cross section of the weld ____________
- Uniform weld size ±1⁄16" (1.6 mm) ____________
- Acceptable weld profile in accordance with Module 29111 *Figure 15* ____________
- Smooth transition with complete fusion at the toes of the weld ____________
- Face reinforcement no greater than 1⁄8" (3.2 mm) ____________
- No pores larger than 3⁄32" (2.4 mm) ____________
- No undercut greater than 1⁄32" (0.8 mm) deep or 10% of the base metal thickness, whichever is less ____________
- No overlap ____________
- No inclusions ____________
- No cracks ____________
- Acceptable guided bend test results (instructor option) ____________

Performance Accreditation Tasks

Module 29111

V-Groove Welds with Backing in the Overhead (4G) Position

Using ³⁄₃₂", ⅛", or ⁵⁄₃₂" (2.4 mm, 3.2 mm, or 4.0 mm) E7018 electrodes, make a V-groove weld with steel backing on carbon steel plate in the overhead position as indicated.

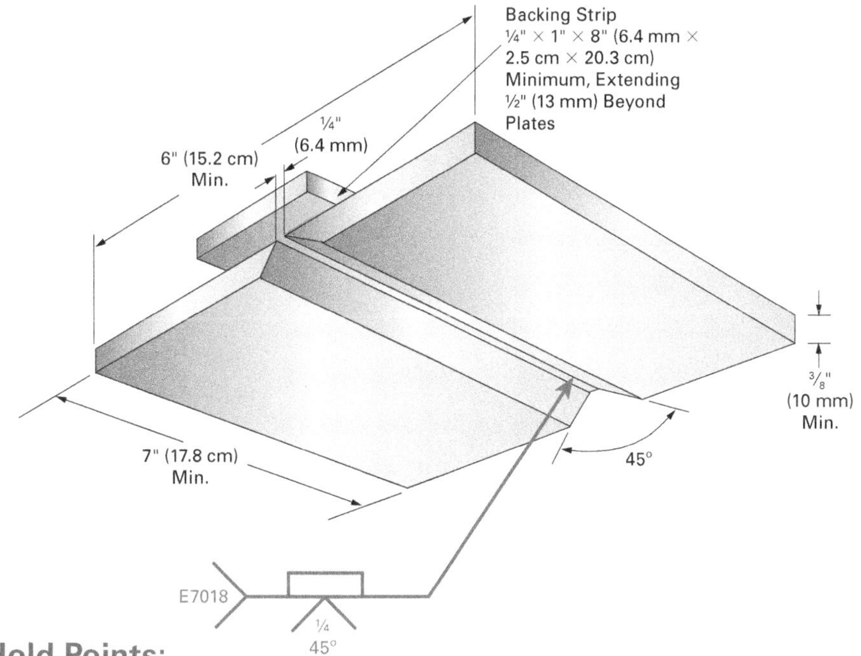

Inspection Hold Points:

- Fit-up ________
- Root pass ________
- Final ________

Criteria for Acceptance:

- No arc strikes outside the weld area ________
- Uniform rippled appearance on the bead face with no valley between the beads and acceptable tie-in ________
- Craters and restarts filled to the full cross section of the weld ________
- Uniform weld size ±¹⁄₁₆" (1.6 mm) ________
- Acceptable weld profile in accordance with Module 29111 *Figure 15* ________
- Smooth transition with complete fusion at the toes of the weld ________
- Face reinforcement no greater than ⅛" (3.2 mm) ________
- No pores larger than ³⁄₃₂" (2.4 mm) ________
- No undercut greater than ¹⁄₃₂" (0.8 mm) deep or 10% of the base metal thickness, whichever is less ________
- No overlap ________
- No inclusions ________
- No cracks ________
- Acceptable guided bend test results (instructor option) ________

APPENDIX 29112A

Performance Accreditation Tasks

The American Welding Society (AWS) School Excelling through National Skills Standards Education (SENSE) program is a comprehensive set of minimum Standards and Guidelines for Welding Education programs. The following Performance Accreditation Tasks (PATs) are aligned with and designed around the SENSE program.

PATs correspond to and support the learning objectives in *AWS EG2.0, Guide for the Training and Qualification of Welding Personnel: Entry-Level Welder*.

Note that to satisfy all learning objectives in *AWS EG2.0*, the instructor must also use the PATs contained in the second level of the NCCER *Welding* curriculum.

PATs 1 through 4 correspond to *AWS EG2.0, Module 4 – Shielded Metal Arc Welding*, Key Indicators 3, 4, and 6. PATs 2 and 3 also correspond to *AWS EG2.0, Module 4 – Shielded Metal Arc Welding*, Key Indicator 7, if the instructor chooses to perform the guided bend test. Refer to NCCER Module 29106, *Weld Quality*, for guided bend acceptance criteria.

PATs provide specific acceptable criteria for performance and help to ensure a true competency-based welding program for students.

The following tasks test your competency in welding open-root single V-groove welds in the 1G, 2G, 3G, and 4G positions. Don't perform these tasks until your instructor directs you to do so. Practice the tasks until you're thoroughly familiar with the procedures.

After you complete each task, take it to your instructor for evaluation.

Performance Accreditation Tasks

Module 29112

Open-Root V-Groove with E6010/E6011 and E7018 Electrodes in the Flat (1G) Position

Using $^{3}/_{32}$" or $^{1}/_{8}$" (2.4 mm or 3.2 mm) E6010/E6011 electrodes for the root pass and $^{3}/_{32}$" or $^{1}/_{8}$" (2.4 mm or 3.2 mm) E7018 electrodes for the fill and cover passes, make an open-root V-groove weld on carbon steel plate in the flat position as indicated.

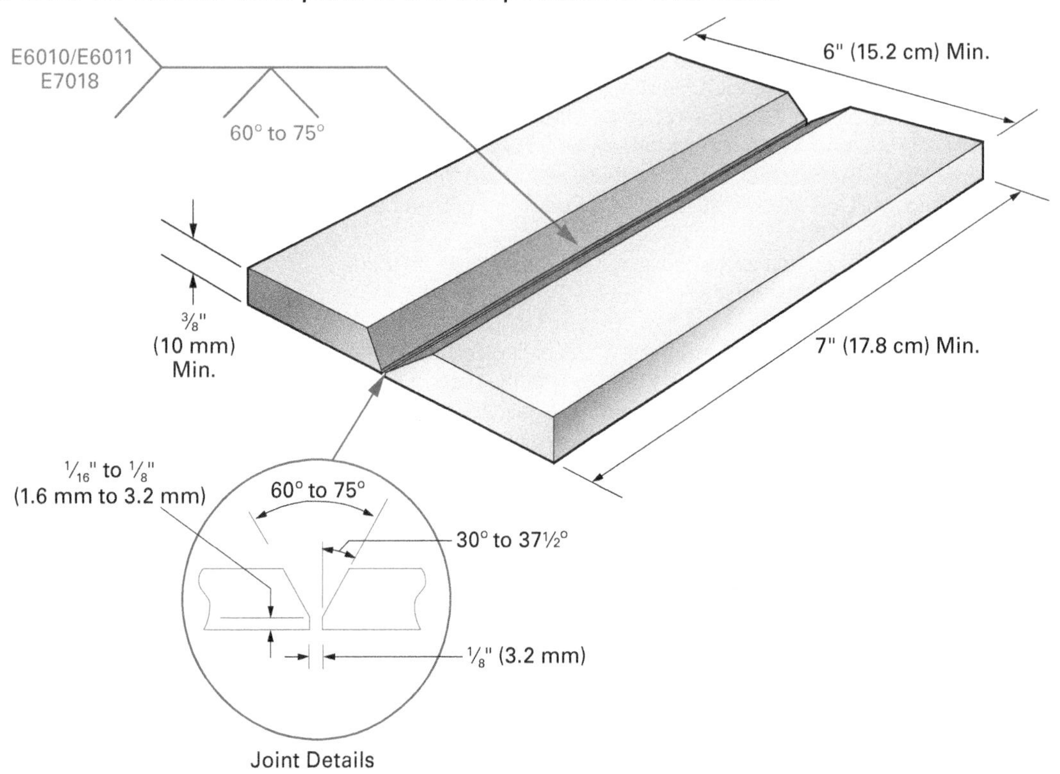

Criteria for Acceptance:

- No arc strikes outside the weld area ________
- Uniform rippled appearance on the bead face with no valley between the beads and acceptable tie-in ________
- Craters and restarts filled to the full cross section of the weld ________
- Uniform weld size $\pm ^{1}/_{16}$" (1.6 mm) ________
- Acceptable weld profile in accordance with Module 29112 *Figure 12* ________
- Complete uniform root penetration at least flush with the base metal to a maximum buildup of $^{1}/_{8}$" (3.2 mm) ________
- Smooth transition with complete fusion at the toes of the weld ________
- Face reinforcement no greater than $^{1}/_{8}$" (3.2 mm) ________
- No pores larger than $^{3}/_{32}$" (2.4 mm) ________
- No undercut greater than $^{1}/_{32}$" (0.8 mm) deep or 10% of the base metal thickness, whichever is less ________
- No overlap ________
- No inclusions ________
- No cracks ________
- Acceptable guided bend test results (instructor option) ________

Performance Accreditation Tasks

Module 29112

Open-Root V-Groove with E6010/E6011 and E7018 Electrodes in the Horizontal (2G) Position

Using $\frac{3}{32}$" or $\frac{1}{8}$" (2.4 mm or 3.2 mm) E6010/E6011 electrodes for the root pass and $\frac{3}{32}$" or $\frac{1}{8}$" (2.4 mm or 3.2 mm) E7018 electrodes for the fill and cover passes, make an open-root V-groove weld on carbon steel plate in the horizontal position as shown.

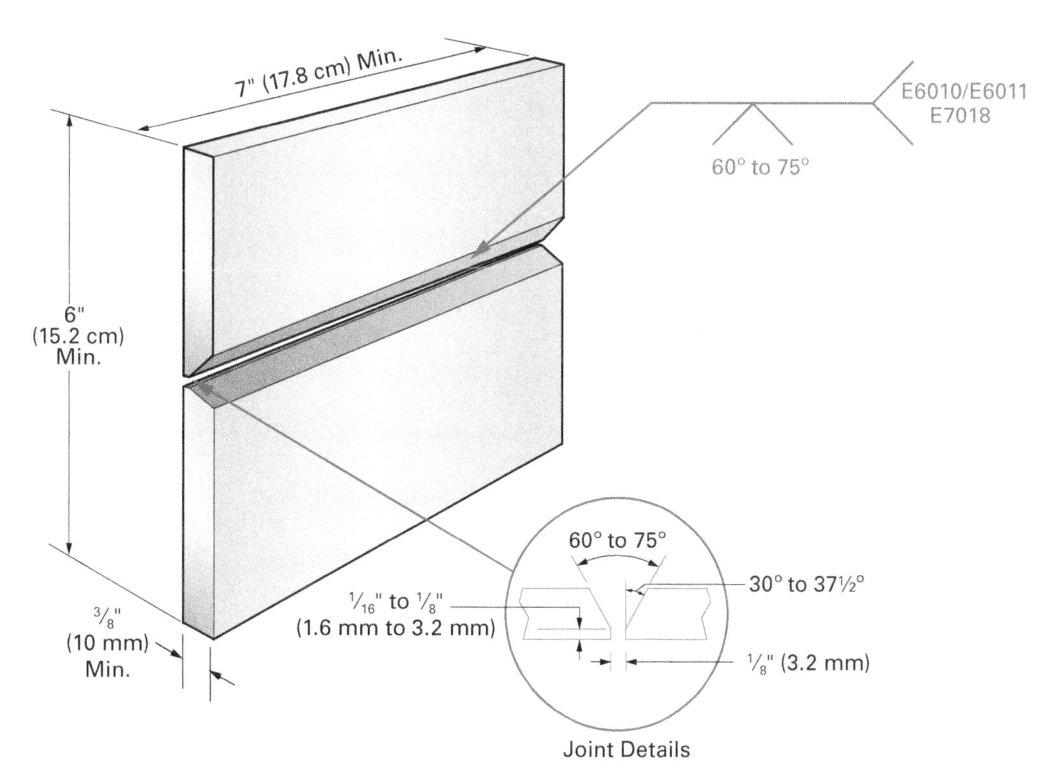

Criteria for Acceptance:

- No arc strikes outside the weld area ____________
- Uniform rippled appearance on the bead face with no valley between the beads and acceptable tie-in ____________
- Craters and restarts filled to the full cross section of the weld ____________
- Uniform weld size ±$\frac{1}{16}$" (1.6 mm) ____________
- Acceptable weld profile in accordance with Module 29112 *Figure 12* ____________
- Complete uniform root penetration at least flush with the base metal to a maximum buildup of $\frac{1}{8}$" (3.2 mm) ____________
- Smooth transition with complete fusion at the toes of the weld ____________
- Face reinforcement no greater than $\frac{1}{8}$" (3.2 mm) ____________
- No pores larger than $\frac{3}{32}$" (2.4 mm) ____________
- No undercut greater than $\frac{1}{32}$" (0.8 mm) deep or 10% of the base metal thickness, whichever is less ____________
- No overlap ____________
- No inclusions ____________
- No cracks ____________
- Acceptable guided bend test results (instructor option) ____________

Performance Accreditation Tasks

Module 29112

Open-Root V-Groove with E6010/E6011 and E7018 Electrodes in the Vertical (3G) Position

Using $\frac{3}{32}$" or $\frac{1}{8}$" (2.4 mm or 3.2 mm) E6010/E6011 electrodes for the root pass and $\frac{3}{32}$" or $\frac{1}{8}$" (2.4 mm or 3.2 mm) E7018 electrodes for the fill and cover passes, make an open-root V-groove weld on carbon steel plate in the vertical position as shown.

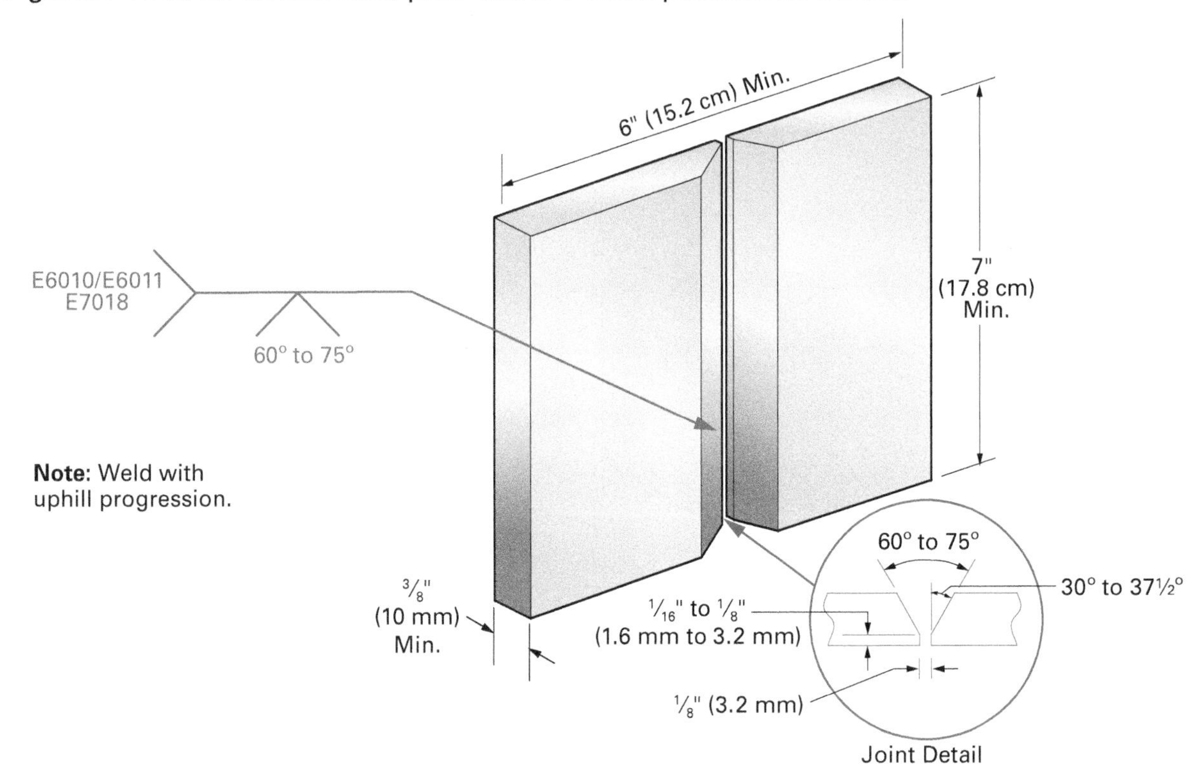

Criteria for Acceptance:

- No arc strikes outside the weld area ____________
- Uniform rippled appearance on the bead face with no valley between the beads and acceptable tie-in ____________
- Craters and restarts filled to the full cross section of the weld ____________
- Uniform weld size ±$\frac{1}{16}$" (1.6 mm) ____________
- Acceptable weld profile in accordance with Module 29112 *Figure 12* ____________
- Complete uniform root penetration at least flush with the base metal to a maximum buildup of $\frac{1}{8}$" (3.2 mm) ____________
- Smooth transition with complete fusion at the toes of the weld ____________
- Face reinforcement no greater than $\frac{1}{8}$" (3.2 mm) ____________
- No pores larger than $\frac{3}{32}$" (2.4 mm) ____________
- No undercut greater than $\frac{1}{32}$" (0.8 mm) deep or 10% of the base metal thickness, whichever is less ____________
- No overlap ____________
- No inclusions ____________
- No cracks ____________
- Acceptable guided bend test results (instructor option) ____________

Performance Accreditation Tasks

Module 29112

Open-Root V-Groove with E6010/E6011 and E7018 Electrodes in the Overhead (4G) Position

Using $\frac{3}{32}$" or $\frac{1}{8}$" (2.4 mm or 3.2 mm) E6010/E6011 electrodes for the root pass and $\frac{3}{32}$" or $\frac{1}{8}$" (2.4 mm or 3.2 mm) E7018 electrodes for the fill and cover passes, make an open-root V-groove weld on carbon steel plate in the overhead position as indicated.

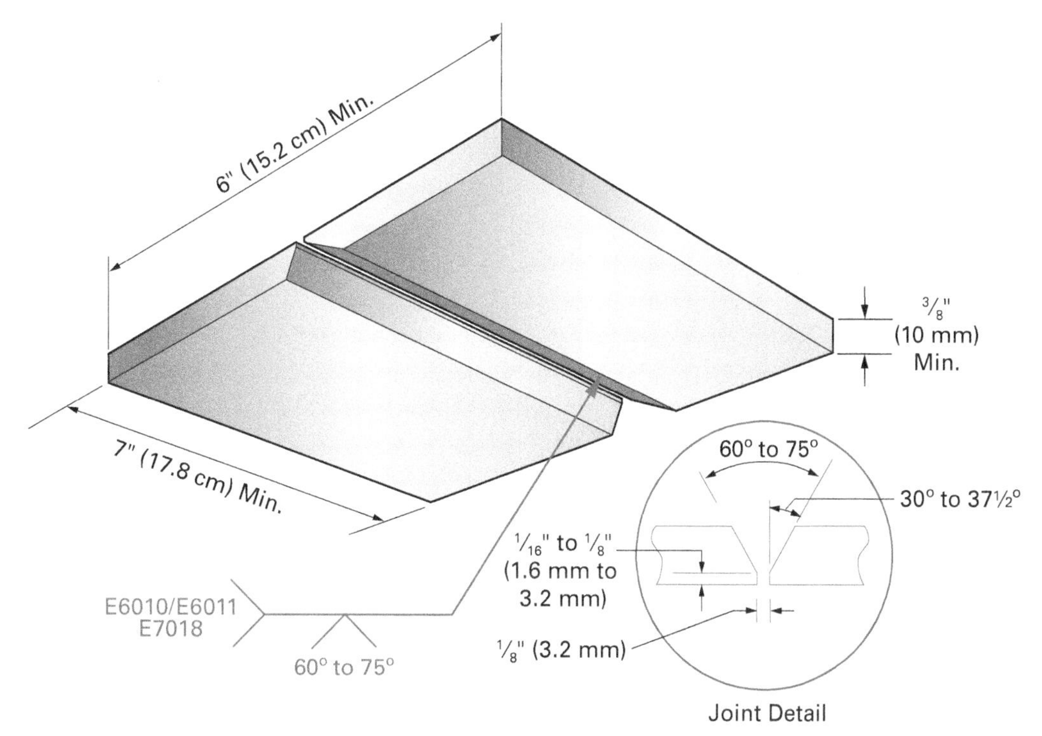

Criteria for Acceptance:

- No arc strikes outside the weld area __________
- Uniform rippled appearance on the bead face with no valley between the beads and acceptable tie-in __________
- Craters and restarts filled to the full cross section of the weld __________
- Uniform weld size $\pm\frac{1}{16}$" (1.6 mm) __________
- Acceptable weld profile in accordance with Module 29112 *Figure 12* __________
- Complete uniform root penetration at least flush with the base metal to a maximum buildup of $\frac{1}{8}$" (3.2 mm) __________
- Smooth transition with complete fusion at the toes of the weld __________
- Face reinforcement no greater than $\frac{1}{8}$" (3.2 mm) __________
- No pores larger than $\frac{3}{32}$" (2.4 mm) __________
- No undercut greater than $\frac{1}{32}$" (0.8 mm) deep or 10% of the base metal thickness, whichever is less __________
- No overlap __________
- No inclusions __________
- No cracks __________
- Acceptable guided bend test results (instructor option) __________

ADDITIONAL RESOURCES

29101 Welding Safety

ANSI/AWS Z49.1, Safety in Welding, Cutting, and Allied Processes. Latest Edition. Miami, FL: American Welding Society.

Arc Welding Safety E205. Latest Edition. Cleveland, OH: Lincoln Electric.

ASTM F2757, Standard Guide for Home Laundering Care and Maintenance of Flame Resistant or Arc Rated Clothing. Latest Edition. West Conshohocken, PA: ASTM International.

AWS F32M/F32, Ventilation Guide for Weld Fumes. Latest Edition. Miami, FL: American Welding Society.

AWS F4.1, Safe Practices for the Preparation of Containers and Piping for Welding, Cutting and Allied Processes. Latest Edition. Miami, FL: American Welding Society.

AWS Safety and Health Fact Sheet No. 12, Contact Lens Wear. **www.aws.org**.

OSHA 29 CFR 1910.134, Respiratory Protection Standard, Occupational Safety and Health Administration, US Department of labor. **www.ecfr.gov**.

OSHA FactSheet: Eye Protection Against Radiant Energy during Welding and Cutting in Shipyard Employment. Latest Edition. Washington DC: United States Department of Labor/OSHA. **www.osha.gov**.

Personal Protective Equipment (PPE) for Welding and Cutting. Miami, FL: American Welding Society.

Safety Topics in Welding, Cutting, and Brazing. **www.osha.gov**.

29102 Oxyfuel Cutting

ANSI/AWS Z49.1, Safety in Welding, Cutting, and Allied Processes. Latest Edition. Miami, FL: American Welding Society.

AWS C4.1, Criteria for Describing Oxygen-Cut Surfaces, and Oxygen Cutting Surface Roughness Gauge. Latest Edition. Miami, FL: American Welding Society.

AWS C4.2, Recommended Practices for Safe Oxyfuel Gas Cutting Torch Operation. Latest Edition. Miami, FL: American Welding Society.

AWS C4.3, Recommended Practices for Safe Oxyfuel Gas Heating Torch Operation. Latest Edition. Miami, FL: American Welding Society.

AWS C4.5M, Uniform Designation System for Oxyfuel Nozzles. Latest Edition. Miami, FL: American Welding Society.

AWS F3.2M/F3.2, Ventilation Guide for Weld Fumes. Latest Edition. Miami, FL: American Welding Society.

AWS F4.1, Safe Practices for the Preparation of Containers and Piping for Welding, Cutting, and Allied Processes. Latest Edition. Miami, FL: American Welding Society.

ESAB®. **www.esabna.com**.

Lincoln Electric. **www.lincolnelectric.com**.

Miller Electric. **www.millerwelds.com**.

OSHA FactSheet: Eye Protection Against Radiant Energy during Welding and Cutting in Shipyard Employment. Latest Edition. Washington DC: United States Department of Labor/OSHA. **www.osha.gov**.

29103 Plasma Arc Cutting

ANSI/AWS Z49.1, Safety in Welding, Cutting, and Allied Processes. Latest Edition. Miami, FL: American Welding Society.

AWS C5.2, Recommended Practices for Plasma Arc Cutting and Gouging. Latest Edition. Miami, FL: American Welding Society.

AWS F3.2M/F3.2, Ventilation Guide for Weld Fumes. Latest Edition. Miami, FL: American Welding Society.

AWS F4.1, Safe Practices for the Preparation of Containers and Piping for Welding, Cutting and Allied Processes. Latest Edition. Miami, FL: American Welding Society.

Hypertherm®. **www.hypertherm.com**.

OSHA FactSheet: Eye Protection Against Radiant Energy during Welding and Cutting in Shipyard Employment. Latest Edition. Washington DC: United States Department of Labor/OSHA. **www.osha.gov**.

Plasma Cutters Handbook: Choosing Plasma Cutters, Shop Safety, Basic Operation, Cutting Procedures, Advanced Cutting Tips, CNC Plasma Cutters, Troubleshooting, and Sample Projects. Eddie Paul. New York, NY: Penguin Group.

29104 Air-Carbon Arc Cutting and Gouging

Air Carbon-Arc Guide, Form Number: 89-250-008. Denton, Texas: Victor Technologies, Inc.

ANSI/AWS Z49.1, Safety in Welding, Cutting, and Allied Processes. Latest Edition. Miami, FL: American Welding Society.

ANSI C5.3, Recommended Practices for Air Carbon Arc Gouging and Cutting. Latest Edition. Miami, FL: American Welding Society.

AWS F32M/F32, Ventilation Guide for Weld Fumes. Latest Edition. Miami, FL: American Welding Society.

AWS F4.1, Safe Practices for the Preparation of Containers and Piping for Welding, Cutting and Allied Processes. Latest Edition. Miami, FL: American Welding Society.

ESAB®. **www.esabna.com**.

Lincoln Electric. **www.lincolnelectric.com**.

Miller Electric. **www.millerwelds.com**.

OSHA FactSheet: Eye Protection Against Radiant Energy during Welding and Cutting in Shipyard Employment. Latest Edition. Washington DC: United States Department of Labor/OSHA. **www.osha.gov**.

29105 Base Metal Prep

American Iron and Steel Institute. **www.steel.org**.

ANSI/AWS Z49.1, Safety in Welding, Cutting, and Allied Processes. Latest Edition. Miami, FL: American Welding Society.

Association for Iron & Steel Technology. **www.aist.org**.

AWS F3.2M/F3.2, Ventilation Guide for Weld Fumes. Latest Edition. Miami, FL: American Welding Society.

AWS F4.1, Safe Practices for the Preparation of Containers and Piping for Welding, Cutting and Allied Processes. Latest Edition. Miami, FL: American Welding Society.

OSHA 1910.252, Welding, Cutting and Brazing, General Requirements. Latest Edition. Washington DC: United States Department of Labor/OSHA.

OSHA FactSheet: Eye Protection Against Radiant Energy during Welding and Cutting in Shipyard Employment. Latest Edition. Washington DC: United States Department of Labor/OSHA. **www.osha.gov**.

Steel Metallurgy for the Non-Metallurgist. Latest Edition. John D. Verhoeven. Materials Park, OH: ASM International.

Welding Handbook. Latest Edition. Miami, FL: The American Welding Society.

29106 Weld Quality

API 1104, Welding of Pipelines and Related Facilities. Latest Edition. Washington, DC: The American Petroleum Institute.

ASME Boiler and Pressure Vessel Code (BVPC). Latest Edition. New York, NY: American Society of Mechanical Engineers.

AWS B1.10M/B1.10, Guide for the Nondestructive Examination of Welds. Latest Edition. Miami, FL: American Welding Society.

AWS D1.1/D1.1M, Structural Welding Code—Steel. Latest Edition. Miami, FL: American Welding Society.

AWS VIW-M,Visual Inspection Workshop Reference Manual (Historical). 2006 Edition. Miami, FL: American Welding Society.

Certification Manual for Welding Inspectors. Latest Edition. Miami, FL: American Welding Society.

Welding Inspection Technology. Latest Edition. Miami, FL: American Welding Society.

www.ansi.org

www.api.org

www.asme.org

www.asnt.org

www.astm.org

www.aws.org

www.iso.org

29107 SMAW – Equipment and Setup

ANSI/AWS Z49.1, Safety in Welding, Cutting, and Allied Processes. Latest Edition. Miami, FL: American Welding Society.

AWS F32M/F32, Ventilation Guide for Weld Fumes. Latest Edition. Miami, FL: American Welding Society.

AWS F4.1, Safe Practices for the Preparation of Containers and Piping for Welding, Cutting and Allied Processes. Latest Edition. Miami, FL: American Welding Society.

Covering the Basics – Engine Driven Welder Maintenance. Al Nystrom, Technical Service Representative. Cleveland, OH, USA: The Lincoln Electric Company.

Lincoln Electric. **www.lincolnelectric.com**.

Miller Electric. **www.millerwelds.com**.

OSHA FactSheet: Eye Protection Against Radiant Energy during Welding and Cutting in Shipyard Employment. Latest Edition. Washington DC: United States Department of Labor/OSHA. **www.osha.gov**.

29108 SMAW Electrodes

ASME Boiler and Pressure Vessel Code (BVPC). Latest Edition. New York, NY: American Society of Mechanical Engineers.

AWS D1.1/D1.1M, Structural Welding Code—Steel. Latest Edition. Miami, FL: American Welding Society.

ESAB®. **www.esabna.com**.

Guidelines for Shielded Metal Arc Welding (SMAW). Latest Edition. Appleton, WI: Miller Electric Mfg. LLC.

Helpful Hints to Basic Welding. Latest Edition. Troy, OH: Hobart Brothers Company.

Hobart Brothers Company.**www.hobartbrothers.com**.

Lincoln Electric. **www.lincolnelectric.com**.

Miller Electric. **www.millerwelds.com**.

Stick Electrode Welding Guide. Latest Edition. Cleveland, OH: Lincoln Electric.

29109 SMAW – Beads and Fillet Welds

ANSI/AWS Z49.1, Safety in Welding, Cutting, and Allied Processes. Latest Edition. Miami, FL: American Welding Society.

AWS D1.1/D1.1M, Structural Welding Code–Steel. Latest Edition. Miami, FL: American Welding Society.

AWS F32M/F32, Ventilation Guide for Weld Fumes. Latest Edition. Miami, FL: American Welding Society.

AWS F4.1, Safe Practices for the Preparation of Containers and Piping for Welding, Cutting, and Allied Processes. Latest Edition. Miami, FL: American Welding Society.

OSHA FactSheet: Eye Protection Against Radiant Energy during Welding and Cutting in Shipyard Employment. Latest Edition. Washington DC: United States Department of Labor/OSHA. **www.osha.gov**.

Welding Handbook. Latest Edition. Miami, FL: The American Welding Society.

29110 Joint Fit-Up and Alignment

ANSI/AWS Z49.1, Safety in Welding, Cutting, and Allied Processes. Latest Edition. Miami, FL: American Welding Society.

AWS D1.1/D1.1M, Structural Welding Code—Steel. Latest Edition. Miami, FL: American Welding Society.

AWS F32M/F32, Ventilation Guide for Weld Fumes. Latest Edition. Miami, FL: American Welding Society.

AWS F4.1, Safe Practices for the Preparation of Containers and Piping for Welding, Cutting and Allied Processes. Latest Edition. Miami, FL: American Welding Society.

OSHA FactSheet: Eye Protection Against Radiant Energy during Welding and Cutting in Shipyard Employment. Latest Edition. Washington DC: United States Department of Labor/OSHA. **www.osha.gov**.

Welding Handbook. Latest Edition. Miami, FL: The American Welding Society.

29111 SMAW – Groove Welds with Backing

ANSI/AWS Z49.1, Safety in Welding, Cutting, and Allied Processes. Latest Edition. Miami, FL: American Welding Society.

AWS D1.1/D1.1M, Structural Welding Code—Steel. Latest Edition. Miami, FL: American Welding Society.

AWS F32M/F32, Ventilation Guide for Weld Fumes. Latest Edition. Miami, FL: American Welding Society.

AWS F4.1, Safe Practices for the Preparation of Containers and Piping for Welding, Cutting and Allied Processes. Latest Edition. Miami, FL: American Welding Society.

OSHA FactSheet: Eye Protection Against Radiant Energy during Welding and Cutting in Shipyard Employment. Latest Edition. Washington DC: United States Department of Labor/OSHA. **www.osha.gov**.

Welding Handbook. Latest Edition. Miami, FL: The American Welding Society.

29112 SMAW – Open-Root Groove Welds (Plate)

ANSI/AWS Z49.1, Safety in Welding, Cutting, and Allied Processes. Latest Edition. Miami, FL: American Welding Society.

AWS D1.1/D1.1M, Structural Welding Code—Steel. Latest Edition. Miami, FL: American Welding Society.

AWS F32M/F32, Ventilation Guide for Weld Fumes. Latest Edition. Miami, FL: American Welding Society.

AWS F4.1, Safe Practices for the Preparation of Containers and Piping for Welding, Cutting and Allied Processes. Latest Edition. Miami, FL: American Welding Society.

OSHA FactSheet: Eye Protection Against Radiant Energy during Welding and Cutting in Shipyard Employment. Latest Edition. Washington DC: United States Department of Labor/OSHA. **www.osha.gov**.

Welding Handbook. Latest Edition. Miami, FL: The American Welding Society.

GLOSSARY

Addendum: Supplementary information that corrects or revises a document.

Alloy: A metal containing additional elements that significantly change its properties.

Alternating current (AC): An electric current that regularly changes direction and size.

Ambient temperature: The temperature of an object's surroundings.

Amperage: The size of an electric current, measured in amperes or *amps* (A).

Annealing: Heating and then gradually cooling a metal to relieve its internal stresses.

Arc blow: Magnetic forces that deflect the welding arc from its intended path.

Arc blow: Magnetic forces that deflect the welding arc from its intended path.

Arc burn: A burn to the skin produced by brief exposure to intense heat and ultraviolet light.

Arc strike: A discontinuity caused by arc initiation and visible as a melted spot outside the weld area.

Austenitic: A nonmagnetic type of stainless steel whose internal crystal structure prevents it from being hardened by heat treatment.

Backfire: A loud snap or pop that happens upon extinguishing a torch.

Backing: A weldable or non-weldable material placed behind a weld's root opening.

Base metal: A metal to be welded, cut, or brazed.

Bonded: Permanently joining metallic parts to form a conductive path that can safely carry a specific electric current.

Brazing: Joining metal by melting a filler metal at a temperature above 450°C (~840°F) over unmelted base metal.

Carbon-graphite electrode: A rod-shaped electrical cutting component made from soft carbon and hard graphite.

Carbon steel: An iron alloy containing a small amount of carbon to increase its strength.

Carburizing flame: A flame burning with an excess amount of fuel. Also called a *reducing flame.*

Casting: A method of making metal objects by pouring molten metal into a mold.

Codes: Documents establishing the minimum requirements of products or processes.

Coefficient of linear thermal expansion: A number indicating by how much a material's length changes in response to a 1°C temperature change.

Cold working: Methods of shaping metal without applying heat.

Concentric cable: A special torch cable that carries electricity through its outer shell and compressed air through its hollow core.

Condensation: Water vapor in the air turning back to its liquid form when it touches a cool surface.

Conductor: A material that can carry an electric current.

Confined space: A potentially hazardous space, not meant to be continuously occupied, that has limited entry and exit routes.

Consumable inserts: Prefabricated filler metal components designed to fuse into the weld root and become part of the weld.

Defect: A discontinuity or other imperfection that makes a weld unacceptable according to code.

Direct current (DC): An electric current that flows in one direction.

Discontinuity: A change or break in a weld's shape or structure.

Distortion: Changes in a welded assembly's shape caused by its parts changing size as the weld joint is heated and cooled.

Drag angle: The travel angle when the electrode points in the direction opposite the bead's progression.

Drag cutting: A cutting method in which the torch tip or shield rests on the workpiece.

Drag lines: Parallel, nearly vertical lines on oxyfuel-cut edges produced by the oxygen stream.

Dross: Oxidized and molten metal expelled by a thermal cutting process. Sometimes called *slag.*

Ductile: Metals that can be significantly stretched without breaking.

Ductility: A material's ability to be bent, shaped, or stretched without breaking.

Duty cycle: The percentage of time a welding or cutting machine can operate without overheating within a 10-minute period.

Electrically grounded: Connected to the earth or a conducting body that behaves like the earth electrically.

Electric arc: An electric current flow across a gap.

Electric current: The electron flow through a circuit. Measured in amperes or amps (A). Sometimes called *amperage*.

Electrode: A metal rod or wire that carries electric current and forms the welding or cutting arc.

Electrode axis: A line drawn along the electrode's length and through its center.

Embrittled: Metal that has become brittle and tends to crack easily.

Essential variables: Elements in a welding procedure specification (WPS) that can't be changed without requalifying the WPS.

Feather: Tapered.

Ferritic: A magnetic type of stainless steel whose internal crystal structure prevents it from being hardened by heat treatment.

Ferrous metals: Metals containing iron.

Flashback: A hissing or whistling sound caused by the flame burning back into the torch, hose, or regulator.

Flash burn: A burn to the eyes produced by brief exposure to intense heat and ultraviolet light. Also called *welder's flash*.

Flux: A material that dissolves or inhibits oxides and protects a weld joint from the atmosphere.

Fume plume: Smoke, gases, and particles produced from the consumables, metals, and metal coatings during welding.

Galvanized steel: Carbon steel dipped in zinc to inhibit corrosion.

Gouging: Cutting a groove into a surface.

Governor: A device that limits an engine's speed.

Hardenable materials: Materials that can become harder by heating and then cooling them.

Hardness: A material's ability to resist penetration.

Heat-affected zone: Unmelted base metal whose structure has been changed by heating.

Hermetically sealed: An airtight seal.

Homogeneity: Having a uniform structure or composition.

Immediately dangerous to life or health (IDLH): As defined by OSHA, an atmosphere that poses an immediate threat to life, would cause irreversible adverse health effects, or would impair an individual's ability to escape from a dangerous atmosphere.

Inclusions: Foreign matter trapped in a weld.

Joint mismatch: Alignment discrepancy in pipe sections or plates. Also called *high-low*.

Joint root: The place in the joint where the weld members are closest together.

Kerf: The slot produced by a cutting process.

Laminations: Separating layers in the base metal.

Level: A horizontal line parallel to the earth's surface. Also, a tool that checks whether a surface is level or plumb.

Loads: Forces applied to a material or structure.

Low-hydrogen: An electrode type manufactured to contain little or no moisture.

Malleable: Metals that can be shaped by hammering or pressure and not crack.

Martensitic: A magnetic type of stainless steel whose internal crystal structure allows it to be hardened by heat treatment.

Melt-through: Complete joint penetration.

Metallurgy: The science of metals, their chemistry, and their mechanical properties.

Neutral flame: A flame burning with the best mix of fuel gas and oxygen.

Non-essential variables: Elements in a welding procedure specification (WPS) that can be changed without requalifying the WPS.

Nondestructive testing (NDT): Testing methods that don't change or damage the test specimen.

Nonferrous metals: Metals that don't contain significant amounts of iron.

Notch toughness: A material's ability to resist breaking at a point of concentrated stress.

Oscillation: A regular, repetitive side-to-side motion.

Oxides: A scale or film on metal surfaces formed by oxygen in the air combining with the metal.

Oxidize: To chemically combine with oxygen either quickly or slowly.

Oxidizing flame: A flame burning with an excess amount of oxygen.

Piercing: To penetrate a metal plate with a cutting torch.

Piping porosity: Porosity approximately perpendicular to the weld face that is deeper than it is wide. Sometimes called *wormholes*.

Plasma: A superheated gas that's electrically charged (ionized).

Plumb: A vertical line perpendicular to the earth's surface.

Polarity: A terminal's electrical charge, either positive or negative, which determines whether current flows into or away from it.

Porosity: Gas pockets, or voids in weld metal.

Procedure qualification record (PQR): A document recording the test results required to qualify a welding procedure specification (WPS).

Purge gases: Inert gases, such as argon, that drive oxygen-containing air away from a weld zone to protect it from defects.

Push angle: The travel angle when the electrode points in the same direction as the bead's progression.

Quenching: Rapidly cooling a metal by plunging it into a liquid.

Radiographic: Imaging technology that passes X-rays or gamma rays through an object.

Rectifier: An electronic device that converts alternating current to direct current.

Residual stress: Heat-induced stress remaining in a weldment at normal temperatures.

Restart: The point where a new weld bead smoothly continues from the previous one. Sometimes called a *tie-in*.

Root face: The flattened area at the end of a groove weld's bevel. Also called the *land*.

Root opening: The distance between the two base metal parts at the weld root.

Severance cut: A cut that divides metal but leaves behind rough edges with significant dross.

Shielding gas: A gas such as argon, helium, or carbon dioxide that protects the welding electrode and weld zone from contamination during certain types of welding.

Single-phase: An alternating current service that delivers a single sine wave to the powered device.

Soapstone: A soft, white stone used to mark metal.

Solenoid valve: An electrically controlled valve that manages gas or liquid flow.

Specifications: Detailed documents defining the work performed and the materials used in specific products or processes.

Stainless steel: An iron alloy that doesn't rust because it contains chromium.

Standards: Documents that implement code requirements.

Step-down transformer: An electromagnetic device that changes an alternating current to one with a lower voltage.

Stringer bead: A weld bead made without any significant oscillating motion.

Supplemental essential variables: Variables affecting a weld's toughness that are relevant when the welding procedure specification (WPS) requires toughness testing.

Surfacing: Welding, brazing, or thermally spraying a layer of metal onto a surface to increase its thickness.

Tack welds: Small welds made to hold a weldment's parts in the proper alignment before welding.

Tempering: Heating and then slowly cooling metal to improve its strength or toughness.

Thermal conductivity value: A number indicating how readily a material transfers heat.

Three-phase: An alternating current service that delivers three overlapping sine waves to the powered device.

Traceability: Verifying through documentation that the correct materials have been used to produce a weld.

Transducer: An instrument that converts energy from one form to another.

Travel angle: The electrode's forward or backward angle with respect to the groove or weld face.

Ultimate tensile strength: The amount of pulling force a material can tolerate before breaking. Often shortened to *tensile strength*.

Ultraviolet (UV) radiation: Invisible light rays that can burn skin and damage the eyes.

Underbead cracking: Base metal cracking near the weld but below its surface.

Undercut: A discontinuity caused by the welder's melting a groove into the base metal at the weld toe.

Ventricular fibrillation: A lethal heart rhythm in which the heart muscle writhes rather than pumps.

Vertical welding: Welding with an upward or downward progression.

Voltage: The electrical force between two terminals or points in a circuit. Measured in volts (V).

Washing: Cutting off bolts, rivets, and other projections from a metal surface.

Weathering steels: Steel alloys that form a dense oxide layer on their surface, which prevents further oxidation.

Weave bead: A weld bead made by oscillating the electrode in one of several patterns.

Weld axis: A line drawn along the weld's length and through its center.

Weld coupons: Metal pieces that welders use for practice or to produce test welds for qualifications.

Welding: Processes that use heat to join materials by softening or melting them, so they blend and form a strong mechanical bond as they cool.

Welding procedure qualifications: Demonstrating through testing that welds made by following a welding procedure specification (WPS) meet prescribed standards.

Welding Procedure Specification (WPS): A document specifying all essential procedural details related to project-specific welds.

Weldment: An assembly made from parts fastened together by welded joints.

Weld root: The point where the back of the weld extends farthest into the joint.

Work angle: The electrode's side-to-side angle with respect to the groove or weld face.

Wrought: Metal shaped by beating with a hammer.